INTERNATIONAL UNION OF THEORETICAL AND APPLIED MECHANICS

IRREVERSIBLE ASPECTS OF CONTINUUM MECHANICS

AND

TRANSFER OF PHYSICAL CHARACTERISTICS IN MOVING FLUIDS

SYMPOSIA VIENNA, JUNE 22—28, 1966

EDITORS

H. PARKUS
VIENNA

L. I. SEDOV
MOSCOW

WITH 87 FIGURES

1968

SPRINGER SCIENCE+BUSINESS MEDIA, LLC

Originally published by Springer-Verlag New York in 1968
Softcover reprint of the hardcover 1st edition

Library of Congress Catalog Card Number 68-11180
ISBN 978-3-7091-5583-7 ISBN 978-3-7091-5581-3 (eBook)
DOI 10.1007/978-3-7091-5581-3

Title No. 9217

Preface

At its meeting on April 23, 1965 in Paris the Bureau of IUTAM decided to have a Symposium on the Irreversible Aspects of Continaum Mechanics held in June 1966 in Vienna. In addition, a Symposium on the Transfer of Physical Characteristics in Moving Fluids which, originally, had been scheduled to take place in Stockholm was rescheduled to be held in Vienna immediately following the Symposium on the Irreversible Aspects of Continuum Mechanics. It was felt that the subjects of the two symposia were so closely related that participants should be given an opportunity to attend both.

Both decisions were unanimously approved by the members of the General Assembly of IUTAM. Prof. H. Parkus, Vienna, was appointed Chairman of the Symposium on the Irreversible Aspects, and Prof. L. I. Sedov, Moscow, was appointed Chairman of the Symposium on the Transfer of Physical Characteristics, with Prof. Parkus being responsible for the local organization of both symposia.

In accordance with the policy set forth by IUTAM, membership of the Symposia was limited and by invitation only. Financial support for covering the costs of organization and for a partial defrayal of the accomodation and traveling expenses of the participants was provided by IUTAM and by the Austrian Ministry of Education.

The Scientific Committee of the Symposium on the Irreversible Aspects of Continuum Mechanics consisted of the following members:

Prof. H. Parkus, Vienna, Chairman
Prof. S. R. de Groot, Amsterdam
Prof. F. K. G. Odqvist, Stockholm
Prof. W. Olszak, Warsaw
Prof. W. Prager, Providence, R. I.
Prof. M. Roy, Paris
Prof. L. I. Sedov, Moscow
Prof. H. Ziegler, Zürich
(Prof. de Groot was unable to attend the Symposium

The primary purpose of the Symposium was to bring together leading scientists from the fields of Continuum Mechanics and Thermodynamics. Of these two fields which, in the past, have developed more or less inde pendently Continuum Mechanics has now reached a stage where further progress appears impossible without the inclusion of thermodynamic concepts. Emphasis was on the basic concepts in both fields, and not on applications. 23 papers were presented. Discussions were unusually lively and highly fruitful. Although unification of the two main schools of thought still seems to be a matter of the distant future, considerable progress has been made at the Symposium towards closing the gap.

The Scientific Committee of the Symposium on the Transfer of Physical Characteristics in Moving Fluids consisted of

Prof. L. I. Sedov, Moscow, Chairman
Prof. H. Parkus, Vienna, Vice-Chairman
Prof. S. B. Berndt, Stockholm
Prof. S. R. de Groot, Amsterdam
Prof. H. W. Liepmann, Pasadena
Prof. M. J. Lighthill, London
Prof. M. Roy, Paris

A free and sapid discussion followed the presentation of the papers.

Vienna and Moscow, September 1967

H. Parkus, L. I. Sedov

Sessions

Wednesday, June 22, 1966

Chairman: B. R. Seth

L. I. Sedov: Variational methods of constructing models of continuous media.

D. C. Drucker: On the continuum as an assemblage of homogeneous elements or states.

Chairman: U. Uhlhorn

H. Ziegler: A possible generalization of Onsager's theory.

C. Truesdell: Thermodynamics for beginners.

Chairman: P. M. Naghdi

B. D. Coleman[1]: Thermodynamics and wave propagation in non-linear materials with memory.

W. Olszak[2]: On thermodynamics of the differential type material.

S. Kaliski: Čerenkov generation of thermal waves for the wave equations of thermo-electro-magneto-elasticity.

Thursday, June 23, 1966

Chairman: J. Hult

E. T. Onat: The notion of state and its implications in thermodynamics of inelastic solids.

J. Meixner: TIP has many faces.

Chairman: G. A. Kluitenberg

J. F. Besseling: A thermodynamic approach to rheology.

Chairman: A. M. Freudenthal

M. Reiner: The influence of dissipated stresswork on the rupture of materials.

J. Kestin: On the application of the principles of thermodynamics to strained solid materials.

R. S. Rivlin[3]: Generalized continuum mechanics.

Friday, June 24, 1966

Chairman: D. C. Drucker

W. Prager: Composite stress-strain relations for elastoplastic solids.

E. H. Lee[4]: Finite strain elastic-plastic theory.

---

[1] Jointly with M. E. Gurtin.

[2] Jointly with P. Perzyna.

[3] Jointly with A. E. Green.

[4] Jointly with D. T. Liu.

Chairman: J. KESTIN

P. M. NAGHDI[1]: A thermodynamic development of elastic-plastic continua.

G. S. SHAPIRO: On conditions of yielding and fracture for viscoelastic-plastic bodies.

Chairman: S. KALISKI

A. M. FREUDENTHAL[2]: Accumulation of second-order strain in workhardening media.

W. NOWACKI: Couple-stresses in the theory of thermoelasticity.

G. HERRMANN[3]: On stress waves in viscoelastic media conducting heat.

Saturday, June 25, 1966

Chairman: E. H. LEE

YU. N. RABOTNOV: Kinetics of creep and creep rupture.

U. UHLHORN: Thermomechanics of a continuous system with internal structure.

J. HULT: Phenomenological aspects of creep irreversibility.

Monday, June 27, 1966

Chairman: S. B. BERNDT

B. R. SETH: Irreversible transition in continuum mechanics.

YU. P. LUNKIN: Vibrational dissociation relaxation in a multicomponent mixture of viscous heat-conducting gases.

Chairman: N. FRÖSSLING

V. N. NIKOLAEVSKII: Transfer phenomena in fluid-saturated porous media.

Tuesday, June 28, 1966

Chairman: M. J. LIGHTHILL

F. N. FRENKIEL[4]: Joint probability density distributions in a turbulent flow.

J. ZEMÁNEK: Die Beeinflussung der Intensität des Transportprozesses durch das dynamische Verhalten der Phasengrenze Flüssigkeit-Gas oder Flüssigkeit-Dampf.

---

[1] Jointly with A. E. GREEN.

[2] Jointly with MARIA RONAY.

[3] Jointly with J. D. ACHENBACH and S. M. VOGEL.

[4] Jointly with P. S. KLEBANOFF.

List of Participants

ADOMEIT, Dr. G., Rheinisch-Westfälische Technische Hochschule Aachen, Templergraben 55, D-51 Aachen, Germany.

AXELRAD, Dr. D. R., Department of Mechanical Engineering, McGill University, Montreal 2, Canada.

BECKERS, Dr. H. L., Koninklijke Shell-Laboratorium, Amsterdam-N, The Netherlands.

BEDNARCZYK, Dr. H., Technische Hochschule Wien, Karlsplatz 13, A-1040 Wien, Austria.

BĚLÍK, Docent Ing. L., CSc., Spojovací 5, Plzeň, Č.S.S.R.

BERNDT, Prof. S. B., Kungliga Tekniska Högskolan, Stockholm 70, Sweden.

BESSELING, Prof. J. F., Technische Hogeschool Delft, Mekelweg 2, Delft, The Netherlands.

BRUN, Prof. L., Laboratoire de Mécanique des Solides, École Polytechnique, 17 Rue Descartes, Paris Ve, France.

COLEMAN, Prof. B. D., Mellon Institute, 4400 Fifth Avenue, Pittsburgh, Pa. 15213, U. S. A.

DEVILLEZ, Ing. C., École Polytechnique, Faculté des Sciences Appliquées, Université Libre de Bruxelles, Bruxelles, Belgium.

DRUCKER, Prof. D. C., Division of Engineering, Brown University, Providence, R. I. 02912, U. S. A.

FENZ, EDITH, Technische Hochschule Wien, Karlsplatz 13, A-1040 Wien, Austria.

FINZI, Prof. L., Istituto di Scienza delle Costruzioni della Facoltà di Architettura, Politecnico Milano, Milano, Italy.

FISZDON, Prof. W., Polskiej Akademii Nauk, ul. Swiętokrzyska 21, Warszawa, Poland.

FRANK, Dr. W., Bundesministerium für Bauten und Technik, Stubenring 1, A-1010 Wien, Austria.

FRENKIEL, Prof. F. N., Applied Mathematics Laboratory, David Taylor Model Basin, Washington, D. C. 20007, U. S. A.

FREUDENTHAL, Prof. A. M., Department of Civil Engineering and Engineering Mechanics, Columbia University, New York, N. Y. 10027, U. S. A.

FRÖSSLING, Prof. N., The Institute of Applied Thermo- and Fluid Dynamics, Chalmers University of Technology, Gibraltargatan 5 M, Gothenburg S, Sweden.

GRIOLI, Prof. G., Seminario Matematico, Università di Padova, Padova, Italy.

HERRMANN, Prof. G., The Technological Institute, Northwestern University, Evanston, Ill. 60201, U. S. A.

HERRNEGGER, F., Institut für Theoretische Physik, Universität Innsbruck, A-6020 Innsbruck, Austria.

HULT, Prof. J., Division of Strength of Materials, Chalmers University of Technology, Gibraltargatan 5 M, Gothenburg S, Sweden.

KALISKI, Prof. S., Polskiej Akademii Nauk, ul. Swiętokrzyska 21, Warszawa, Poland.

KESTIN, Prof. J., Division of Engineering, Brown University, Providence, R. I. 02912, U. S. A.

KLUITENBERG, Dr. G. A., Technische Hogeschool te Eindhoven, Insulindelaan 2, Eindhoven, The Netherlands.

KOITER, Prof. W. T., Technische Hogeschool Delft, Mekelweg 2, Delft, The Netherlands.

KRÖNER, Prof. E., Institut für Theoretische Physik, Bergakademie Clausthal, D-3392 Clausthal-Zellerfeld, Germany.

LEE, Prof. E. H., Division of Engineering Mechanics, Stanford University, Stanford, Calif. 94305, U. S. A.

LIGHTHILL, Prof. M. J., Imperial College of Science and Technology, University of London, South Kensington, London S. W. 7, England.

LUN'KIN, Prof. YU. P., c/o Prof. G. K. MIKHAILOV, U. S. S. R. National Committee on Theoretical and Applied Mechanics, Leningrad Avenue 7, Moscow A-40, U. S. S. R.

MAIER, Prof. G., Istituto di Scienza delle Costruzioni della Facoltà di Architettura, Politecnico Milano, Milano, Italy.

MANDL, Dr. G., Volmerlaan 6, Rijswijk Z. H., The Netherlands.

MAZUR, Prof. P., Instituut Lorentz, Steenschuur, Leiden, The Netherlands.

MEIXNER, Prof. J., Institut für Theoretische Physik, Rheinisch-Westfälische Technische Hochschule Aachen, Templergraben 55, D-51 Aachen, Germany.

MIKHAILOV, Prof. G. K., U. S. S. R. National Committee on Theoretical and Applied Mechanics, Leningrad Avenue 7, Moscow A-40, U. S. S. R.

MÜLLER, Dr. I., Institut für Theoretische Physik, Rheinisch-Westfälische Technische Hochschule Aachen, Templergraben 55, D-51 Aachen, Germany.

NAGHDI, Prof. P. M., Division of Applied Mechanics, Room 6141, Etcheverry Hall, University of California, Berkeley, Calif. 94720, U. S. A.

NIKOLAEVSKII, Prof. V. N., Institute of Physics of the Earth, U. S. S. R. Academy of Sciences, Bol. Gruzinskaya 10, Moscow G-242, U. S. S. R.

NIORDSON, Prof. F. I., Department of Applied Mechanics, The Technical University of Denmark, Rigensgade 13, Copenhagen K, Denmark.

NOWACKI, Prof. W., Polskiej Akademii Nauk, ul. Swiętokrzyska 21, Warszawa, Poland.

ODQVIST, Prof. F. K. G., The Royal Institute of Technology, Stockholm 70, Sweden.

OLESIAK, Prof. Z., ul. F. Joliot Curie 7 m. 24, Warszawa 87, Poland.

OLSZAK, Prof. W., Polskiej Akademii Nauk, ul. Swiętokrzyska 21, Warszawa, Poland.

ONAT, Prof. E. T., Yale University, Winchester Hall, 15 Prospect Street, New Haven, Conn. 06520, U. S. A.

PARKUS, Prof. H., Technische Hochschule Wien, Karlsplatz 13, A-1040 Wien, Austria.

PERZYNA, Doz. Dr. P., Polskiej Akademii Nauk, ul. Swiętokrzyska 21, Warszawa, Poland.

PLOCHOCKI, Doz. Dr. Z., Polskiej Akademii Nauk, ul. Swiętokrzyska 21, Warszawa, Poland.

PRAGER, Prof. W., Department of the Aerospace and Mechanical Engineering Sciences, P. O. Box 109, University of California, La Jolla, Calif. 92037, U. S. A.

PŘEVRÁTIL, Ing. R., Department of Mechanics, Slovak Technical University, Tolstého 1, Bratislava, Č. S. S. R.

RABOTNOV, Prof. YU. N., Institute of Mathematics and Mechanics, Moscow State University, Moscow B-234, U. S. S. R.

RADENKOVIC, Prof. D., Laboratoire de Mécanique des Solides, École Polytechnique, 17 Rue Descartes, Paris Ve, France.

REINER, Prof. M., Technion — Israel Institute of Technology, Haifa, Israel.

RIMROTT, Prof. F. P. J., Department of Mechanical Engineering, University of Toronto, Toronto 5, Canada.

RIVLIN, Prof. R. S., Center for the Application of Mathematics, Lehigh University, Bethlehem, Pa. 18015, U. S. A.

ROY, Prof. M., 55 Boulevard Malesherbes, Paris VIIIe, France.

SAVE, Prof. M., Département de Mécanique, Faculté Polytechnique de Mons, Rue de Houdain, Mons, Belgium.

SAVIN, Prof. G. N., Institute of Mechanics, Academy of Sciences of the Ukrainian S. S. R., ul. Chkalora-55b, Kiev, U. S. S. R.

SCHWARZL, Dr. F. R., Centraal Laboratorium t. n. o., Postbox 217, Delft, The Netherlands.

SEDOV, Prof. L. I., U. S. S. R. National Committee on Theoretical and Applied Mechanics, Leningrad Avenue 7, Moscow A-40, U. S. S. R.

SETH, Prof. B. R., Dibrugarh University, Dibrugarh, Assam, India.

SHAPIRO, Prof. G. S., Institute of Mechanics, U. S. S. R. Academy of Sciences, Leningrad Avenue 7, Moscow A-40, U. S. S. R.

TANI, Prof. I., Institute of Space and Aeronautical Science, University of Tokyo, Komaba, Meguro-ku, Tokyo, Japan.

THOMANN, Dr. H., The Aeronautical Research Institute of Sweden (FFA), Bromma 11, Sweden.

TORRE, Dr. C., Bundesministerium für Landesverteidigung, A-1024 Wien, Austria.

TRUESDELL, Prof. C., The Johns Hopkins University, Baltimore, Md. 21218, U. S. A.

UHLHORN, Prof. U., Tekniska Högskolan i Lund, Lund, Sweden.

ZEMAN, Dr. J. L., Technische Hochschule Wien, Karlsplatz 13, A-1040 Wien, Austria.

ZEMÁNEK, Dipl. Ing. J., CSc., SVUSS Staatliches Forschungsinstitut für Maschinenbau, Běchovice-Prag, Č. S. S. R.

ZIEGLER, Dr. F., Technische Hochschule Wien, Karlsplatz 13, A-1040 Wien, Austria.

ZIEGLER, Prof. H., Eidgenössische Technische Hochschule, Zürich, Switzerland.

Contents

On Stress Waves in Viscoelastic Media Conducting Heat

By

J. D. Achenbach, S. M. Vogel and **G. Herrmann**
Evanston, Ill. (U.S.A.)

Summary

The theory of propagating surfaces of discontinuity is used to study one-dimensional stress waves in viscoelastic media which conduct heat. The constitutive laws for stress and entropy are derived from the specific free energy which is taken in the form of a functional expansion to the second order in strain and to the third order in temperature change. With the aid of these laws, relations are developed between the kinematical, mechanical and thermodynamical quantities at the wave front. Explicit expressions are established for both the velocity and the change of magnitude of the stress discontinuity as it propagates through the medium. The results include the effects of viscous damping, temperature-dependent material properties and thermo-mechanical coupling.

1. Introduction

The role of non-equilibrium thermodynamics in wave propagation in non-linear dissipative materials has received considerable attention recently in a series of papers by Coleman and Gurtin [1, 2 and 3]. In [1] the velocity and the growth and decay of one-dimensional acceleration waves are discussed for definite conductors, as well as for media which do not conduct heat (non-conductors). In [2] some of the findings of [1] are extended to three-dimensional acceleration waves. In the third paper, [3], one-dimensional shock waves in a material with memory are discussed, but the analysis is restricted to non-conductors. It is thus recognized in [1, 2 and 3] that the physical properties of many viscoelastic materials are sensitive to temperature changes, and such temperature dependence is included in a very general fashion. By considering shock waves in non-conductors the authors exclude, however, part of the effect of thermomechanical coupling, namely, the influence of temperature on deformation.

In this paper we discuss the propagation of stress waves (shocks) in a viscoelastic material with temperature dependent properties and thermomechanical coupling. The term 'wave' has been used with several meanings; in this paper a stress wave shall denote a discontinuity in stress across a propagating surface (singular surface). The analysis is concerned with small strains and small changes in temperature. The wave is induced in the body by the sudden application of surface tractions. The body is homogeneous and, prior to loading, at rest, undeformed and at a uniform temperature[1]. We determine the velocity at which the stress discontinuity propagates and we show how the dissipative nature of a viscoelastic material in conjunction with thermomechanical coupling affects the magnitude of such a wave. In addition, the relation between relevant mechanical and thermodynamic quantities at the wave front is established.

A set of thermomechanically coupled equations for the propagation of stress, strain and temperature fields in viscoelastic solids, was derived earlier by HUNTER [4]. HUNTER's analysis is restricted, however, to a particular dependence of the relaxation function on temperature (thermorheologically simple solids), and he also assumes that the specific heat and the thermal expansion coefficient are algebraic factors rather than integral operators.

After including in the next section a brief statement of the balance or conservation laws, we turn to a discussion of the constitutive relations in Section 3. Certain of COLEMAN's findings [5, 6] on the thermodynamics of materials with memory are used there to obtain specific forms for the stress and entropy functionals. Section 4 is devoted to the presentation of some well-known kinematical jump conditions and the derivation of additional discontinuity relations from the balance laws and constitutive equations. These jump conditions are then employed (Section 5) in the computation of the velocity and the attenuation of the propagating discontinuity.

2. Balance Laws

We are concerned here with problems in which the stress σ, the absolute temperature T (which is always > 0), the specific internal energy ε, the specific entropy η, and the heat flux q are functions of x and t only. It is assumed that the displacement gradient is sufficiently small, such that it is unnecessary to distinguish between its values at the position of a particle before or after deformation. Both extrinsic body forces and extrinsic heat supplies are excluded from the analysis. The fields σ, T, η, ε and q, as functions of x and t, must obey, in addition to the Clausius-Duhem inequality, the laws of balance of momentum

[1] These assumptions could be relaxed somewhat under appropriate restrictions.

and balance of energy. The integral forms of these laws are

$$\frac{d}{dt}\int_{x_1}^{x_2} \varrho v \, dx = \sigma(x_2, t) - \sigma(x_1, t) \tag{2.1}$$

$$\frac{d}{dt}\int_{x_1}^{x_2} \left(\frac{v^2}{2} + \varepsilon\right) \varrho \, dx = \sigma(x_2, t)\, v(x_2, t) - \sigma(x_1, t)\, v(x_1, t) - $$
$$- q(x_2, t) + q(x_1, t) \tag{2.2}$$

where the constant ϱ is the mass density, and $v = \partial u/\partial t$ is the particle velocity. Equations (2.1) and (2.2) are assumed to hold at all times t and for any pair of points x_1, x_2 in the body.

The specific free energy φ is defined by

$$\varphi = \varepsilon - T\eta . \tag{2.3}$$

The substitution of (2.3) into (2.2) changes the form of the energy balance equation to

$$\frac{d}{dt}\int_{x_1}^{x_2} \left(\frac{v^2}{2} + \varphi + T\eta\right) \varrho \, dx = \sigma(x_2, t)\, v(x_2, t) - \sigma(x_1, t)\, v(x_1, t) - $$
$$- q(x_2, t) + q(x_1, t) . \tag{2.4}$$

In the absence of any specific extrinsic heat supply, the Clausius-Duhem inequality may be written as

$$\frac{d}{dt}\int_{x_1}^{x_2} \eta \varrho \, dx + \frac{q(x_2, t)}{T(x_2, t)} - \frac{q(x_1, t)}{T(x_1, t)} \geq 0 . \tag{2.5}$$

3. Constitutive Relations

When thermodynamics is included, a simple material [5,7] is a material for which the stress at time t is determined by the temporal histories of deformation gradient and temperature. Instead of the deformation gradient we use the small strain $\gamma = \partial u/\partial x$. Coleman, in his thermodynamic theory, shows that a simple material is characterized by four functionals, $\boldsymbol{\varphi}$, $\boldsymbol{\sigma}$, $\boldsymbol{\eta}$, $\boldsymbol{q}$. The material functionals $\boldsymbol{\varphi}$, $\boldsymbol{\sigma}$, and $\boldsymbol{\eta}$, give the present values of φ, σ, and η, when the strain history γ^t and the temperature history T^t are specified. The histories γ^t and T^t are defined by

$$\gamma^t(x, s) = \gamma(x, t - s) \qquad 0 \leq s \leq t$$
$$T^t(x, s) = T(x, t - s) \qquad 0 \leq s \leq t .$$

The functional $\boldsymbol{q}$ gives the present value of q when $\partial T(x, t)/\partial x$, as well as γ^t and T^t, are given.

Suppose that γ^t and T^t are known, then one writes for the free energy

$$\varphi(x, t) = \underset{s=0}{\overset{t}{\boldsymbol{\varphi}}}\left(\gamma^t(x, s), T^t(x, s)\right) \tag{3.1}$$

or equivalently

$$\varphi(x, t) = \underset{s=0}{\overset{t}{\boldsymbol{\varphi}}}\left(\gamma_r^t(x, s),\ \theta_r^t(x, s);\ \gamma(x, t),\ \theta(x, t)\right) \tag{3.2}$$

where $\theta = T - T_0$ is the small change in temperature from some uniform reference temperature T_0, and

$$\gamma_r^t(x, s) = \gamma(x, t - s) \qquad 0 < s \leq t$$

$$\theta_r^t(x, s) = \theta(x, t - s) \qquad 0 < s \leq t.$$

Equation (3.2) states that for each value of $\gamma(x, t)$ and $\theta(x, t)$, $\boldsymbol{\varphi}$ is a functional of $\gamma_r^t(x, \cdot)$ and $\theta_r^t(x, \cdot)$, while for each past history $\gamma_r^t(x, \cdot)$ and $\theta_r^t(x, \cdot)$, $\boldsymbol{\varphi}$ is an ordinary function of $\gamma(x, t)$ and $\theta(x, t)$. We should have used different symbols for the functionals appearing in (3.1) and (3.2), but following COLEMAN, GURTIN and HERRERA [8], we rely on the presence or absence of the semicolon to indicate the appropriate point of view.

An important part of COLEMAN's theory is the principle of fading memory, which is discussed in [1, 5, 6, 8 and 9]. One of the implications of the fading memory assumption is the existence and continuity of derivatives $D_\gamma\boldsymbol{\varphi}$, $\delta_\theta\boldsymbol{\varphi}$, $D_\gamma D_\theta\boldsymbol{\varphi} = D_\theta D_\gamma\boldsymbol{\varphi}$, $D_\gamma\delta_\theta\boldsymbol{\varphi} = \delta_\theta D_\gamma\boldsymbol{\varphi}$, $\delta_\theta\delta_\gamma\boldsymbol{\varphi} = \delta_\gamma\delta_\theta\boldsymbol{\varphi}$, etc., of the functional $\boldsymbol{\varphi}$. These derivatives are themselves functionals and are defined in [1]. For illustrative purposes, a few of these definitions are reproduced:

$$D_\gamma \underset{s=0}{\overset{t}{\boldsymbol{\varphi}}}(\gamma^t, \theta^t) \equiv \frac{\partial}{\partial\gamma} \underset{s=0}{\overset{t}{\boldsymbol{\varphi}}}(\gamma_r^t, \theta_r^t; \gamma, \theta) \tag{3.3a}$$

$$\delta_\gamma \underset{s=0}{\overset{t}{\boldsymbol{\varphi}}}(\gamma^t, \theta^t \mid f) \equiv \left\{\frac{\partial}{\partial\alpha} \underset{s=0}{\overset{t}{\boldsymbol{\varphi}}}(\gamma_r^t + \alpha f, \theta_r^t; \gamma, \theta)\right\}_{\alpha=0} \tag{3.3b}$$

$$\begin{aligned}\delta_\gamma D_\gamma \underset{s=0}{\overset{t}{\boldsymbol{\varphi}}}(\gamma^t, \theta^t \mid f) &\equiv \left\{\frac{\partial^2}{\partial\alpha\,\partial\gamma} \underset{s=0}{\overset{t}{\boldsymbol{\varphi}}}(\gamma_r^t + \alpha f, \theta_r^t; \gamma, \theta)\right\}_{\alpha=0} \\ &= \frac{\partial}{\partial\gamma}\left\{\frac{\partial}{\partial\alpha} \underset{s=0}{\overset{t}{\boldsymbol{\varphi}}}(\gamma_r^t + \alpha f, \theta_r^t; \gamma, \theta)\right\}_{\alpha=0} \equiv D_\gamma\delta_\gamma \underset{s=0}{\overset{t}{\boldsymbol{\varphi}}}(\gamma^t, \theta^t \mid f).\end{aligned} \tag{3.3c}$$

In (3.3b, c) f is an arbitrary function over the interval $(0, t]$. These new functionals are jointly continuous in all of their arguments.

The main new result of the investigation into the thermodynamics of materials with memory presented in [5] is a theorem relating the stress and entropy functionals to the specific free energy functional. This theorem states that

$$\sigma(x, t) = \varrho D_\gamma \overset{t}{\underset{s=0}{\boldsymbol{\varphi}}} \left(\gamma^t(x, s), \theta^t(x, s)\right) \tag{3.4}$$

and

$$\eta(x, t) = -D_\theta \overset{t}{\underset{s=0}{\boldsymbol{\varphi}}} \left(\gamma^t(x, s), \theta^t(x, s)\right). \tag{3.5}$$

Thus, once $\boldsymbol{\varphi}$ is known, one can determine the functionals $\boldsymbol{\sigma}$ and $\boldsymbol{\eta}$ by taking the partial derivatives of $\boldsymbol{\varphi}$ with respect to the present values γ and θ, respectively, and keeping the past histories γ_r^t and θ_r^t fixed.

Because of the smoothness properties given to $\boldsymbol{\varphi}$ by the assumption of fading memory, $\boldsymbol{\varphi}$ can be represented to any desired order by its "Taylor's expansion" in terms of the D and δ operators. Since we are assuming that the strains and temperature changes are small but that the physical properties of the material are temperature-dependent, we carry out the expansion of the specific free energy to the second order in strain and third order in temperature change. A discussion of this expansion in terms of the D and δ operators is given in the Appendix. Because only regular homogeneous functionals of the first, second and third degree appear in eq. (A. 2) of the Appendix, these functionals can be represented by multiple integrals [10]. The expansion for $\boldsymbol{\varphi}$, using the integral representation, then becomes

$$\begin{aligned}\varrho\boldsymbol{\varphi} = A_0 + A_1\gamma + A_2\gamma\theta + A_3\gamma\theta^2 + A_4\gamma^2 + A_5\gamma^2\theta + \\ + B_1\theta + B_2\theta^2 + B_3\theta^3,\end{aligned} \tag{3.6}$$

where

$$\begin{aligned}A_0 &= a + R_{22}*\theta*\theta^{(1)} + R_{11}*\gamma*\gamma^{(1)} + R_{12}*\gamma*\theta^{(1)} + \\ &\quad + R_{222}*\theta*\theta^{(1)}*\theta^{(1)} + R_{112}*\gamma*\gamma^{(1)}*\theta^{(1)} + R_{122}*\gamma*\theta^{(1)}*\theta^{(1)}\end{aligned} \tag{3.7 a}$$

$$A_1 = E + G_1'*\gamma + M_1'*\theta + M_2'*\theta*\theta^{(1)} + G_2'*\gamma*\theta^{(1)} \tag{3.7 b}$$

$$A_2 = M_1(0) + K_2'*\theta + K_1'*\gamma \tag{3.7 c}$$

$$A_3 = K_2(0) \tag{3.7 d}$$

$$A_4 = \frac{1}{2}G_1(0) + N'*\theta \tag{3.7 e}$$

$$A_5 = K_1(0) \tag{3.7 f}$$

$$B_1 = H + L_1' * \theta + L_2' * \gamma + L_{11}' * \theta * \theta^{(1)} + L_{22}' * \gamma * \gamma^{(1)} + L_{12}' * \gamma * \theta^{(1)} \quad (3.7\,\mathrm{g})$$

$$B_2 = \frac{1}{2} L_1(0) + \frac{1}{2} F_1' * \theta + \frac{1}{2} F_2' * \gamma \quad (3.7\,\mathrm{h})$$

$$B_3 = \frac{1}{3} F_1(0). \quad (3.7\,\mathrm{i})$$

The asterisk appearing in (3.7) is a shorthand notation for the following types of integrals:

$$G_1' * \gamma = \int_0^t G_1'(t - s)\, \gamma(x, s)\, ds \quad (3.8\,\mathrm{a})$$

$$G_2' * \gamma * \theta^{(1)} = \int_0^t \int_0^t G_2'(t - s, t - \xi)\, \gamma(x, s)\, \frac{\partial \theta(x, \xi)}{\partial \xi}\, ds\, d\xi. \quad (3.8\,\mathrm{b})$$

The prime in expressions (3.8a, b) denotes differentiation with respect to $t - s$. The lower limit of integration is the time at which deformation of the body is initiated.

The stress and entropy are found by substituting (3.6) into eqs. (3.4) and (3.5). Thus

$$\sigma = A_1 + A_2 \theta + A_3 \theta^2 + 2 A_4 \gamma + 2 A_5 \gamma \theta, \quad (3.9)$$

and

$$-\eta \varrho = B_1 + 2 B_2 \theta + 3 B_3 \theta^2 + A_2 \gamma + 2 A_3 \gamma \theta + A_5 \gamma^2. \quad (3.10)$$

If the terms containing temperature change are neglected in eq. (3.9), this equation becomes the classical constitutive law of isothermal linear viscoelasticity. On the other hand, if the integral terms are assumed absent, eq. (3.9) reduces to the stress-strain law of thermoelasticity with material properties that depend linearly on temperature.

To complete the characterization of a simple material, we need to specify the heat flux functional. Here we assume that Fourier's law is valid, i.e.,

$$q = -k \frac{\partial T}{\partial x} \quad (3.11)$$

where k is the thermal conductivity, which is considered constant. We observe that a material obeying (3.11) is a definite conductor if and only if $k > 0$. Such a material is a non-conductor if $k \equiv 0$.

4. Jump Conditions

Suppose that a discontinuity propagates through a medium, and suppose that the position of the wave front is given by

$$x = y(t). \quad (4.1)$$

The velocity of the wave relative to the material may then be written as

$$c = dy/dt. \tag{4.2}$$

Consider a function $f(x, t)$, which, as a function of x, has a jump discontinuity at $x = y(t)$. The jump in $f(x, t)$ across the wave front is then defined by

$$[f](t) = f_1(t) - f_2(t) = \lim_{x \to y^-(t)} f(x, t) - \lim_{x \to y^+(t)} f(x, t). \tag{4.3}$$

If the derivatives $\partial f/\partial t$ and $\partial f/\partial x$ are continuous functions of x and t jointly, except for jump discontinuities across the wave front, the kinematical condition of compatibility [11] yields

$$\frac{d}{dt}[f](t) = \left[\frac{\partial f}{\partial t}\right] + \left[\frac{\partial f}{\partial x}\right] c. \tag{4.4}$$

For a function $f(x, t)$ continuous in all x with jump discontinuities in $\partial f/\partial t$ and $\partial f/\partial x$ at $x = y(t)$, eq. (4.4) reduces to

$$\left[\frac{\partial f}{\partial t}\right] = -c\left[\frac{\partial f}{\partial x}\right]. \tag{4.5}$$

We call $y(t)$ the wave front of a stress wave if the field $u(x, t)$ is a continuous function of x and t jointly for all x and t while $\dot{u}$ and γ have at most jump discontinuities across the wave front but are continuous in x and t everywhere else. Since the jump in strain is finite, the propagating discontinuity acts as a moving source generating a certain amount of heat. It can be shown that such a moving source in a material obeying Fourier's law of heat conduction leads to a continuous temperature distribution. The quantities $\partial\theta/\partial t$ and $\partial\theta/\partial x$ can, however, have jump discontinuities at the wave front. The occurrence of a temperature discontinuity at the wave front is found in more pathological cases as, for example, when a concentrated body force is suddenly applied in an infinite medium [12].

With $y(t)$ between x_1 and x_2, we rewrite eq. (2.1) in the form

$$\frac{d}{dt}\left\{\int_{x_1}^{y(t)} \varrho v(x, t)\, dx + \int_{y(t)}^{x_2} \varrho v(x, t)\, dx\right\} = \sigma(x_2, t) - \sigma(x_1, t). \tag{4.6}$$

Upon performing the differentiation and letting $x_1 \to y^-(t)$ and $x_2 \to y^+(t)$, we obtain from (4.6), using (4.2), the following expression for $[\sigma]$:

$$[\sigma] = -\varrho c[v]. \tag{4.7}$$

In a similar manner, the jump relation corresponding to the balance of energy law is obtained as

$$-\varrho c \left\{\frac{1}{2}[v^2] + [\varphi] + T[\eta]\right\} = [\sigma v] - [q]. \tag{4.8}$$

Applying the limiting process once more, this time to the Clausius-Duhem inequality (2.5), we find at the wave front

$$\varrho c[\eta] - \frac{1}{T}[q] \geq 0. \tag{4.9}$$

5. The Propagation and Decay of Discontinuities

For a stress wave the constitutive laws (3.9) and (3.10) yield across $x = y(t)$

$$[\sigma] = 2(A_4 + A_5\theta)[\gamma], \tag{5.1}$$

$$-\varrho[\eta] = (A_2 + 2A_3\theta)[\gamma] + A_5[\gamma^2]. \tag{5.2}$$

Thus $[\eta]$ is $O([\gamma])$. By contrast, the jump in specific entropy was shown to be $O([\gamma]^3)$ when shock waves propagate in non-conductors [3].

The compatibility relation, $[v] = -c[\gamma]$, eq. (4.5), when substituted into eq. (4.7), yields another relation between $[\sigma]$ and $[\gamma]$, namely,

$$[\sigma] = \varrho c^2[\gamma]. \tag{5.3}$$

Combining eqs. (5.1) and (5.3), the speed of propagation of stress waves is obtained as

$$c = [2(A_4 + A_5\theta)/\varrho]^{1/2}. \tag{5.4}$$

Equation (5.4) shows that c depends on both the past history and the present value of the temperature change.

We now assert

$$-\varrho c \left\{\frac{1}{2}[v^2] + [\varphi]\right\} = [\sigma v]. \tag{5.5}$$

The validity of (5.5) can be established as follows: From definition (4.3), the compatibility relation $[v] = -c[\gamma]$, and eq. (4.8) we obtain

$$\begin{aligned}[\sigma v] + \frac{1}{2}\varrho c[v^2] &= \sigma_1 v_1 - \sigma_2 v_2 + \frac{1}{2}\varrho c(v_1^2 - v_2^2)\\ &= \frac{1}{2}(\sigma_1 + \sigma_2)(v_1 - v_2)\\ &= -\frac{c}{2}(\sigma_1 + \sigma_2)[\gamma].\end{aligned} \tag{5.6}$$

Since only γ is discontinuous across the wave, we find from eq. (3.9)

$$\frac{\sigma_1 + \sigma_2}{2} = (A_1 + A_2\theta + A_3\theta^2) + (A_4 + A_5\theta)(\gamma_1 + \gamma_2). \qquad (5.7)$$

Hence eq. (5.6) becomes

$$[\sigma v] + \frac{1}{2}\varrho c[v^2] = -c(A_1 + A_2\theta + A_3\theta^2)[\gamma] - c(A_4 + A_5\theta)[\gamma^2]. \qquad (5.8)$$

At $x = y(t)$, the expression for φ as given by (3.6) assumes the form

$$\varrho[\varphi] = (A_1 + A_2\theta + A_3\theta^2)[\gamma] + (A_4 + A_5\theta)[\gamma^2]. \qquad (5.9)$$

A comparison of (5.8) and (5.9) reveals that eq. (5.5) is indeed valid.

As a consequence of (5.5), the jump condition corresponding to the energy balance eq. (4.9) becomes

$$\varrho c T[\eta] = [q]. \qquad (5.10)$$

Substitution of (5.10) into (4.9) shows that the latter inequality becomes an equality.

The constitutive equation for the heat flux, eq. (3.11), provides yet another jump condition at $x = y(t)$. This relation is

$$[q] = -k\left[\frac{\partial T}{\partial x}\right] = -k\left[\frac{\partial \theta}{\partial x}\right]. \qquad (5.11)$$

Employing (5.2) and (5.11) in (5.10) yields the formula

$$\left[\frac{\partial \theta}{\partial x}\right] = \frac{cT}{k}\{(A_2 + 2A_3\theta)[\gamma] + A_5[\gamma^2]\}. \qquad (5.12)$$

The next objective is to derive the differential equation governing the magnitude of the stress discontinuity as it propagates through the medium. From eq. (4.4), with f replaced first by γ and then by v, we have

$$\frac{d[\gamma]}{dt} = [\dot{\gamma}] + c\left[\frac{\partial \gamma}{\partial x}\right] \qquad (5.13)$$

and

$$\frac{d[v]}{dt} = [\dot{v}] + c\left[\frac{\partial v}{\partial x}\right] = [\dot{v}] + c[\dot{\gamma}]. \qquad (5.14)$$

The elimination of $[\dot{\gamma}]$ between (5.13) and (5.14) yields

$$\frac{d[v]}{dt} - [\dot{v}] = c\frac{d[\gamma]}{dt} - c^2\left[\frac{\partial \gamma}{dx}\right]. \qquad (5.15)$$

By differentiating the compatibility relation $[\gamma] = -[v]/c$ with respect to time, we obtain

$$c\frac{d[\gamma]}{dt} = -c\frac{d}{dt}\left(\frac{[v]}{c}\right) = -\frac{d[v]}{dt} + \frac{1}{c}[v]\frac{dc}{dt}. \tag{5.16}$$

Substitution of (5.16) into (5.15) yields

$$2\frac{d[v]}{dt} - \frac{[v]}{c}\frac{dc}{dt} = [\dot{v}] - c^2\left[\frac{\partial\gamma}{\partial x}\right]. \tag{5.17}$$

The law of balance of momentum (2.1) implies that for $x \neq y(t)$: $\partial\sigma/\partial x = \varrho\dot{v}$, while for $x = y(t)$

$$\left[\frac{\partial\sigma}{\partial x}\right] = \varrho[\dot{v}]. \tag{5.18}$$

Another expression for $[\partial\sigma/\partial x]$ can be obtained from the constitutive law (3.9):

$$\left[\frac{\partial\sigma}{\partial x}\right] = \left[\frac{\partial}{\partial x}(A_1 + A_2\theta + A_3\theta^2)\right] + 2\left[\gamma\frac{\partial}{\partial x}(A_4 + A_5\theta)\right] + \varrho c^2\left[\frac{\partial\gamma}{\partial x}\right]. \tag{5.19}$$

It follows from (5.17), (5.18) and (5.19)

$$2\frac{d[v]}{dt} - \frac{[v]}{c}\frac{dc}{dt} = \frac{1}{\varrho}\left[\frac{\partial}{\partial x}(A_1 + A_2\theta + A_3\theta^2)\right] + \frac{2}{\varrho}\left[\gamma\frac{\partial}{\partial x}(A_4 + A_5\theta)\right] \tag{5.20}$$

or

$$\frac{2}{\varrho c}\frac{d[\sigma]}{dt} - \frac{3[\sigma]}{\varrho c^2}\frac{dc}{dt} = \frac{1}{\varrho c}\left[\frac{\partial}{\partial t}(A_1 + A_2\theta + A_3\theta^2)\right] - \\ - \frac{2}{\varrho}\left[\gamma\frac{\partial}{\partial x}(A_4 + A_5\theta)\right], \tag{5.21}$$

where use is made of eqs. (4.5) and (4.7). By employing the expressions for the A's, as given by eqs. (3.7), in eq. (5.21), we find that the differential equation for the magnitude of the stress wave becomes

$$\frac{2}{\varrho c}\frac{d[\sigma]}{dt} - \frac{3[\sigma]}{\varrho c^2}\frac{dc}{dt} = \frac{1}{\varrho c}\Big\{G_1'(0)[\gamma] + (M_2'{*}\theta)\left[\frac{\partial\theta}{\partial t}\right] + (G_2'{*}\gamma)\left[\frac{\partial\theta}{\partial t}\right] + \\ + (G_2'{*}\theta^{(1)})[\gamma] + K_1'(0)\,\theta[\gamma] + A_2\left[\frac{\partial\theta}{\partial t}\right] + 2K_2(0)\,\theta\left[\frac{\partial\theta}{\partial t}\right]\Big\} - \\ - \frac{2}{\varrho}\Big\{(N'{*}\theta^{(1)})[\gamma] + K_1(0)\left[\frac{\partial\theta}{\partial x}\gamma\right]\Big\}. \tag{5.22}$$

Equation (5.22), as it stands, cannot be solved easily. The assumption that ahead of the stress wave the quantities u and θ, as well as their gradients, are negligible suffices, however, to make the equation more tractable. Since, in the absence of thermomechanical coupling these quantities are indeed zero, the above assumption appears to be justified. Also, it has been experimentally observed [13] that in a viscoelastic body the term involving the time derivative of the mechanical dilatation has only a small influence on the temperature[1]. With this simplification, c becomes a constant and eq. (5.22) reduces to

$$\frac{2}{\varrho c}\frac{d[\sigma]}{dt} = \frac{G_1'(0)}{\varrho c}[\gamma] + \frac{M_1(0)}{\varrho c}\left[\frac{\partial \theta}{\partial t}\right] - \frac{2}{\varrho} K_1(0)\left[\frac{\partial \theta}{\partial x}\right][\gamma]. \tag{5.23}$$

By substitution of the jump conditions (5.3) and (5.12) and by application of (4.5), eq. (5.23) may be written as

$$\frac{d[\sigma]}{dt} = C_1[\sigma] + C_2[\sigma]^2 + C_3[\sigma]^3, \tag{5.24}$$

where

$$C_1 = \frac{G_1'(0)}{2\varrho c^2} - \frac{M_1^2(0)\, T_0}{2k}, \tag{5.25}$$

$$C_2 = -\frac{3 M_1(0)\, K_1(0)\, T_0}{2k\varrho c^2}, \tag{5.26}$$

$$C_3 = -\frac{K_1^2(0)\, T_0}{k\varrho^2 c^4}. \tag{5.27}$$

With the initial condition, $[\sigma] = \sigma_0$ at $t = 0$,[2] the solution of eq. (5.24) is

$$\ln\left\{\frac{[\sigma]^2}{\sigma_0^2}\,\frac{C_1 + C_2\sigma_0 + C_3\sigma_0^2}{C_1 + C_2[\sigma] + C_3[\sigma]^2}\right\} - \frac{C_2}{\sqrt{C_2^2 - 4C_1C_3}} \times$$
$$\times \ln\left\{\left(\frac{2C_3[\sigma] + C_2 - \sqrt{C_2^2 - 4C_1C_3}}{2C_3[\sigma] + C_2 + \sqrt{C_2^2 - 4C_1C_3}}\right)\left(\frac{2C_3\sigma_0 + C_2 - \sqrt{C_2^2 - 4C_1C_3}}{2C_3\sigma_0 + C_2 - \sqrt{C_2^2 - 4C_1C_3}}\right)\right\} = 2C_1 t. \tag{5.28}$$

The solution of (5.24) can be simplified considerably if the ratio of the amplitude of the stress discontinuity to the instantaneous relaxation modulus $G_1(0)$ is $\ll 1$. The term $C_3[\sigma]^3$ is then negligible in comparison

[1] It should be noted that the influence of thermomechanical coupling at and behind the wave front is retained, and $[\eta]$ is still $O([\gamma])$.

[2] This initial condition implies a tension wave.

to $C_2[\sigma]^2$ and can be ignored. As a result eq. (5.24) becomes

$$\frac{d[\sigma]}{dt} = C_1[\sigma] + C_2[\sigma]^2. \tag{5.29}$$

The solution of (5.29) is

$$[\sigma] = \frac{\beta}{\left(\frac{\beta}{\sigma_0} + 1\right) e^{-C_1 t} - 1}, \tag{5.30}$$

where

$$\beta = C_1/C_2. \tag{5.31}$$

Since $G_1'(0) < 0$, eq. (5.30) states that the magnitude of the discontinuity always decays.

Further simplification of eq. (5.24) can be achieved by introducing the assumption that the material functions are temperature-independent. Then $K_1(0) \equiv 0$, and eq. (5.24) reduces to a linear first order ordinary differential equation with the solution

$$[\sigma] = \sigma_0 \exp\left\{\frac{G_1'(0)}{2G_1(0)} - \frac{M_1^2(0)\,T_0}{2k}\right\} t. \tag{5.32}$$

If we also neglect the effect of thermomechanical coupling, then $M_1(0) \equiv \equiv 0$ also and eq. (5.32) becomes

$$[\sigma] = \sigma_0 \exp\left\{\frac{G_1'(0)}{2G_1(0)}\right\} t. \tag{5.33}$$

Equation (5.33) is a well-known result, see, for example, the papers by VALANIS [14] and CHU [15]. By setting $G_1'(0) \equiv 0$ in (5.32) and (5.33) the corresponding results for the propagation of stress waves in elastic materials are obtained. Equation (5.33) is then obvious and eq. (5.32) agrees with earlier results [16] on the propagation of stress discontinuities according to the coupled equations of thermoelasticity.

Acknowledgment

This work was supported in part by the Advanced Research Project Agency of the U.S. Department of Defense through the Northwestern University Materials Research Center, and in part by the Office of Naval Research.

Appendix

The expansion of $\boldsymbol{\varphi}$ as given in Section 3 is an alternate representation of the "Taylor's expansion" of this functional in terms of the D and δ differential operators. The existence of such an expansion is guaranteed by the smoothness properties given to $\boldsymbol{\varphi}$ by the fading memory assumption.

Let $0^{\dagger}(s)$ be a zero function defined by

$$0^{\dagger}(s) = 0, \qquad 0 \leq s \leq t,$$

where 0 is the scalar zero. Furthermore, let

$$\gamma_0^{\dagger}(\ ) = \theta_0^{\dagger}(\ ) = 0^{\dagger}(\ ).$$

Since we assumed that the body was initially undeformed and at a uniform temperature, we shall expand $\boldsymbol{\varphi}$ about the zero functions $\gamma_0^{\dagger}$ and $\theta_0^{\dagger}$. In terms of the D and δ differential operators, the "Taylor's expansion" of φ to the third order in θ and second order in γ has the form (see [1], [8])

$$\boldsymbol{\varphi} = A_0 + A_1\gamma + A_2\gamma\theta + A_3\gamma\theta^2 + A_4\gamma^2 + A_5\gamma^2\theta + \\ + B_1\theta + B_2\theta^2 + B_3\theta^3, \tag{A.1}$$

where

$$A_0 = \overset{t}{\underset{s=0}{\boldsymbol{\varphi}}}(-) + \frac{1}{2}\delta^2_{\theta_0}\overset{t}{\underset{s=0}{\boldsymbol{\varphi}}}(-\,|\,\theta^t_r, \theta^t_r) + \frac{1}{2}\delta^2_{\gamma_0}\overset{t}{\underset{s=0}{\boldsymbol{\varphi}}}(-\,|\,\gamma^t_r, \gamma^t_r) + \\ + \delta_{\gamma_0}\delta_{\theta_0}\overset{t}{\underset{s=0}{\boldsymbol{\varphi}}}(-\,|\,\gamma^t_r, \theta^t_r) + \frac{1}{6}\delta^3_{\theta_0}\overset{t}{\underset{s=0}{\boldsymbol{\varphi}}}(-\,|\,\theta^t_r, \theta^t_r, \theta^t_r) + \\ + \frac{1}{2}\delta^2_{\gamma_0}\delta_{\theta_0}\overset{t}{\underset{s=0}{\boldsymbol{\varphi}}}(-\,|\,\gamma^t_r, \gamma^t_r, \theta^t_r) + \frac{1}{2}\delta_{\gamma_0}\delta^2_{\theta_0}\overset{t}{\underset{s=0}{\boldsymbol{\varphi}}}(-\,|\,\gamma^t_r, \theta^t_r, \theta^t_r), \tag{A.2a}$$

$$A_1 = D_{\gamma_0}\overset{t}{\underset{s=0}{\boldsymbol{\varphi}}}(-) + D_{\gamma_0}\delta_{\gamma_0}\overset{t}{\underset{s=0}{\boldsymbol{\varphi}}}(-\,|\,\gamma^t_r) + D_{\gamma_0}\delta_{\theta_0}\overset{t}{\underset{s=0}{\boldsymbol{\varphi}}}(-\,|\,\theta^t_r) + \\ + \frac{1}{2}D_{\gamma_0}\delta^2_{\theta_0}\overset{t}{\underset{s=0}{\boldsymbol{\varphi}}}(-\,|\,\theta^t_r, \theta^t_r) + D_{\gamma_0}\delta_{\gamma_0}\delta_{\theta_0}\overset{t}{\underset{s=0}{\boldsymbol{\varphi}}}(-\,|\,\gamma^t_r, \theta^t_r), \tag{A.2b}$$

$$A_2 = D_{\gamma_0}D_{\theta_0}\overset{t}{\underset{s=0}{\boldsymbol{\varphi}}}(-) + D_{\gamma_0}D_{\theta_0}\delta_{\theta_0}\overset{t}{\underset{s=0}{\boldsymbol{\varphi}}}(-\,|\,\theta^t_r) + D_{\gamma_0}D_{\theta_0}\delta_{\gamma_0}\overset{t}{\underset{s=0}{\boldsymbol{\varphi}}}(-\,|\,\gamma^t_r), \tag{A.2c}$$

$$A_3 = \frac{1}{2}D_{\gamma_0}D^2_{\theta_0}\overset{t}{\underset{s=0}{\boldsymbol{\varphi}}}(-), \tag{A.2d}$$

$$A_4 = \frac{1}{2}D^2_{\gamma_0}\overset{t}{\underset{s=0}{\boldsymbol{\varphi}}}(-) + \frac{1}{2}D^2_{\gamma_0}\delta_{\theta_0}\overset{t}{\underset{s=0}{\boldsymbol{\varphi}}}(-\,|\,\theta^t_r), \tag{A.2e}$$

$$A_5 = \frac{1}{2}D^2_{\gamma_0}D_{\theta_0}\overset{t}{\underset{s=0}{\boldsymbol{\varphi}}}(-), \tag{A.2f}$$

$$B_1 = D_{\theta_0} \overset{t}{\underset{s=0}{\boldsymbol{\varphi}}} (-) + D_{\theta_0} \delta_{\theta_0} \overset{t}{\underset{s=0}{\boldsymbol{\varphi}}} \left(- \mid \theta_r^t\right) + D_{\theta_0} \delta_{\gamma_0} \overset{t}{\underset{s=0}{\boldsymbol{\varphi}}} \left(- \mid \gamma_r^t\right) + \\ + \frac{1}{2} D_{\theta_0} \delta_{\theta_0}^2 \overset{t}{\underset{s=0}{\boldsymbol{\varphi}}} \left(- \mid \theta_r^t, \theta_r^t\right) + \frac{1}{2} D_{\theta_0} \delta_{\gamma_0}^2 \overset{t}{\underset{s=0}{\boldsymbol{\varphi}}} \left(- \mid \gamma_r^t, \gamma_r^t\right) + \\ + D_{\theta_0} \delta_{\gamma_0} \delta_{\theta_0} \overset{t}{\underset{s=0}{\boldsymbol{\varphi}}} \left(- \mid \gamma_r^t, \theta_r^t\right), \tag{A.2g}$$

$$B_2 = \frac{1}{2} D_{\theta_0}^2 \overset{t}{\underset{s=0}{\boldsymbol{\varphi}}} (-) + \frac{1}{2} D_{\theta_0}^2 \delta_{\theta_0} \overset{t}{\underset{s=0}{\boldsymbol{\varphi}}} \left(- \mid \theta_r^t\right) + \frac{1}{2} D_{\theta_0}^2 \delta_{\gamma_0} \overset{t}{\underset{s=0}{\boldsymbol{\varphi}}} \left(- \mid \gamma_r^t\right), \tag{A.2h}$$

$$B_3 = \frac{1}{6} D_{\theta_0}^3 \overset{t}{\underset{s=0}{\boldsymbol{\varphi}}} (-), \tag{A.2i}$$

and $\overset{t}{\underset{s=0}{\boldsymbol{\varphi}}} (-)$ represents $\overset{t}{\underset{s=0}{\boldsymbol{\varphi}}} \left(\gamma_0^\dagger, \theta_0^\dagger\right)$.

Each term on the right in eqs. (3.7) is the equivalent representation of the term which occupies the same position in eqs. (A.2).

In eq. (A.2a), the terms $\delta_{\gamma_0} \boldsymbol{\varphi}$ and $\delta_{\theta_0} \boldsymbol{\varphi}$ have been omitted. It follows from the Clausius-Duhem inequality [5] that whenever γ^t and θ^t are the (constant) functions $\gamma_0^\dagger$ and $\theta_0^\dagger$ these functionals must vanish.

References

[1] COLEMAN, B. D., and M. E. GURTIN: Waves in Materials with Memory. III. Thermodynamic Influences on the Growth and Decay of Acceleration Waves. Arch. Ratl Mech. Anal. **19**, 266—298 (1965).

[2] COLEMAN, B. D., and M. E. GURTIN: Waves in Materials with Memory. IV. Thermodynamics and the Velocity of General Acceleration Waves. Arch. Ratl Mech. Anal. **19**, 317—338 (1965).

[3] COLEMAN, B. D., and M. E. GURTIN: Thermodynamics and One-Dimensional Shock Waves in Materials with Memory. Proc. Roy. Soc. **A-292**, 562—574 (1966).

[4] HUNTER, S. C.: Tentative Equations for the Propagation of Stress, Strain, and Temperature Fields in Viscoelastic Solids. J. Mech. Phys. Solids **9**, 39—51 (1961).

[5] COLEMAN, B. D.: Thermodynamics of Materials with Memory. Arch. Ratl Mech. Anal. **17**, 1—46 (1964).

[6] COLEMAN, B. D.: On Thermodynamics, Strain Impulses, and Viscoelasticity. Arch. Ratl Mech. Anal. **17**, 230—254 (1964).

[7] NOLL, W.: A Mathematical Theory of the Mechanical Behavior of Continuous Media. Arch. Ratl Mech. Anal. **2**, 197—226 (1958).

[8] COLEMAN, B. D., M. E. GURTIN, and I. HERRERA R.: Waves in Materials with Memory. I. The Velocity of One-Dimensional Shock and Acceleration Waves. Arch. Ratl Mech. Anal. **19**, 1—19 (1965).

[9] COLEMAN, B. D., and W. NOLL: Foundations of Linear Viscoelasticity. Rev. Modern Phys. **33**, 239—249 (1961).

[10] VOLTERRA, V.: Theory of Functionals and of Integral and Integro-Differential Equations. New York: Dover Publications. 1959.

[11] TRUESDELL, C., and R. A. TOUPIN: The Classical Field Theories. In: Encyclopedia of Physics. Vol. III/1, § 180. Ed. by S. FLÜGGE. Berlin-Göttingen-Heidelberg: Springer. 1960.
[12] ACHENBACH, J. D.: Approximate Transient Solutions for the Coupled Equations of Thermoelasticity. J. Acoust. Soc. Amer. **36**, 10–18 (1964).
[13] SCHAPERY, R. A.: Effect of Cyclic Loading on the Temperature in Viscoelastic Media with Variable Properties. AIAA Journal **2**, 827–835 (1964).
[14] VALANIS, K. C.: Propagation and Attenuation of Waves in Linear Viscoelastic Solids. J. Math. Phys. **44**, 227–239 (1965).
[15] CHU, B. T.: Stress Waves in Isotropic Linear Viscoelastic Materials. J. Mécanique **1**, 439–462 (1962).
[16] ACHENBACH, J. D.: The Propagation of Stress Discontinuities According to the Coupled Equations of Thermoelasticity. Acta Mechanica **3**, 4, 342–351 (1967).

A Thermodynamic Approach to Rheology

By

J. F. Besseling

Delft (Netherlands)

Summary

A continuum theory of deformation and flow is presented, that is based on the principle of conservation of mass, the first and second law of thermodynamics, the concepts of a local thermodynamic state and of a local geometric natural reference state, a principle of determinism and on a postulate concerning the production of entropy. As special cases are considered gases, elastic materials, simple solids and liquids. Stress is in this theory a thermodynamically defined quantity.

Introduction

The mathematical theories of deformation and flow of matter deal essentially with the gross properties of a medium. Addition of heat and performance of mechanical work are considered as distinct means for a change of the state of the medium. The resulting phenomena in any particular material are, however, not unrelated and a thermodynamical treatment of the foundations of the theory of flow and deformation seems the appropriate, and indeed the obvious approach, but then full account should be given of the rate of change of all quantities that determine the thermodynamic state.

A theory of deformation and flow, founded on thermodynamic concepts, should not make use of stress as a primitive concept and of the equations of motion, but the stress concept and the equations of motion should follow as mathematical consequences of the theory. It is to be understood, however, that within the context of a thermodynamic theory the concept of stress is limited by the nature of the state variables, that are being considered. Thus in a theory, in which the effect of changes of geometry on the state of a medium is solely represented by the mass density as a state variable, the concept of stress will only arise as a hydrostatic pressure. One might agree therefore with TRUESDELL's statement, that the equations of mechanics describe a wider range of phenomena than do the equations of thermodynamics (TRUESDELL 1952), but it should be realized, that the equations of mechanics are lacking in physical content as long as no constitutive equations for the stresses are given.

It is the conviction of the author that stress as a thermodynamically defined quantity is a sufficiently broad concept to cope with the observed phenomena in the deformation and flow of matter, though admittedly in many instances the equations may in an approximation be simplified in a way, that obscures their thermodynamical content. This paper is an attempt to show that a continuum theory of deformation and flow can be based on the principle of conservation of mass, the first and second law of thermodynamics, the concepts of a local thermodynamic state and a local geometric natural reference state, a principle of determinism and on a postulate concerning the production of entropy.

The concept of a local geometric natural reference state in connection with inelastic phenomena was first introduced by ECKART (1948). His use of a reference rate tensor, not subjected to kinematical compatibility equations, has unduly been criticized by TRUESDELL (1952, 1960), because the latter did not recognize the physical content of this concept. The local, generally non-Euclidean metric of the natural reference state is to be considered as a thermodynamic state variable, as it is done for instance by COLEMAN and NOLL (1959), and its rate of change is to be determined by a constitutive equation for the production of entropy.

The principle of determinism put forward in this paper is in the spirit of classical mechanics. It contends that the process of flow and deformation shall be fully determined, if for a body, free in space, at a any initial time the thermodynamic state and, independently of this state, the velocity field inside the body are prescribed. From the arbitrariness of the velocity field with respect to the initial thermodynamic state then follow the equations of motion, while the requirement that interaction between various regions in the body shall be such, that the equation expressing the principle of conservation of energy is identically satisfied, leads to the equation for the local rate of energy dissipation.

For the postulate concerning the production of entropy no other justification can be given than that it may be regarded as another consequence of the principle of determinism if one refuses to complicate the theory further by the addition of other new principles.

It will be shown how the theory, put forward in this paper, reduces the problem of rheological constitutive equations to the determination of scalar functions of state variables, which define the internal energy and the rate of energy dissipation. It should be observed that in this theory, there is no place for pure viscosity, while volume change is always a purely thermo-elastic phenomenon. It may well be argued however, that this is by no means a limitation of the theory in view of the available data on actual behaviour of materials. The theory leads to the Navier-Stokes equations of viscous flow if the restriction of quasi-stationary flow is introduced. It is this restriction that obscures the thermodynamic nature of the shear stresses in this case.

Also in the case of time-independent plasticity, particularly for the rigid-perfectly plastic model, the thermodynamic content of the equations is obscured by the limiting process, that is to be applied to the equations of the general theory.

Still the author realizes that the thermodynamic approach to rheology, presented in this paper, is open to many questions. In particular the extension of the theory to electro-magnetic phenomena is by no means clear, while further the rigid deterministic approach to rheology may well fail to describe adequately the infinite complexity of actual material behaviour.

Notation

Throughout the paper index notation with summation convention and Kronecker delta is used.

The theory is formulated in terms of rectangular Cartesian coordinates and Cartesian tensors.

We employ the superposed dot for the material timederivative in the moving continuum:

$$\dot{A} \doteq \frac{\partial A}{\partial t} + \frac{\partial A}{\partial x_i}\,\dot{x}_i .$$

If necessary the extent of application of the dot will be indicated by an overbar, for instance in $\dot{\overline{\varrho\, d V}}$.

The superposed -1 indicates the inverse of a transformation, defined by

$$\overset{-1}{b}_{i\alpha} b_{\alpha j} = \delta_{ij} .$$

Orthogonal transformations are indicated by R_{ij}, where

$$\overset{-1}{R}_{ij} = R_{ji} .$$

The superposed $\frac{1}{2}$ denotes the square root of a symmetric tensor, defined by

$$\overset{1/2}{C}_{\alpha\beta} = \sum_p \sqrt{C^{(p)}}\, n_\alpha^{(p)} n_\beta^{(p)} .$$

where $C^{(p)}$ is a principal value of the tensor $C_{\alpha\beta}$ and where $n_\alpha^{(p)}$ are the direction cosines of the corresponding principal direction.

Conservation of Mass

In classical mechanics and thermodynamics mass is recognized as an indestructible property of any closed system. The motion of elementary particles of a system can in general not be described by the motion of a continuum, in which each particle occupies a material point. Furthermore the system may consist of different types of particles. If, however, at any instant the probability for a particle to be of a certain type is independent of its location in the system, the average motion of mass inside the system can be represented by the motion $\dot{x}_i = \dot{x}_i(x_1, x_2, x_3, t)$ of a continuum with a mass density $\varrho = \varrho(x_1, x_2, x_3, t)$. We shall consider only systems for which this continuum approach is permitted and which are of finite extent, with a non-vanishing mass density ϱ inside a moving volume V. This volume with constant mass will be denoted as a body.

We shall assume that for the body the velocities $\dot{x}_i$ and the mass density ϱ are continuous differentiable functions of the coordinates. The principle of conservation of mass is then expressed by

$$\dot{\overline{\varrho\, d V}} = \left(\dot{\varrho} + \varrho \frac{\partial \dot{x}_i}{\partial x_i}\right) d V = 0,$$

or

$$\frac{\partial \varrho}{\partial t} + \frac{\partial \varrho}{\partial x_i} \dot{x}_i + \varrho \frac{\partial \dot{x}_i}{\partial x_i} = 0. \tag{1}$$

First Law of Thermodynamics

In classical mechanics and thermodynamics it is asserted that with respect to each of certain preferred coordinate systems, the so-called inertial systems, the quantity energy,

$$U + \frac{1}{2} \int\limits_V \varrho \dot{x}_i \dot{x}_i d V,$$

is a second indestructible property of any closed system, where the internal energy U is independent of the average mass velocity $\dot{x}_i$ and is an unique functional of appropriate state variables for the material under consideration. The class of coordinate systems, with respect to which the kinetic energy $\frac{1}{2} \int\limits_V \varrho \dot{x}_i \dot{x}_i d V$ is to be defined, is determined by the group of Galilean transformations

$$x_i' = V_i t + \overline{R}_{ij}(x_j - V_j t), \overline{R}_{ij} \overline{R}_{kj} = \delta_{ik}, \tag{2}$$

where V_i represents an arbitrary constant velocity and where $\overline{R}_{ij}$ is a constant orthogonal matrix. Since U is to be independent of $\dot{x}_i$, it follows that U must be invariant under the transformation (2).

Second Law of Thermodynamics

Changes of state of a material system can be due to external and internal causes. For a closed system, where the state can only change by internal causes, it is a fact of experience that the system will tend to an equilibrium state, characterized by the vanishing of the timederivatives of all state variables. This concept of irreversibility of the changes of state is by the second law of thermodynamics formulated in terms of a functional of state variables, the entropy S, that for a closed system satisfies the inequality.

$$\dot{S} \geq 0. \tag{3}$$

It may be observed that in an equilibrium state holds $\dot{U} = 0$ and $\dot{S} = 0$, since both U and S are functionals of state variables only.

The Concept of a Local Thermodynamic State

A closed material system may be broken into a number of interacting subsystems. Now it is postulated that, like the kinetic energies, the internal energies as well as the entropies are additive. This implies that the internal energy and the entropy of a subsystem are functionals of state variables, which refer to this subsystem only. It is clear that there will be a limit as to the smallness of a subsystem to which this postulate still applies. The thermodynamic approach is a-priori restricted to systems, where statistical averages are meaningful. However, the definition of a local internal energy u and *a* local entropy s, both taken per unit mass, for a system in a homogeneous state a mathematical triviality, is a lawful artifice in the description of non-homogeneous states of bodies, where u- and s-distributions may represent the internal energy and the entropy of all subsystems up to the smallest system, that permits a statistical averaging procedure. If furthermore the variation of state variables is small within a region that is large compared to the smallest thermodynamic system, then we may consider the values of u and s in a point as functions of locally defined state variables. However, since the entropy has been introduced to express the irreversibility of the changes of state of a closed system, the entropy per unit mass, s, may itself, without loss of generality, be treated as a parameter specifying the local state, together with other appropriate state variables for which the rate of change is not subjected to any inequality. The scalar quantity s is invariant under the transformation (2) and the internal energy per unit mass, u, may therefore be any arbitrary function of s.

In order to indicate, that the notion of the smallest thermodynamic system presents under normal circumstances no restriction to the validity of the concept of a local thermodynamic state, it can be observed that the number of particles in air at sealevel amounts to $2 \cdot 7 \cdot 10^{19}\ \mathrm{cm}^{-3}$ and in iron to $8 \cdot 6 \cdot 10^{22}\ \mathrm{cm}^{-3}$.

According to the second law of thermodynamics for any closed system the rate of change of s must satisfy the inequality (3)

$$\dot{S} = \int_V \varrho \dot{s} \, d V \geq 0 .$$

Locally, however, $\dot{s}$ may be negative due to the interaction of various subsystems in the body. Because of the additive property of the entropy we can, besides an entropy production per unit mass $\overset{+}{s} \geq 0$, define an entropy flux $\dot{h}_i = \dot{h}_i(x_1, x_2, x_3, t)$, that determines the distribution of entropy, such that

$$\varrho \dot{s} = -\frac{\partial \dot{h}_i}{\partial x_i} + \varrho \overset{+}{s}, \quad \overset{+}{s} \geq 0 . \tag{4}$$

If the body as a whole is a closed system, its boundary is a surface through which no interaction takes place, and hence on this surface shall hold $\dot{h}_i n_i = 0$, where n_i is the outward unit normal. The inequality for the local rate of entropy production, $\overset{+}{s} \geq 0$, ensures that the second law of thermodynamics is satisfied, since

$$\dot{S} = \int_V \left(-\frac{\partial \dot{h}_i}{\partial x_i} + \varrho \overset{+}{s} \right) d V = -\int_A \dot{h}_i n_i \, d A + \int_V \varrho \overset{+}{s} \, d V .$$

Later on we shall require that the flux of interaction $\dot{h}_i$ is such, that the equation expressing the principle of conservation of energy shall be satisfied identically, thereby establishing the connection between the flux $\dot{h}_i$ and the flux of energy, known as the heatflux. Inequality (4) then becomes equivalent to the Clausius-Duhem inequality of thermodynamics.

Insofar as the rate of change of state variables is not defined in terms of the velocity field of the moving continuum, representing the average mass motion, this rate of change shall be determined by thermodynamic constitutive equations, which are an expression of the properties of the material as a thermodynamic system. They constitute the essence of the theory of irreversible thermodynamics.

The Geometric Natural Reference State

Let the properties of the material under consideration be defined in some reference state. The internal energy u may now be considered as a function of state variables, that define changes with respect to this reference state. Thus in the following s will represent the change of entropy and will be taken equal to zero in the reference state.

It will be assumed that the physical properties of the material in the reference state, that determine the functional dependence of u on the state variables, are preserved in any subsequent process, in particular any anisotropy, that the material may possess. Anisotropic properties can be defined with reference to orthogonal triads in each point of the body, which without rotation are carried along by the motion of the continuum. It is now postulated that the geometrical configuration of the material at any instant in each point of the body is related to the configuration in the reference state by a invertible transformation of line elements

$$d a_\alpha = b_{\alpha i} d x_i, \quad \det (b_{\alpha i}) > 0 . \tag{5}$$

Here the $d a_\alpha$'s define line elements in the local coordinate system, determined by the orthogonal triads, which, with their fixed directions in space, form a base of reference for the anisotropic properties of the material.

By the theorem on the unique polar decomposition of a tensor with a non-vanishing determinant the above transformation of lineelements may be decomposed into a rigid rotation followed by a deformation with a symmetric transformation matrix, as illustrated for the two-dimensional case in Fig. 1. It should be realized that the configuration of the natural reference state in a point of the moving continuum with respect to the orthogonal triads remains always the same, though the identity of the material particles, that make up this configuration, is in general subject to change. This is an allowable concept since the motion of the continuum describes only the average motion of mass, but not the motion of the individual particles. So, if we consider for instance a crystal lattice, in which a process of slip has taken place, then a unit cell can be made up of atoms, that before slip took place belonged with other atoms to different unit cells. The unit cell, illustrated in Fig. 1, in a point of the continuum, covering the lattice, is at any instant the same, but the identity of the atoms, that make up the unit cell, may be subject to change.

It follows from the above that it is in general not permissible to consider the system (5) as exact differentials for the corresponding functions of the coordinate points in the space x_1, x_2, x_3, because the conditions of integrability will only be satisfied in special cases (i.e. in thermo-elasticity). The tensor $b_{\alpha i}$ is a physical concept, like the metric tensor of the natural reference state, as introduced by ECKART (1948), and it is not subject to kinematic compatibility equations derived from the

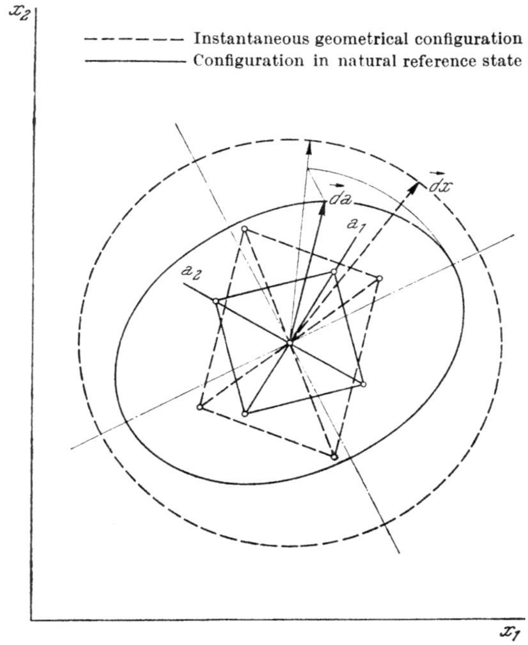

Fig. 1. Two-dimensional illustration of the transformation $d a_\alpha = b_{\alpha i} \, d x_i$

motion of the continuum, which represents only the average motion of mass inside the body.

State variables in the internal energy function u shall, like the internal energy itself, be invariant under the group of Galilean transformation (2). Since the orthogonal triads in each point of the body are connected with the fixed natural reference state, they are not affected by the transformations (2) and hence the $d a_\alpha$'s are invariant under this transformation. It follows that

$$d a_\alpha = b_{\alpha i} \, d x_i = b_{\alpha i} \overline{R}_{ji} \, d x'_j = b'_{\alpha j} \, d x'_j ,$$

or

$$b_{\alpha i} = b'_{\alpha j} \overline{R}_{ji} . \tag{6}$$

Since the components of the tensor $b_{\alpha i}$ are not invariant under the transformation (2), they are not thermodynamic state variables. They define, however, a tensor $C_{\alpha\beta}$

$$C_{\alpha\beta} = b_{\alpha i} b_{\beta i} = b'_{\alpha j} \overline{R}_{ji} b'_{\beta k} \overline{R}_{ki} = b'_{\alpha j} b'_{\beta j}, \tag{7}$$

which does possess the required property of invariance under the group of Galilean transformations. We shall show in the following that a broad theory of rheology can be based upon an internal energy, that is an unique function of the state variables s and $C_{\alpha\beta}$.

$$u = u(s, C_{\alpha\beta}). \tag{8}$$

The Principle of Determinism

From classical mechanics it is known that if, independently of each other, initial positions and initial velocities of mass points are given, the subsequent motion is determined. It is in the spirit of this principle of determinism, that it is postulated here, that an initial thermodynamic state and an initial velocity field, prescribed independently of each other, shall suffice to determine the subsequent motion and accompanying thermodynamic processes. It follows that to all possible thermodynamic states and to all velocity fields shall correspond definite rates of change of all state variables.

It was pointed out before that in sofar as the rate of change of state variables is not defined in terms of the velocity field of the moving continuum, this rate of change shall be determined by the material as a thermodynamic system. However, an element of volume dV, imbedded in the moving continuum, does in general not represent one and the same thermodynamic system at subsequent times, since the velocity field defines only the average motion of mass. This difficulty does not arise, if it is postulated that the thermodynamic processes depend only on the instantaneous values of the state variables, because at any time t an element of volume dV does comprise a thermodynamic system within the limitations imposed by the concept of a local thermodynamic state.

From here on therefore, we shall consider the rate of entropy production $\overset{+}{s}$ and the flux of interaction $\dot{h}_i$ to be at any instant uniquely determined by the thermodynamic state, as characterized by the values of the state variables.

With respect to the geometric natural reference state it can be observed that its rate of change, characterized by $\dot{\overline{da_\alpha}}$, may be expressed in terms of quantities, determined by a local thermodynamic process through the values of the state variables, as well as in terms of the rate of change of the tensor $b_{\alpha i}$ and the velocity field of the moving continuum:

$$\overline{da_\alpha}^{\,\cdot} = p_{\alpha i} dx_i = \left(\dot{b}_{\alpha i} + b_{\alpha j}\frac{\partial \dot{x}_j}{\partial x_i}\right) dx_i,$$

or

$$\dot{b}_{\alpha i} = p_{\alpha i} - b_{\alpha j}\frac{\partial \dot{x}_j}{\partial x_i}, \tag{9}$$

where the tensor $p_{\alpha i}$ shall now be uniquely determined by the values of the state variables. This concept of a rate of change of the natural reference state determined by a thermodynamic process, independent of the concept of a moving continuum, is due to ECKART (1948), though by him it was formulated in a more restricted sense in terms of a rate of change of the reference metric.

With the aid of (9) the rate of change of the state variables $C_{\alpha\beta}$ can be expressed by

$$\dot{C}_{\alpha\beta} = \dot{b}_{\alpha i} b_{\beta i} + b_{\alpha i}\dot{b}_{\beta i} = p_{\alpha i} b_{\beta i} + b_{\alpha i} p_{\beta i} - 2 b_{\alpha i} b_{\beta j} d_{ij}, \tag{10}$$

where d_{ij} is the rate of deformation tensor, defined by

$$d_{ij} = \frac{1}{2}\left(\frac{\partial \dot{x}_i}{\partial x_j} + \frac{\partial \dot{x}_j}{\partial x_i}\right).$$

Now, on the basis of an internal energy function $u = u(s, C_{\alpha\beta})$, a number of conclusions can be drawn from the first law of thermodynamics and the principle of determinism.

The rate of change of u is given by

$$\dot{u} = \frac{\partial u}{\partial C_{\alpha\beta}}\dot{C}_{\alpha\beta} + \frac{\partial u}{\partial s}\dot{s},$$

or, according to expressions (4) and (10), by

$$\varrho\dot{u} = 2\varrho\frac{\partial u}{\partial C_{\alpha\beta}} b_{\alpha i} p_{\beta i} - 2\varrho\frac{\partial u}{\partial C_{\alpha\beta}} b_{\alpha i} b_{\beta j} d_{ij} + \varrho\frac{\partial u}{\partial s}\dot{\bar{s}} - \frac{\partial u}{\partial s}\frac{\partial \dot{h}_i}{\partial x_i}. \tag{11}$$

We shall define the following quantities, which are determined by the thermodynamic state through the internal energy function u:

$$t_{ij} = -2\varrho\frac{\partial u}{\partial C_{\alpha\beta}} b_{\alpha i} b_{\beta j}, \tag{12}$$

$$T = \frac{\partial u}{\partial s}. \tag{13}$$

The tensor t_{ij} will be denoted as stress tensor and the quantity T as temperature, because these quantities will appear to possess all the characteristics, usually attributed to quantities with these names.

For a body, that constitutes a closed system, conservation of energy is expressed by

$$\dot{U} + \int_V \varrho \ddot{x}_i \dot{x}_i \, dV = \int_V \varrho (\dot{u} + \ddot{x}_i \dot{x}_i) dV$$

$$= \int_V \left[t_{ij} d_{ij} - t_{ij} \overset{-1}{b}_{j\alpha} \dot{p}_{\alpha i} + T \varrho \overset{+}{\dot{s}} - T \frac{\partial \dot{h}_i}{\partial x_i} + \varrho \ddot{x}_i \dot{x}_i \right] dV = 0 . \quad (14)$$

By means of the divergence theorem, and observing that on the bounding surface of the body as a closed system holds $\dot{h}_i n_i = 0$, we can transform eq. (14) into

$$\int_V \left[\left(-\frac{\partial t_{ij}}{\partial x_j} + \varrho \ddot{x}_i \right) \dot{x}_i + T \varrho \overset{+}{\dot{s}} + \frac{\partial T}{\partial x_i} \dot{h}_i - t_{ij} \overset{-1}{b}_{j\alpha} \dot{p}_{\alpha i} \right] dV +$$

$$+ \int_A t_{ij} n_j \dot{x}_i \, dA = 0 . \quad (15)$$

According to the principle of determinism, formulated above, any possible thermodynamic state and any independently chosen velocity field constitute allowable and sufficient initial conditions for the mathematical description of the subsequent motion with the accompanying thermodynamic processes. Therefore eq. (15) must hold for arbitrary velocity fields $\dot{x}_i$, while all other quantities, ϱ, $\ddot{x}_i$, and the quantities determined by the thermodynamic state, are independent of $\dot{x}_i$. From the principal theorem of the calculus of variations then follows the important conclusion, that

$$\frac{\partial t_{ij}}{\partial x_j} - \varrho \ddot{x}_i = 0 \text{ in } V \quad (16)$$

and

$$t_{ij} n_j = 0 \quad \text{on } A. \quad (17)$$

Equations (16) are to be identified as the equations of motion of Cauchy, which in our case are obtained without reference to the principle of balance of momentum, be it in the limited sense of a thermodynamically defined stress tensor. Further this stress tensor is symmetric, not by virtue of the principle of balance of moment of momentum, but because of the symmetry of the tensor $C_{\alpha\beta}$ in (12).

By eqs. (16) and (17) eq. (15), expressing the conservation of energy, is reduced to

$$\int_V \left(T \varrho \overset{+}{\dot{s}} + \frac{\partial T}{\partial x_i} \dot{h}_i - t_{ij} \overset{-1}{b}_{j\alpha} \dot{p}_{\alpha i} \right) dV = 0 . \quad (18)$$

The flux $\dot{h}_i$ has been introduced to represent the interaction between various regions in the body. So far no restriction has been imposed on this interaction. We shall now require that the flux of interaction $\dot{h}_i$ is such, that the eq. (18) is satisfied identically, i.e.

$$T\varrho \overset{+}{\dot{s}} + \frac{\partial T}{\partial x_i}\dot{h}_i - t_{ij}\overset{-1}{b}_{j\alpha}p_{\alpha i} = 0. \tag{19}$$

Let us now consider a closed region of volume $V_{\text{int.}}$, bounded by a surface $A_{\text{int.}}$ inside the body. For the rate of change of the energy, contained in the volume $V_{\text{int.}}$, is found

$$\int_{V_{\text{int.}}} \varrho(\dot{u} + \ddot{x}_i\dot{x}_i)\,dv = \int_{A_{\text{int.}}} (t_{ij}n_j\dot{x}_i - T\dot{h}_i n_i)\,dA. \tag{20}$$

Hence the interaction between the region $V_{\text{int.}}$ and the rest of the body can be interpreted as a flux of mechanical energy (i.e. defined with the aid of the velocity field of the continuum), $t_{ij}n_j\dot{x}_i$, and a flux of non-mechanical energy, $-T\dot{h}_i n_i$, through the surface $A_{\text{int.}}$. The flux $q_i = T\dot{h}_i$ is then to be identified as the heatflux.

While according to (17) the mechanical interaction vanishes, if the thermodynamic state is such that $t_{ij}n_j = 0$, the non-mechanical interaction by the very definition of $\dot{h}_i$, vanishes only if $\dot{h}_i n_i = 0$. Interaction without energy exchange is beyond the scope of thermodynamics. Hence the temperature T must be unequal to zero for all thermodynamic states, where a non-mechanical interaction is possible. Therefore the scale for T, hitherto left undefined, must satisfy the condition $T \geq 0$; the state with $T = 0$ being defined as the limiting case of a state where any non-mechanical thermodynamic interaction becomes impossible. Inequality (4), after substitution of the heatflux, $q_i = T\dot{h}_i$, with the so-defined absolute temperature, is known as the Clausius-Duhem inequality.

Irreversible Thermodynamics

For the problem of rheology, as it is being considered here, the theory of irreversible thermodynamics pertains to the determination of constitutive equations for the quantities $\overset{+}{\dot{s}}$, $\dot{h}_i$ and $p_{\alpha i}$, which occur in eq. (19).

First we shall show that the tensor $p_{\alpha i}$ must satisfy a specific condition in order to ensure, that the logical requirement is fulfilled, that the mass density, corresponding to the geometric natural reference state, is constant and equal to a given value ϱ_0.

The length of a line element in the natural reference state is according to (5) defined by

$$dl_0^2 = da_\alpha\, da_\alpha = b_{\alpha i}\, b_{\alpha j}\, dx_i\, dx_j. \tag{21}$$

The tensor

$$g_{ij} = b_{\alpha i} b_{\alpha j}, \; g = \det(g_{ij}) \neq 0, \tag{22}$$

is the metric tensor, that defines the generally non-Euclidean metric of the natural reference state in the x_1, x_2, x_3 space. By virtue of the following equalities

$$\overline{\ln g}^{\,\cdot} = \overset{-1}{g}_{ij} \dot{g}_{ij} = \overset{-1}{g}_{ij} \left[\left(p_{\alpha i} - b_{\alpha k} \frac{\partial \dot{x}_k}{\partial x_i}\right) b_{\alpha j} + b_{\alpha i}\left(p_{\alpha j} - b_{\alpha k} \frac{\partial \dot{x}_k}{\partial x_j}\right)\right],$$

$$\overline{\ln g}^{\,\cdot} = \overset{-1}{g}_{ij} (p_{\alpha i} b_{\alpha j} + b_{\alpha i} p_{\alpha j}) - 2 \frac{\partial \dot{x}_k}{\partial x_k},$$

and with the aid of eq. (1) in the form

$$\frac{\dot{\varrho}}{\varrho} = \overline{\ln \frac{\varrho}{\varrho_0}}^{\,\cdot} = - \frac{\partial \dot{x}_k}{\partial x_k}$$

we arrive at

$$\overset{-1}{g}_{ij} (p_{\alpha i} b_{\alpha j} + b_{\alpha i} p_{\alpha j}) = 2 \, \overline{\ln \frac{\varrho_0 \sqrt{g}}{\varrho}}^{\,\cdot}.$$

Since

$$\overset{-1}{g}_{ij} = (b_{\alpha i} b_{\alpha j})^{-1} = \overset{-1}{b}_{i\alpha} \overset{-1}{b}_{j\alpha},$$

we have

$$\overset{-1}{g}_{ij} (p_{\alpha i} b_{\alpha j} + b_{\alpha i} p_{\alpha j}) = 2 \overset{-1}{b}_{i\alpha} p_{\alpha i} = 2 \, \overline{\ln \left(\frac{\varrho_0 \sqrt{g}}{\varrho}\right)}^{\,\cdot}. \tag{23}$$

Locally the metric tensor g_{ij} defines the relation between a material element of volume dV in the x_1, x_2, x_3 space and the same material element of volume dV_0 in the natural reference state by

$$dV_0 = \sqrt{g}\, dV.$$

Because the mass content of dV and dV_0 is the same, we have

$$\varrho_0 \, dV_0 = \varrho_0 \sqrt{g} \, dV = \varrho \, dV$$

$$\frac{\varrho_0 \sqrt{g}}{\varrho} = 1. \tag{24}$$

Hence according to eqs. (23) and (24) the tensor $p_{\alpha i}$ must satisfy the condition

$$2 \overset{-1}{b}_{i\alpha} p_{\alpha i} = 0. \tag{25}$$

If the stress tensor is decomposed into its deviator s_{ij} and its pressure component p,

$$t_{ij} = s_{ij} - p\delta_{ij}, \; s_{\alpha\alpha} = 0, \tag{26}$$

we can observe, that because of condition (25), the pressure component does not contribute to the rate of energy dissipation. According to eq. (19)

$$T\varrho\overset{+}{s} = -\frac{\partial T}{\partial x_i}\dot{h}_i + s_{ij}\overset{-1}{b}_{j\alpha}p_{\alpha i} - p\overset{-1}{b}_{j\alpha}p_{\alpha j},$$

or by (25)

$$T\varrho\overset{+}{s} = -\frac{\partial T}{\partial x_i}\dot{h}_i + s_{ij}\overset{-1}{b}_{j\alpha}p_{\alpha i}. \tag{27}$$

As all three quantities, T, ϱ, and $\overset{+}{s}$, are non-negative, we know from what has been postulated in the preceding paragraphs, that the rate of energy dissipation $T\varrho\overset{+}{s}$ is a positive, semi-definite function of state variables.

We note that the rate of energy dissipation is always the scalar product of a vector or tensor of quantities, determined by the thermodynamic state, and a dissipation vector or tensor. The relative magnitudes of the components of the dissipation vector or tensor shall depend on the dissipation mechanism. We shall now *postulate* that sufficient information on the dissipation mechanism is contained in the energy dissipation function, such as to ensure that the relative magnitudes of the components of the dissipation vector or tensor depend on the state variables solely through the functional dependence of the energy dissipation function on those state variables, that are associated with the dissipation vector or tensor in the expression for the rate of energy dissipation.

The quantities $\frac{\partial T}{\partial x_i}$ and s_{ij} in eq. (27) are no state variables, since they are not invariant under the transformation (2). We shall show, however, that they are related to state variables through orthogonal transformations.

According to the theorem on the unique polar decomposition of a tensor with a non-vanishing determinant, we have

$$b_{\alpha i} = \overset{1/2}{C}_{\alpha\beta}R_{\beta i}, \quad R_{\alpha i}R_{\beta i} = \delta_{\alpha\beta}, \tag{28}$$

where $\overset{1/2}{C}_{\alpha\beta}$ is invariant and where $R_{\alpha i}$ obeys the transformation rule

$$R'_{\alpha i} = R_{\alpha j}\overline{R}_{ij}. \tag{29}$$

Hence the quantities

$$t_{ij} R_{\alpha i} R_{\beta j} = -2\varrho \frac{\partial u}{\partial C_{\mu\nu}} \overset{1/2}{C}_{\mu\alpha} \overset{1/2}{C}_{\nu\beta} \tag{30}$$

are proper state variables.

Further it can be observed that by virtue of the invariance of the quantity

$$\frac{\partial T}{\partial x_i} d x_i = \frac{\partial T}{\partial x_i} \overline{R}_{ji} \, d x_j'$$

the transformation rule for $\frac{\partial T}{\partial x_i}$ reads

$$\left(\frac{\partial T}{\partial x_i}\right)' = \frac{\partial T}{\partial x_j} \overline{R}_{ij} . \tag{31}$$

Therefore the quantities

$$R_{\alpha i} \frac{\partial T}{\partial x_i} = R_{\alpha j} \overline{R}_{ij} \frac{\partial T}{\partial x_k} \overline{R}_{ik} = \left(R_{\alpha i} \frac{\partial T}{\partial x_i}\right)' \tag{32}$$

are invariant under the transformation (2) and are proper state variables.

The orthogonal transformations applied to t_{ij} and $\frac{\partial T}{\partial x_i}$ in (30) and (32) are of importance in the case of anisotropy. Then the rate of energy dissipation is not a function of the invariants of these quantities and a rotation is necessary to bring the material directions in concordance with the orthogonal triads, specifying the local anisotropy.

We shall now assume that the rate of energy dissipation due to spatial interaction, represented by the first term on the right hand side of eq. (27), is independent of the local rate of energy dissipation, represented by the second term. We then have

$$T \varrho \overset{+}{s} = D_1 + D_2 \tag{33}$$

with

$$D_1 = -\frac{\partial T}{\partial x_i} \dot{h}_i = D_1\left(p, T, R_{\alpha i} \frac{\partial T}{\partial x_i}\right) \geq 0, \tag{34}$$

$$D_2 = s_{ij} \overset{-1}{b}_{j\alpha} p_{\alpha i} = D_2(p, T, s_{ij} R_{\alpha i} R_{\beta j}) \geq 0. \tag{35}$$

According to what has been postulated above the vector h_i, as a dissipation vector in $\frac{\partial T}{\partial x_i}$-space, is determined by

$$\dot{h}_i = -\lambda_1 \frac{\partial D_1}{\partial \left(\frac{\partial T}{\partial x_i}\right)}, \tag{36}$$

and $\overset{-1}{b}_{j\alpha} p_{\alpha i}$, as a dissipation tensor in s_{ij}-space, by

$$\overset{-1}{b}_{j\alpha} p_{\alpha i} = \lambda_2 \frac{\partial D_2}{\partial s_{ij}},$$

or

$$p_{\alpha i} = b_{\alpha j} \overset{-1}{b}_{j\beta} p_{\beta i} = \lambda_2 b_{\alpha j} \frac{\partial D_2}{\partial s_{ij}}. \tag{37}$$

Thus the directions of the dissipation vector and tensor satisfy the so-called normality condition with respect to surfaces of constant rate of energy dissipation in the $\frac{\partial T}{\partial x_i}$- and s_{ij}-space respectively. This condition has by ZIEGLER been interpreted as a generalization to the nonlinear case of the Onsager relations (ZIEGLER 1961). Here we have simply postulated this condition on the basis of deterministic reasoning. It leads to constitutive equations of a special nature, which have the advantage that they are sufficiently specific to permit, if experimentally verified, the definition of rheological properties of materials in terms of an empirically determinable scalar function for the rate of energy dissipation.

The material derivative of the tensor $b_{\alpha i}$ and of the state varibles $C_{\alpha\beta}$, given by (9) and (10), can now be expressed in terms of the velocity field of the continuum and the function D_2

$$\dot{b}_{\alpha i} = b_{\alpha j} \left[\lambda_2 \frac{\partial D_2}{\partial s_{ij}} - \frac{\partial \dot{x}_j}{\partial x_i} \right], \tag{38}$$

$$\dot{C}_{\alpha\beta} = \lambda_2 \left[b_{\alpha j} \frac{\partial D_2}{\partial s_{ij}} b_{\beta i} + b_{\alpha i} b_{\beta j} \frac{\partial D_2}{\partial s_{ij}} \right] - 2 b_{\alpha i} b_{\beta j} d_{ij},$$

or

$$\dot{C}_{\alpha\beta} = 2 b_{\alpha i} b_{\beta j} \left(\lambda_2 \frac{\partial D_2}{\partial s_{ij}} - d_{ij} \right). \tag{39}$$

The values of the scalar factors λ_1 and λ_2, that determine the magnitude of the dissipation quantities, are found by substitution of (36) and (37) into the expressions (34) and (35)

$$\lambda_1 = \left[\frac{\partial T}{\partial x_i} \frac{\partial D_1}{\partial \left(\frac{\partial T}{\partial x_i} \right)} \right]^{-1} D_1, \tag{40}$$

$$\lambda_2 = \left[s_{ij} \frac{\partial D_2}{\partial s_{ij}} \right]^{-1} D_2. \tag{41}$$

The theory is now complete, since the rate of change of all quantities, that have been introduced, is determined as soon as an initial thermodynamic state and an initial velocity field are given. The material properties and the constitutive equations are fully defined in terms of the function for the internal energy, $u = u(C_{\alpha\beta}, s)$, and the functions $D_1 = D_1\left(p, T, R_{\alpha i}\frac{\partial T}{\partial x_i}\right)$ and $D_2 = D_2(p, T, s_{ij}R_{\alpha i}R_{\beta j})$. that determine the rate energy dissipation in the irreversible thermodynamic processes.

It should be observed that the function $u = u(C_{\alpha\beta}, s)$ need only to define the change of the internal energy per unit mass with respect to some known reference state, characterized by a mass density ϱ_0, an absolute temperature T_0, and a stress tensor $(t_{ij})_0$. The absolute value of the internal energy per unit mass is irrelevant in the theory of rheology; this in contrast to the absolute values of the dissipation functions D_1 and D_2, which determine the magnitudes of the dissipation quantities.

Summary of Rheological Equations

The rheological problem, according to the theory that has been formulated in this paper, is described by the following equations:

$$\frac{\partial\varrho}{\partial t} + \frac{\partial\varrho}{\partial x_i}\dot{x}_i + \varrho\frac{\partial\dot{x}_i}{\partial x_i} = 0; \tag{1}$$

$$\varrho\left(\frac{\partial s}{\partial t} + \frac{\partial s}{\partial x_i}\dot{x}_i\right) = -\frac{\partial\dot{h}_i}{\partial x_i} + \frac{D_1 + D_2}{T}; \tag{4), (33}$$

$$\dot{h}_i = -\lambda_1\frac{\partial D_1}{\partial\left(\frac{\partial T}{\partial x_i}\right)}; \tag{36}$$

$$\frac{\partial b_{\alpha i}}{\partial t} + \frac{\partial b_{\alpha i}}{\partial x_k}\dot{x}_k = b_{\alpha j}\left(\lambda_2\frac{\partial D_2}{\partial s_{ij}} - \frac{\partial\dot{x}_j}{\partial x_i}\right); \tag{38}$$

$$\frac{\partial C_{\alpha\beta}}{\partial t} + \frac{\partial C_{\alpha\beta}}{\partial x_i}\dot{x}_i = 2b_{\alpha i}b_{\beta j}\left[\lambda_2\frac{\partial D_2}{\partial s_{ij}} - \frac{1}{2}\left(\frac{\partial\dot{x}_i}{\partial x_j} + \frac{\partial\dot{x}_j}{\partial x_i}\right)\right]; \tag{39}$$

$$R_{\alpha i} = \overset{-1/2}{C}_{\alpha\beta}b_{\beta i}; \tag{28}$$

$$t_{ij} = -2\varrho\frac{\partial u}{\partial C_{\alpha\beta}}b_{\alpha i}b_{\beta j} = s_{ij} - p\delta_{ij}, \quad p = -\frac{1}{3}t_{ii}; \tag{12}$$

$$T = \frac{\partial u}{\partial s}, \quad T \geq 0; \tag{13}$$

$$\frac{\partial t_{ij}}{\partial x_j} = \varrho\ddot{x}_i = \varrho\left(\frac{\partial\dot{x}_i}{\partial t} + \frac{\partial\dot{x}_i}{\partial x_k}\dot{x}_k\right); \tag{16}$$

$$u = u(C_{\alpha\beta}, s), \quad u = 0 \text{ for } \varrho = \varrho_0, \quad T = T_0, \quad t_{ij} = (t_{ij})_0; \tag{8}$$

$$D_1 = D_1\left(p, T, R_{\alpha i}\frac{\partial T}{\partial x_i}\right), \quad \lambda_1 = \left[\frac{\partial T}{\partial x_i}\frac{\partial D_1}{\partial\left(\dfrac{\partial T}{\partial x_i}\right)}\right]^{-1} D_1; \qquad (34), (40)$$

$$D_2 = D_2(p, T, s_{ij} R_{\alpha i} R_{\beta i}), \quad \lambda_2 = \left[s_{ij}\frac{\partial D_2}{\partial s_{ij}}\right]^{-1} D_2. \qquad (35), (41)$$

These equations define a system of 30 field equations for the 30 dependent variables ϱ, $\dot{x}_i$, s, $\dot{h}_i$, T, $b_{\alpha i}$, $C_{\alpha\beta}$, and t_{ij}. Proper initial conditions for this system are of the form

$$\begin{aligned} t = 0: \quad b_{\alpha i} &= b_{\alpha i}(x_1, x_2, x_3), \\ s &= s(x_1, x_2, x_3), \\ \dot{x}_i &= \dot{x}_i(x_1, x_2, x_3). \end{aligned}$$

The initial mass density follows then from (24)

$$\varrho = \varrho_0 \sqrt{g}, \quad g = \det(g_{ij}), \quad g_{ij} = b_{\alpha i} b_{\alpha j},$$

while the initial values of $C_{\alpha\beta}$ are determined by $C_{\alpha\beta} = b_{\alpha i} b_{\beta i}$.

For a body that constitutes a closed thermodynamic system, the boundary conditions are given by

$$t_{ij} n_j = 0, \quad \dot{h}_i n_i = 0 \quad \text{on } A.$$

Proper boundary conditions for non-closed thermodynamic systems are

$$t_{ij} n_j = p_i^0 \quad \text{or} \quad \dot{x}_i = \dot{x}_i^0 \quad \text{on } A$$

and

$$T = T^0 \quad \text{or} \quad \dot{h}_i n_i = \dot{h}^0 \quad \text{on } A.$$

Here the super index 0 indicates a prescribed function of time.

It is clear that the boundary conditions impose certain restrictions on the initial conditions in the boundary layer of the body. It should be realized, however, that prescribed initial boundary conditions are an artificial concept in the description of natural phenomena occuring in infinite space, where all systems interact with adjoining systems and where this interaction is never a one-way traffic.

Special Cases

In order to illustrate the theory we shall now consider a number of special constitutive equations, corresponding to special forms for the functions u, D_1 and D_2,

a) Gases

Let us consider the case that the internal energy per unit mass depends only on the absolute temperature, $u = f(T)$, or more specifically

$$u = c(T - T_0), \quad c = \text{constant}. \tag{42}$$

With the aid of (13) we arrive at the following equation

$$c \frac{\partial u}{\partial s} - u = c T_0,$$

which determines the dependence of u on s

$$u = k \exp\left(\frac{s}{c}\right) - c T_0, \quad k = k(C_{\alpha\beta}). \tag{43}$$

First we shall assume now that the effect of changes of geometry on the state of the medium is solely represented by the mass density. According to (24) we may then put $k = k(g)$. More specifically we shall consider ($u = 0$ for $g = 1$ and $s = 0$)

$$u = c T_0 \left[g^\gamma \exp\left(\frac{s}{c}\right) - 1\right]. \tag{44}$$

The stress tensor and the temperature are then given by

$$\begin{aligned} t_{ij} &= -2\varrho \frac{\partial u}{\partial C_{\alpha\beta}} b_{\alpha i} b_{\beta i} = -2\varrho \frac{\partial u}{\partial g} g \delta_{ij} \\ &= -2 c \gamma \varrho T_0 g^\gamma \exp\left(\frac{s}{c}\right) \delta_{ij}, \end{aligned} \tag{45}$$

$$T = \frac{\partial u}{\partial s} = T_0 g^\gamma \exp\left(\frac{s}{c}\right). \tag{46}$$

Stress arises in this case only as a hydrostatic pressure, which by elimination of the entropy from (46) and (45) is found to obey the equation of state for a perfect gas.:

$$p = 2 c \gamma \varrho T.$$

The constant c is to be interpreted as the specific heat at constant volume, c_v, while the introduction of the specific heat at constant pressure, c_p, through the enthalpy or heat function h.

$$dh = d\left(u + \frac{p}{\varrho}\right) = (c_v + 2 c_v \gamma)\, dT = c_p\, dT,$$

shows that the constant $2 c_v \gamma$ is the gas constant $R = c_p - c_v$.

The simplest example of an equation for u of the type (43), that leads to shear stresses, is offered by

$$u = cT_0\left[\frac{C_{\alpha\alpha}}{3}\exp\left(\frac{s}{c}\right) - 1\right]. \tag{47}$$

For the stresses and the temperature follows

$$t_{ij} = -2\varrho\frac{\partial u}{\partial C_{\alpha\beta}}b_{\alpha i}b_{\beta j} = -\frac{2}{3}\varrho cT_0\exp\left(\frac{s}{c}\right)g_{ij}, \tag{48}$$

$$T = \frac{\partial u}{\partial s} = \frac{T_0}{3}C_{\alpha\alpha}\exp\left(\frac{s}{c}\right). \tag{49}$$

Since

$$g_{ii} = C_{\alpha\alpha},$$

the expression for the pressure component may, after elimination of s, be written in the form

$$p = -\frac{1}{3}t_{ii} = \frac{2}{9}\varrho cT_0\exp\left(\frac{s}{c}\right)g_{ii} = \frac{2}{3}c\varrho T. \tag{50}$$

This is the equation of state for a mono-atomic gas with

$$R = \frac{2}{3}c = c_p - c_v = \frac{2}{3}c_v.$$

For a complete description of the mono-atomic gas the dissipation functions D_1 and D_2 must be specified. We shall examine the simplest forms, that satisfy the conditions $D_1 \geq 0$, $D_2 \geq 0$.

$$D_1 = \alpha\frac{\partial T}{\partial x_i}\frac{\partial T}{\partial x_i}, \quad \alpha = \alpha(p, T) > 0, \tag{51}$$

$$D_2 = \beta s_{ij}s_{ij}, \qquad \beta = \beta(p, T) > 0. \tag{52}$$

We then have according to (36) a heatflux $q_i = T\dot{h}_i$, determined by

$$q_i = -\alpha T\frac{\partial T}{\partial x_i}, \tag{53}$$

which is equivalent to Fourier's law of heat conduction. The dissipation function D_2 can be used to determine the stress rate

$$\dot{t}_{ij} = t_{ij}\left(\frac{\dot{s}}{c} + \frac{\dot{\varrho}}{\varrho}\right) - \frac{2}{3}cT_0\varrho\exp\left(\frac{s}{c}\right)\dot{g}_{ij}$$

or, according to (1) and (38)

$$\dot{t}_{ij} = t_{ik}\frac{\partial\dot{x}_k}{\partial x_j} + t_{kj}\frac{\partial\dot{x}_k}{\partial x_i} + t_{ij}\frac{\partial\dot{x}_k}{\partial x_k} - t_{ij}\frac{\dot{s}}{c} = \beta(t_{ik}s_{kj} + t_{jk}s_{ki}). \tag{54}$$

This rather unwieldy expression may be replaced by a considerably simpler constitutive equation in the case of quasi-stationary, small distortions with respect to the natural reference state.

Small distortions with respect to the natural reference state can be defined with

$$\frac{C_{\alpha\beta}}{g^{\frac{1}{3}}} = \delta_{\alpha\beta} + e_{\alpha\beta}, \quad |e_{\alpha\beta}| \ll 1. \tag{55}$$

We note that eq. (39) can be written in the form

$$\overline{g^{\frac{1}{3}} \frac{C_{\alpha\beta}}{g^{\frac{1}{3}}}}^{\textstyle\cdot} = \dot{\overline{g^{\frac{1}{3}}}} \left(\frac{C_{\alpha\beta}}{g^{\frac{1}{3}}}\right) + g^{\frac{1}{3}} \left(\overline{\frac{C_{\alpha\beta}}{g^{\frac{1}{3}}}}^{\textstyle\cdot}\right) =$$

$$= -\frac{2}{3} g^{\frac{1}{3}} \frac{\partial \dot{x}_k}{\partial x_k} \left(\frac{C_{\alpha\beta}}{g^{\frac{1}{3}}}\right) + 2 b_{\alpha i} b_{\beta j} \left[\lambda_2 \frac{\partial D_2}{\partial s_{ij}} - d_{ij} + \frac{1}{3} d_{kk} \delta_{ij}\right].$$

But according to eqs (1) and (24) we have

$$\dot{\overline{g^{\frac{1}{3}}}} + \frac{2}{3} g^{\frac{1}{3}} \frac{\partial \dot{x}_k}{\partial x_k} = 0.$$

Hence the rate of distortion is given by

$$\left(\overline{\frac{C_{\alpha\beta}}{g^{\frac{1}{3}}}}^{\textstyle\cdot}\right) = \frac{2 b_{\alpha i} b_{\beta j}}{g^{\frac{1}{3}}} \left[\lambda_2 \frac{\partial D_2}{\partial s_{ij}} - d_{ij} + \frac{1}{3} d_{kk} \delta_{ij}\right], \tag{56}$$

or, after substitution of the expressions (48), (49), (50), and (52)

$$\left(\overline{\frac{C_{\alpha\beta}}{g^{\frac{1}{3}}}}^{\textstyle\cdot}\right) = -6 p \beta \left(\frac{C_{\alpha\beta}}{g^{\frac{1}{3}}}\right) \left(\frac{C_{\beta\gamma} - \frac{1}{3} C_{\delta\delta} \delta_{\beta\gamma}}{C_{\alpha\alpha}}\right) -$$

$$- \frac{2 b_{\alpha i} b_{\beta j}}{g^{\frac{1}{3}}} \left(d_{ij} - \frac{1}{3} d_{kk} \delta_{ij}\right). \tag{57}$$

Introduction of the tensor $e_{\alpha\beta}$ from (55) leads to

$$\left(\overline{\frac{C_{\alpha\beta}}{g^{\frac{1}{3}}}}^{\textstyle\cdot}\right) = \dot{e}_{\alpha\beta} = -2 p \beta (\delta_{\alpha\gamma} + e_{\alpha\gamma}) \left(\frac{e_{\beta\gamma} - \frac{1}{3} e_{\delta\delta} \delta_{\beta\gamma}}{1 + \frac{1}{3} e_{\alpha\alpha}}\right) -$$

$$- \frac{2 b_{\alpha i} b_{\beta j}}{g^{\frac{1}{3}}} \left(d_{ij} - \frac{1}{3} d_{kk} \delta_{ij}\right). \tag{58}$$

If we neglect the tensor $e_{\alpha\gamma}$ with respect to $\delta_{\alpha\gamma}$, because the distortions are assumed to be small, we have

$$\dot{e}_{\alpha\beta} + 2p\beta\left(e_{\alpha\beta} - \frac{1}{3}\, e_{\gamma\gamma}\delta_{\alpha\beta}\right) \approx -\frac{2b_{\alpha i}\, b_{\beta j}}{g^{\frac{1}{3}}}\left(d_{ij} - \frac{1}{3}\, d_{kk}\,\delta_{ij}\right). \tag{59}$$

It follows from the theory of differential equations that

$$|\dot{e}_{\alpha\beta}| \ll \left|\frac{2b_{\alpha i}\, b_{\beta j}}{g^{\frac{1}{3}}}\left(d_{ij} - \frac{1}{3}\, d_{kk}\,\delta_{ij}\right)\right|, \tag{60}$$

provided in the gas flow the following conditions are satisfied

$$\left|\frac{d^n}{dt^n}\left[\frac{2b_{\alpha i}\, b_{\beta j}}{g^{\frac{1}{3}}}\left(d_{ij} - \frac{1}{3}\, d_{kk}\,\delta_{ij}\right)\right]\right| \ll (2p\beta)^n \left|\frac{2b_{\alpha i}\, b_{\beta j}}{g^{\frac{1}{3}}}\left(d_{ij} - \frac{1}{3}\, d_{kk}\,\delta_{ij}\right)\right|. \tag{61}$$

By virtue of (60) we may then put according to (56)

$$d_{ij} - \frac{1}{3}\, d_{kk}\,\delta_{ij} \approx \lambda_2 \frac{\partial D_2}{\partial s_{ij}} = \beta s_{ij}. \tag{62}$$

Thus we are led to the equations of viscosity of Navier-Stokes without volume viscosity. The shear stresses are in our case still thermodynamically defined quantities, though their thermodynamic origin is fully obscured in the eqs. (62).

In most gas flows the conditions (61) will be satisfied. Shock waves may be an exception, and it is just in connection with the calculated and measured thickness of shock waves, that the need for the existence of volume viscosity was felt (GILBARG and PAOLUCCI 1953). Perhaps the explanation for the discrepancy between theory and experiment must in this case be sought in the fact, that for shock waves the conditions (61) are not satisfied. The decisive quantity is here the value of

$$2p\beta = \frac{\varrho R T}{\mu}, \tag{63}$$

where μ represents the viscosity $\left(s_{ij} = 2\mu(d_{ij} - {}^1/_3 d_{kk}\delta_{ij})\right)$.

b) Elastic Materials

An elastic material will be defined here as a material, for which the dissipation tensor $p_{\alpha i}$ (9) vanishes. The geometric natural reference state can then always be defined such that the system (5), $da_\alpha = b_{\alpha i}\, dx_i$, represents exact differentials for functions $a_\alpha = a_\alpha(x_1, x_2, x_3)$, or

$$b_{\alpha i} = \frac{\partial a_\alpha}{\partial x_i}. \tag{64}$$

Hence the motion of elementary particles in an elastic body can be described by the motion of a continuum, in which each particle occupies a material point. The displacement tensor $\frac{\partial a_\alpha}{\partial x_i}$ describes the change, which the displacement from the state of the body at time t to the reference state induces in the relative position of any two points as these points draw together.

For the material of a body, that as a system free in space will not desintegrate, the stress-free state must correspond to a thermodynamic equilibrium state with non-vanishing mass density. Besides on the temperature, the internal energy function u for such a material will depend on the inter-particle distances, and thereby on $g = (d V_0/d V)^2$. The simplest form of such a function is presented by

$$u = cT + cT_0(g^{-\gamma} - 2)$$

or, according to

$$u - c\frac{\partial u}{\partial s} = cT_0(g^{-\gamma} - 2),$$

by

$$u = cT_0\left[k(C_{\alpha\beta})\, e^{\frac{s}{c}} + g^{-\gamma} - 2\right]. \tag{65}$$

Let us consider an elastic body for which there exists a state, to be taken as reference state, which is completely stress-free. Any choice for $k(C_{\alpha\beta})$ in the function u, defined with respect to this reference state, must satisfy the following requirements:

1⁰ the stresses t_{ij} vanish for $T = T_0$ and $C_{\alpha\beta} = \delta_{\alpha\beta}$,
2⁰ u is positive definite for all isentropic deformations ($s = 0$) from the stress-free reference state.

If we restrict ourselves to isotropic materials, and if the first invariant of the inverse of $C_{\alpha\beta}$ is denoted by $\overset{-1}{C}_{\alpha\alpha}$, the following expression for u satisfies the requirements, formulated above.

$$u = cT_0\left[\frac{1}{3}\left\{\left(\frac{1}{2} - \beta\right) g^{-\frac{1}{3}} C_{\alpha\alpha} + \left(\frac{1}{2} + \beta\right) g^{\frac{1}{3}} \overset{-1}{C}_{\alpha\alpha}\right\} \times \times\, g^{\gamma} \exp\left(\frac{s}{c}\right) + g^{-\gamma} - 2\right]. \tag{66}$$

Here c, β, and γ represent material constants with the restriction: $-\frac{1}{2} \leq \beta \leq \frac{1}{2}$.

It can be observed that, since $\operatorname{Det}\left(g^{-\frac{1}{3}} C_{\alpha\beta}\right) = 1$, the tensor $g^{-\frac{1}{3}} C_{\alpha\beta}$ is a measure for distortion at constant volume, while g represents the influence of change of volume.

Expression (66) appears to lead to some of the thermo-elastic effects observed in rubber-like materials, where the "compression modulus", $4\varrho c \gamma^2 T_0$, is large compared to the "shear modulus", $\frac{2}{3}\varrho c T_0$. For $\gamma \to \infty$ we arrive at the Mooney-Rivlin theory for rubber, in which the pressure component is treated as a thermodynamically undefined quantity and the deformations are subjected to the subsidiary condition $g = 1$.

However, the three material constants in expression (66) prove to be inadequate to cope with the experimental data on rubber-like materials. A better quantitative description is expected from a five constants theory, based on the following expressions for u:

$$u = c(T - T_0) + C_1(g^{-\gamma} - 1) \tag{67}$$

and

$$u = C_2\left[\frac{1}{3}\left\{\left(\frac{1}{2} - \beta\right) g^{-\frac{1}{3}} C_{\alpha\alpha} + \left(\frac{1}{2} + \beta\right) g^{\frac{1}{3}} \overset{-1}{C}_{\alpha\alpha}\right\} - 1\right] e^{\frac{s}{c}} + \\ + c T_0 \left(e^{\frac{s}{c}} - 1\right) + C_1\left[(g^{\gamma} - 1) e^{\frac{s}{c}} + (g^{-\gamma} - 1)\right]. \tag{68}$$

Here the isothermal bulkmodulus of elasticity, C, the shear modulus of elasticity, G, and the coefficient of cubic thermal expansion, α, all defined with respect to small deviations from the natural reference state, are related to the specific heat at constant volume, c, and to the constants C_1, C_2, and γ by

$$C = 8\varrho\gamma^2 C_1\left[1 - \frac{C_1}{2cT_0}\right], \quad G = \frac{2}{3}\varrho C_2,$$

$$\alpha = \frac{1}{4\gamma T_0\left[1 - \frac{C_1}{2cT_0}\right]}. \tag{69}$$

c) Simple Solids

A material will be denoted here as a solid, if the stress-free state corresponds to a thermodynamic equilibrium state. As it already follows from this definition of a solid, and as it will be shown in the following, it may be difficult to distingiush between a solid and a liquid.

The adjective "simple" will be used here to indicate that the same functions u, D_1 and D_2 apply to the material in all points of the body Unfortunately all real solids are to some extent non-simple solids, as for instance all metals with their polycristalline structure.

In contrast to an elastic material, in a general solid the dissipation tensor $p_{\alpha i}$ shall vanish only under special conditions of stress and temperature; and hence the system (5), $da_\alpha = b_{\alpha i}\, dx_i$, after an arbitrary thermodynamic history from a certain reference time t_0, will not represent exact differentials for functions $a_\alpha = a_\alpha(x_1, x_2, x_3)$.

For the internal energy function u of a simple solid we could consider expression (68). However, since for most solids, as for instance all metals, only small deviations from the natural reference state can be realised, it usually suffices to consider the theory for this case.

We shall first consider the theory for small deviations from the natural reference state as it is applied in the field of structural analysis. There the theory is usually further restricted by the condition that the deviations of the configuration in the natural reference state from a certain reference configuration of the continuum remain small.

If we denote the coordinates of the material points of the continuum in a certain reference geometry by ξ_1, ξ_2, ξ_3, then we can define the motion of the continuum by three single-valued functions

$$x_i = x_i(\xi_1, \xi_2, \xi_3, t). \tag{70}$$

We can now write

$$d a_\alpha = b_{\alpha i}\, d x_i = b_{\alpha i} \frac{\partial x_i}{\partial \xi_k} d\xi_k. \tag{71}$$

If we put

$$b_{\alpha i} \frac{\partial x_i}{\partial \xi_k} = \delta_{\alpha k} + \beta_{\alpha k}, \tag{72}$$

then we can express the condition, that the deviations of the configuration in the natural reference state from the reference configuration of the continuum remain small, by

$$|\beta_{\alpha k}| \ll 1. \tag{73}$$

Here it should be observed that we wish the quantities $\beta_{\alpha k}$ to be invariant under the Galilean transformation (2) in order that condition (73) may be fulfilled, irrespective of the choice of the inertial system. The coordinates ξ_1, ξ_2, ξ_3 should then be invariant under the transformation (2), which implies that the axes of the coordinate system ξ_i must concur with fixed material directions in the reference geometry of the body, like the orthogonal triads, that serve as a base for the da_α's.

If we introduce as a measure of deformation of the continuum the Lagrangian strain tensor

$$E_{ij} = \frac{1}{2}\left(\frac{\partial x_k}{\partial \xi_i} \frac{\partial x_k}{\partial \xi_j} - \delta_{ij}\right), \tag{74}$$

and if we take as geometric state variables the components of an "elastic" strain tensor, defined by

$$E'_{\alpha\beta} = \frac{1}{2}\left(\overset{-1}{C}_{\alpha\beta} - \delta_{\alpha\beta}\right) = \frac{1}{2}\left(\overset{-1}{b}_{i\alpha}\, \overset{-1}{b}_{i\beta} - \delta_{\alpha\beta}\right), \tag{75}$$

then we have the equality

$$E'_{\alpha\beta}\, b_{\alpha i} \frac{\partial x_i}{\partial \xi_k}\, b_{\beta j} \frac{\partial x_j}{\partial \xi_l} = \frac{1}{2}\left(2 E_{kl} + \delta_{kl} - b_{\alpha i} \frac{\partial x_i}{\partial \xi_k}\, b_{\alpha j} \frac{\partial x_j}{\partial \xi_l}\right).$$

From (72) and (73) we conclude that

$$E_{ij} - E''_{ij} = E'_{\alpha\beta}\delta_{\alpha i}\delta_{\beta j} = E'_{ij}, \qquad |E''_{ij}| = \frac{1}{2}\,|\beta_{\alpha i}\delta_{\alpha j} + \beta_{\alpha j}\delta_{\alpha i} + \beta_{\alpha i}\beta_{\alpha j}| \ll 1, \tag{76}$$

where E''_{ij} is called the inelastic or plastic strain tensor,

The state of the material is now at any time t determined by the variables E'_{ij} and s, while according to (74) and (76) the tensor E'_{ij} is defined in terms of the three functions (70) and the six independent components of the symmetric tensor E''_{ij}. The internal energy per unit mass may then be given by

$$u = u(E'_{ij}, s). \tag{77}$$

If we let the natural reference state at some instant coincide with the reference geometry of the continuum, then the mass density is the same for the natural reference state and for the reference geometry of the continuum and is equal to ϱ_0. Accordingly, we have

$$\operatorname{Det}\left(b_{\alpha i} \frac{\partial x_i}{\partial \xi_k}\, b_{\alpha j} \frac{\partial x_j}{\partial \xi_l}\right) = 1, \quad \text{or } E''_{ii} = 0. \tag{78}$$

In this case, by taking ξ_1, ξ_2, ξ_3 as the independent space variables and performing all calculations in the reference geometry of the body, we satisfy the principle of conservation of mass automatically, while the principle of conservation of energy for the body as a closed system is expressed by

$$\int_{V_0} \varrho_0(\dot{u} + \ddot{x}_i \dot{x}_i)\, dV$$

$$= \int_{V_0} \left[\varrho_0 \frac{\partial u}{\partial E'_{ij}} \dot{E}_{ij} - \varrho_0 \frac{\partial u}{\partial E'_{ij}} \dot{E}''_{ij} + \varrho_0 \frac{\partial u}{\partial s} \dot{s} + \varrho_0 \ddot{x}_i \dot{x}_i\right] dV = 0$$

We introduce the temperature T by (13) and, instead of the stress tensor t_{ij} (12) of Cauchy, the pseudo stress tensor of Kirchhoff, σ_{ij}

$$\sigma_{ij} = \varrho_0 \frac{\partial u}{\partial E'_{ij}}. \tag{79}$$

Further we have according to (4) and (70)

$$\varrho_0 \dot{s} = -\frac{\varrho_0}{\varrho} \frac{\partial \dot{h}_i}{\partial \xi_k} \frac{\partial \xi_k}{\partial x_i} + \varrho_0 \dot{\bar{s}} . \tag{80}$$

On the surface of the body as a closed system $\dot{h}_i n_i = 0$ holds. With the aid of the divergence theorem the equation for conservation of energy can now be written in the form

$$\int\limits_{V_0} \left[\left\{ -\frac{\partial}{\partial \xi_j} \left(\frac{\partial x_i}{\partial \xi_k} \sigma_{kj} \right) + \varrho_0 \ddot{x}_i \right\} \dot{x}_i + T \varrho_0 \dot{\bar{s}} + \frac{\partial}{\partial \xi_j} \left(\frac{\varrho_0}{\varrho} \frac{\partial \xi_j}{\partial x_i} T \right) \dot{h}_i - \sigma_{ij} \dot{E}''_{ij} \right] d V +$$
$$+ \int\limits_{A_0} \frac{\partial x_i}{\partial \xi_k} \sigma_{kj} n_j \dot{x}_i \, d A = 0 . \tag{81}$$

By similiar reasoning as applied to eq. (15) we may deduce from eq. (81) the following equations

$$\frac{\partial}{\partial \xi_j} \left(\frac{\partial x_i}{\partial \xi_k} \sigma_{kj} \right) - \varrho_0 \ddot{x}_i = 0 \qquad \text{in } V_0 , \tag{82}$$

$$T \varrho_0 \dot{\bar{s}} + \frac{\partial}{\partial \xi_j} \left(\frac{\varrho_0}{\varrho} \frac{\partial \xi_j}{\partial x_i} T \right) \dot{h}_i + \sigma_{ij} \dot{E}''_{ij} = 0 \quad \text{in } V_0 , \tag{83}$$

$$\frac{\partial x_i}{\partial \xi_k} \sigma_{kj} n_j = 0 \qquad \text{on } A_0 . \tag{84}$$

If we rewrite the expression (83) for the rate of energy dissipation in the form

$$T \varrho_0 \dot{\bar{s}} = -\frac{\partial}{\partial \xi_j} \left(\frac{\varrho_0}{\varrho} \frac{\partial \xi_j}{\partial x_i} T \right) \frac{\partial x_i}{\partial \xi_k} \frac{\partial \xi_k}{\partial x_l} \dot{h}_l + \sigma_{ij} \dot{E}''_{ij} ,$$

we observe, that the rate of energy dissipation is given by the sum of two scalar products. One is the product of the vector of state variables, invariant under the transformation (2),

$$-\frac{\partial}{\partial \xi_j} \left(\frac{\varrho_0}{\varrho} \frac{\partial \xi_j}{\partial x_i} T \right) \frac{\partial x_i}{\partial \xi_k} ,$$

and the dissipation vector $\frac{\partial \xi_k}{\partial x_l} \dot{h}_l$. The other is the scalar product of the stress tensor σ_{ij} and the dissipation tensor $\dot{E}''_{ij}$. After introducing

$$s^*_{ij} = \sigma_{ij} + p^* \delta_{ij} , \qquad p^* = -\frac{1}{3} \sigma_{\alpha\alpha} , \tag{85}$$

$$\frac{\partial T^*_{ji}}{\partial \xi_j} = \frac{\partial}{\partial \xi_j} \left(\frac{\varrho_0}{\varrho} \frac{\partial \xi_j}{\partial x_i} T \right) , \tag{86}$$

and using further the notation and assumptions, underlying (33) through (37), we may write

$$D_1 = -\frac{\partial}{\partial \xi_j}\left(\frac{\varrho_0}{\varrho}\frac{\partial \xi_j}{\partial x_i} T\right) \dot{h}_i = D_1\left(p^*, T, \frac{\partial T^*_{ji}}{\partial \xi_j}\frac{\partial x_i}{\partial \xi_k}\right) \geq 0, \tag{87}$$

$$D_2 = \sigma_{ij} \dot{E}''_{ij} = s^*_{ij} \dot{E}''_{ij} = D_2(p^*, T, s^*_{ij}) \geq 0, \tag{88}$$

$$\dot{h}_i = -\lambda_1 \frac{\partial D_1}{\partial\left(\frac{\partial T^*_{ji}}{\partial \xi_j}\right)}, \quad \lambda_1 = \left[\frac{\partial T^*_{ji}}{\partial \xi_j}\frac{\partial D_1}{\partial\left(\frac{\partial T^*_{ki}}{\partial \xi_k}\right)}\right]^{-1} D_1, \tag{89}$$

$$\dot{E}''_{ij} = \lambda_2 \frac{\partial D_2}{\partial s^*_{ij}}, \quad \lambda_2 = \left[s^*_{ij}\frac{\partial D_2}{\partial s^*_{ij}}\right]^{-1} D_2. \tag{90}$$

The rheological problem for a certain material is again reduced to the determination of three scalar functions: the internal energy function, u, and the two dissipation functions D_1 and D_2. The restriction that the inelastic strains shall remain small, (73), is indeed a natural one in a theory of deformation, where all changes are a priori to be defined with respect to a geometric reference state of a body. The internal energy function can no longer be considered as a function of the Langrangian strain tensor if the condition $|\beta_{\alpha k}| \ll 1$ is not fulfilled. Consequently the geometric reference state of a body then loses its meaning with respect to the thermodynamic state of the material. A theory for large inelastic strains is a theory of flow rather than of deformation.

It can be shown that for small deviations from the natural reference state of the material $\left(|E'_{ij}| \ll 1, \left|\frac{s}{c_v}\right| \ll 1\right)$ the expression for the internal energy for an isotropic material is given by (BESSELING 1960)

$$\varrho_0 u = T_0 \varrho_0 s + \frac{1}{2} C\left(1 + \frac{C\alpha^2 T_0}{\varrho_0 c_v}\right) E^2_{ii} + G\left(E'_{ij} - \frac{1}{3}E_{kk}\delta_{ij}\right) \times$$
$$\times \left(E'_{ij} - \frac{1}{3}E_{kk}\delta_{ij}\right) - \frac{C\alpha T_0}{c_v} E_{ii} s + \frac{\varrho_0 T_0}{2 c_v} s^2. \tag{91}$$

Here is C the isothermal bulk modulus of elasticity, G the shear modulus of elasticity, α the coefficient of cubic thermal expansion, and c_v the specific heat per unit mass at constant volume.

Though for $|E'_{ij}| \ll 1$ the difference between ϱ_0 and ϱ may be neglected, the geometric non-linearities due to a deviation of $\frac{\partial x_i}{\partial \xi_j}$ from δ_{ij} play an essential role in the determination of the stability of structures. If we neglect the deviations of $\frac{\partial x_i}{\partial \xi_j}$ from δ_{ij} with respect to δ_{ij} and put $\varrho = \varrho_0$, we arrive for the isentropic and the isothermal case with $E''_{ij} = 0$ at the classic linear theory of elasticity.

A non-vanishing dissipation function D_2 leads to a theory of creep, from which the theory of plasticity may be derived by a limiting process, as indicated by ZIEGLER (1962).

From the experimental fact that even for small strains the stresses s^*_{ij} cannot exceed certain values, depending on the material, we may conclude that with increasing stresses s^*_{ij} the rate of energy dissipation, given by D_2, tends to infinity. As the relation between rate of dissipation and stress is more strongly non-linear, surfaces for equal difference in rate of dissipation will lie closer together in s^*_{ij}-space. In the limit a surface of indefinite rate of energy dissipation may be conceived, that separates the region of rate of dissipation $= 0$ from the region of rate of dissipation $= \infty$. It is the yield surface of the so-called elastic-ideally plastic material. It can be defined by an equation of the form

$$\varphi(p^*, T, s^*_{ij}) = 0. \tag{92}$$

For values of s^*_{ij} inside the yield surface the material is thermo-elastic ($\dot{E}''_{ij} = 0$). At the yield surface, as a surface of indefinite rate of energy dissipation, the rate of inelastic strain is given by

$$\dot{E}''_{ij} = \lambda_2 \frac{\partial \varphi}{\partial s^*_{ij}}. \tag{93}$$

The positive scalar λ_2, which determines the magnitude of the rate of energy dissipation, now depends on the rate of total deformation, $\dot{E}_{ij}$, or follows from the rate of stress. In order that the state of stress does not leave the yield surface, which would imply $\dot{E}''_{ij} = 0$, the following condition must be satisfied.

$$\dot{\varphi} = \frac{\partial \varphi}{\partial s^*_{ij}} \dot{s}^*_{ij} + \frac{\partial \varphi}{\partial p^*} \dot{p}^* + \frac{\partial \varphi}{\partial T} \dot{T} = 0 \tag{94}$$

or

$$\dot{\varphi} = \frac{\partial \varphi}{\partial s^*_{ij}} 2G \left[\dot{E}_{ij} - \frac{1}{3} \dot{E}_{kk} \delta_{ij} - \dot{E}''_{ij}\right] + \frac{\partial \varphi}{\partial p^*} \dot{p}^* + \frac{\partial \varphi}{\partial T} \dot{T} = 0. \tag{95}$$

By substitution of (93) into eq. (95) we arrive at the following expression for λ_2

$$\lambda_2 = \left[2G \frac{\partial \varphi}{\partial s^*_{ij}} \frac{\partial \varphi}{\partial s^*_{ij}}\right]^{-1} \left[\frac{\partial \varphi}{\partial s^*_{ij}} 2G \left(\dot{E}_{ij} - \frac{1}{3} \dot{E}_{kk} \delta_{ij}\right) + \frac{\partial \varphi}{\partial p^*} \dot{p}^* + \frac{\partial \varphi}{\partial T} \dot{T}\right], \tag{96}$$

subject to the condition $\lambda_2 \geq 0$.

If the thermal effects, and the influence of the hydrostatic component of stress, p^*, on the yield condition, may be neglected, the plasticity equations given above are equivalent to the Prandtl-Reuss theory of plasticity (PRANDTL 1924; REUSS 1930).

In the mechanics of metal-forming we have to deal with large inelastic strains and, as it has already been pointed out above, the theory will then be a theory of flow rather than a theory of deformation. In principle the general rheological equations, derived in this paper, present such a theory of flow. They are, however, so complicated that rigorous solutions of practical problems seem to be out of the question, even if the deviations from the natural reference state remain small. If, however, the latter condition is combined with the condition of isotropy of the material, a simplified version of the theory can be formulated.

We consider as a measure of deviation from the natural reference state the quantities.

$$\varepsilon'_{ij} = \frac{1}{2}(\delta_{ij} - g_{ij}), \tag{97}$$

which we subject to the condition

$$|\varepsilon'_{ij}| \ll 1. \tag{98}$$

Because of (24) and (98) we may neglect the quantity $\frac{\varrho - \varrho_0}{\varrho_0}$ with respect to unity.

Though the components of ε'_{ij} are not invariant under the transformation (2) and are, therefore, themselves not state variables, the invariants of e'_{ij} do possess the required invariance properties of state variables for an isotropic material.

If we introduce the deviator of the tensor ε'_{ij},

$$e'_{ij} = \varepsilon'_{ij} - \frac{1}{3}\varepsilon'_{kk}\,\delta_{ij}, \tag{99}$$

and if for small deviations from the natural reference state $\left(|\varepsilon'_{ij}| \ll 1,\ \left|\frac{s}{c_v}\right| \ll 1\right)$ the expression for the internal energy is limited to terms quadratic in e'_{ij} and $\frac{s}{c_v}$, we arrive for an isotropic material at the following result

$$\varrho u = T_0 \varrho s + \frac{1}{2} C\left(1 + \frac{C\alpha^2 T_0}{\varrho c_v}\right)\varepsilon'^2_{ii} + G e'_{ij} e'_{ij} - \frac{C\alpha T_0}{c_v}\varepsilon'_{ii} s + \frac{\varrho T_0}{2 c_v} s^2. \tag{100}$$

Here again C is the isothermal bulk modulus of elasticity, G the shear modulus of elasticity, α the coefficient of cubic thermal expansion, and c_v the specific heat per unit mass at constant volume. It should be observed that $\frac{C}{\varrho}$ and $\frac{G}{\varrho}$ are the proper material constants, but since $\frac{\varrho - \varrho_0}{\varrho_0} \ll 1$, C and G are themselves nearly constant.

We observe that, because of (98), ε'_{ij} may be neglected with respect to δ_{ij}. Consequently we have

$$t_{ij} = -2\varrho \frac{\partial u}{\partial C_{\alpha\beta}} b_{\alpha i} b_{\beta j} = \varrho \frac{\partial u}{\partial \varepsilon'_{kl}} \overset{1/2}{g}_{ki} \overset{1/2}{g}_{lj} \approx \varrho \frac{\partial u}{\partial \varepsilon'_{ij}} .$$

The expressions for the stresses and the temperature, that can be deduced from (100), read then as follows

$$t_{ij} = \varrho \frac{\partial u}{\partial \varepsilon'_{ij}} = 2 G e'_{ij} + C \left(1 + \frac{C \alpha^2 T_0}{\varrho c_v}\right) \varepsilon'_{kk} \delta_{ij} - \frac{C \alpha T_0}{c_v} s \delta_{ij}, \qquad (101)$$

$$T = \frac{\partial u}{\partial s} = T_0 - \frac{C \alpha T_0}{\varrho c_v} \varepsilon'_{ii} + \frac{T_0}{c_v} s . \qquad (102)$$

The rate of change of the tensor ε'_{ij} is according to eqs. (22) and (38) determined by

$$\dot{\varepsilon}'_{ij} = -\frac{1}{2} \dot{g}_{ij} = -\frac{1}{2} (\dot{b}_{\alpha i} b_{\alpha j} + b_{\alpha i} \dot{b}_{\alpha j}),$$

$$\dot{\varepsilon}'_{ij} = -\frac{1}{2} g_{kj} \left(\lambda_2 \frac{\partial D_2}{\partial s_{ik}} - d_{ik}\right) - \frac{1}{2} g_{ik} \left(\lambda_2 \frac{\partial D_2}{\partial s_{kj}} - d_{kj}\right) -$$
$$- \frac{1}{2} g_{kj} \frac{1}{2} \left(\frac{\partial \dot{x}_i}{\partial x_k} - \frac{\partial \dot{x}_k}{\partial x_i}\right) + \frac{1}{2} g_{ik} \frac{1}{2} \left(\frac{\partial \dot{x}_k}{\partial x_j} - \frac{\partial \dot{x}_j}{\partial x_k}\right),$$

or, since ε'_{ij} may be neglected with respect to δ_{ij},

$$\dot{\varepsilon}'_{ij} = d_{ij} - \lambda_2 \frac{\partial D_2}{\partial s_{ij}} + \varepsilon'_{kj} \frac{1}{2} \left(\frac{\partial \dot{x}_i}{\partial x_k} - \frac{\partial \dot{x}_k}{\partial x_i}\right) - \varepsilon'_{ik} \frac{1}{2} \left(\frac{\partial \dot{x}_k}{\partial x_j} - \frac{\partial \dot{x}_j}{\partial x_k}\right). \qquad (103)$$

The skew-symmetric tensor

$$\omega_{ij} = \frac{1}{2} \left(\frac{\partial \dot{x}_i}{\partial x_j} - \frac{\partial \dot{x}_j}{\partial x_i}\right) \qquad (104)$$

is known as the spin tensor. We can now rewrite eqs. (103) as follows

$$\dot{\varepsilon}'_{ii} = d_{ii} . \qquad (105\,\text{a})$$

$$\dot{e}'_{ij} - e'_{kj} \omega_{ik} + e'_{ik} \omega_{kj} = d_{ij} - \frac{1}{3} d_{kk} \delta_{ij} - \lambda_2 \frac{\partial D_2}{\partial s_{ij}} . \qquad (105\,\text{b})$$

The left-hand side of eq. (105 b) represents the so-called Jaumann derivative (JAUMANN 1911) of the tensor e'_{ij}. It differs from the ordinary material derivative only in case $|\omega_{ij}|$ is by an order of magnitude larger than $\left| d_{ij} - \frac{1}{3} d_{kk}\delta_{ij} - \lambda_2 \frac{\partial D_2}{\partial s_{ij}} \right|$.

In order to complete the theory of flow with small deviations from the natural reference state expressions should be given for the dissipation functions D_1 and D_2. Since the material must be isotropic we have

$$D_1 = D_1\left(p, T, \frac{\partial T}{\partial x_i} \frac{\partial T}{\partial x_i}\right), \tag{106}$$

$$D_2 = D_2(p, T, s_{ij}s_{ij}, s_{ij}s_{jk}s_{ki}). \tag{107}$$

An expression of the form (51) leads again to Fourier's heat conduction law, while $D_2 = D_2(T, s_{ij}s_{ij})$ furnishes the creep equations, proposed by ODQVIST (1936). When we consider time-independent plasticity again as a limiting case of creep, a yield surface

$$\varphi(T, s_{ij}s_{ij}) = 0 \tag{108}$$

leads to equations of the form

$$\dot{e}'_{ij} - e'_{kj}\omega_{ik} + e'_{ik}\omega_{kj} = d_{ij} - \frac{1}{3} d_{kk}\delta_{ij} - \lambda s_{ij}, \quad \lambda \geq 0, \tag{109}$$

which are a generalization of the Prandtl-Reuss theory of plasticity.

In view of the difficulties, encountered when we try to solve in practical plasticity problems the equations for small deviations from the natural reference state, the theory is often simplified further by neglecting these deviations from the natural reference state altogether. The constitutive equations are then reduced to

$$d_{ii} = 0,$$

$$d_{ij} - \frac{1}{3} d_{kk}\delta_{ij} = \lambda s_{ij}, \tag{110}$$

where

$$\lambda \geq 0 \text{ if } \varphi = 0, \dot{\varphi} = 0,$$

$$\lambda = 0 \text{ if } \varphi < 0 \text{ or } \varphi = 0 \text{ and } \dot{\varphi} < 0. \tag{111}$$

In this theory for the rigid-ideally plastic model the thermodynamic origin of the stresses is fully obscured. The pressure component is an undefined quantity, subjected only to boundary conditions and the equations of motion for the stresses (16), while the deviator components of stress have to satisfy the boundary conditions and the equations of motion, as well as the flow eqs. (110) in those regions, where the yield condition (108) is satisfied.

d) Liquids

Liquids are usually defined as materials in which shear stresses are necessarily accompanied by motion of the continuum. However every material with a dissipation function D_2, that vanishes only for $s_{ij} = 0$, complies with this definition, in particular those materials, which we have called simple solids, if they are subject to creep for non-vanishing s_{ij}. The distinction between a solid and a liquid thus can only be a quantitative one in terms of the time constant in the dissipation function D_2.

Since liquids offer great resistance against compression, while shear stresses remain small, a theory of flow for liquids, based upon the assumption that the deviations from the natural reference state remain small, will usually be adequate. For the internal energy, the stresses and the temperature we may then employ the expressions (100), (101) and (102), as liquids will generally be isotropic. In many flow problems the constitutive eqs. (105) can for liquids be simplified considerably, however.

If we consider a dissipation function of the type

$$D_2 = \beta s_{ij} s_{ij},$$

then we have according to (56) the equation

$$\overline{\left(\frac{C_{\alpha\beta}}{g^{\frac{1}{3}}}\right)}^{\raisebox{0pt}{$\cdot$}} = \frac{2 b_{\alpha i} b_{\beta j}}{g^{\frac{1}{3}}} \left[\beta s_{ij} - d_{ij} + \frac{1}{3} d_{kk} \delta_{ij}\right].$$

From the equality

$$(C_{\alpha\beta} - \delta_{\alpha\beta}) R_{\alpha i} R_{\beta j} = -2 \varepsilon'_{ij}$$

and the condition (98) we can conclude that

$$|C_{\alpha\beta} - \delta_{\alpha\beta}| \ll 1,$$

and

$$e_{\alpha\beta} R_{\alpha i} R_{\beta j} \approx -2 e'_{ij}.$$

Hence, according to $s_{ij} = 2 G e'_{ij}$ and (28)

$$\dot{e}_{\alpha\beta} \approx -2 G_\beta \frac{\overset{1/2}{C}_{\alpha\gamma} \overset{1/2}{C}_{\beta\delta}}{g^{\frac{1}{3}}} e_{\gamma\delta} - 2 \frac{\overset{1/2}{C}_{\alpha\gamma} R_{\gamma i} \overset{1/2}{C}_{\beta\delta} R_{\delta j}}{g^{\frac{1}{3}}} \left(d_{ij} - \frac{1}{3} d_{kk} \delta_{ij}\right)$$

or, if $C_{\alpha\beta} - \delta_{\alpha\beta}$ is neglected with respect to $\delta_{\alpha\beta}$,

$$\dot{e}_{\alpha\beta} + 2 G \beta e_{\alpha\beta} \approx -2 R_{\alpha i} R_{\beta j} \left(d_{ij} - \frac{1}{3} d_{kk} \delta_{ij}\right). \tag{112}$$

If the flow is such, that the following condition is satisfied

$$\left|\frac{d^n}{dt^n}\left[2 R_{\alpha i} R_{\beta j}\left(d_{ij}-\frac{1}{3} d_{kk}\delta_{ij}\right)\right]\right| \ll (2G\beta)^n \left|2 R_{\alpha i} R_{\beta j}\left(d_{ij}-\frac{1}{3} d_{kk}\delta_{ij}\right)\right|, \tag{113}$$

then it follows from the theory of differential equations that

$$|\dot{e}_{\alpha\beta}| \ll \left|2 R_{\alpha i} R_{\beta j}\left(d_{ij}-\frac{1}{3} d_{kk}\delta_{ij}\right)\right|. \tag{114}$$

In that case the eqs. (109) may be replaced by

$$-G\beta e_{\alpha\beta} R_{\alpha i} R_{\beta j} = \beta s_{ij} = d_{ij} - \frac{1}{3} d_{kk}\delta_{ij}. \tag{115}$$

These are the usual constitutive equations for shear viscosity of an isotropic liquid, in which the thermodynamic origin of the stresses is fully obscured.

Whether a flow problem may be treated as a problem of liquid flow with the aid of eqs. (115) or should be treated as flow of a solid with the aid of eqs. (105b) will depend on whether or not in the solution of the problem the conditions (113) are satisfied. However, in the decisive quantity

$$2G\beta = \frac{G}{\mu},$$

where μ is the viscosity $\left(s_{ij} = 2\mu\left(d_{ij}-\frac{1}{3} d_{kk}\delta_{ij}\right)\right)$, appears the shear modulus of elasticity, which does not lend itself to a direct measurement for liquids with a small value of μ, just because for such liquids the conditions (113) will be satisfied in practically all flow phenomena.

Concluding Remarks

Whether stress as a thermodynamically defined quantity, subjected to equations of motion as a consequence of the principle of determinism, is a sufficiently broad concept in the theory of rheology, should ultimately be decided by experiments. Mathematical abstractions in physical theories can only find their justification in so far as they provide a basis for the correlation of observable phenomena.

In author's opinion a sharp distinction should be made between theories, in which distribution of stress vectors is introduced as the action of the material outside upon the material inside a closed surface, and theories in which a description is given of a material as a thermodynamic system. Since the response of the material will in general depend on the history of the external action, in theories of the first kind materials have to be characterized by a response functional. On the other hand in theories of the second kind the state of the material is at any moment supposed to be fully determined by the values of the state variables and the response of the material will be known as soon as the internal energy function is known and constitutive equations are given for the rate of change of the state variables, where these equations are independent of the history of the process, if the set of thermodynamic state variables is complete for the phenomena under consideration.

In some special cases the theories of the first and second kind are equivalent. Such is the case in the theory for ideal creep and ideal plasticity of an isotropic material with small deviations from the natural thermodynamic reference state. Then the thermodynamic state is assumed to be fully determined by the temperature and the state of stress (BESSELING 1966). The constitutive equations of the theory of the second kind may then be interpreted as response functions in a theory of the first kind. The theory of elasticity on the other hand must be considered as a theory of the second kind, if it is based upon GREEN's concept of a natural state of the continuum (TRUESDELL 1952). Recent papers on materials with fading memory (TRUESDELL 1965) deal again with theories of the first kind. Though they present a perfectly valid approach of the rheological problem, it must be doubted, whether it will ever be possible to determine for a specific material by laboratory experiments the tensorial functionals, that appear in these theories.

In author's opinion a more promising approach is offered by theories of the second kind, in which new state variables are introduced as the need for them arises, as for instance in the case of non-simple solids. Though in principle one could try to treat a polycristalline metal as a conglomeration of differently orientated anisotropic bodies of a simple solid, where the physical properties might even vary from one body to another, this approach is only feasible if the configuration of the various bodies is simplified to such an extent, that it becomes doubtful, whether the analysis has some bearing on the behaviour of the actual material. Author prefers a representation of the non-simple solid, in which the internal energy distribution is the weighted sum of a small number of distributions, each depending on its own set of state variables and representative for a portion of the material imbedded in the moving continuum together with all other portions. By assigning different dissipation functions to the various portions we can obtain a description of phenomena like primary creep, creep-recovery, anisotropic strain hardening and Bauschinger effect (BESSELING 1953, 1958).

There remains the difficult and laborious task of the experimentalist to show which is the proper theory for the analysis of rheological problems for real materials.

References

[1] BESSELING, J. F.: Nat. Aero. Res. Inst. (Amsterdam) Report S-410, 1953.
[2] — J. Appl. Mech. **25**, 4, 529 (1958).
[3] — Aeronautics and Astronautics **1960**, 247—266.
[4] — In: The Folke Odqvist Volume. Stockholm: Almqvist and Wiksell/Gebers. 1966.
[5] COLEMAN, B. D., and W. NOLL: Arch. Ratl Mech. Anal. **4**, 97—128 (1959).
[6] ECKART, C.: Phys. Rev. **73**, 2, 373—382 (1948).
[7] GILBARG, D., and D. PAOLUCCI: J. Ratl Mech. Anal. **2**, 617—642 (1953).
[8] JAUMANN, G.: Sitzungs-Ber. Akad. Wiss. (Wien) **110**, 120 (1911).
[9] PRANDTL, L.: Proc. 1st Int. Congr. Appl. Mech. (Delft) 43 (1924).
[10] REUSS, A.: Z. A. M. M. **10**, 266 (1930).
[11] TRUESDELL, C.: J. Ratl Mech. Anal. **1**, 1, 125—300; **1**, 1, 593—616 (1952).
[12] — and R. A. TOUPIN: Handbuch der Physik. Vol. III/1. Berlin—Göttingen—Heidelberg: Springer. 1960.
[13] — and W. NOLL: Handbuch der Physik. Vol. III/3. Berlin—Göttingen—Heidelberg: Springer. 1965.
[14] ZIEGLER, H.: Ing. Arch. **30**, 410 (1961).
[15] — Rheologica Acta **2** (3), 230—235 (1962).

Discussion

ONAT: The number of state variables $C_{\alpha\beta}$ which you have in your theory is six?

BESSELING: Six, because you eliminate three rotations. But the three rotations are always there because you have to return to the natural reference state. So there are really nine quantities geometrically and one for the thermodynamics.

ONAT: It seems to me that at least in linear viscoelasticity six would be too small a number for state variables.

BESSELING: I agree, of course. This is just the simplest case and it leads only to the theory of secondary creep and ideal plasticity. But you can improve on the representation by taking more state variables.

NAGHDI: If I understood you correctly, you defined the stress tensor in terms of internal energy as partial derivative of . . .

BESSELING: Yes.

NAGHDI: That is a constitutive assumption. It is the same point as discussed yesterday that constitutive equations should be separated from the basic field equations.

BESSELING: I only look at it differently. You start out from an a-priori concept of stress.

NAGHDI: The stress tensor is introduced (in the usual way) through the stress vector which acts across a surface of the body.

BESSELING: I do not introduce this stress vector.

NAGHDI: But you use it!

BESSELING: I call it stress tensor. That is just a quantity that appears in my equations but I nowhere introduce the equation of balance of momentum or something like that where you have to introduce the a-priori concept of stress.

NAGHDI: In your energy equation you have to state what is the rate of work done by surface forces.

BESSELING: No, I have no surface forces, I consider a closed system in an empty universe. And of course, inside this closed system I can then turn to a non-closed system. It would be more usual to introduce the stress at the surface but then, at a later point, you find the stress in terms of the internal energy and somewhere you have used two definitions of stress, I think.

NAGHDI: No.

BESSELING: Maybe you can say I have not introduced the more general concept of stress. I consider a special case, in which the stress is related to the internal energy.

NAGHDI: The constitutive assumption for the stress tensor is separate from the basic field equations.

BESSELING: Yes, you may introduce stress as separate from the basic theory — but I don't see any real objections against not doing that.

KESTIN: Perhaps it is worth remarking here that the derivation of Cauchy's equations of motion from the assumptions on the slides was achieved about three years ago by GREEN and RIVLIN. The assumption of the invariance with respect to a Galilean transformation together with the additivity of internal and kinetic energy yields the equations of motion. It is also worth mentioning, I think, that the argument can be inverted and is normally inverted to state that the use of the equations of motion together with the principle of invariance with respect to Galilean transformation yields the statement that internal energy is additive to kinetic energy.

BESSELING: With respect to the paper mentioned I think there is a difference. GREEN and RIVLIN introduce stress as a stress vector at the surface. Then, from that the stress tensor inside is derived. I think, in that argument you are not allowed to do the step which is done in the paper to say that the equations shall be satisfied for any volume and therefore also for a volume dV. I think, in order to have these arguments you should really introduce stress as a quantity, determined by the internal state, and not as some kind of abstract a-priori concept.

NAGHDI: GREEN and RIVLIN start with a balance of energy together with invariance requirements under superposed rigid body motions. Then, they derive Cauchy's equations of motion and the equation for conservation of mass. You objected to the introduction of the idea of the stress vector. The stress vector is introduced as part of the balance of energy and the arguments they use about the integrand are the usual arguments based on the continuity of the functions. This time it is applied to the balance of energy.

BESSELING: You have introduced the internal energy and you have not stated what things this internal energy depends on.

NAGHDI: It does not matter, as long as they state that it is invariant under superposed rigid body motions. This is precisely an argument that one would use in the construction of any type of constitutive equations.

BESSELING: I think, we should talk about it in private.

KLUITENBERG: You do not derive the entropy production but you suggest it. Can you generalize this idea, so that you have also volume-viscosity?

BESSELING: No. In this approach there is no place for volume-viscosity. But I am still wondering whether there are convincing experimental data that there is volume-viscosity. As long as this is not settled I don't see why we should complicate things by introducing it. Now, of course, I know of some instances where you can get better correspondence between theory and experiments by introducing volume-viscosity. But on the other hand maybe you may equally improve on the correspondence by introducing the kind of relaxation effects that are associated with highly non-stationary states. As long as this point is not settled I would like to consider volume-viscosity as not existent.

KLUITENBERG: I think, there is a paper by KARIM and ROSENHEAD[1] where they agree, that volume-viscosity does exist.

ONAT: But certainly there is relaxation phenomena observed in linear visco-elastic solids in the presence of pure volume changes.

BESSELING: Of course, you can get relaxation effects by the thermodynamic phenomena of heat conduction and convection.

KESTIN: It can be shown that if a system performs work in a reversible way which is described by the pressure multiplied into the total derivative of volume and does not possess internal variables of state then the bulk-viscosity is zero. This is the normal assumption that we make when we study gases. In the case of liquids the statement carries through but it becomes somewhat trivial by the fact that in liquids we normally also introduce the assumption that the density remains constant. Now, in a solid, in a viscoelastic solid, we have the case where internal variables operate and under those conditions bulk-viscosity has a non-vanishing value.

BESSELING: I agree that by introducing certain internal variables it will be possible to obtain a non-vanishing bulk-viscosity.

---

[1] L. M. KARIM and L. ROSENHEAD, Rev. Mod. Phys. 24 (1952), 108.

Thermodynamics and Wave Propagation in Non-Linear Materials with Memory

By

B. D. Coleman and **M. E. Gurtin**
Pittsburgh, Pa. (U. S. A.)

Preface

In the past year several articles have appeared on the theory of propagating singular surfaces in non-linear dissipative materials[1]. Since non-equilibrium thermodynamics plays a basic role in this infant subject, we have prepared for this symposium a survey of its most developed part: the theory of one-dimensional waves in materials which do not conduct heat but have long-range viscoelastic memory.

Our analysis rests upon the constitutive assumption that the stress, temperature, and specific internal energy at a material point depend on the temporal histories of the strain and specific entropy at the point. The functionals expressing this dependence are not assumed to have any special properties other than those implied by the laws of thermodynamics and the principle of fading memory in the form proposed by COLEMAN and NOLL[2]. Particular emphasis is laid on a theorem in the thermodynamics of materials with memory relating the stress and temperature functionals to the internal-energy functional [COLEMAN, 1964, 1]. After generalizing to materials with memory results obtained by DUHEM and TRUESDELL for acceleration waves and mild discontinuities in elastic bodies[3], we give explicit and exact formulae showing how non-linearity

---

[1] COLEMAN, GURTIN and HERRERA [1965, 1], COLEMAN and GURTIN [1965, 2—7], [1966, 1], VARLEY [1965, 9], PIPKIN [1966, 3], COLEMAN, GREENBERG and GURTIN [1966, 2].

[2] COLEMAN and NOLL [1960, 1; 1961, 1]. For a survey see TRUESDELL and NOLL [1965, 8, §§ 38—41].

[3] The classical theorems of Duhem are extended, explained, and given modern proofs by TRUESDELL [1961, 2]. For surveys of the theory of homentropic and homothermal waves in elastic materials, see TRUESDELL and NOLL [1965, 8, particularly §§ 71 and 72], and COLEMAN and GURTIN [1965, 5].

and internal dissipation affect the amplitude of these higher-order waves when they propagate into homogeneous regions at equilibrium. In our analysis of shock waves we observe that much of the classical theory of Hugoniot curves can be extended to dissipative materials.

We make many assertions here without giving details of the proofs. Some of the omiitted demonstrations will appear shortly in an article on shock waves prepared for the Proceedings of the Royal Society[1]; most of the others will be found in the book "Wave Propagation in Dissipative Materials"[2].

1. Balance Laws

As in the first three articles in the series Waves in Materials with Memory [1965, 1—3], the mechanical and thermodynamic concepts used here are one-dimensional. A motion is described by a scalar function $x = x(X, t)$ giving the place x at time t of the material point which has the position X in a fixed reference configuration R with mass density ϱ_R. We find it convenient to identify each material point with its position X in R. The stress (or "tension") T, the absolute temperature $\Theta\,(\Theta > 0)$, the specific internal energy ε (in units of energy per unit mass), and the specific entropy η are all taken to be functions of X and t. We assume that the material does not conduct heat and that both the extrinsic body force and the extrinsic heat supply are zero; thus, for the laws of balance of momentum and energy we write:

$$\frac{d}{dt}\int_{X_\alpha}^{X_\beta} \dot{x}\varrho_R\, dX = T(X_\beta, t) - T(X_\alpha, t), \tag{1.1}$$

$$\frac{d}{dt}\int_{X_\alpha}^{X_\beta} \left(\varepsilon + \frac{1}{2}\,\dot{x}^2\right) \varrho_R\, dX = T(X_\beta, t)\,\dot{x}\,(X_\beta, t) - T(X_\alpha, t)\,\dot{x}\,(X_\alpha, t). \tag{1.2}$$

A superposed dot denotes the material time derivative; i. e., $\dot{x} = \partial_t x\,(X, t)$ is the velocity. Equations (1.1) and (1.2) hold for all times t and for every interval $[X_\alpha, X_\beta]$ in R.

The deformation gradient F at a fixed material point X at a given time t is the number defined by $F(X, t) = \partial_X x(X, t)$. When ϱ_R is specified and the present theory is applied to longitudinal motions of a three-dimensional body, F determines the mass density ϱ at time t

[1] Now available as [1965, 6].

[2] COLEMAN, GURTIN, HERRERA, and TRUESDELL [1965, 7]. This book is a reprint of the following five articles: [1961, 2], [1965, 1—4].

through the relation

$$\varrho = \frac{\varrho_R}{|F|}, \tag{1.3}$$

which is then equivalent to the law of conservation of mass.

Since the heat conduction and extrinsic heat supply are zero for the materials and processes we consider here, the Clausius-Duhem inequality takes the simple form

$$\frac{d}{dt} \int_{X_\alpha}^{X_\beta} \eta \varrho_R \, dX \geq 0. \tag{1.4}$$

2. Constitutive Assumptions

The histories up to time t of the deformation gradient and the entropy at X are the real-valued functions F^t and η^t defined by

$$F^t(s) = F(X, t-s), \quad \eta^t(s) = \eta(X, t-s), \quad 0 \leq s < \infty. \tag{2.1}$$

In our present theory the material at X is characterized by three constitutive functionals, $\mathfrak{e}$, $\mathfrak{T}$, and $\mathfrak{t}$; these functionals give the present values of ε, T, and Θ at X whenever the histories F^t and η^t at X are specified:

$$\left.\begin{aligned} \varepsilon(t) &= \mathfrak{e}(F^t, \eta^t), \\ T(t) &= \mathfrak{T}(F^t, \eta^t), \\ \Theta(t) &= \mathfrak{t}(F^t, \eta^t). \end{aligned}\right\} \tag{2.2}$$

We suppose that our constitutive functionals obey the principle of fading memory. To state this principle in our present context we let h denote an influence function, i. e., a positive, monotone[1], continuous function on $[0, \infty)$ decaying fast enough to be square integrable, and we define the norm $\|\Lambda^t\|$ of a pair of histories $\Lambda^t = (F^t, \eta^t)$ by[2]

$$\|\Lambda^t\|^2 = F^t(0)^2 + \eta^t(0)^2 + \int_0^\infty (F^t(s)^2 + \eta^t(s)^2)\, h(s)^2 \, ds. \tag{2.3}$$

We take the common domain of $\mathfrak{e}$, $\mathfrak{T}$, and $\mathfrak{t}$ to be an open subset D in the cone $\mathfrak{C}$ of all pairs Λ^t of positive-valued functions F^t, η^t with $\|\Lambda^t\|$

---

[1] COLEMAN and MIZEL [1966, 1] have recently shown that conditions on h weaker than monotonicity suffice for most of the theory of fading memory.

[2] Cf., COLEMAN and NOLL [1960, 1; 1961, 1].

finite, and we assume that $\mathfrak{e}$, $\mathfrak{T}$, and $\mathfrak{t}$ are several times continuously differentiable throughout D with respect to the norm $||\cdot||$. To be more specific, let $\mathfrak{f}$ be a functional representing $\mathfrak{e}$, $\mathfrak{T}$, or $\mathfrak{t}$. We are assuming here that, for each Λ^t in D, $\mathfrak{f}$ has nth order Fréchet derivatives $d^n\mathfrak{f}(\Lambda^t|\cdot, \ldots, \cdot)$ for $n = 1, \ldots, M$; these derivatives are bounded, symmetric, n-linear forms with the property that for all functions Ω in $\mathfrak{C}$ with $\Lambda^t + \Omega$ in D

$$\mathfrak{f}(\Lambda^t + \Omega) = \mathfrak{f}(\Lambda^t) + \sum_{n=1}^{M} \frac{1}{n!}\, d^n\mathfrak{f}(\Lambda^t \mid \overbrace{\Omega, \ldots, \Omega}^{n-times}) + o(||\Omega||^M), \tag{2.4}$$

where

$$\lim_{||\Omega|| \to 0} \frac{o(||\Omega||^M)}{||\Omega||^M} = 0. \tag{2.5}$$

Each functional $d^n\mathfrak{f}(\cdot \mid \cdot, \ldots, \cdot)$ is assumed to be jointly continuous in all its $n + 1$ arguments. For a theory of wave propagation, we do not need Fréchet derivatives of all orders. When we discuss a wave of order $N \geq 2$ it suffices to assume that $M = N + 1$ for $\mathfrak{e}$, and $M = N$ for $\mathfrak{T}$ and $\mathfrak{t}$. In our treatment of shock waves $(N = 1)$ we shall assume that $M = 5$ for $\mathfrak{e}$ and $M = 4$ for $\mathfrak{T}$ and $\mathfrak{t}$.

In this essay we shall call a material obeying (2.2) and the smoothness assumption (2.4) a *non-conductor*; this is an abbreviation for "a simple material which does not conduct heat but which has a long-range, non-linear, viscoelastic memory compatible with the principle of fading memory."

Let T_0 be the stress the material would exhibit had it always been at rest with some particular fixed specific entropy η_0 and with the fixed deformation gradient F_0, and let T^* be the stress corresponding to the history $\Lambda^t = (\mathrm{F}^t, \eta^\dagger)$ where F^t is arbitrary but $\eta^\dagger$ is the constant history $\eta^\dagger(s) \equiv \eta_0$, $0 \leq s < \infty$. Using (2.4) with $\mathfrak{f} = \mathfrak{T}$ and $M = 2$, one can easily show that[1]

$$T^* - T_0 = G(0)\, \Psi(t) + \int_0^\infty \psi(t-s) \frac{d}{ds} G(s)\, ds + \mathrm{O}(\delta^2), \tag{2.6}$$

where

$$\psi(t-s) = F^t(s) - F_0 = F(t-s) - F_0, \tag{2.7}$$

[1] See COLEMAN and NOLL [1961, 1] and COLEMAN [1964, 1, § 7] for the general argument for three-dimensional bodies. The present one-dimensional case is discussed in detail in [1965, 3, p. 289].

is called the *relative strain*, and δ, given by

$$\delta = \sup_{0<s<\infty} |\psi(t-s)|, \tag{2.8}$$

is small if $|\psi|$ has always been small. The function G, called the *stress-strain relaxation function* corresponding to F_0 and η_0, is determined when the functional $\mathfrak{T}$ is specified. For a given material, different choices of F_0 and η_0 give rise to different functions G. The equation (2.6) states that if the entropy is held fixed and the relative strain is kept small, the stress increment $T^* - T_0$ for a non-linear material with memory is given, approximately, by the constitutive equation of a linearly viscoelastic material.

3. Some Results from the Thermodynamics of Materials with Memory

Let F^t be a history and ΔF a number. The function $F^t \wedge \Delta F$, defined by

$$(F^t \wedge \Delta F)(s) = \begin{cases} F^t(0) + \Delta F, & s = 0, \\ (F^t(s), & 0 < s < \infty, \end{cases} \tag{3.1}$$

is called the *jump continuation of F^t with jump ΔF* [COLEMAN, 1964, 1, pp. 238—240]. An equation analogous to (3.1) defines the history $\eta^t \wedge \Delta\eta$. When the constitutive functionals are known and the pair F^t, η^t is specified, the values of ε, T, and Θ corresponding to $F^t \wedge \Delta F$ and $\eta^t \wedge \Delta\eta$ are given by functions of the jumps ΔF and $\Delta\eta$:

$$\left.\begin{aligned} \varepsilon^*(\Delta F, \Delta\eta) &= \mathfrak{e}(F^t \wedge \Delta F, \ \eta^t \wedge \Delta\eta), \\ T^*(\Delta F, \Delta\eta) &= \mathfrak{T}(F^t \wedge \Delta F, \ \eta^t \wedge \Delta\eta), \\ \Theta^*(\Delta F, \Delta\eta) &= \mathfrak{t}(F^t \wedge \Delta F, \ \eta^t \wedge \Delta\eta). \end{aligned}\right\} \tag{3.2}$$

The functions ε^*, T^*, and Θ^* defined here are called the *instantaneous response functions corresponding to the underlying histories F^t, η^t*. Of course, these instantaneous response functions depend not only on the material under consideration but also on F^t and η^t.

Working in a more general theory which applies to arbitrary, rather than one-dimensional processes and which allows for the presence of heat conduction, COLEMAN [1964, 1, Thm. 1, p. 19 and Thm. 6, p. 33] has found the complete set of restrictions which the constitutive functionals must obey in order that the entropy production never be negative in a thermodynamic process. Among these restrictions are the following: ε^* must determine T^* through the *stress-relation*,

$$T^*(\varDelta F, \varDelta \eta) = \varrho_R \frac{\partial}{\partial \varDelta F} \varepsilon^*(\varDelta F, \varDelta \eta), \tag{3.3}$$

and Θ^* through the *temperature relation*,

$$\Theta^*(\varDelta F, \varDelta \eta) = \frac{\partial}{\partial \varDelta \eta} \varepsilon^*(\varDelta F, \varDelta \eta). \tag{3.4}$$

The stress and temperature relations have an elementary physical interpretation [1964, 1, pp. 240—241]: Let us take a material with arbitrary total history $\varLambda^t = (F^t, \eta^t)$, and then change F and η at time t so fast that in comparison with $\varLambda^t$ the new history so generated appears to be discontinuous[1]. The values of F, η, ε, T, and Θ at each "instant" of this new process will be $F^t(0) + \varDelta F$, $\eta^t(0) + \varDelta \eta$, $\varepsilon^*(\varDelta F, \varDelta \eta)$, $T^*(\varDelta F, \varDelta \eta)$, and $\Theta^*(\varDelta F, \varDelta \eta)$. The equations (3.3) and (3.4) state that $T^*(\varDelta F, \varDelta \eta)$ and $\Theta^*(\varDelta F, \varDelta \eta)$ are determinable from the function ε^* by precisely the rules that give T and Θ in a perfectly elastic material (i. e., the rules used in statical situations). That the function ε^* depends on the previous total history $\varLambda^t$ means that the nature of the "perfectly elastic material" of the rapid investigation depends on $\varLambda^t$: Corresponding to each previous total history we have a new elastic material. This conclusion is exemplified by considering the behavior of a glass in fast experiments. A glass is but a fluid obeying the principle of fading memory. When studied rapidly, a glass appears to be a perfectly elastic solid with a stored energy function; the various moduli of the "elastic solid" depend, however, on the previous history of the glass. COLEMAN's derivation of (3.3) and (3.4) leads to the conclusion that, when studied sufficiently rapidly, any material obeying (2.2) and (2.4) must exhibit, as a glass does, elastic behavior with a stored energy function.

In our present notation the "D-operators" introduced by COLEMAN [1964, 1] are given by

$$\begin{aligned} D_F \mathfrak{f}(F^t, \eta^t) &= \frac{\partial}{\partial \varDelta F} f^*(\varDelta F, 0)|_{\varDelta F=0}, \\ D_\eta \mathfrak{f}(F^t, \eta^t) &= \frac{\partial}{\partial \varDelta \eta} f^*(0, \varDelta \eta)|_{\varDelta \eta=0}, \end{aligned} \tag{3.5}$$

where $\mathfrak{f}$ stands for $\mathfrak{e}$, $\mathfrak{T}$, or $\mathfrak{t}$, and f^* is the corresponding instantaneous response function ε^*, T^*, or Θ^*. Thus the linear operators D_F, D_η represent partial differentiation with respect to present values holding the past history fixed. In terms of these operators the stress relation and temperature relation may be written in the compact form [COLEMAN 1964, 1, p. 33]

$$\mathfrak{T} = \varrho_R D_F \mathfrak{e}, \quad \mathfrak{t} = D_\eta \mathfrak{e}. \tag{3.6}$$

---

[1] In [1965, 2, § 8] we give a method of approximating jump continuations by continuous histories.

Corresponding to any specification of F^t, η^t in the domain of our constitutive functionals, we may consider the *instantaneous stress-strain modulus* E defined by[1]

$$E = D_F \mathfrak{T}(F^t, \eta^t) = \frac{\partial}{\partial \Delta F} T^*(\Delta F, 0)|_{\Delta F = 0} \tag{3.7}$$

and the *instantaneous second-order stress-strain modulus*:

$$\tilde{E} = D_F^2 \mathfrak{T}(F^t, \eta^t) = \frac{\partial^2}{\partial \Delta F^2} T^*(\Delta F, 0)_{\Delta F = 0}. \tag{3.8}$$

These moduli, which depend of course not only on the material but also on the total history $\Lambda^t = (F^t, \eta^t)$, play an important role in the theory of wave propagation. They measure, respectively, the initial slope and initial curvature of the stress-strain curve $T^*(\Delta F, 0)$ for instantaneous response to isentropic strain impulses, of intensity ΔF, superimposed on F^t, η^t at time t. In the special case in which F^t and η^t are the constant histories $F^\dagger$ and $\eta^\dagger$ defined by

$$F^\dagger(s) = F_0 = \text{const.}, \quad \eta^\dagger(s) = \eta_0 = \text{const.}, \quad 0 \leq s < \infty, \tag{3.9}$$

E equals the initial value $G(0)$ of the stress-strain relaxation function G discussed in (2.6) — (2.8).

The quantity[2]

$$\alpha = D_\eta \mathfrak{T}(F^t, \eta^t) = \frac{\partial}{\partial \Delta \eta} T^*(0, \Delta \eta)|_{\Delta \eta = 0}, \tag{3.10}$$

is called the *instantaneous stress-entropy modulus* corresponding to F^t, η^t. It follows from the familiar rule $\partial_x \partial_y f(x, y) = \partial_y \partial_x f(y, x)$ for C^2 functions f, that the operators D_F and D_η commute. In particular we have $D_F D_\eta \mathfrak{e} = D_\eta D_F \mathfrak{e}$, and (3.6) yields the following analogue of Maxwell's relation, familiar in equilibrium thermodynamics:

$$D_\eta \mathfrak{T} = \varrho_R D_F \mathfrak{t}. \tag{3.11}$$

Thus, another formula for α is

$$\alpha = \varrho_R D_F \mathfrak{t}(F^t, \eta^t) = \varrho_R \frac{\partial}{\partial \Delta F} \Theta^*(\Delta F, 0)|_{\Delta F = 0}. \tag{3.12}$$

---

[1] In [1965, 3] our present E and $\tilde{E}$ are written $\hat{E}_t$ and $\hat{\tilde{E}}_t$, respectively, and are called instantaneous moduli "at fixed entropy" to distinguish them from the corresponding stress-strain moduli at fixed present temperature.

[2] Our present α is written $\hat{\alpha}_t$ in [1965, 3].

4. Classification of Waves

In continuum physics the word *wave* is used with several distinct meanings. To some a wave is a sinusoidal disturbance, to some it is any member of a certain class of solutions to a hyperbolic equation, and to others it is a propagating singular surface. Here we follow CHRISTOFFEL, DUHEM, HADAMARD, and HUGONIOT and use the word in the last sense. Of course in our one-dimensional theory the singular surface becomes a singular point. When we say that this singularity *propagates*, we mean that it moves relative to the material. We let Y_t be the material point, labeled by its position in the reference configuration R, at which the wave is to be found at time t. The *intrinsic velocity* of the wave is defined by

$$U(t) = \frac{d}{dt} Y_t. \tag{4.1}$$

For definiteness we shall assume that $U(t) > 0$.

The location in space of the wave at time t is the place

$$y_t = x(Y_t, t). \tag{4.2}$$

The *wave velocity* u is the rate of advance of the wave as seen by an observer at rest:

$$u(t) = \frac{d}{dt} y_t = \frac{d}{dt} x(Y_t, t). \tag{4.3}$$

If $f(X, t)$ is continuous in X except for a jump discontinuity at $X = Y_t$, we write

$$f_1 = f_1(t) \stackrel{\text{def}}{=} \lim_{X \to Y_t^+} f(X, t), \quad f_2 = f_2(t) \stackrel{\text{def}}{=} \lim_{X \to Y_t^-} f(X, t), \tag{4.4}$$

and define the jump in f across the wave to be

$$[f] = [f](t) \stackrel{\text{def}}{=} f_2(t) - f_1(t). \tag{4.5}$$

We shall assume throughout this article that $x(X, t)$ is a continuous function of X, t for all X, t and that the fields $\eta(X, t)$, $\partial_X^m \partial_t^n \eta(X, t)$, $\partial_X^m \partial_t^n x(X, t)$, for all $m, n = 0, 1, 2, \ldots$, are continuous at all points away from Y_t, t and suffer nothing worse than jump discontinuities across the wave.

We say that a wave is of order N if the N'th-order derivatives of $x(X, t)$ suffer non-zero jumps upon the wave while all derivatives of $x(X, t)$ of order less than N and all derivatives of $\eta(X, t)$ of order less than $N-1$ are continuous across the wave. Thus, at a wave of order N

$$[\partial_X^m \partial_t^n x] \neq 0 \quad \text{whenever } m+n=N, \tag{4.6a}$$

but

$$[\partial_X^m \partial_t^n x] = 0 \quad \text{whenever } 0 \leq m+n < N, \tag{4.6b}$$

and

$$[\partial_X^m \partial_t^n \eta] = 0 \quad \text{whenever } 0 \leq m+n < N-1. \tag{4.6c}$$

If a wave of order N is such that

$$[\partial_X^m \partial_t^n \eta] = 0 \quad \text{whenever } m+n = N-1, \tag{4.7}$$

then we say that the wave is *homentropic*[1].

Across a *shock wave*

$$[x] = 0, \quad [\dot{x}] \neq 0 \quad [F] \neq 0, \tag{4.8}$$

and it is possible to have $[\eta] \neq 0$. Thus a shock wave is a wave of order 1.

Across an *acceleration wave*

$$[x] = [\dot{x}] = [F] = [\eta] = 0, \tag{4.9a}$$

but

$$[\ddot{x}] \neq 0, \quad [\dot{F}] \neq 0, \quad [\partial_X F] \neq 0. \tag{4.9b}$$

Thus, an acceleration wave is a wave of order 2. If $[\partial_X \eta] = [\dot{\eta}] = 0$, the acceleration wave is *homentropic*.

A wave of order $N \geq 3$ is called a mild *discontinuity*.

When we discuss a wave of order N, our basic assumption is that given in (4.6). It is obvious, however, that in a theory which accounts for long range memory, some assumption must be made to limit the wildness of past histories. The following assumption is more than sufficient for the present theory: We suppose that in the past the wave has not passed through any material point more than a countable number of times and that if the order of the wave is N, then whenever $1 \leq m + n \leq N$ and $0 \leq p + q \leq N-1$ the fields $\partial_X^m \partial_\tau^n x(X,\tau)$ and $\partial_X^p \partial_\tau^q \eta(X,\tau)$ are for each t uniformly bounded over the set $\{X, \tau \mid Y_t - \delta < X < Y_t + \delta, -\infty < \tau < t\}$ for some $\delta > 0$. This assumption will be met in almost all applications. In those few applications where $F(X,\tau) \to \infty$ as $\tau \to -\infty$, other, weaker regularity assumptions can be given which will also suffice for the present theory[2].

We shall discuss first the theory of waves of order $N \geq 2$, for their theory is simpler than the theory of shock waves.

---

[1] Cf. TRUESDELL and NOLL [1965, 8, §§ 72 and 73].

[2] The minimal regularity conditions on past histories are discussed in [1965, 1—4], particularly [1965, 3].

5. Acceleration Waves and Mild Discontinuities

First we note that arguments given by DUHEM[1] for acceleration waves in elastic materials, by COLEMAN and GURTIN [1965, 3, § 4] for acceleration waves in materials with memory, and by TRUESDELL [1961, 2, § 13] for mild discontinuities in elastic materials, may be combined to prove the following remark, provided one uses the relations (3.6) for materials with memory: *In a non-conductor all waves of order* $N \geq 2$ *are homentropic and hence obey* (4.7).

The main reason for discussing in one section all waves of order $N \geq 2$ lies in the fact that such waves share a common expression for the velocity[2]. Let F_Y^t and η_Y^t denote the functions F^t and η^t at the wave, i. e., the histories of F and η up to time t at the point Y_t:

$$F_Y^t(s) = F(Y_t, t - s), \quad \eta_Y^t(s) = \eta(Y_t, t - s), \quad 0 \leq s < \infty. \tag{5.1}$$

The intrinsic velocity $U(t)$ *of any wave of order* $N \geq 2$ *propagating in a non-conductor satisfies*[3]

$$U(t)^2 = \frac{E_Y}{\varrho_R}, \tag{5.2}$$

with E_Y *the instantaneous stress-strain modulus evaluated at the wave*:

$$E_Y = D_F \mathfrak{T}(F_Y^t, \eta_Y^t). \tag{5.3}$$

Since F is continuous across a wave of order 2 or higher, application of a familiar chain rule to (4.2) yields, upon use of the definitions (4.1) and (4.3),

$$u(t) = F_Y U(t) + \dot{x}_Y, \tag{5.4}$$

where $F_Y = F(Y_t, t)$ and $\dot{x}_Y = \dot{x}(Y_t, t)$ are the deformation gradient and material velocity at the wave. The velocity of the wave relative to the material point instantaneously situated at it is

$$u^r = u - \dot{x}_Y. \tag{5.5}$$

---

[1] For a discussion of DUHEM's theorems on waves in elastic materials, see TRUESDELL [1961, 2, § 13].

[2] This result, long known to hold for perfect fluids, was derived for isotropic incompressible elastic materials in three-dimensional situations by ERICKSEN [1953, 1] and for general elastic materials by TRUESDELL [1961, 2].

[3] For the proof of eq. (5.2), accounting for both memory effects and thermodynamic influences, see COLEMAN and GURTIN [1965, 3, § 9; 1965, 4, § 2].

We call u^r the *relative velocity*. It follows from (5.4) that $u^r = F_Y U$. Therefore, (5.2) tells us that *in a non-conductor the relative velocity of a wave of order* $N \geq 2$ *is given by*

$$u^r = \pm F_Y \sqrt{\frac{E_Y}{\varrho_R}}. \tag{5.6}$$

It follows from (5.2) and (5.3) that both $U(t)$ and $u^r(t)$ depend on the history of the material point at which the wave is located. We consider now a wave of order $N \geq 2$ which since time $t = 0$ has been propagating into a region which previously had always been at rest in a fixed homogeneous configuration with constant and uniform specific entropy. Thus, we suppose that $U(t) > 0$ and that for $X > Y_t$ and $t \geq 0$,

$$x = x(X, t - s) = F_0 X + X_0, \quad 0 \leq s < \infty, \tag{5.7}$$

$$\eta = \eta(X, t - s) = \eta_0, \quad 0 \leq s < \infty.$$

Equation (5.3) now becomes

$$E_Y = D_F \mathfrak{T}(F^\dagger, \eta^\dagger) = G(0) \tag{5.8}$$

with $F^\dagger$ and $\eta^\dagger$ defined in (3.9), and $G(0)$ the initial value of G for that linearly viscoelastic material which approximates the general simple material in small isentropic deformations from the state (5.7), as explained in (2.6) — (2.8). Of course we here have $\dot{x}_Y = 0$ and $F_Y = F_0$, and, therefore, (5.6) yields the following assertion: *When a wave of order* $N \geq 2$ *travels in a homogeneous region of a non-conductor which has been at equilibrium, as shown in* (5.7), *the wave velocity* u, *the relative velocity* u^r, *and the intrinsic velocity* U *are constants given by*

$$u = u^r = F_0 U = F_0 \sqrt{\frac{G(0)}{\varrho_R}}. \tag{5.9}$$

Although all waves of order greater than 1 travel at a common speed, a fundamental distinction between acceleration waves and mild discontinuities arises as soon as we consider the problem of reinforcement and attenuation. We may define the *amplitude* $a(t)$ of a wave of order N to be

$$a(t) = [\partial_N^t x(X, t)]; \tag{5.10}$$

e. g., the amplitude of an acceleration wave is just the jump $[\ddot{x}]$ in the acceleration, while for a wave of order 3 we have $a = [\dddot{x}]$.

If we are given a at some time $t = 0$, then it is possible to derive expressions for a at later times; the form of these expressions will depend on the total history of the region traversed by the wave in the time interval $[0, t]$[1]. A discussion of this problem in full generality, though not intrinsically difficult, seems to require the development of an elaborate notation. In the interest of simplicity, we here give results only for waves propagating into homogeneous equilibrium regions of the type (5.7). We continue to let $G(s)$ be the stress-strain relaxation function for that linearly viscoelastic material which approximates our non-linear material in small isentropic departures from $F_0^\dagger$, $\eta_0^\dagger$, and we put

$$G'(0) = \frac{d}{ds} G(s)|_{s=0}. \tag{5.11}$$

We always assume $G(0) > 0$; let us now add the assumption that

$$G'(0) < 0 \tag{5.12}$$

and further assume that the instantaneous second-order stress-strain modulus at the wave is non-zero:

$$\tilde{E}_Y = D_F^2 \mathfrak{T}(F^\dagger, \eta^\dagger) \neq 0. \tag{5.13}$$

Our principal result is the following: *Consider a wave of order* $N \geq 2$ *propagating, since* $t = 0$, *into a non-conductor of infinite extent. Assume that previous to arrival of the wave the non-conductor had always been at rest with its deformation gradient* F_0 *and specific entropy* η_0 *constant in space and time. If* $N = 2$, *then*[2]

$$a(t) = \frac{\lambda}{\left(\frac{\lambda}{a(0)} - 1\right) e^{\gamma t} + 1}, \tag{5.14}$$

where

$$\gamma = -\frac{G'(0)}{2G(0)} = \text{const.}, \tag{5.15}$$

$$\lambda = \frac{G'(0)\,U}{\tilde{E}_Y} = \text{const}. \tag{5.16}$$

If $N \geq 3$, *then*

$$a(t) = a(0)e^{-\gamma t}, \tag{5.17}$$

with γ *again given by* (5.15).

---

[1] The amplitude of acceleration waves in materials with memory is discussed in [1965, 3] taking account of thermodynamics; a more detailed analysis which, though it neglects thermodynamic influences, includes waves of all orders greater than 1 is given by COLEMAN, GREENBERG and GURTIN in [1966, 2].

[2] Equations (5.14)—(5.16) for acceleration waves were obtained in our present thermodynamical theory by COLEMAN and GURTIN [1965, 3, Thm. 11.1, p. 297].

In spite of their simplicity, eqs. (5.14) — (5.17) are exact and hold for waves of arbitrary initial amplitude $a(0)$, provided only that the wave has been entering a homogeneous region in equilibrium. Equations of the general form (5.14) and (5.17) may be derived also in theories which neglect the influence of thermodynamic variables on the stress[1]. However, to obtain these results in a theory starting with the general constitutive eqs. (2.2), one must use Coleman's relations (3.6), and hence the second law of thermodynamics.

It is interesting to note that for an acceleration wave $|\lambda|$ plays the role of a critical amplitude. Since we assume (5.12), equation (5.14) has the following properties: *If* $|a(0)| < |\lambda|$ *or if* $\operatorname{sgn} a(0) = \operatorname{sgn} \tilde{E}$, *then* $a(t) \to 0$ *monotonically as* $t \to \infty$. *If* $a(0) = \lambda$, *then* $a(t) \equiv a(0)$. *On the other hand, if both* $|a(0)| > |\lambda|$ *and* $\operatorname{sgn} a(0) = -\operatorname{sgn} \tilde{E}$, *then* $|a(t)| \to \infty$ *monotonically in a finite time* t_∞ *given by*

$$t_\infty = -\frac{1}{\gamma} \ln\left(1 - \frac{\lambda}{a(0)}\right). \tag{5.18}$$

One is tempted to suppose that the approach of $[\ddot{x}](t)$ to ∞ as $t \to t_\infty$ is indicative of the formation of a shock wave at time t_∞, but a proof to this effect is lacking. In any event, it is clear from the results in hand that the presence of internal damping, manifested by a strictly negative value of $G'(0)$, does not always imply that an acceleration wave moving into a homogeneous region must be damped out.

The implications of (5.17) are much simpler: *If we assume* (5.12), *then for a mild discontinuity propagating into a homogeneous region at equilibrium,* $a(t) \to 0$ *exponentially as* $t \to \infty$.

Of course, if $\tilde{E}_Y$ were zero, then in place of (5.14) we should have (5.17) holding also for acceleration waves. When we use (1.3) we are assuming that our one-dimensional theory is being applied to plane longitudinal motions; however, the present theory may be applied also to shearing motions in incompressible materials[2]. In view of this it is worthwhile to note that although (5.13) should be expected to hold for longitudinal waves, $\tilde{E}$ is zero for a plane shear wave propagating along a principal axis of strain into an incompressible isotropic material [1965, 2, § 6].

---

[1] An equation of the form (5.14) was obtained for materials of integral type by Varley [1965, 9] and for more general materials with fading memory by Coleman and Gurtin [1965, 2]. A result of the form (5.17) was derived by Coleman, Greenberg and Gurtin [1966, 2] under neglect of thermodynamic influences. The observation that (5.17) holds in the present theory is new.

[2] Cf. [1965, 1, § 6].

If we assume that $G'(0) = 0$, then eq. (5.14) for acceleration waves assumes the form[1]

$$a(t) = \frac{a(0)}{1 + \nu a(0) t}, \qquad \nu = \frac{\tilde{E}}{2UG(0)} \tag{5.19}$$

and eq. (5.17) for mild discontinuities reduces to[2]

$$a(t) \equiv a(0). \tag{5.20}$$

We have already observed that every acceleration wave in a non-conductor is homentropic, i. e., has $[\dot{\eta}] = [\partial_X \eta] = 0$; when, however, the acceleration wave is propagating into an equilibrium region of the type (5.7), the following stronger condition on the entropy change can be derived [1965, 3, Remark 11.3]:

$$[\ddot{\eta}] = [\partial_X \dot{\eta}] = [\partial_X^2 \eta] = 0. \tag{5.21}$$

Our one-dimensional theory may be applied to the study of longitudinal waves proceeding along the axis of a cylinder composed of an isotropic or transversely isotropic non-conductor with memory[3]. In this application X and $x = x(X, t)$ are distances along the cylindrical axis, and $F(X, t)$, which we take to be positive, determines the density through (1.3). Hence, for a longitudinal acceleration wave $\varrho(X, t)$ is continuous for all X and t, while $\dot{\varrho}(X, t)$ and $\frac{\partial}{\partial X}\varrho(X, t)$ have jump discontinuities at $X = Y_t$ but are continuous everywhere else. We call the wave *compressive* at time t if $U(t)[\dot{\varrho}](t) > 0$ and *expansive at* t if $U(t)[\dot{\varrho}](t) < 0$. Combining (1.3) with the compatibility condition[4]

$$[\ddot{x}] = -U[\dot{F}], \tag{5.22}$$

which holds for all acceleration waves, we obtain

$$a = \varrho_R \varrho^{-2} U[\dot{\varrho}]. \tag{5.23}$$

---

[1] Equation (5.19) was derived by THOMAS [1957, 1] for longitudinal waves in elastic fluids which do, of course, have $G'(0) = 0$. See also the recent work of GREEN [1964, 2] and VARLEY and DUNWOODY [1965, 10].

[2] GREEN [1964, 2] has recently derived (5.20) for elastic materials.

[3] The theoretical basis of this statement and some of the experimental difficulties involved in achieving purely longitudinal motions in actual three-dimensional specimens are discussed in § 6 A of [1965, 1]. Our present longitudinal waves are usually called *plane longitudinal waves* but are sometimes also called *plane dilatational waves*.

[4] Vid. [1960, 2, § 190] or [1965, 1, p. 12].

Let us put

$$\lambda_* = \frac{G'(0)\varrho_0^2}{\tilde{E}\varrho_R}, \qquad \varrho_0 = \frac{\varrho_R}{F_0},$$

and consider a longitudinal acceleration wave which has been propagating since time $t = 0$ into a homogeneous region of a non-conductor which previously had been at rest with its density ϱ_0 and entropy η_0 constant in space and time. From our analysis of (5.14) — (5.16) we obtain the following observations: If *either* (i) $|[\dot{\varrho}]| < |\lambda_*|$, or (ii) the wave is initially compressive with $\tilde{E} > 0$, or (iii) the wave is initially expansive with $\tilde{E} < 0$, *then* $[\dot{\varrho}] \to 0$ monotonically as $t \to \infty$. If $[\dot{\varrho}](0) = \lambda_*$, then $[\dot{\varrho}(t)] \equiv [\dot{\varrho}](0)$. But, if (i*) $|[\dot{\varrho}](0)| > |\lambda_*|$ *and* either (ii*) the wave is initially compressive with $\tilde{E} < 0$, or (iii*) the wave is initially expansive with $\tilde{E} > 0$, *then* $|[\dot{\varrho}]| \to \infty$ monotonically and in the finite time

$$t_\infty = \frac{2G(0)}{G'(0)} \ln\left(1 - \frac{\lambda_*}{[\dot{\varrho}](0)}\right) \tag{5.24}$$

Common experience with shock tubes yields a qualitative verification of the conclusions just reached. The motion of a gas in a shock tube can be assumed to be longitudinal. For most gases, including ideal gases, $\tilde{E} < 0$. Therefore, our theory tells us that an acceleration wave moving into an equilibrium region in a shock tube should grow into a shock wave only if it is compressive, and this is just what is observed. The theory states further that a shock wave should not form unless $|[\dot{\varrho}]|$ is initially greater than $|\lambda_*|$. We do not know whether this critical jump $|\lambda_*|$ has been measured precisely, but it surely exists in experience; otherwise every sudden burst of sound, no matter how faint, as it moves down a long tube of air would grow into a shock wave.

We do not wish to give the impression that we regard the present theory as particularly applicable to gases. In fact, although (2.2) should hold well for a gas at high density, they clearly cannot apply to rarefied gases for which the thermodynamic state depends on gradients of the density. We expect that the interested experimenter will have no difficulty in finding examples of solids and liquids of low thermal conductivity, exhibiting long-range viscoelastic memory and a variety of values of $\tilde{E}$, some positive and some negative, with which to test our conclusions.

6. Shock Waves

At a shock wave there is a jump in the velocity $\dot{x}$ and the deformation gradient F. Since we assume that the intrinsic velocity (4.1) is positive, it follows from (4.3) and (4.4) that the velocity of the shock wave relative to the material just ahead of the wave is

$$u_1^r = u - \dot{x}_1, \tag{6.1a}$$

and the velocity of the wave relative to the material immediately behind it is

$$u_2^r = u - \dot{x}_2. \tag{6.1b}$$

Furthermore, since (4.3) and (4.4) also imply that

$$u = F_1 U + \dot{x}_1 = F_2 U + \dot{x}_2, \tag{6.2}$$

we have the following easily interpreted expressions for the intrinsic velocity

$$U = \frac{u_1^r}{F_1} = \frac{u_2^r}{F_2}. \tag{6.3}$$

When these expressions are combined with (1.3), we obtain the familiar formulae

$$u_1^r \varrho_1 = u_2^r \varrho_2 = U \varrho_R \stackrel{\text{def}}{=} m. \tag{6.4}$$

Equation (6.2) also yields the compatibility condition

$$[\dot{x}] = -U[F]. \tag{6.5}$$

It follows from the assumptions laid down in Section 4 and the smoothness (2.4) of the constitutive functionals in (2.2), that η, ε, T, Θ, and their time derivatives $\dot{\eta}$, $\dot{\varepsilon}$, $\dot{T}$, $\dot{\Theta}$ are continuous in X, t except for jump discontinuities at Y_t, t. Once this is noted, arguments completely analogous to those used in the classical theory of perfect fluids[1] may be applied to the balance laws (1.1), (1.2), and the entropy inequality (1.4) to obtain the following *jump conditions*:

$$m[\dot{x}] + [T] = 0, \tag{6.6}$$

$$m\left(\frac{1}{2}[\dot{x}^2] + [\varepsilon]\right) + [T\dot{x}] = 0, \tag{6.7}$$

$$m[\eta] \geq 0. \tag{6.8}$$

Interpreting our one-dimensional theory as a description of a longitudinal motion of a three-dimensional body we call a shock wave *compressive* if $U[\varrho] > 0$ and expansive if $U[\varrho] < 0$. Since we assume $U > 0$, it follows from (1.3), (6.5), and (6.6) that

$$\operatorname{sgn}[\varrho] = -\operatorname{sgn}[F] = \operatorname{sgn}[\dot{x}] = -\operatorname{sgn}[T]. \tag{6.9}$$

---

[1] E. g. [1948, 1, p. 124] or [1959, 1, p. 219].

Let $\Lambda^t = (F^t, \eta^t)$ be the total history of Y_t just before arrival of the wave; then, by (2.2) and (4.4),

$$\varepsilon_1 = \mathfrak{e}(F^t, \eta^t), \quad T_1 = \mathfrak{T}(F^t, \eta^t), \quad \Theta_1 = \mathfrak{t}(F^t, \eta^t), \tag{6.10}$$

and, by (3.1) and (4.5),

$$\varepsilon_2 = \mathfrak{e}(F^t \wedge [F], \eta^t \wedge [\eta]), \quad T_2 = \mathfrak{T}(F^t \wedge [F], \eta^t \wedge [\eta]),$$

$$\Theta_2 = \mathfrak{t}(F^t \wedge [F], \eta^t \wedge [\eta]). \tag{6.11}$$

Thus, if ε^*, T^*, and Θ^* are the instantaneous response functions corresponding to F^t, η^t, we have

$$\varepsilon_1 = \varepsilon^*(0, 0), \quad T_1 = T^*(0, 0), \quad \Theta_1 = \Theta^*(0, 0), \tag{6.12}$$

$$\varepsilon_2 = \varepsilon^*([F], [\eta]), \quad T_2 = T^*([F], [\eta]), \quad \Theta_2 = \Theta^*([F], [\eta]). \tag{6.13}$$

Straightforward algebraic manipulation of eqs. (6.5) — (6.8) yields a formula generalizing the Hugoniot relation to non-conductors with memory:

$$\varepsilon_2 - \varepsilon_1 = \frac{1}{2\varrho_R}(T_1 + T_2)(F_2 - F_1). \tag{6.14}$$

Substituting (6.13) into (6.14) we obtain the following equation relating the jumps $[F]$ and $[\eta]$:

$$H([F], [\eta]) \stackrel{\text{def}}{=} \varepsilon^*([F], [\eta]) - \varepsilon_1 - \frac{1}{2\varrho_R}\{T^*([F], [\eta]) + T_1\}[F] = 0. \tag{6.15}$$

It follows from $(6.11)_3$ and the temperature relation (3.4) that

$$\frac{\partial}{\partial[\eta]} H(0, [\eta])\bigg|_{[\eta]=0} = \Theta_1 > 0, \tag{6.16}$$

and the implicit function theorem tells us that in some neighborhood of $[F] = 0$ we can solve (6.15) to obtain $[\eta]$ as a function of $[F]$:

$$[\eta] = \eta_H([F]). \tag{6.17}$$

The function T_H defined by

$$T_H([F]) \stackrel{\text{def}}{=} T^*\big([F], \eta_H([F])\big) \tag{6.18}$$

may be called the *instantaneous Hugoniot stress-strain function*; it gives the stress just behind the wave as a function of the jump $[F]$ in the strain; i. e.,

$$T_2 = T_H([F]). \tag{6.19}$$

Of course, the form of T_H will depend on the histories $(F,^t \eta^t)$ just before arrival of the wave. The *instantaneous isentropic stress-strain* function T_I defined by

$$T_I(\Delta F) \stackrel{\text{def}}{=} T^*(\Delta F, 0) \tag{6.20}$$

also depends on the original underlying histories F^t, η^t.[1] As in the theory of perfect fluids, of fundamental importance here is the attainment of an understanding of the relation between T_H and T_I.

Arguments completely analogous to those used to obtain related results in the classical theory may here be applied to prove the following theorem, which for materials with memory rests heavily on the stress relation (3.3) and the temperature relation (3.4): *The instantaneous Hugoniot stress-strain function and the instantaneous isentropic stress-strain function corresponding to a given pair F^t, η^t of underlying histories not only touch at $[F] = 0$ but also have there the same slope and curvature; in fact, as $[F] \to 0$ we have,* [1965, 6, § 3],

$$[T] = E_1[F] + \frac{1}{2}\tilde{E}_1[F]^2 + O([F]^3), \tag{6.21}$$

where E_1 and *$\tilde{E}_1$ are the first and second-order instantaneous stress-strain moduli*

$$\left.\begin{aligned} E_1 &= \left.\frac{d T_I(\Delta F)}{d\Delta F}\right|_{\Delta F=0} = D_F \mathfrak{T}(F^t, \eta^t), \\ \tilde{E}_1 &= \left.\frac{d^2 T_I(\Delta F)}{d\Delta F^2}\right|_{\Delta F=0} = D_F^2 \mathfrak{T}(F^t, \eta^t). \end{aligned}\right\} \tag{6.22}$$

In view of equations (5.2) and (5.3), the number

$$C_1 = \sqrt{\frac{E_1}{\varrho_R}} \tag{6.23}$$

may be called the intrinsic sound speed for the material immediately ahead of the wave. Since (6.4)—(6.6) yield

$$\varrho_R U^2 = \frac{[T]}{[F]}, \tag{6.24}$$

eq. (6.21) gives us the following formula relating the intrinsic velocity of a shock wave to the intrinsic speed C_1 any higher-order wave would exhibit upon entering the same region:

$$U^2 = C_1^2 + \frac{1}{2\varrho_R}\tilde{E}[F] + O([F]^2). \tag{6.25}$$

---

[1] Note that our use of the word *isentropic* signifies only $\Delta\eta = 0$, *not* $\eta^t(s) =$ const.; the history η^t is arbitrary.

In the course of proving (6.21), one also obtains the following result, which is perhaps more enlightening and which also rests on (3.3) and (3.4): *For small* $[F]$ *the jump in the entropy obeys*[1]

$$[\eta] = \frac{\tilde{E}_1}{12\varrho_R \Theta_1} [F]^3 + O([F]^4). \tag{6.26}$$

On combining this result with (6.8) and (6.9) we see that if $\tilde{E}_1 \neq 0$, a shock wave of small amplitude with $U > 0$ can exist only if $\tilde{E}_1$ and $[\varrho]$ have opposite sign. In a few moments we shall observe that this conclusion, which is in agreement with the conditions for the growth of acceleration waves given at the end of the previous section, can be extended to shock waves of arbitrary intensity.

Continuing to let (F^t, η^t) be the total history immediately ahead of the wave, we call the number C_2 defined by

$$C_2 = \sqrt{\frac{E_2}{\varrho}}, \quad E_2 = \frac{\partial}{\partial \Delta F} T^*(\Delta F, [\eta])\Big|_{\Delta F = [F]} = D_F \mathfrak{T}(F^t \wedge [F], \eta^t \wedge [\eta]), \tag{6.27}$$

the intrinsic sound speed for the material immediately behind the wave. We of course assume that E is always positive, i. e., that

$$\frac{\partial}{\partial \Delta F} T^*(\Delta F, \Delta \eta) > 0; \tag{6.28}$$

let us now assume further that the instantaneous stress-entropy modulus α of (3.10) and (3.12) is non-zero:

$$\frac{\partial}{\partial \Delta \eta} T^*(\Delta F, \Delta \eta) \neq 0. \tag{6.29}$$

Of course, (6.29) implies that either

$$\frac{\partial T^*}{\partial \Delta \eta} < 0, \tag{6.30}$$

or

$$\frac{\partial T^*}{\partial \Delta \eta} > 0. \tag{6.31}$$

The inequality (6.30) is the more common case as it holds not only for gases, which have $T = -p$ with p increasing as η increases, but also for most solids. Although the equilibrium analogue of (6.31) holds for

[1] [1965, 6, § 3, Thm. 5]. Our eq. (6.26) generalizes to materials with memory a result obtained by JOUGUET [1920, 1, p. 906] for elastic materials; for the form taken by (6.26) for perfect fluids see [1959, 1, footnote to p. 225].

elastomers such as rubber in slow measurements, (6.31) itself involves rapid behavior which seems to be always "glassy". In deciding the appropriateness of (6.30) or (6.31) for a given real material, it is useful to note that, as we saw in (3.12), the stress and temperature relations yield

$$\frac{\partial T^*}{\partial \Delta\eta} = \varrho_R \frac{\partial \Theta^*}{\partial \Delta F}, \tag{6.32}$$

which relation, when combined with (6.9) and the theorem containing (6.21), justifies the following assertion: *If* $\begin{Bmatrix}(6.30)\\(6.31)\end{Bmatrix}$ *holds, a weak compressive shock wave results in an* $\begin{Bmatrix}\textit{increase}\\\textit{decrease}\end{Bmatrix}$ *in temperature, and a weak expansive shock wave results in a* $\begin{Bmatrix}\textit{decrease}\\\textit{increase}\end{Bmatrix}$ *in temperature.*

We are now prepared to state a generalization to materials with memory of BETHE [1942, 1] and WEYL's [1949, 1] theorem for finite amplitude shock waves in perfect fluids[1]: *If in addition to* (6.28) *and* (6.29) *we assume that*

$$\frac{\partial^2}{\partial \Delta F^2} T^*(\Delta F, \Delta\eta) \neq 0, \tag{6.33}$$

then the intrinsic velocity is supersonic relative to the material ahead of the wave and subsonic relative to the material behind the wave; i. e.,

$$C_2 > U > C_1. \tag{6.34}$$

If $\tilde{E} < 0$, i. e., *if*

$$\frac{\partial^2}{\partial \Delta F^2} T^*(\Delta F, \Delta\eta) < 0, \tag{6.35}$$

then the shock wave must be compressive: $[F] < 0$. If $\tilde{E} > 0$, i. e., *if*

$$\frac{\partial^2}{\partial \Delta F^2} T^*(\Delta F, \Delta\eta) > 0, \tag{6.36}$$

then the shock wave must be expansive: $[F] > 0$.

In terms of the relative sound speeds

$$c_1 = F_1 C_1, \quad c_2 = F_2 C_2, \tag{6.37}$$

the inequality (6.34) may be written in the equivalent and more familiar form

$$c_2 > u_2^r, \quad u_1^r > c_1. \tag{6.38}$$

---

[1] It is not difficult to translate into a proof of this theorem the arguments given by COURANT and FRIEDRICHS [1948, 1, pp. 141–148] for the analogous problem in the theory of perfect fluids.

Acknowledgement

The support of this research by the Air Force Office of Scientific Research and the Office of Naval Research is gratefully acknowledged.

References

1920 [1] JOUGUET, E.: C. R. Acad. Sci. Paris **171**, 461—464, 512—515, 789—791, 904—907.
1942 [1] BETHE, H.: Office of Scientific Research and Development, Division B, Report No. 545.
1948 [1] COURANT, R., and K. O. FRIEDRICHS: Supersonic Flow and Shock Waves. New York: Interscience.
1949 [1] WEYL, H.: Comm. Pure. Appl. Math. **2**, 103—122.
1953 [1] ERICKSEN, J. L.: J. Ratl Mech. Anal. **2**, 329—337.
1957 [1] THOMAS, T. Y.: J. Math. Mech. **6**, 455—469.
1959 [1] SERRIN, J.: Mathematical Principles of Classical Fluid Mechanics. In: Encyclopedia of Physics, Vol. VIII/1, ed. by S. FLÜGGE. Berlin—Göttingen—Heidelberg: Springer.
1960 [1] COLEMAN, B. D., and W. NOLL: Arch. Ratl Mech. Anal. **6**, 355—370.
[2] TRUESDELL, C., and R. A. TOUPIN: The Classical Field Theories. In: Encyclopedia of Physics, Vol. III/1, ed. by S. FLÜGGE. Berlin—Göttingen—Heidelberg: Springer.
1961 [1] COLEMAN, B. D., and W. NOLL: Reviews Mod. Phys. **33**, 239—249 (1961); erratum: ibid. **36**, 1103 (1964).
[2] TRUESDELL, C.: Arch. Ratl Mech. Anal. **8**, 263—296.
1964 [1] COLEMAN, B. D.: Arch. Ratl Mech. Anal. **17**, 1—46, 230—254.
[2] GREEN, W. A.: Arch. Ratl Mech. Anal. **16**, 79—89.
1965 [1] COLEMAN, B. D., M. E. GURTIN and I. HERRERA R.: Arch. Ratl Mech. Anal. **19**, 1—19.
[2] — — Arch. Ratl Mech. Anal. **19**, 239—265.
[3] — — Arch. Ratl Mech. Anal. **19**, 266—298.
[4] — — Arch. Ratl Mech. Anal. **19**, 317—338.
[5] — — Thermodynamics and Wave Propagation. Report No. 100, Contract Nonr 562(40), Division of Applied Mathematics, Brown University, April 1965.
[6] — — Thermodynamics and One-Dimensional Shock Waves in Materials with Memory. Report No. 3, Contract Nonr 562(40), Division of Applied Mathematics, Brown University, August, 1965. = Proc. Royal Soc. **A 292**, 562—574.
[7] — — I. HERRERA R. and C. TRUESDELL: Wave Propagation in Dissipative Materials. New York: Springer.
[8] TRUESDELL, C., and W. NOLL: The Non-Linear Field Theories of Mechanics. In: Encyclopedia of Physics, Vol. III/2, ed. by S. FLÜGGE. Berlin—Göttingen—Heidelberg: Springer.
[9] VARLEY, E.: Arch. Ratl Mech. Anal. **19**, 215—225.
[10] — and J. DUNWOODY: J. Mech. Phys. Solids **13**, 17—28.
1966 [1] COLEMAN, B. D., and V. J. MIZEL: Arch. Ratl Mech. Anal. **23**, 87—123.
[2] — J. M. GREENBERG, and M. E. GURTIN: Arch. Ratl Mech. Anal. **22**, 333 to 354.
[3] PIPKIN, A. C.: Quart. Appl. Math. **23**, 297—303.

Discussion

BRUN: In this theory you make use of a space of histories of deformations which are infinite. The beginning may be very remote — as remote as we want. The consequences of this hypothesis bound with the fact that the functionals are continuous lead to somewhat stronger consequences as the hypothesis where we only consider finite deformation histories. I want your opinion about that fact — it might be as good as your point of view.

COLEMAN: One can consider histories defined on finite intervals $[t - \tau, t]$ and assume that the material was in equilibrium up to time $t - \tau$. Since such histories are everywhere dense in the domain of the continuous functionals we discuss here, knowledge of the response to such histories determines our functionals. Thus, "infinite histories" enter our theory naturally. On the other hand, omission of "infinite histories" from a memory theory renders impossible any discussion of steady flows and the manner in which they may be achieved asymptotically. [For a discussion of the "infinite histories" characterizing such steady flows as Couette flow and Poiseuille flow, see the book "Viscometric Flows of Non-Newtonian Fluids, Theory and Experiment" by COLEMAN, MARKOVITZ, and NOLL (Springer, 1966).]

ONAT: I would like to make a remark on Dr. BRUN's question. I think many times we indeed work with histories defined on finite time intervals — on the observation interval — and it is quite possible to do this, provided that at zero time you can have a set of identical specimens. Now, working with histories based on finite intervals has its advantages. One of them is one does not have to introduce the so-called principle of fading memory which for someone who works in plasticity, like myself, is unacceptable because metals remember things done to them in the distant past. Another, I think, important thing is that some materials like concrete age as time goes on. Its properties change over the course of time. Now, this type of material could be taken care of if one works with finite histories based on finite time-intervals whereas in COLEMAN's framework one cannot cope with such situations.

NAGHDI: Would you object if I replace that and say: this development considered a class of materials subject to the requirement of fading memory, and just that.

COLEMAN: I agree with you. The principle of fading memory is a constitutive assumption, not a law of physics.

MEIXNER: You start from some axioms and you do certainly very excellent mathematics. But now, when I want to apply these things what is, for instance, your Θ? How do you realize the Θ or how do you measure it? How do you define it in connection with an experiment on a real material?

COLEMAN: When the thermodynamics of materials with memory is applied to gases with internal molecular relaxation, Θ represents the *translational temperature*. This is true, in particular, in the theory of acceleration waves and shock waves in such gases. Experimenters tell us that in ordinary rheological experiments done, for example, on viscoelastic polymers undergoing oscillatory shear, the temperature may be measured with mercury thermometers, resistance thermometers, or thin copper-constantin thermocouples.

MEIXNER: Well, they measure with thermometers which are scaled according to the thermodynamic scale. How do you prove that your scale is a thermodynamic scale?

COLEMAN: We believe that experimenters often can find thermometers whose constitutive equations are unaffected by the material flowing past them and whose response time is rapid relative to the time scale of the experiment. Perhaps I should also remark that the temperature occurring in the Clausius-Duhem inequality obeys KELVIN's formulation of thermodynamics.

NAGHDI: You have some results in the case of a Newtonian fluid. Does that come out naturally or do you have to do with slow motion? (Professor NAGHDI was referring to the general theory of the thermodynamics of materials with memory, not to wave propagation.)

COLEMAN: When we apply the Clausius-Duhem inequality *directly* to linearly viscous fluids we obtain restrictions that are in accord with experience (cf. COLEMAN and MIZEL [1966, 1]). On the other hand, a general material with fading memory is approximated by a linearly viscous material in the limit of slow motions (COLEMAN and NOLL [1960, 1]). If one *first* finds the implications of the Clausius-Duhem inequality for a material obeying the principle of fading memory and *then* takes the asymptotic limit of slow motions, one derives restrictions *equivalent* to those obtained directly for linearly viscous fluids (Cf. COLEMAN [1964, 1]). In both cases, departures from the equilibrium caloric equation of state are found to be of order 2 in the time scale (or the rate of strain) whereas the linearly viscous correction to the equilibrium stress-strain law is, of course, of order 1 in the time scale.

On the Continuum as an Assemblage of Homogeneous Elements or States[1]

By

D. C. Drucker[2]

Providence, R.I. (U.S.A.)

Summary

Some of the implications are explored of considering a continuum or a "point" in a material as an assemblage of simple models. A discussion is given of the need for and the distinction between a thermodynamics based upon a highly specialized or restricted model and one which is applicable to a wide class of material behavior. A thermodynamic statement developed on the basis of a model of material cannot be a valid *general* principle if it is not applicable to the combined response of two or more such models. Combinations of viscous models and combinations of elastic — perfectly plastic models are discussed in these terms. The significance of reversibility, or the ability to restore the initial state through mechanical deformation and moderate temperature changes alone, is related to dislocation concepts. Materials which work-harden with strain cycling are contrasted with those which work-soften and those which have been stabilized and do neither. The distinction between frictional and plastic behavior is discussed in thermodynamic terms along with the related questions of the degree of path independence in a space of controllable state variables and the order of the infinity of independent state variables.

Introduction

Considerable confusion exists at the present time on the essential features of the application of thermodynamics to the inelastic behavior of materials. In fact, the statement has been made [1] that "... for materials with perfect memory, no thermodynamic structure can be expected to hold". Elastic-plastic materials, which are time-independent by definition, have perfect memory in the sense that prior plastic defor-

---

[1] The results in this paper were obtained in the course of research sponsored by the Office of Naval Research under Contract Nonr 562(20) with Brown University. Much of the approach described is complementary to that of Professor MEIXNER and that of Professor KESTIN with whom I had so many valuable discussions.

[2] L. Herbert Ballou University Professor, Brown University.

mation produces an effect which never is wiped out by any amount of time spent on paths in the purely elastic range. Viscoelastic materials very far from their equilibrium configurations pose problems of similar conceptual difficulty. On the other hand, the feeling has been expressed that "... to fit the behavior of solid, strained materials into the framework of classical thermodynamics ... all that is required is a careful interpretation of experimental results in terms of quite familiar ideas" [2].

Confusion is compounded by the natural desire on the part of some investigators to generalize all results which are obtained so that they are applicable in the realm of large strains and rotations, and the equally natural desire on the part of others to restrict themselves to uniaxial tension and compression or to simple shear. Sometimes an attempt is made to carry over the concepts of classical thermodynamics or of the thermodynamics of linear irreversible processes with as little modification as possible. At other times, attention is directed toward broad and hopefully useful classifications of material with some well-defined thermodynamic or quasi-thermodynamic properties.

The confusion is well warranted by the difficulty of the subject and the present limited knowledge of its fundamental features [3—8]. Indeed there may be no such features of general or even of specialized utility as distinguished from correct statements of thermodynamics which lead to no new results. Consequently, every effort should be made to clarify the bases of the various approaches already proposed and those to be advanced in the future. If a crucial starting assumption is made, it should stand out clearly and should not be obscured by a complex generalization. There is little point to building an elegant mathematical edifice on an oversimplified picture of the real world. The simplest of all possible situations should be examined which are adequate to establish and test the basic premises. Therefore, small rather than large strains and rotations should be discussed first. However, even in the small displacement range, it is necessary to avoid drawing an unwarranted generalization from an oversimplified situation.

Assemblages of homogeneous elements or states are useful models of material or continuum behavior [9—16]. They also serve as excellent examples of the many sources of confusion and the ease of drawing unacceptable generalizations. Well-known models help in the study of the value and desirability of developing a complete thermodynamics based upon a simple idealization or very restricted set of state variables [2, 3], as opposed to starting with a high order of infinity of state variables and writing thermodynamic restrictions or rules for broad classifications of material behavior or combinations of strong idealization and broad classification [17—19].

On the Validity of Thermodynamic Statements about Material Systems

If a material system is sufficiently well-defined or idealized, its state variables will be known or computable. Straightforward application of the laws of thermodynamics or of special postulates of thermodynamic nature then will lead to certain logical consequences or thermodynamic statements. The question which arises is how broadly valid they are. Clearly, the thermodynamic statements will be correct for the system studied, provided no error or inadmissible assumption has been made. This is a rather unsatisfactory limitation unless the system studied is of great generality or at least includes a number of examples of practical or conceptual importance. The alternative of treating the stress problem and the thermal problem separately is available instead. Because the temperature terms in expressions for internal energy or entropy tend to overwhelm the stress terms, an inadequate thermodynamics may be worse than none at all.

When a material system is studied, and a rather well-known explicit but drastic idealization is chosen, the hope often is that the result will transcend the limitations of its origin. Sometimes the hope is taken as certainty. Sometimes the limitations are not fully appreciated, and unacceptable new rules of behavior are proposed for some or all materials. Unfortunately it is far easier to be destructive than constructive in an evaluation. A negative approach only can be offered here, a generalization of an earlier counterexample [7].

The suggestion here is that a thermodynamic (or mechanics) statement be considered of deficient validity or generality if it applies individually to each of two material systems but not to a meaningful mechanical combination of the two. If the combination is also a member of the original class, an error in assumption or logic is indicated. If not, and the statement does not apply to the combination, the conclusion is not that the thermodynamic statement is incorrect, but rather that it cannot be counted upon as a principle of behavior in any but the special case for which it was developed. A proper but limited thermodynamics is consistent and may be of value for particular problems or particular materials. It is its generality and broad implications which cannot be depended upon.

Two examples, one of each type, will be mentioned here briefly for illustration. The first example shows how difficult it is to avoid falling into a trap. A careful study of thermodynamics, checked by a thorough look at one subclass of materials and the limiting case of a perfectly plastic material [5, 20], indicated that surfaces of constant rate of entropy production or rate of dissipation of energy should occupy a central position in the theory of non-linear viscous materials. However, a linear combination of any two such materials, each with a fundamentally different functional form governing its dissipation rate, also is a non-linear viscous material. On the one hand, therefore, the surface of constant dissipation rate for the combination should play a central role. On the other, the combination of two non-linear components with different degrees of non-linearity leads to a governing functional form which is not the rate of dissipation except for special states of stress. Therefore, the conclusion must be that the hypothesis itself is inadmissible [7]. Furthermore, general plasticity theory is not a limiting case of visco-elasticity theory or vice versa. Had one-dimensional states of stress only been considered, these contradictions and difficulties would not have appeared and a false generality might have been accepted.

The second example is quite different in nature. As a first step beyond the assumption of perfect plasticity for an elastic-plastic material [3, 11, 12], it is not unreasonable to choose the plastic strain ε_{ij}^p at the reference state of stress and temperature as the additional state variable (or as 6 or 9 state variables if each component is counted separately). The combination in parallel of two elastic-plastic materials, each of which has its own ε_{ij}^p as an additional state variable is not of the same class, because ε_{ij}^p for the composite does not determine the state uniquely at any reference stress and temperature. This fact does not invalidate the original hypothesis. It does, however, show the hypothesis to be of rather restrictive nature and so likely to be an indifferent approximation to the fundamental behavior of almost all real materials. Again it is worth repeating that the solution of particular problems with restrictive hypotheses is well worthwhile. The cautionary flag is raised only against the tendency to assume general validity for any thermodynamic statements, or perhaps against the use of a consistent but restricted thermodynamics in preference to the approximate but well understood direct treatment of the decoupled stress and temperature problem.

Isothermal Elastic-Plastic Behavior

Simple examples of a trivial nature concerned with the validity of hypothesis or generalization appear in the usual considerations of elastic-plastic behavior. They are so obvious that they are dismissed immediately in isothermal plasticity. Yet their counterparts are all too likely to appear in the thermodynamics of plasticity because so little is known about the thermomechanical behavior of materials, and their thermodynamics is just developing to the threshold of a useful theory.

For example, suppose the clever and extremely valuable concept of linear elastic behavior followed by perfectly plastic behavior was thought to be a general principle of elastic-plastic material behavior. Its generality could be tested by considering the combination of two or more linear elastic — perfectly plastic elements with different yield points in tension. The combination in series would still be linear elastic — perfectly plastic. The combination in parallel would not be, although the difference would appear only in the intermediate range when at least one element was plastic while at least one remained elastic (Fig. 1).

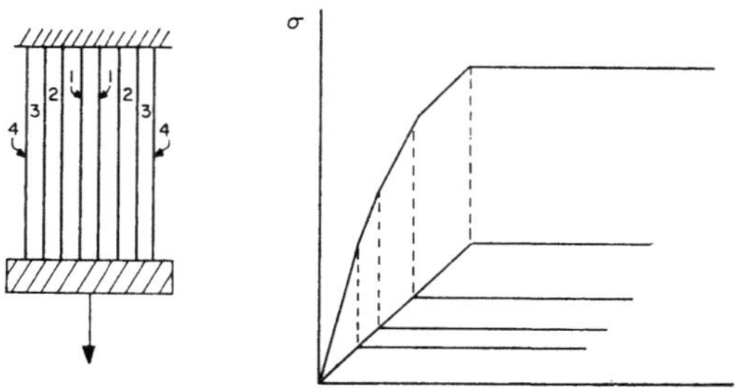

Fig. 1. Parallel combination of linear elastic-perfectly plastic elements in tension

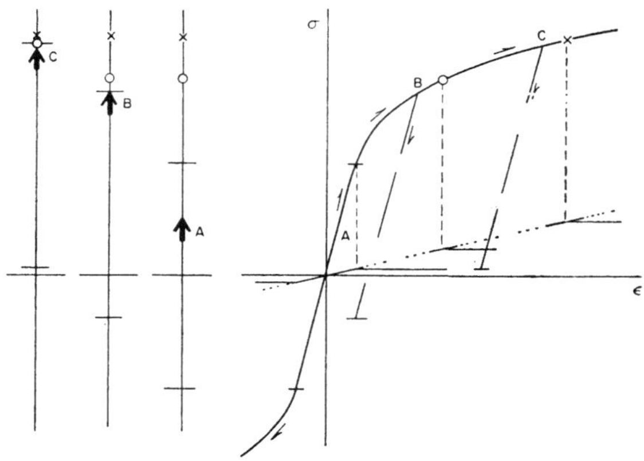

Fig. 2. The elastic range remains constant with a finite or infinite assemblage, each of whose perfectly plastic elements individually has the same response in tension and compression

Carrying this trivial case still further makes it a little more instructive. Suppose an infinite number of elements is assumed so that any tension stress-strain curve can be matched completely. This satisfaction of an infinite number of conditions with an infinite number of elements still does not guarantee generality. Quite the contrary, the behavior in compression following tension will exhibit a purely elastic range of fixed magnitude, regardless of the tensile strain imposed, when each of the elements is perfectly plastic with the same yield stress in tension and compression (Fig. 2). Also, much of the prior history is wiped out if a path of repeated compressive and tensile loading is followed. In fact, with infinite patience it is possible to restore the original state completely. This last point is especially troublesome both on the macroscale and the microscale. At moderate and low temperatures, the number, type, and configuration of dislocations cannot possibly revert to the original state when the metal starts off in the fully annealed condition. Such extended reversibility is unlikely for almost any given initial state. The yield strength and the flow curve of the metal after many cycles of deformation will be very different from the corresponding properties at the start of the loading history.

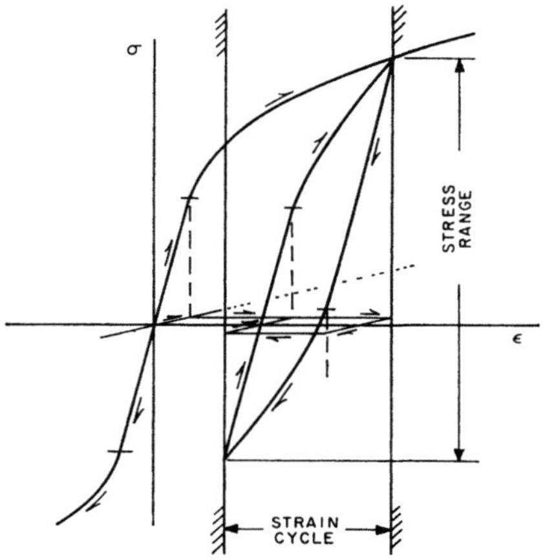

Fig. 3. An immediate steady state under cyclic strain for assemblage of Fig. 2

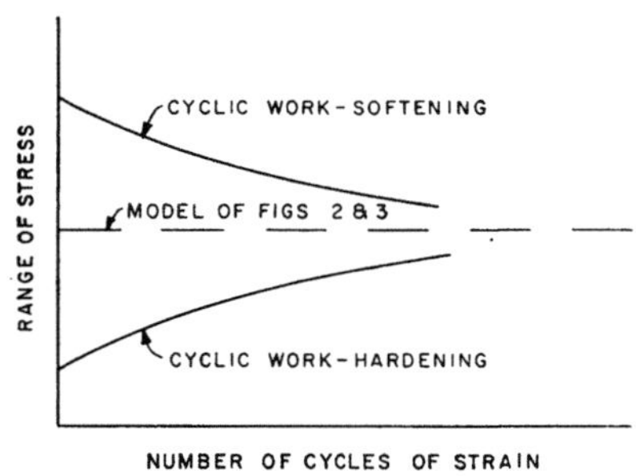

Fig. 4. Hardening or softening under cyclic strain (schematic)

An assemblage in parallel of a finite number or an infinite number of linear elastic — perfectly plastic elements with the same behavior in tension and compression will settle down immediately to a steady state under cyclic strain (Fig. 3). Neither the cyclic strain-hardening of annealed metals, nor the cyclic strain-softening of metals over-hardened by heat treatment or neutron irradiation (Fig. 4) can appear in this model of material behavior.

The ability of the model to revert to its initial state, and inability to alter its elastic range or to cyclically harden or soften, do not violate any fundamental laws. Therefore the model is permissible. The question is simply how general and how useful it is. Any rule devised on the basis of so special a set of properties is unlikely to have general validity, if it depends heavily upon them.

Assemblages of perfectly plastic elements can be devised which overcome some of the shortcomings of simulation of the behavior of real materials. A combination in parallel of two pairs of elements, in each of which there is an element with a much larger yield stress in tension than compression and an element in which the reverse is true, serves as illustration (Fig. 5). Now the elastic range is increased by plastic deformation, but the model is too simple to show a cyclic strain-hardening.

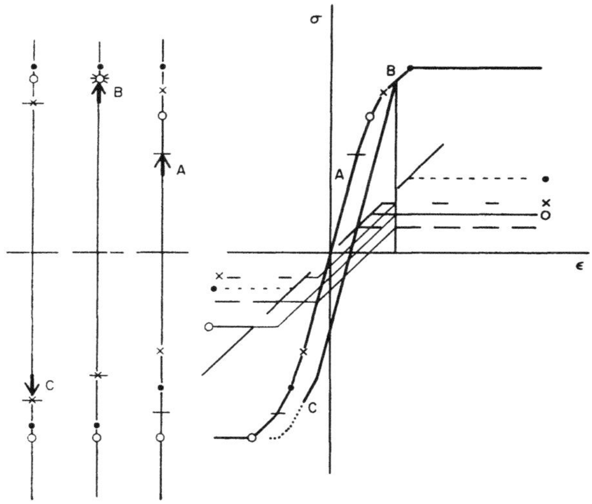

Fig. 5. Parallel assemblage of elements with unequal yield points in tension and compression has an increasing elastic range

More elaborate series and parallel combinations, however, can do a remarkable job of simulating materials which harden with cyclic strain.

These assemblages are not returnable to their original state by mechanical working. However each individual element is returnable if some way can be found to get inside the model and to pull and push in the right amount and sequence. Therefore, although the behavior is very different, the state of the model is well-defined in essentially the same terms as for initially symmetric elements.

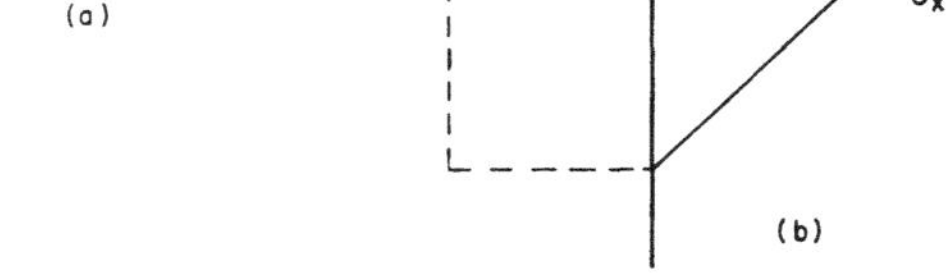

Fig. 6. Illustration of an extension of Fig. 5 to two or more dimensions of stress. The Tresca criterion is chosen for clarity; (a) behavior of each of the four elastic-perfectly plastic elements, (b) initial yield surface for assemblage of the four elements [drawn to same scale as (a)]

Fig. 6 illustrates how this picture of intertwined off-center perfectly plastic elements in combination can be extended to two or more dimensions of stress. The Tresca criterion of maximum shear stress and a

σ_x, σ_y space have been chosen for clarity of presentation. Note that the initial yield surface remains a Tresca hexagon in this special example but that two of the sides belong to one of the component elements, two to another, and one each to the remaining two elements. Just as for Fig. 5, when plastic deformation occurs, the yield domain increases in size as it moves about in stress space and becomes unsymmetric. This expansion process stops as soon as the four elements are brought into coincidence, at which stage the yield surface again is a symmetric hexagon but is 1.5 times the original size.

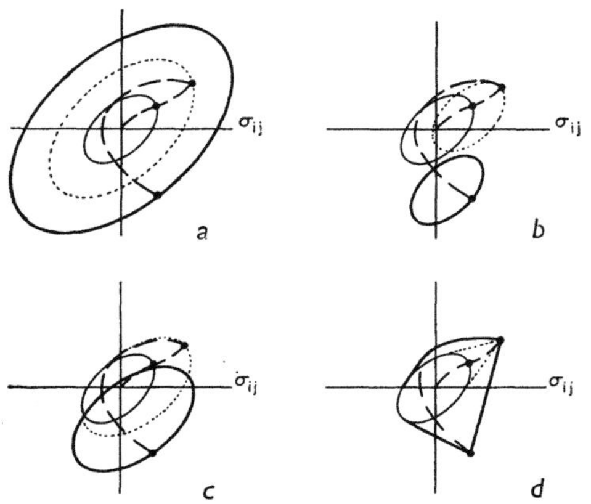

Fig. 7. Schematic representation of a few examples of the infinte variety of isothermal behavior. *a* isotropic hardening; *b* kinematic hardening; *c* an intermediate hardening; *d* independent plane loading surfaces

Fig. 7 shows a few highly idealized models of work-hardening materials. The important consideration here is not the advantages or shortcomings of each, but rather that the behavior of real materials is infinitely more complex than any one of them can predict. Some feeling for the complexity can be obtained by modifying the model [13—16] of plane loading surfaces (Fig. 7d) to include interaction among the loading surfaces. In this manner, the initial yield surface can expand as in isotropic hardening (Fig. 7a), translate as in kinematic hardening (Fig. 7b), do both as in Fig. 7c, or even shrink in size to give cyclic work-softening.

Infinitely complex in infinitely many ways, actual materials really are submerged in a sea of mechanical irreversibility. In dislocation terms, there are essentially an infinite number of dislocations which can be rearranged in an infinite number of independent ways which are only partially reversible at most. In continuum terms, the actual initial and subsequent yield surfaces are infinitely complex in their positions and shape. Every change in position or shape represents a change in state. Any finite assemblage of models can go only part way in this direction. For each of the orders of infinity individually, the match will be no better than the simulation of a real stress-strain curve by the elements of Fig. 1. The simulation in combination will be far worse, yet may still be very useful in practice. After all, metal structures are designed on the drastic assumption of perfect plasticity and the associated limit theorems.

State Variables and Fixed Parameters in Functional Forms

At each stage in its history, an elastic-plastic material at a given state of stress and temperature will have a set of properties or state parameters. These include, among many others, the elastic moduli, the coefficient(s) of expansion, the specific heat, the yield stress for simple tension or compression or shear, and the form of the current yield function. The time-independence implicit in the term "plastic" presumes that there is no effect of time spent at the given state [21]. Chemical changes at temperature, which include such realities as annealing, ageing, changes of grain size, and precipitation from solution are ignored.

The state variables within the elastic range are the temperature Θ and the stress σ_{ij} or equivalently the elastic strain ε^e_{ij}. Elastic moduli and other state parameters may vary with the state variables, but the changes are determined uniquely by them when the values at a given stage are known. In general, of course, the functional forms governing the plastic deformation as well as particular yield values will change with plastic deformation (Fig. 7). If the yield surface at each temperature is fixed (unaffected by increments of plastic strain $\delta\varepsilon^p_{ij}$), the material is perfectly plastic and may be modeled by a single element whose state variables are θ and σ_{ij} or ε^e_{ij}.

When the material is modeled by an assemblage of elastic — perfectly plastic elements of the intertwined type (Figs. 5 and 6) or the non-overlapping form (Fig. 2), the behavior depends strongly on the parameters and functional forms for each element. However, the state variables are just the θ and σ_{ij} or ε^e_{ij} for the individual elements because the functional forms do not change and the parameters are determined uniquely by the σ_{ij} and θ. Within the current elastic domain of the entire assemblage, the applied or total stress which is a controllable variable also is a state variable and may be substituted for any one of the individual stress states. If desired, the state variables may be chosen as the total stress, the temperature(s), and the location of the origin or any other recognizable point of the individual yield surfaces in total stress space.

When the model consists of just two elastic — perfectly plastic elements in parallel, one of which remains below yield, it is permissible to choose the controllable variable of total stress and the measurable variable of plastic strain as the mechanical state variables. This kinematic hardening model (Fig. 7b) is self-consistent and useful for solving boundary value problems but clearly lacks generality. Thermodynamic statements based strongly upon it should not be expected to hold for real materials whose change in state for a given plastic strain will vary enormously with prior history. In particular, a return to zero plastic strain will not restore their initial state. With rare exceptions, materials do not have the reversibility of this simple model or even of more elaborate two element models whose component elements harden isotropically.

Isotropic hardening (Fig. 7a) displays none of this reversibility or restoration of state and cannot be modeled by any assemblage of independent perfectly plastic elements. It too is self-consistent and useful for solving problems but lacks generality. The yield stress in simple tension may be used as a parameter of state. Within the current yield surface, behavior is fully elastic and the state variables are θ and σ_{ij} or ε''_{ij}. Plastic strain increases the yield stress in tension or the diameter of the elastic domain and so alters the state of the material. Perfect plasticity is a limiting case of isotropic hardening in which the rate of hardening approaches zero; subsequent yield surfaces shrink toward the initial yield surface.

Level of Phenomenological Abstraction of Inhomogeneity of Material

Once broad principles have been laid down for the choice of a proper model representation of a particular material in a particular application, the choice to be made will be governed in part by convenience and in part by the phenomenological aspects to be simulated. A single element for each point of a continuum is the simplest representation and may well be forced by the need to obtain answers to practical problems. If an assemblage of elements is employed, each is at the temperature of the point of the body. The load-deflection-heat transfer response of the entire body or of any portion subjected to inhomogeneous stress and temperature then is the direct equivalent of the constitutive relations for the next level of refinement in the modeling of material behavior.

At this next level of representation, each "point" of the continuum is recognized as a region containing inhomogeneities on the microscale. Each substructural component, such as a grain or inclusion, is considered to have its own constitutive relation of the usual phenomenological type. Each is modeled by a single element or by an assemblage of simple elements at the temperature of the substructural component. The system then has a well-defined set of state variables and parameters whose changes can be followed fully, in principle at least.

Time is needed for an equilibrium temperature to be established among the components at the "point". Therefore, time effects due to temperature change will appear as a primary creep in the stress-strain relations for a model whose individual elements are all time-independent of the usual elastic-plastic type with temperature dependence of the yield surface. The heat transfer now involves distances of the order of the grain size of a metal and temperature changes which may be very high locally in the microstructure. Corresponding effects in the load-deflection relations for the body as a whole involve the very much larger dimensions of the body and very much smaller (average) temperature changes at a point in a supposedly microscopically homogeneous continuum.

The complexity of response to the imposed loads and temperature variations of a boundary value problem is enormous at this level of representation, despite the well-defined set of state variables. A further step downward in scale from grain size to the dislocation level of dimensions can be visualized but there are subtle as well as obvious difficulties with the continuum approach at that level [22]. Regardless of the level selected, it is important to keep in mind that all choices of well-defined elements or sub-elements are not equivalent, nor are they equally good in the simulation of a particular aspect of real material behavior. A single element properly selected for a particular material and problem can be superior to an infinite array of the wrong type of elements.

Despite the sense of immense complexity exhibited by the model approach, or perhaps because of it, the lack of special importance of the stress-free state is clear. A load-free body in thermal equilibrium is likely to have large residual or initial stresses on the continuum scale, and even larger stresses on the microscale. Departure from zero load therefore represents departure from a wide variety of initial states of stress. Consequently, if material behavior does have any general characterization, this should be exhibited by the behavior over paths or cycles of stress or of strain [23] starting from any point inside the elastic domain [12, 17—19]. This is a partial justification only for considering the work done by an external agency which adds to and then removes a set of stresses or loads or deformations from the existing set, in preference to considering the work of the existing set.

In the general situation, the modeling of the material requires an infinite array of infinite elements with various degrees of intertwining and interlocking. The concept of state remains clear but the number of state variables needed for an accurate representation becomes unmanageable. Some other approach than the ad hoc combination of simple models is required to define and develop the meaning of a good approximation and an appropriate thermodynamics.

Path Independence in the Small

A thermodynamic approach which includes all possible classes of material response can lead only to restatements of the first and second law which will appear either trivial or well-known. The alternative to the simple model approach, with its limited and uncertain generality, is to devise reasonably rather than excessively broad classifications of material response without restriction on the number of variables and parameters of state. It is reasonable to assume thermodynamic stability for structural materials which do not alter spontaneously within the time scale of interest. A stronger postulate of mechanical stability under isothermal conditions also appears appropriate for these materials [12, 17,

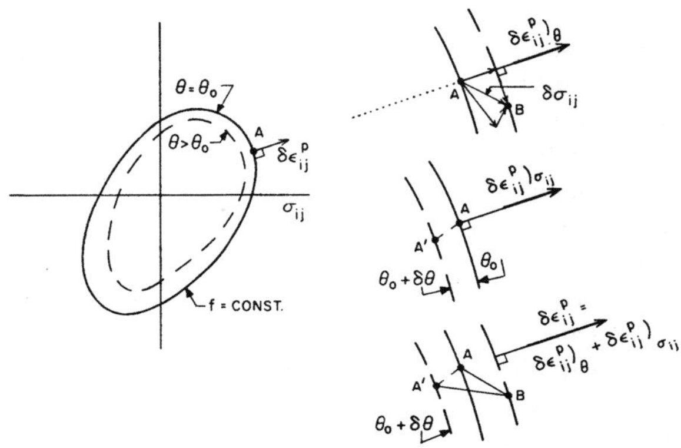

Fig. 8. Path independence in the small for no unloading at A

$$\delta\varepsilon_{ij}^{p} = \delta\varepsilon_{ij}^{p})_{\theta} + \delta\varepsilon_{ij}^{p})_{\sigma_{ij}} = h_{ijmn}\,\delta\sigma_{mn} + H_{ij}\,\delta\theta = \left(G\,\frac{\partial f}{\partial\sigma_{mn}}\,\delta\sigma_{mn} + M\,\frac{\partial f}{\partial\theta}\,\delta\theta\right)\frac{\partial f}{\partial\sigma_{ij}}$$

18]. However, a correspondingly strong thermomechanical stability with variable temperature is far too restrictive for solids and liquids. An increase in temperature lowers yield and flow strengths, and so increases plastic or viscous dissipation. They in turn tend to increase the temperature still more. This tendency toward instability is balanced locally or globally by such factors as work-hardening or stress redistribution and heat transfer [17, 24], just as geometric instability of structural elements is counteracted by work-hardening and the redistribution of stress throughout the structure [25]. However, a degree of instability is fundamental and should not be completely ruled out explicitly or implicitly in a thermodynamic treatment of materials.

A suitable path independence in the small (Fig. 8) somewhat analogous to the microscopic reversibility involved in Onsager's relations, may be a worthwhile point of departure [7]. All paths of continued loading from a smooth point $A(\sigma_{ij}, \theta_0)$ to a neighboring point $B(\sigma_{ij} + \delta\sigma_{ij}, \theta_0 + \delta\theta)$ are assumed to give the same increment of strain and require the same increment of work per unit volume to order $\delta\sigma_{ij}\delta\varepsilon_{ij}$ when the material is elastic-plastic. Linearity in the small is included but is a weaker assumption. Normality of $\delta \varepsilon_{ij}^{p}$ follows as a consequence. Convexity of individual yield surfaces in each isothermal stress space follows from the added assumption for either mechanically stable or unstable material [19] that positive work is done by an external agency in any cycle of stress which produces $\delta \varepsilon_{ij}^{p}$. If the yield surface has a corner at A, the paths of continued loading may be restricted to narrow regions of stress and temperature space or may not exist at all. Unless the corner is intrinsic, as for infinitely many independent plane loading surfaces (Fig. 7d) normality holds at points adjacent to the corner and no real difficulty of definition arises. When the corner is intrinsic, there should be normality of the components of the plastic strain increment to the individual loading surfaces associated with each [13—16].

Although some aspects of path independence were explored in Ref. 7 for materials in which time plays an essential role, much more remains to be done on the level of definition as well as logical consequence.

Dry Friction vs. Perfect Plasticity

The stress-strain curve in simple tension or compression or shear for a perfectly plastic idealization of material is the same as the horizontal force-displacement curve for a block resting on a horizontal plane with a constant coefficient of friction. Neither plastic deformation of the material nor sliding of the block alters its state. On these grounds alone it might seem reasonable to suppose that dry friction and perfect plasticity are analogous. In addition, when metal slides on metal, the frictional force is due to the plastic deformation of the little peaks of metal in contact with each other over a very small fraction of the total contact area. Despite this physical confirmation, the fact is that the idealizations of dry friction and of perfect plasticity are very different indeed.

The difference shows up when two or more components of stress or of force are considered and the yield surface for a perfectly plastic material is taken to coincide with the limiting surface for a frictional material. A limiting frictional surface is based on the linear increase of shear force on the surface of sliding with the normal force on that surface. This normal stress effect does not exist in the usual shear stress criteria of yielding such as the Tresca or Mises forms. However, it is included in a more general criterion of perfect plasticity [26]. Normality of $\delta \varepsilon_{ij}^{p}$ to the yield surface in a theory of plasticity now must be contrasted with the lack of normality in the frictional case (Fig. 9). Frictional sliding occurs without changes in length or volume; plastic deformation in shear must be accompanied by a volume increase if the normality requirement of the theory of perfect plasticity is to be satisfied.

This difference is large and significant. Considerably less energy appears as heat in the plastic system because part or all of the work done by the tangential forces which produce the shearing or sliding goes into potential energy of strain or of position. In the frictional system all of this work appears as heat. The frictional system therefore is far less stable than the plastic. Laws of thermodynamics apply to both classes of systems but they are in different thermodynamic classifications.

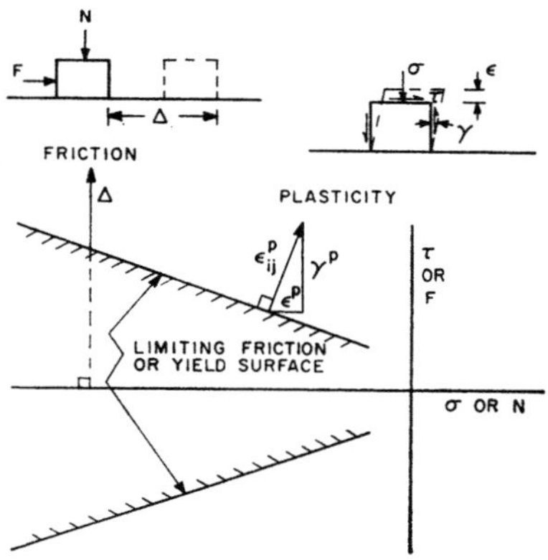

Fig. 9. Friction vs. plasticity (normality)

How then can the plastic shearing of asperities of metal upon metal not be in the same class as plastic behavior? The main point here is a matter of idealization. The metal itself obeys a shear law of some form and does not show a visible expansion in volume upon shearing. Therefore blocks slide over each other without being pushed apart. The normal force effect in friction is explained easily by the greater area of contact at higher pressure than at lower due to the flattening of asperities and the bringing of more of them into contact. This is a significant geometry change on the microscale which affects the equations of equilibrium in the sense that the local contact stresses are not at all proportional to the normal force, as they would be if the area of contact remained fixed. The need to take into account this significant but hidden geometry change in the domain of small overall displacements places this plastic system which represents frictional behavior in a very different class from ordinary plasticity.

Of course, the two classes of behavior can be brought together by considering just one class in which both large and small strains and rotations are taken into account [3, 4]. Pride in any success of this approach would have to be tempered a little by the realization that the normality condition is lost and the resulting thermodynamic statements will be more general but less useful.

Concluding Remarks

One of the questions touched upon only peripherally is whether it is necessary to develop a thermodynamic theory for irreversible materials in order to understand or to solve problems of analysis and design. If the answer is no, as suggested here, the purpose of a thermodynamic treatment is to elucidate the basic features of the behavior of real materials, not to develop a consistent theory for an arbitrary idealization. The thermodynamic statements which are sought go beyond the predictions of the first and second laws for all materials. Therefore useful classifications of materials must be devised which abstract essential features of material behavior in at least one important practical situation. Degrees of path independence and degrees of stability seem far more relevant for this purpose than special assemblages of models or their equivalents.

On the other hand, when a particular problem of analysis or design must be solved, success is likely to require immediate recourse to a highly idealized model. Most often the equations of heat transfer will be solved first without consideration of the stress field. Then the equations of stress analysis will be solved as uncoupled equations into which the local temperatures found previously are substituted or otherwise taken into account. When, as in metal cutting, this procedure breaks down because the plastic deformation of the metal is the major source of heat, a successive approximation technique is more likely to be successful than a full thermodynamic treatment because the thermal terms tend to overwhelm the stress terms.

References

[1] TRUESDELL, C. A.: Thermodynamics of Deformation, In Non-Equilibrium Thermodynamics, Variational Techniques, and Stability. Chicago, Ill.: Univ. of Chicago. 1966. p. 110 referring to B. D. COLEMAN.

[2] KESTIN, J.: On the Application of the Principles of Thermodynamics to Strained Solid Materials. In: IUTAM Symposia 1966, 177. Wien-New York: Springer. 1968.

[3] GREEN, A. E., and P. M. NAGHDI: A General Theory of an Elastic Plastic Continuum. Arch. Ratl Mech. Anal. **18**, 251–281 (1965).

[4] PIPKIN, A. C., and R. S. RIVLIN: Mechanics of Rate Independent Materials. ZAMP **16**, 313–327 (1965).

[5] ZIEGLER, H.: An Attempt to Generalize Onsager's Principle. ZAMP **9b**, 748 to 763 (1958).

[6] MEIXNER, J.: Consequences of an Inequality in Nonequilibrium Thermodynamics. J. Appl. Mech. **33** (1966). See also IUTAM Symposia 1966, 236. Wien-New York: Springer. 1968.

[7] DRUCKER, D. C.: Stress-Strain-Time Relations and Irreversible Thermodynamics. Proc. 1962 IUTAM Symposium on Second Order Effects in Elasticity, Plasticity, and Fluid Dynamics. Eds. D. ABIR and M. REINER. pp. 331–350. New York-London: Pergamon Press. 1964.

[8] SEDOV, L. I.: Introduction to the Mechanics of a Continuous Medium. Reading, Mass.: Addison-Wesley. 1965. (Translation of 1962 Edition.)

[9] BOHNENBLUST, H. F., and P. DUWEZ: Some Properties of a Mechanical Model of Plasticity. J. Appl. Mech. **15**, Trans. ASME **70**, 222–225 (1948).

[10] WHITE, G. N. Jr.: Application of the Theory of Perfectly Plastic Solids to Stress Analysis of Strain Hardening Solids. Brown Univ. Tech. Report All/51 (Aug. 1950).

[11] Prager, W.: Introduction to Plasticity. Reading, Mass.: Addison-Wesley. 1959. Models of Plastic Behavior. Also Proceedings 5th U. S. National Congress of Applied Mechanics. pp. 435—450. ASME 1966. Refers to work of Ilyushin, Ishlinsky, Ivlev, Kachanov, Kadashevich and Novozhilov, Mandel, Shield, and Ziegler.

[12] Koiter, W. T.: General Theorems for Elastic-Plastic Solids. Progress in Solid Mechanics **1**. Amsterdam: North-Holland Publishing Co. 1960.

[13] Naghdi, P. M.: Stress-Strain Relations in Plasticity and Thermoplasticity. Second Symposium on Naval Structural Mechanics „Plasticity". Eds. E. H. Lee and P. S. Symonds, pp. 121—167. New York: Pergamon Press. 1960.

[14] Sanders, J. L.: Plastic Stress-Strain Relations Based on Infinitely Many Plane Loading Surfaces. Proc. 2nd U. S. National Congress of Applied Mechanics. pp. 455—460. ASME. 1954.

[15] Hodge, P. G., Jr.: Plastic Analysis of Structures. New York: McGraw-Hill. 1959.

[16] Budiansky, B., and T. T. Wu: Theoretical Prediction of Plastic Strains of Polycrystals. Proc. 4th U. S. National Congress of Applied Mechanics. v. 2, pp. 1175—1185. ASME. 1962.

[17] Drucker, D. C.: Extension of the Stability Postulate with Emphasis on Temperature Changes. Second Symposium on Naval Structural Mechanics "Plasticity". Eds. E. H. Lee and P. S. Symonds, pp. 170—184. New York: Pergamon Press. 1960.

[18] Drucker, D. C.: On the Postulate of Stability of Material in the Mechanics of Continua. J. Mécanique **3**, 235—249 (1964).

[19] Palmer, A. C., G. Maier, and D. C. Drucker: Convexity of Yield Surfaces and Normality Relations for Unstable Materials or Structural Elements. J. Appl. Mech. **34**, Trans. ASME **89**, 464—470 (1967).

[20] Calladine, C. R., and D. C. Drucker: Nesting Surfaces of Constant Rate of Energy Dissipation in Creep. Q. Appl. Math. **20**, 79—84 (1962).

[21] Drucker, D. C.: On Time Independent Plasticity and Metals under Combined Stress at Elevated Temperature. Recent Progress in Applied Mechanics. The Folke Odqvist Volume. Stockholm: Almqvist and Wiksell/Gebers. 1966.

[22] Drucker, D. C.: The Continuum Theory of Plasticity on the Macroscale and the Microscale. 1966 ASTM Marburg Lecture. J. Materials **1**, 873—910 (1966).

[23] Ilyushin, A. A.: On the Postulate of Plasticity. Prikl. Mat. Mekh. **25**, 503—507 (1961).

[24] Rubin, D., and D. C. Drucker: On Stability of Viscoplastic Systems with Thermo-mechanical Coupling. Reiner Anniversary Volume. New York: Pergamon Press. 1967.

[25] Maier, G., and D. C. Drucker: Elastic-Plastic Continua Containing Unstable Elements Obeying Normality and Convexity Relations. Schweizerische Bauzeitung **84** (23), 447—450 (1966).

[26] Drucker, D. C., and W. Prager: Soil Mechanics and Plastic Analysis or Limit Design. Q. Appl. Math. **10**, 157—165 (1952).

Discussion

Naghdi: What does the term thermodynamic instability mean?

Drucker: By thermodynamic instability I mean instability under both temperature change and mechanical loading. Consider an increase in temperature in a body which, in the case of a viscous material, is deforming under a given set of loads or, in the case of a plastic material, is at the point of yielding. Then, of course, additional yielding or flow will occur. This additional yielding will generate heat. The additional heat further lowers the flow strength and, therefore, one has a degree of instability. This instability is balanced in real materials and real bodies by transfer of heat to the outside world, and by transfer of heat to neighboring points of lower stress. Unless the system were stable, or nearly so, we would not be able to observe it under almost stationary conditions. However, it is stable only because of the transfer terms; the material itself is not stable at the point.

LEE: As I understood it, you are willing to admit thermomechanical instability, but not purely mechanical instability. But isn't it true that materials like mild steel exhibit mechanical instability?

DRUCKER: One can devise a theory for mechanical instability as well. However, if you include in your material behavior all possible mechanically and thermally stable and unstable behavior, with the geometric instability effects you can get in large strain problems, you end up simply with the first and second laws. The statements may not look trivial, because they will be so very elaborate, but they are.

At this meeting, we have seen two opposite thermodynamic approaches. One is a general approach, which in a formal sense attempts to extend thermodynamics to the most complex of cases. The other makes a very simple restrictive physical or mathematical assumption. It seems to me that neither of these two extremes is truly useful in developing a thermodynamics of materials. We must decide not to develop a thermodynamics for all materials, but first to develop a thermodynamics for mechanically stable materials, then perhaps a thermodynamics for mechanically unstable, and so on. I suspect even this will soon turn out to be too general.

Prof. SETH pointed out that there are certain classes of problems, such as the shock-wave-problem, where you are not going to be able to get rid of the thermomechanical interaction. However, this is not necessarily so for static problems and the ordinary dynamic problems. You are better off dealing with a decoupled, well understood, system of sufficient generality than a coupled but over-idealized system.

LEE: You mentioned that in usual design problems one would not be involved in a coupled thermomechanical problem. It seems to me this very much depends on the problem you are concerned with. If, for example one is concerned with dissipative bodies, which are loaded cyclicly, as in, say, solid-rocket-propellants, then with continued dissipation of mechanical energy to heat, the actual temperature value becomes exceedingly important and to analyse this without thermomechanical coupling would lead to completely wrong results.

DRUCKER: Let me explain what I mean by decoupled systems. I don't mean that heat generation is ignored, but rather that the coupled equations in the full sense are not considered. The thermal problem is solved separately, and the mechanical problem is solved separately. However, the heat generated in oscillation is taken into account to a first approximation in the determination of the local temperature which in turn determines the behavior of the material at the point. If you try to do the fully coupled thermodynamics, I am afraid you will be forced to select too strong an idealization of material and so obtain an incorrect result for design purposes.

Joint Probability Density Distributions in a Turbulent Flow

By

François N. Frenkiel and **Philip S. Klebanoff**
Washington, D.C. (U.S.A.)

Often in order to resolve the mathematical formulation of a physical problem there is a need for further simplifications or assumptions, and not too seldom such simplifications may obscure the essential physical behavior. The statistical theory of turbulence is particularly vulnerable to such pitfalls. It is therefore important to check such assumptions by appropriate experimental measurements or indeed to use experimental data as a basis for reasonable simplifying assumptions.

In fact such close connection between experiment and theory has been the basis for much of the progress in the statistical theory of turbulence, however the existing experimental data with respect to the statistical behavior of turbulence is somewhat limited. An intensive study has therefore been initiated using high-speed digital computing methods to provide more complete data on the statistical characteristics of turbulence in particular higher-order correlation coefficients and joint-probability density distributions of turbulent velocities.

The simultaneous turbulent velocity components parallel to the mean velocity $u_1' = u'(x_1, y_1, z_1, t)$ and $u_2' = u'(x_2, y_2, z_2, t)$ at two points located downstream of a turbulence producing grid were recorded on magnetic tape using hot-wire anemometry. The recorded data were digitized and the appropriate statistical characteristics were obtained using high-speed computing techniques [6].

Measurements of one-dimensional probability density distribution of turbulent velocity components

$$\text{Prob}\,[\xi < u' < \xi + d\xi] = p(u')du',$$

and of the (two-dimensional) joint probability density distribution

$$\text{Prob}\,[\xi < u_1' < \xi + d\xi,\ \eta < u_2' < \eta + d\eta] = p(u_1', u_2')du_1'du_2',$$

are presented and compared with the Gaussian probability distribution

$$p_0(u_1', u_2') = \frac{1}{2\pi[\overline{u_1'^2}\ \overline{u_2'^2}\ (1 - R^2)]^{1/2}}$$

$$\cdot \exp\left[-\frac{1}{2(1 - R^2)}\left(\frac{u_1'^2}{\overline{u_1'^2}} - 2R\,\frac{u_1' u_2'}{(\overline{u_1'^2}\ \overline{u_1'^2})^{1/2}} + \frac{u_2'^2}{\overline{u_2'^2}}\right)\right]$$

where

$$R(x_2 - x_1, y_2 - y_1, z_2 - z_1) = \frac{\overline{u_1' u_2'}}{(\overline{u_1'^2}\ \overline{u_2'^2})^{1/2}},$$

is a correlation coefficient between simultaneous turbulent velocities at the two points [4].

The probability distributions $p(u')$ and $p(u_1', u_2')$ were obtained for two points located in a plane parallel to the mean velocity ($x_2 - x_1 = x$; $y_2 - y_1 = 0$; $z_2 - z_1 = 0$) and in a plane perpendicular to the mean velocity ($x_2 - x_1 = 0$; $y_2 - y_1 = y$; $z_2 - z_1 = 0$). Although the one-dimensional probability distributions agreed well with the Gaussian law, as has been similarly observed by others [7], the measured two-dimensional joint probability distribution shows a clear departure from Gaussianity [4]. It should be noted that in the case of a Gaussian probability distribution all the *skewness factors*

$$S^{(n)} = \frac{\overline{(u_2' - u_1')^n}}{[\overline{(u_2' - u_1')^2}]^{n/2}} \qquad (n \text{ odd})$$

would have been equal to zero. Measurement of the skewness factor $S^{(3)}$ (the results obtained for $S^{(3)}$ agree with previous measurements [8]), super-skewness factor $S^{(5)}$, and hyper-skewness factor $S^{(7)}$ are presented[1] and the values obtained for these factors indicate the departure from Gaussianity. The relation between these factors is shown [3, 6] to correspond very well to the case when a non-Gaussian probability density distribution of the form

$$p(u_1', u_2') = p_0(u_1', u_2') \sum A_{j,k} H_{j,k}(u_1', u_2'),$$

involving Hermite polynomials of two variables [1], $H_{j,k}$, is used. The coefficients $A_{j,k}$ of the polynomials are given as functions of the measured higher-order correlation coefficients

$$R^{m,n} = \frac{\overline{u_1'^m\ u_2'^n}}{(\overline{u_1'^2})^{m/2}\ (\overline{u_2'^2})^{n/2}}.$$

[1] Moments of 3rd, 5th, and 7th order have been referred to by Frenkiel and Klebanoff as the "shichi go san" moments from the Japanese name 七五三 representing the benevolent numbers 7, 5, 3 given to a Japanese children holiday (15 November).

Measurements of flatness factors

$$Q^{(n)} = \frac{\overline{(u_2' - u_1')^n}}{[\overline{(u_2' - u_1')^2}]^{n/2}} \qquad (n \text{ even})$$

up to the eighth order are also presented and it is shown that their behavior is characterized better by the non-Gaussian assumption than by assuming Gaussianity. Higher even-order correlations ($R^{m,n}$ for $m + n = 4, 6, 8$) although agreeing quite well with the assumption of Gaussianity are nevertheless also in better agreement with the non-Gaussian probability distribution [6]. It is noted that the odd-order moments are very sensitive to the departures from Gaussianity however this is not the case for the even orders.

Space-time correlations of turbulent velocities along the direction of the mean velocity

$$R_{x,t}^{m,n} = \frac{\overline{u'^m(0,0,0,t)\, u'^n(x,0,0,t+h)}}{[\overline{u'^2}(0,0,0)]^{m/2}\,[\overline{u'^2}(x,0,0)]^{n/2}}$$

of even order $m + n$ up to the eighth order and of odd order up to the fifth are presented. The applicability of TAYLOR's space-time approximation [9], i. e. the concept of a frozen pattern of turbulence moving with the mean velocity, is examined. In addition to the second-order space-time correlations, also measured previously by others [2], the difference between the behavior of the odd-order space-time correlations and of the even-order space-time correlations is demonstrated [5].

References

[1] APPELL, P., and J. KAMPÉ DE FÉRIET: Fonctions hypergéométriques et hypersphériques. Polynomes d'Hermite. Paris: Gauthier-Villars. 1926.

[2] FAVRE, A., T. GAVIGLIO and R. DUMAS: Quelques mesures de corrélation dans le temps et l'espace en soufflerie. Rech. aéronautique, No. 32 (1953).

[3] FRENKIEL, F. N., and P. S. KLEBANOFF: Les asymétries d'ordre supérieur dans un écoulement turbulent. C. R. Acad. Sci. (Paris) **260**, 6026 (1965).

[4] — — Two-dimensional Probability Density Distribution in a Turbulent Field. Phys. of Fluids 8, 2291 (1965).

[5] — — Space-time Correlations in Turbulence. In: Dynamics of Fluids and Plasmas, ed. by S. I. PAI. p. 257. New York: Academic Press. 1966.

[6] — — Higher-order Correlations in a Turbulent Field. Phys. of Fluids **10**, 507 (1967).

[7] SIMMONS, L. F. G., and C. SALTER: An Experimental Determination of the Spectrum of Turbulence. Proc. R. Soc. (London) **A 165**, 74 (1938).

[8] STEWART, R. W.: Triple Velocity Correlations in Isotropic Turbulence. Proc. Camb. Phil. Soc. 47, 146 (1951).

[9] TAYLOR, G. I.: The Spectrum of Turbulence. Proc. R. Soc. (London) **A 164**, 476 (1938).

Accumulation of Second-Order Strain in Workhardening Media

By

A. M. Freudenthal and **Maria Ronay**
New York, N.Y. (U.S.A.)

Summary

Observations of accumulating axial extension produced by cyclic torsion of cylindrical bars and tubes of some workhardening metals are presented. The phenomenon is interpreted as an irreversible second order effect arising from the quadratic term in the constitutive equation of such media required by the condition of tensorial invariance.

1. Introduction

Combined torsion and tension of thin-walled tubes or cylindrical bars represents a multi-axial state of stress with easily variable directions of principal shear that can be realized with moderate experimental difficulties. It has therefore been widely used for the experimental investigation of various aspects of the theories of plasticity and strain-hardening, such as St. Venant's assumption of coincidence of the directions of principal shear velocities with the directions of principal shear stress [1], the difference between deformation and flow theories [2] and the distortion and motion of yield-surfaces produced by deformation [3]. In these experiments tensile force and torque were applied consecutively, in different orders, and only minor deviations from St. Venant's assumption were observed beyond the range of small strains; they were ascribed to strain-anisotropy developing during deformation. Significant deviations from this assumption under large torsion strains ($\gamma > 0.5$) were reported by Swift [4], who observed permanent axial elongations produced by severe torsion of cylindrical bars and tubes of mild steel, copper and brass, in the latter of an order magnitude of about 3 percent of the shear strain over the greater part of the plastic range. The intermediate strain is therefore not zero, which implies a dissimilarity between Mohr's stress- and strain circles, attributed again to strain-anisotropy associated with the large shear strains. Similar observations on Armco iron were made by L'HERMITE and DAWANCE [5], who, however, investigated mainly the combined effect of severe torsion and moderate tension in producing axial elongation.

In recent years a rather striking phenomenon, usually referred to as "cyclic creep", was reported by several experimenters engaged in research on metal fatigue in combined torsion and tension [6, 7]. When in a cylindrical bar or tube cyclic torsion of moderate amplitude was superimposed on an axial stress of moderate intensity, the bar started to

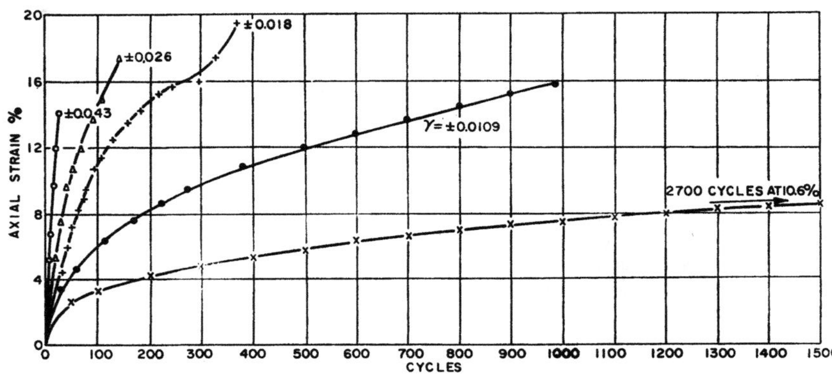

Fig. 1

elongate at an unexpectedly high rate. The observed functions of elongation versus time (cycles) are quite similar in form to "creep-functions", consisting of the characteristic three stages and are terminated by fracture. Figure 1 shows a typical set of observations on mild steel tubes [7] under an axial stress of 21.500 psi (1535 kg/cm²) and various amplitudes of cyclic torsion between $\gamma = \pm 0.006$ and $\pm 0{,}026$; similar diagrams were recorded for copper tubes [6]. A related phenomenon has also been reported [8]: cyclic creep in torsion could be produced by superposition of reversed cyclic axial strain on steady torsion.

No satisfactory explanation of these phenomena, which have significant technical implications, has so far been proposed. One of the interpretations has been that the creep is "genuine" i. e. time-dependent and produced by the axial force, and arises because the strain-hardening due to extension is offset by recovery accompanied by softening due to the cyclic strain [9]; alternatively, the phenomenon has been considered to be a "mechanically activated" creep process [6].

A phenomenological explanation of the interaction, in which the nature of the apparent "creep" as an accumulation of plastic axial strain increments is considered, can be based on the theory of kinematic hardening in which the Bauschinger effect is associated with a residual stress field [10]: if the initial yield surface is displaced by cyclic torsion, first in one direction then in the other (Fig. 2), an axial stress which is below the

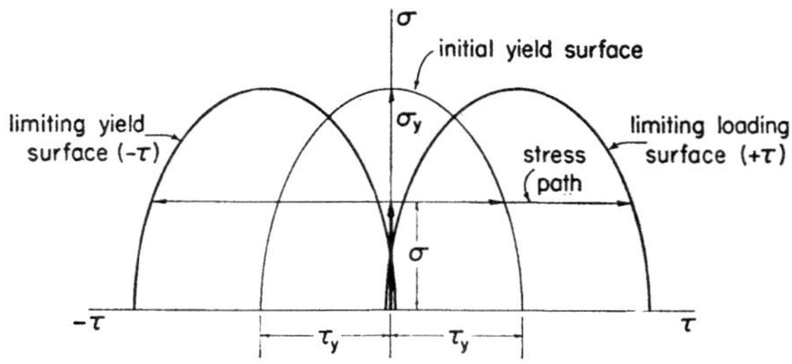

Fig. 2

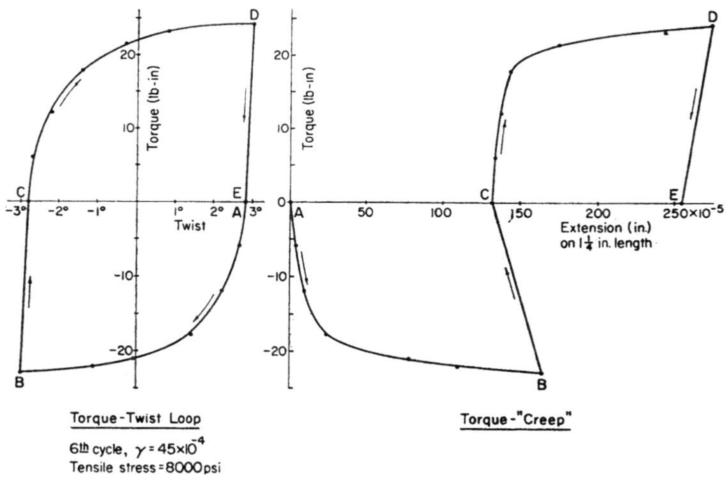

Fig. 3

yield stress in tension will penetrate the moving yield surface twice in each cycle, thus producing the axial strain increments which make up the accumulating axial elongation [11]. A detailed observation of the deformation process in a single torsion cycle with axial tension seems to support this explanation (Fig. 3). It can be clearly observed that most of the axial strain increment accumulates in the final parts of the loading and reversed loading range (when the intercept of the moving yield surface on the normal stress-axis drops rapidly below the axial stress intensity) while the shear strain axceeds the elastic range practically over the whole range of loading as a result of the displacement of the yield surface associated with the Bauschinger effect.

The above explanation is, nevertheless, not entirely satisfactory since it does not account for the fact that substantial "cyclic creep" can be produced by cyclic torsion alone, without superimposed tension.

2. Accumulation of Axial Strain in Cyclic Torsion at Room Temperature

In all experiments in which "cyclic creep" has been observed as a result of cyclic torsion superimposed on constant axial tension either the amplitudes of the applied torsion strain were not very small, (between 0.5 and several percent) or the sustained axial stress was a substantial

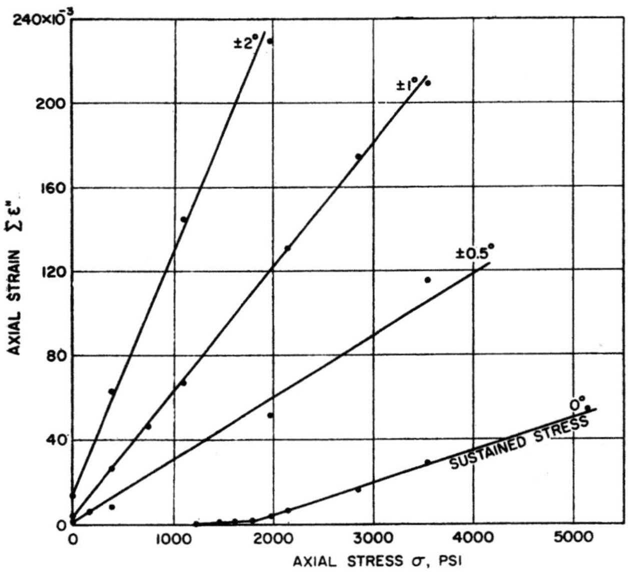

Fig. 4. Axial strain accumulation $\Sigma\varepsilon''(N)$ in superpurity aluminum at the start of crack propagation under combined axial tension and cyclic reversed torsion as a function of the applied tension σ for $\varphi = \pm 0.5°$, 1° and 2°

fraction (one third or more) of the yield stress in tension. In an attempt to explore the lower limits of the interaction between sustained tension and cyclic torsion, a systematic study of the range of small cyclic strain-amplitudes ($0.00075 < \gamma < 0.0045$) and very low axial stress intensities was undertaken, using super-pure aluminium. The results are shown in Figs. 4 and 5 and indicate not only the existence of a strong interaction even in this range but, more important, the accumulation of axial strain under cyclic torsion of small amplitude without axial load. These observations demonstrate the inadequacy of the interpretation of the

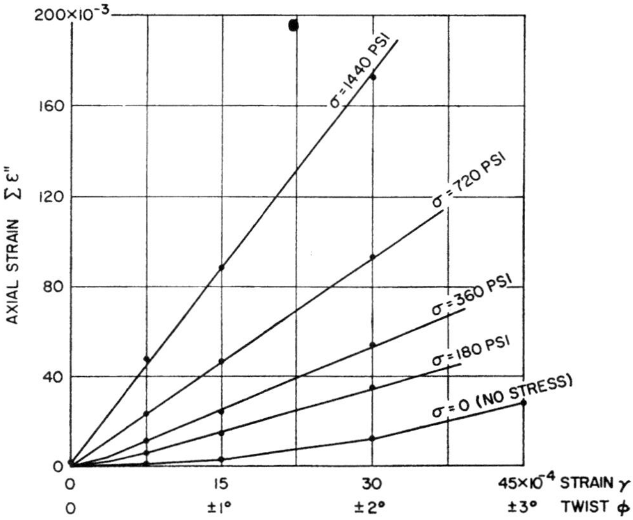

Fig. 5

phenomenon of "cyclic creep" with the aid of either of the concepts of "mechanically activated" true creep or "kinematic hardening".

In order to establish the nature of axial strain accumulation in pure cyclic torsion, the phenomenon was studied experimentally in considerable detail. Some of the results of this study are shown in Figs. 6 and 7 for aluminum bars and in Fig. 8 for copper tubes. The form of these curves, obtained in cyclic torsion at a frequency of 1750 cycles per minute, is similar to that of the cyclic creep curves obtained by superposition of cyclic torsion on steady tension. Therefore it seems at least as reasonable to base the interpretation of "cyclic creep" on Fig. 5 as on Fig. 4: in other words, instead of considering the phenomenon to result from an "activation" by cyclic torsion of the steady state deformation produced by sustained stress (curve labelled 0° in Fig. 4), it might be considered to result from an amplification, by the axial stress, of the axial strain accumulation produced by cyclic torsion (curve labelled $\sigma = 0$ in Fig. 5).

Plotting this accumulation in aluminum bars at various specified numbers of torsion cycles as a function of the applied cyclic strain

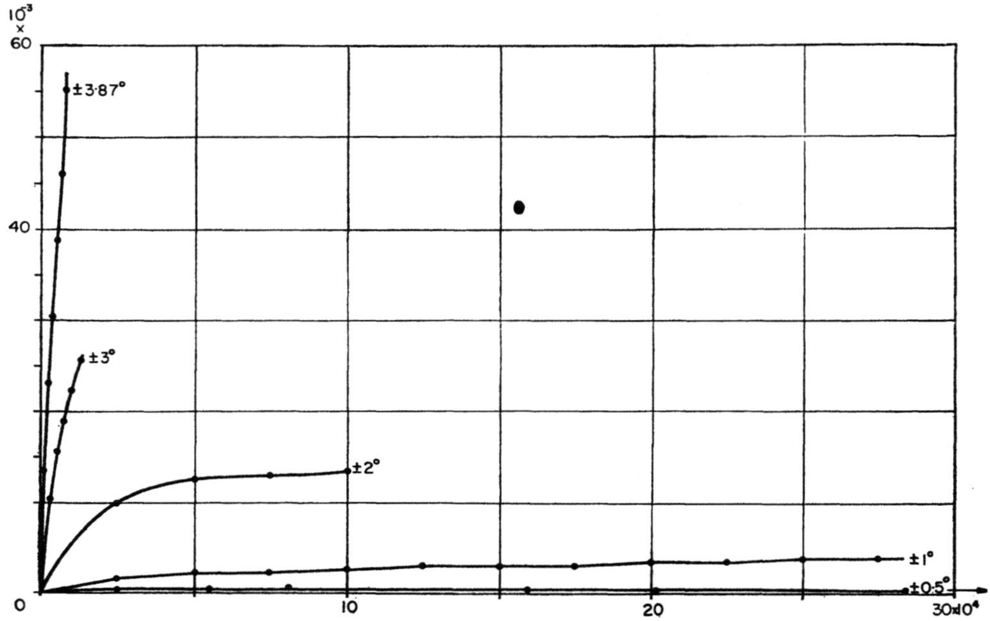

Fig. 6. Second-order axial strain accumulation $\Sigma\varepsilon''(N)$ in high purity aluminum specimens subject to reversed torsion cycles at several imposed angles $0.5° \leq \varphi \leq 3.87°$ (to start off crack propagation)

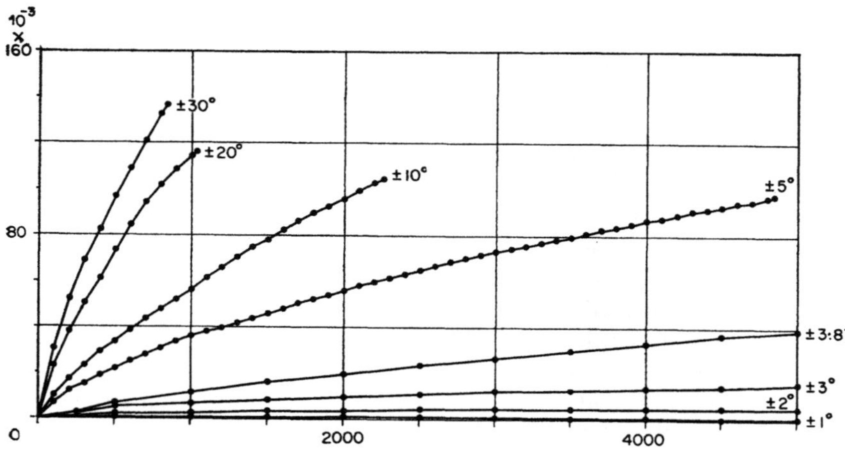

Fig. 7

amplitude (or angle) the observations can be fitted by a quadratic relation within the range of small strains (Fig. 9); for large strains the fitted

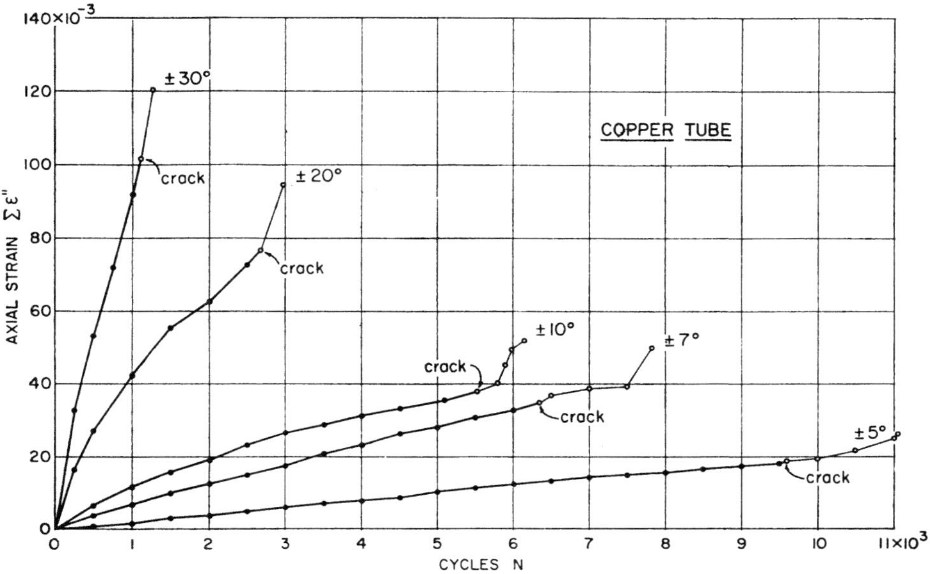

Fig. 8

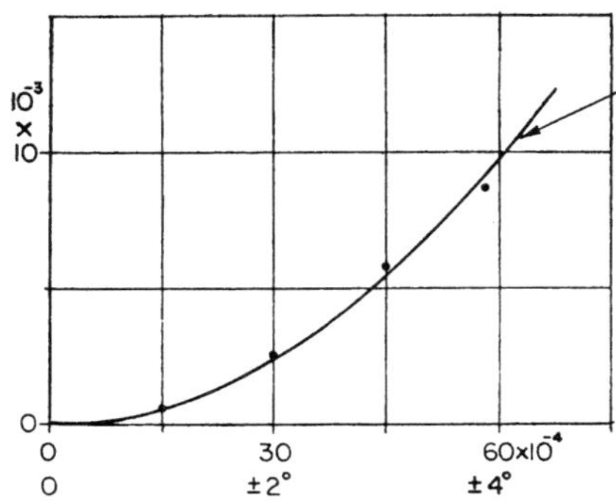

Fig. 9. Second-order axial strain accumulation $\Sigma\varepsilon''(N)$ at $N = 750$ cycles of reversed torsion as a function of the torsion angle: $0.5° \leq \varphi \leq 3.87°$ (quadratic relation)

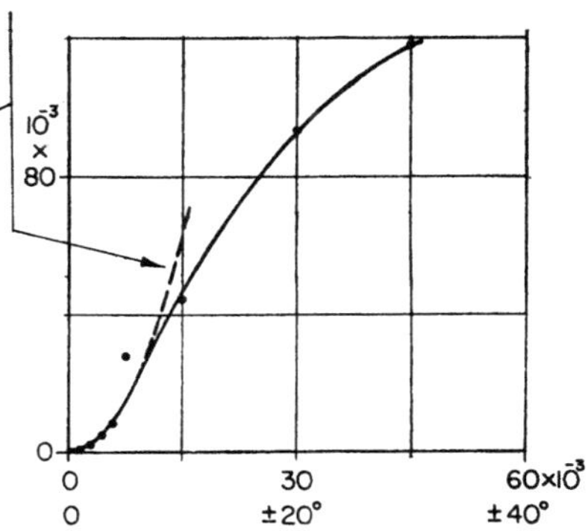

Fig. 10. Second-order axial strain accumulation $\Sigma\varepsilon''(N)$ at $N = 750$ cycles of reversed torsion as a function of the torsion angle: $0.5° \leq \varphi \leq 30°$

curves have a tendency to flatten out (Fig. 10). Similar results were obtained for copper, brass and titanium tubes (Fig. 11). The shape of the curves obtained under conditions of very small cyclic torsion strains is similar to that obtained by SWIFT for large unidirectional torsion strains (Fig. 12).

The existence of quadratic relations between the accumulated axial elongation at a certain number of cycles and the applied small shear strain amplitudes suggests that the axial strain accumulation is a manifestation of an irreversible second-order effect, of similar origin as the Poynting effect [12] observed in torsion of elastic wires and the Weissenberg effect [13] accompanying shear flow of polymeric fluids and due to the existence of a quadratic term in the constitutive equation. However, the Poynting effect is very small and reversible, vanishing at zero strain, while the Weissenberg effect vanishes at zero velocity. Their observation requires therefore rather elaborate experimental techniques and high accuracy. An irrecoverable second-order strain increment, on the other hand, can be amplified arbitrarily within the range of extensional instability or of fatigue fracture, by increasing the number of applied torsion cycles; the resulting second-order strain is therefore easily observed and measured. Figure 13 illustrates the order of magnitude of the accumulated axial extension in an aluminum bar by comparison of its extended to its initial length.

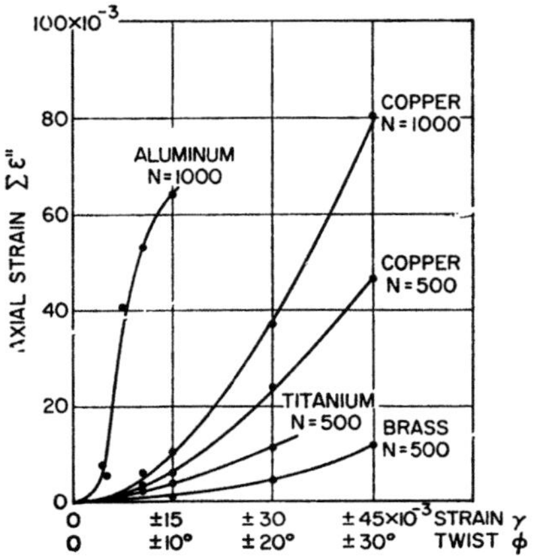

Fig. 11

The Weissenberg and the Poynting effects have been identified as second-order effects by REINER on the basis of his development of invariant constitutive equations for viscous [14] and elastic [15] media; they have been extensively studied in recent years [16, 17]. The existence of an irrecoverable second-order axial strain or strain-increment accompanying the shear strain produced by an applied torque and particularly its accumulation under cyclic torsion has, however, only recently been reported [11, 18]. Analytically this second-order effect is already implied in the stress-strain relations proposed by PRAGER [19] for an incompressible strain-hardening medium with yield condition $f(J_2, J_3) = 0$

$$\varepsilon'' = \alpha_0 \mathbf{1} + \alpha_1 \mathbf{s} + \alpha_2 \boldsymbol{t} \tag{1}$$

and in the flow relation proposed by HANDELMAN, LIN and PRAGER [20]

$$d\varepsilon'' = \alpha_0 \mathbf{1} + (\alpha_1 \mathbf{s} + \alpha_2 \boldsymbol{t}) \cdot df \qquad \text{for } df > 0 \tag{1a}$$

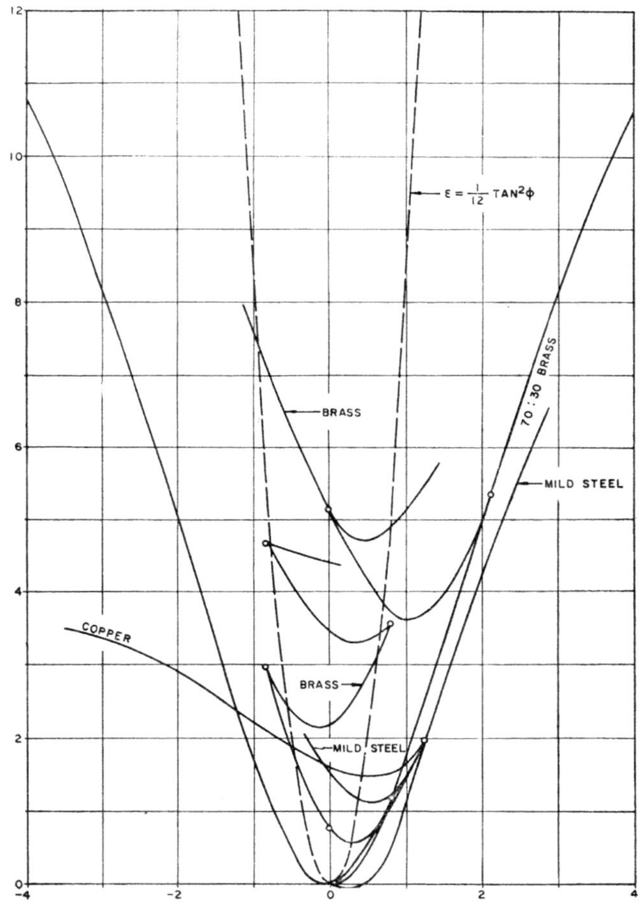
12
10
8
6
4
2
0
-4
-2
0
2
4
$\varepsilon = \frac{1}{12} \tan^2\phi$
BRASS
70 : 30 BRASS
MILD STEEL
COPPER
BRASS
MILD STEEL

Fig. 12

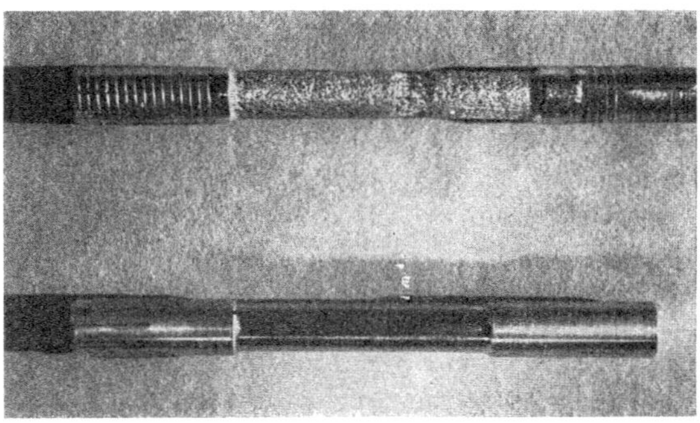

provided the symmetry restrictions imposed by their authors are removed; $\alpha_0(J_2, J_3)$, $\alpha_1(J_2, J_3)$ and $\alpha_2(J_2, J_3)$ are polynomial functions in the scalar invariants J_2 and J_3 of the stress deviation $\mathbf{s}$ and $\boldsymbol{t} = \mathbf{s}^2 - \frac{2}{3} J_2\mathbf{1}$.

According to either of the above equations a torque applied to a thin-walled tube producing the shear stress components $s_{r\theta} = s_{\theta r} = \tau$ causes irreversible shear strain or strain increments (since torsion constitutes "radial" loading, stress-strain relations can be used) $\gamma''_{r\theta} = \gamma''_{\theta r} = \gamma'' = \alpha_1\tau$, as well as radial and circumferential strains $\varepsilon''_{rr} = \varepsilon''_{\theta\theta} = \alpha_2\tau^2$ the sign of which depends on the sign of the parameter α_2. In an incompressible material these strain components produce the axial strain $\varepsilon_{22}'' = \varepsilon'' = -\frac{2}{3}\alpha_2\tau^2$. Since $\tau = \alpha_1^{-1}\gamma$, the relation between ε'' and γ'' is of the form

$$\varepsilon'' = -\frac{2}{3}\alpha_2\alpha_1^{-2}(\gamma'')^2 \tag{2}$$

which represents the observed quadratic relation (Fig. 9), provided α_1 and α_2 are nearly enough true constants, as might be assumed within the range of small strains. Since the observed axial strains are extensions, the parameter $\alpha_2 < 0$.

The circumferential strain $\varepsilon''_{\theta\theta}$ is therefore compressive; when the axial strain accumulates under applied cyclic strain, the longitudinal

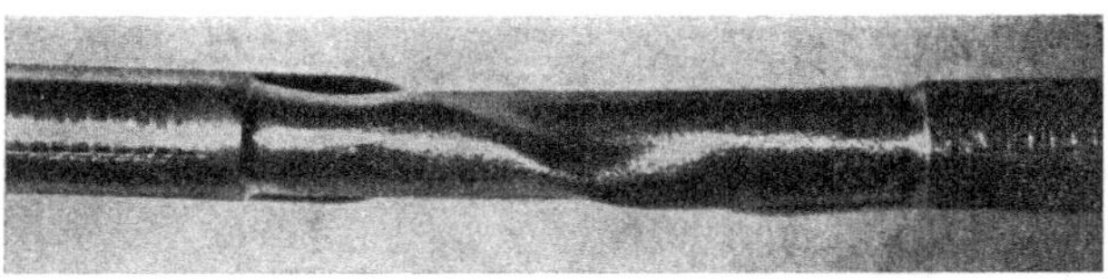

Fig. 14a

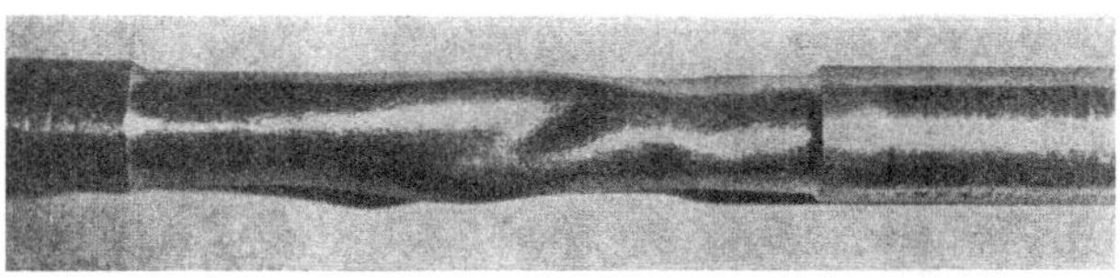

Fig. 14b

extension is accompanied by circumferential contraction, similar to that produced by external pressure. In the limiting case this second-order contraction of a thin-walled tube causes the type of instability that is characteristic of buckling of a cylinder under external pressure (Fig. 14).

In the range of large strains in strain-hardening metals the parameter α_1 increases with τ and therefore with γ''. Unless α_2 were to increase much faster, an assumption for which no reason exists, the second-order extension can therefore be expected to increase with increasing strain more slowly than with $(\gamma'')^2$ a conclusion that is in agreement with the observations (Figs. 10 and 12).

According to eq. (1a) irreversible axial strain accumulates only over those portions of each cycle for which the condition of "loading" $df > 0$ is satisfied, while $f = 0$; in the case of torsion with $J_2 = \tau^2$, $J_3 = 0$ this is equivalent to $\tau d\tau > 0$. Hence the axial strain accumulates during those parts of the first and third quarters of the full torsion cycle in which the condition $f = 0$ is satisfied. This is shown in Fig. 15 in which the

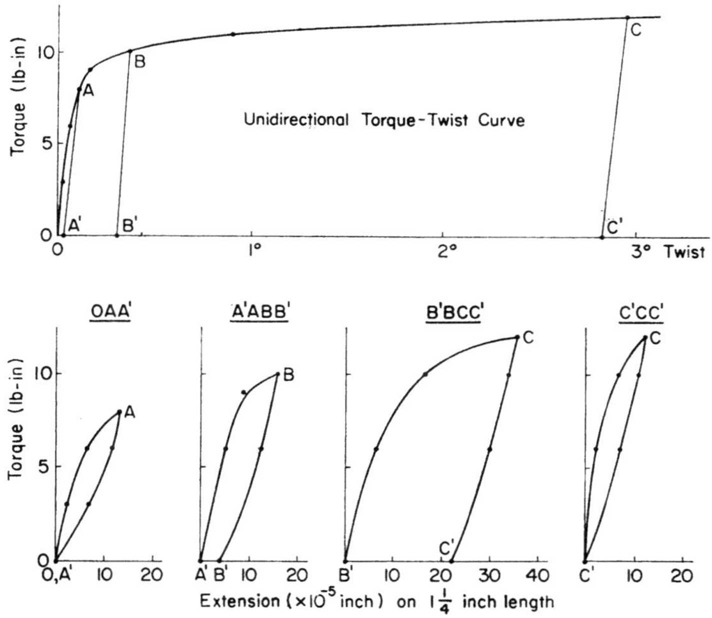

Fig. 15

second-order axial extension in three half-cycles of torsion straining at three different strain amplitudes has been plotted as a function of the applied torque. The recoverable and irrecoverable axial strain can be clearly seen; the observed recoverable strain is much larger than the possible magnitude of the Poynting effect. The form of the curves suggests that the irrecoverable part of the axial extension starts to accumulate as soon as $\tau d\tau > 0$; this implies that in reversed torsion, due to the Bauschinger effect, the condition $f = 0$ is satisfied at any value $\tau > 0$, an observation that is compatible with the concept of kinematic hardening illustrated in Fig. 2.

In Fig. 16 the accumulation of second-order axial strain in relation to the shear strain cycle is schematically illustrated under the assumption of simple superposition of elastic and plastic strain. As the material strain-

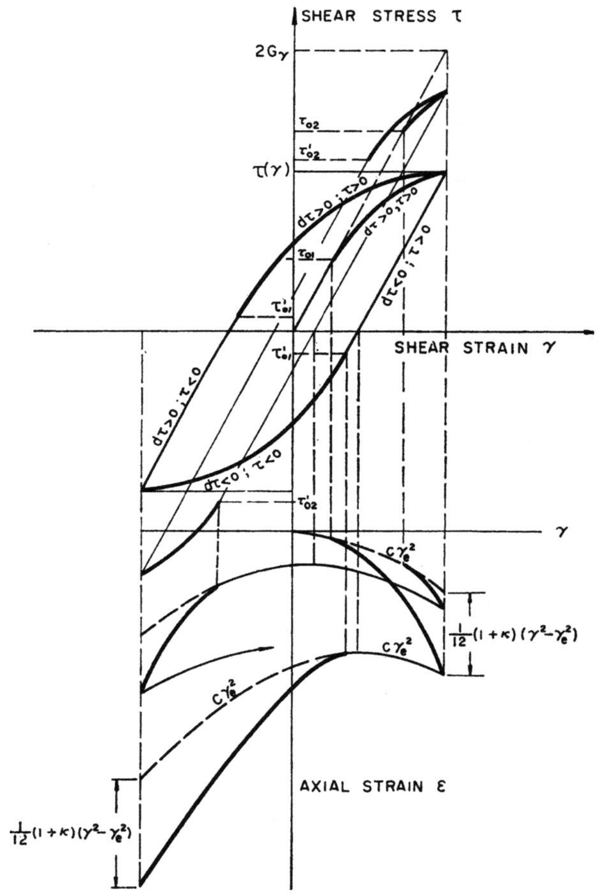

Fig. 16

hardens in the course of the strain cycling (Fig. 17), the rate of axial strain accumulation decreases, as observed in Figs. 6, 7 and 8.

Within the unloading range $df = \tau d\tau < 0$ and $f < 0$; therefore $\varepsilon'' = 0$, while some part of the second-order axial strain is recovered, Fig. 15 clearly shows that this part is much larger than could be expected if only the Poynting effect were involved. According to Poynting's observations on copper wire [12] the elastic axial strains ε' associated with an elastic shear strain of $\gamma' = 2.4 \times 10^{-3}$ is $\varepsilon' = 4.25 \times 10^{-6}$, which implies a quadratic relation of the form $\varepsilon' = 0.74\,(\gamma')^2$. The observed values of the recovered axial strain associated with a cyclic (total) strain of $(\gamma' + \gamma'') = 4.5 \times 10^{-3}$ are, however, of the order of magnitude of $\varepsilon' = 12 \times 10^{-5}$. Considering that for pure copper $\gamma' \ll \gamma''$,

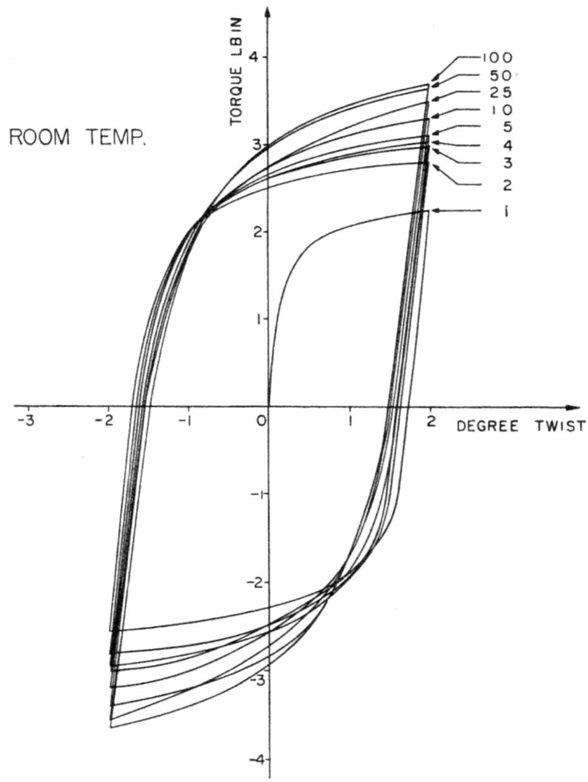

Fig. 17. Hysteresis loops in torsion for the first 100 cycles at room temperature

so that presumably the elastic shear strain $\gamma' < 10^{-3}$, it appears that the recoverable axial strain is about two orders of magnitude larger than the Poynting strain. It is therefore measurable in spite of the short gauge length used (1.5 in), while measurement of the Poynting effect required wire lenght of several feet.

The existence of the large recoverable second-order strain cannot be accounted for unless it is assumed that the interaction between the elastic and the strain-hardening response, as a result of which "latent" elastic energy (energy of micro-residual stresses) is stored within the material, produces an additional second-order effect which is associated with the existence of a recoverable non-elastic shear strain demonstrated by the existence of a closed hysteresis loop in one-directional loading and unloading in a strain-hardened medium.

To describe this phenomenon analytically eq. (1a) has to be extented to include the elastic strain by specifying that the total strain-increment is produced by both the elastic stress deviation and the plastic "overstress" deviation. In terms of eq. (1a) and considering only first and second-order terms in the Rivlin-Ericksen generalization of the Cayley-Hamilton theorem [21] for incompressible elastic strain-hardening media.

$$d\varepsilon = \alpha_1' \, d\boldsymbol{s} + \alpha_2' \, d\boldsymbol{t} + (\alpha_1'' \boldsymbol{s} + \alpha_2'' \boldsymbol{t}) \, df + \alpha_3 (\boldsymbol{s} \cdot d\boldsymbol{s} + d\boldsymbol{s} \cdot \boldsymbol{s}) \qquad (3)$$

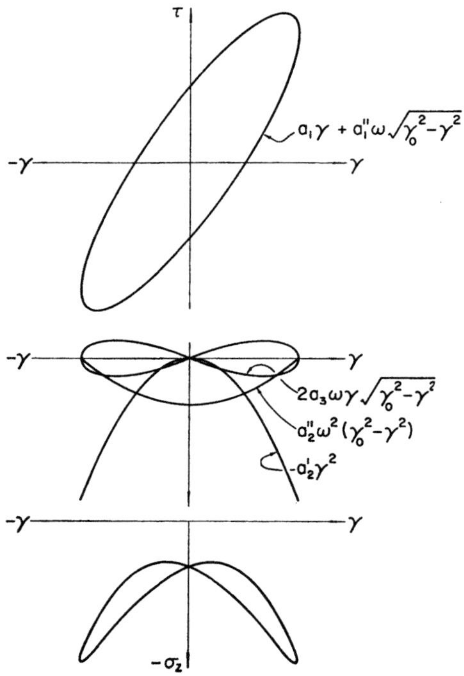

Fig. 18

where α_1' and α_2' are elastic constants, $\alpha_1''(J_2, J_3)$ and $\alpha_2''(J_2, J_3)$ are polynomial functions in the scaler invariants J_2 and J_3 of the "overstress" deviation, $\alpha_3(J_2, J_3, J_2', J_3')$ is a polynomial in the invariants J_2 and J_3 and the invariants J_2' and J_3' of the elastic stress deviation, and $d\boldsymbol{t} = (d\boldsymbol{s})^2 - \frac{2}{3} J_2' \mathbf{1}$. Equation (3) represents a generalized form of the Prandtl-Reuss equation including all second-order effects.

The coefficients $\alpha_1'' \, df$ and $\alpha_2'' \, df$ are non-zero only in the "loading" ranges where $df > 0$ and $f = 0$; the coefficients α_1', α_2' and α_3 are non-zero in both the loading and unloading ranges although α_3 may have a different form in the loading and unloading ranges. The coefficient α_3 governs the recoverable second-order effect that apparently arises from the interaction of elastic and plastic strain. In torsion this term produces additional radial and circumferential strain increments $d\varepsilon_{rr}' = d\varepsilon_{\theta\theta}' = 2\alpha_3 \tau \, d\tau$ with associated axial strain increment $d\varepsilon_{zz}' = d\varepsilon' = -\frac{4}{3} \alpha_3 \cdot \tau \, d\tau = -\frac{2}{3} \alpha_3 d(\tau^2)$ which is reversed when $\tau \, d\tau < 0$; it is of similar symmetric form as the Poynting effect. The reversible second-order strain is thus the sum of two terms governed, respectively, by the parameters α_2' and α_3, the first being the Poynting effect proper, the second an additional (and much larger) effect observable in strain-hardening media.

The relation between the total strain-increment matrix and the elastic and plastic stress matrices expressed by eq. (3) is similar in form to the relation between the stress matrix and the kinematic matrices specifying a generalized retarded elastic (Voigt) body

$$\sigma = a_1'\varepsilon + a_2'\varepsilon^2 + a_1''\dot{\varepsilon} + a_2''\dot{\varepsilon} + a_3(\varepsilon\dot{\varepsilon} + \dot{\varepsilon}\,\varepsilon) \tag{4}$$

which can be used to interpret the results of experiments in cyclic torsion performed in a Weissenberg Rheogoniometer on short cylindrical tube specimens of rubber filled with solid particles [22]. In this case the second-order effects in cyclic torsion with $\gamma = \gamma_0 \sin \omega t$ and therefore $\dot{\gamma} = \gamma_0 \omega \cos \omega t = \omega \sqrt{\gamma_o^2 - \gamma^2}$ consist also of three terms: the elastic term

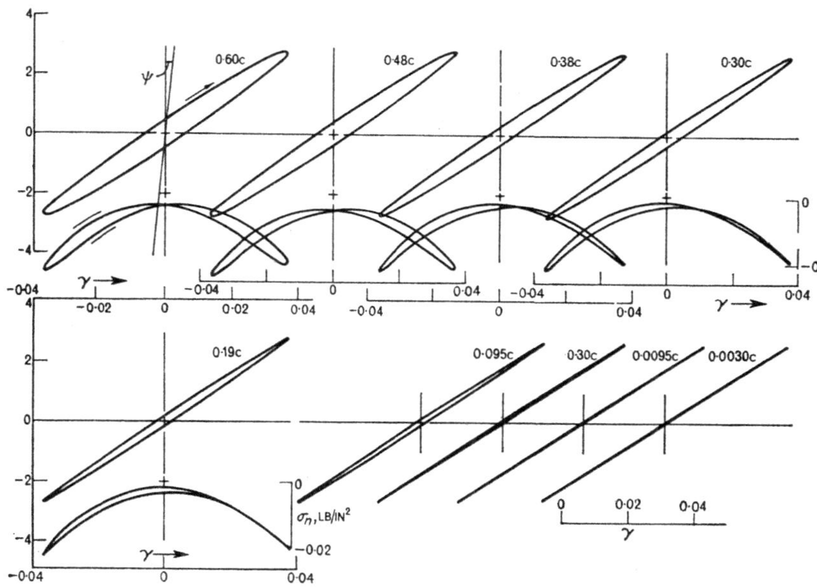

Fig. 19

$\sigma_{zz}' = a_2'\gamma_0^2 \sin^2 \omega t = 1/2\, a_2'\gamma_0^2\,(1 - \cos 2\omega t)$, the viscous term $\sigma_{zz}'' = a_2''\gamma_0^2\omega^2 \cos^2 \omega t = 1/2\, a_2''\gamma_0^2\omega^2\,(1 + \cos 2\omega t)$ and the visco-elastic interaction term $\sigma_{zz}''' = \gamma_0^2\omega \sin 2\omega t$; in terms of the shear strain $\sigma_{zz}' + \sigma_{zz}''' = a_2'\gamma^2 + 2a_3\omega\gamma\sqrt{\gamma_0^2 - \gamma^2}$, while $\sigma_z'' = a_2''\omega^2\,(\gamma_0^2 - \gamma^2)$. Fig. 18 illustrates the three second-order terms under specific assumptions for the ratio between the coefficients, as well as the first order stress-strain relation $\tau = a_1'\gamma + a_1''\omega\sqrt{\gamma_0^2 - \gamma^2}$. Figures 19 and 20 show these functions as actually recorded in experiments on an unfilled natural rubber and a highly filled polyurethane rubber. The existence of the three second-order effects in a visco-elastic medium is clearly demonstrated. The difference between eqs. (4) and (3) is in the existence of a

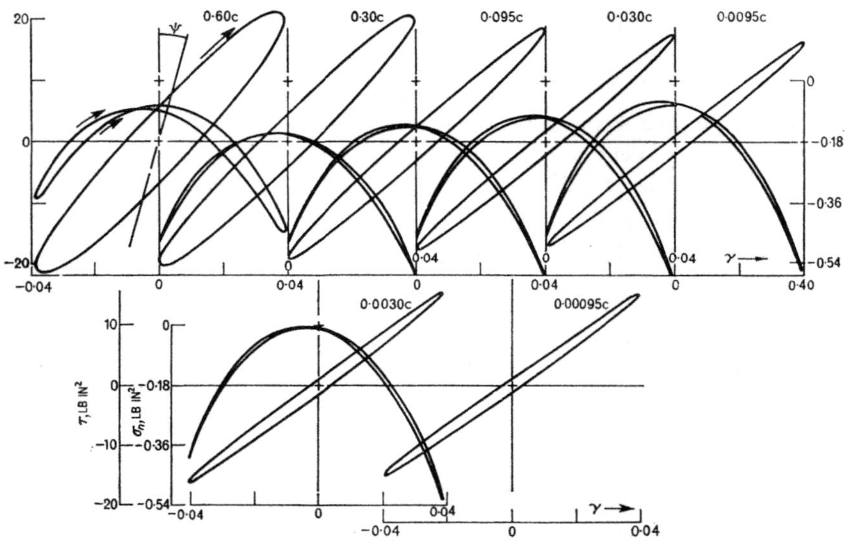

Fig. 20

phase shift between γ and $\dot{\gamma}$ in eq. (4) while τ and $d\tau$ in eq. (3) are in phase.

It seems therefore reasonable, to attribute the observed large recoverable second-order axial strain in the strain-hardening medium (Fig. 15) to a similar tensorial interaction between the elastic stress and the "overstress" producing strain-hardening. The formal similarity between the constitutive equations of the general visco-elastic and the generalized "rate independent" (plastic) medium which has recently been pointed out by PIPKIN and RIVLIN [23] is thus experimentally demonstrated.

3. Second-Order Strain Accumulation at Elevated Temperatures

Observations of the accumulation of axial strain at elevated temperatures indicate a significant increase with temperature of the rate of such accumulation. This may be partly due to the absence at elevated temperatures of strain-hardining in cyclic straining, indicated in Fig. 21 for aluminum at a temperature level of 150 °C. The axial strain-increment,

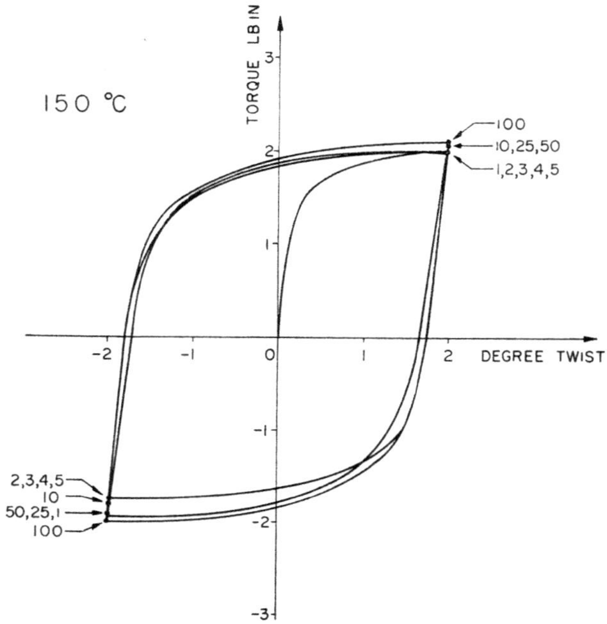

Fig. 21. Hysteresis loops in torsion for the first 100 cycles at 150 °C temperature

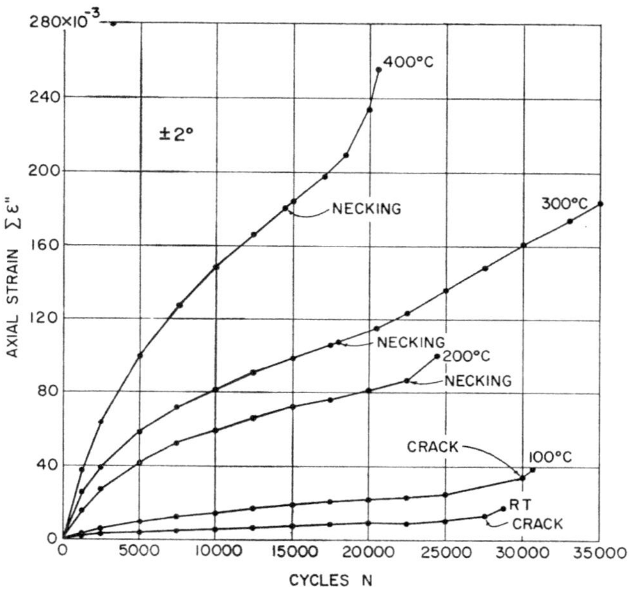

Fig. 22

therefore, does not decrease as much with increasing number of cycles as at room temperature, with resulting larger total accumulation at a specified number of cycles (Fig. 22). Moreover preliminary observations at different frequencies of strain cycling show that the magnitude of the axial strain increment per cycle increases with decreasing frequency, which suggests the existence of an additional viscous response in the medium which would produce additional viscous or combined viscoplastic second-order axial strain accumulation of considerable temperature- and frequency sensitivity. The observations of axial strain accumulation at the constant frequency of 1750 cycles per minute and at two specified numbers of cycles are summarized in Figs. 23 and 24.

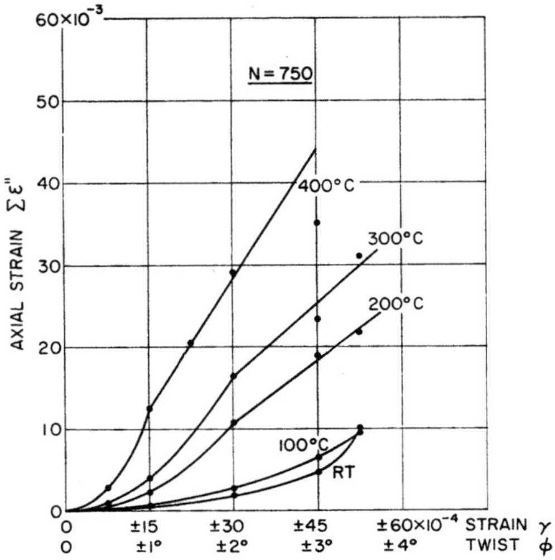

Fig. 23

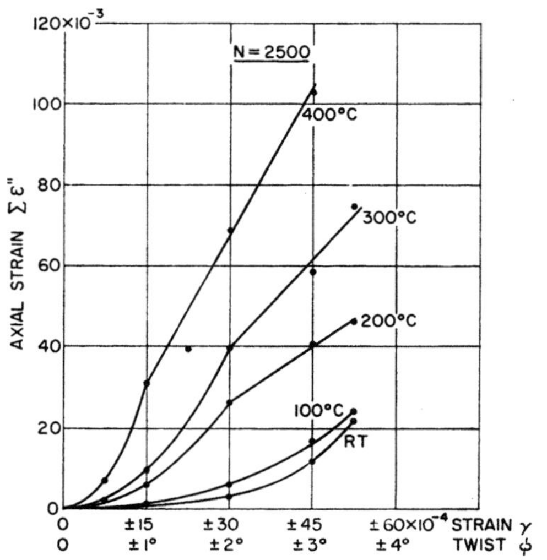

Fig. 24

The significant effect of temperature is thus clearly demonstrated. Studies of the effect of frequency as well as of axial strain recovery at elevated temperatures are now in progress. In the case of the existence of a combined visco-plastic response at elevated temperatures eq. (1) governing the irrecoverable strain in the strain hardening medium will have to be replaced by an equation of the form of eq. (3) by which the dependence of the strain increment on the plastic and viscous response is expressed.

4. Conclusions

Second-order strain accumulation in strain-hardening media seems to be the physically and technically most significant second-order effect. While it is most pronounced and most easily observable in cyclic torsion, it obviously exists in other types of cyclic deformation such as tension and bending where irrecoverable strain accumulation under axial restraint causes apparent "instability effects" which have been reported as such in the literature [24].

References

[1a] TAYLOR, G. I., and H. QUINNEY: Phil. Trans. Roy. Soc. London A **230**, 323–362 (1939).
[1b] HSU, T. C.: J. Basic Engng, Trans. Am. Soc. Mech. Engrs (Ser. D) **88**, 61–70 (1966).
[1c] FEIGEN, M.: Proc. Second U.S. Congress Appl. Mech. 1954. pp. 469–476. New York: Am. Soc. Mech. Engng 1955.
[2] MANDEL, J., P. HABIB, and H. D. BUI: Proc. Fourth Internat. Congres Rheology, 1963, **3**, 97–117. New York: Interscience. 1965.
[3a] IVEY, H.: J. Mech. Engng Sci., **3**, 15–31 (1961).
[3b] NAGHDI, P. M., F. ESSENBURG, and W. J. KORFF: J. Appl. Mech. **25**, 201–209 (1958).
[4] SWIFT, H. W.: Engineering **1947**, 253–257.
[5] L'HERMITE, R., and G. DAWANCE: Circul. Inst. Tech. Batiment, Serie G, No. 16 (1947).
[6a] BENDLER, H. M., and W. A. WOOD: Trans. Met. Soc. AIME **224**, 18 (1962).
[6b] WOOD, W. A., and S. MCK. COUSLAND: Inst. Conf. Creep (London) **1**, Paper 3–25 (1963).
[6c] MOYAR, G. J., and C. M. SINCLAIR: Inst. Conf. Creep (London) **1**, Paper 2–47 (1963).
[7] BENHAM, P. P.: Int. J. Mech. Sci. **7**, 81–86 (1965).
[8] COFFIN, L. F.: J. Basic Engng, Trans. Am. Soc. Mech. Engrs (Ser. D) **82**, 671 to 682 (1960).
[9] FELTNER, C. E.: Acta Met. **11**, 817 (1963).
[10] KADASHEVICH, I. I., and V. V. NOVOZHILOV: Prikl. Math. Mech. **22**, 78–89 (1958).
[11] FREUDENTHAL, A. M., and M. RONAY: Proc. Roy. Soc. (London) A **292**, 14–50 (1966).
[12] POYNTING, J. H.: Proc. Roy. Soc. (London) A **82**, 564–559 (1909); A **86**, 534–561 (1912).
[13] WEISSENBERG, K.: Nature **159**, 310–311 (1947).
[14] REINER, M.: Amer. J. Math. **67**, 350–362 (1945).
[15] REINER, M.: Amer. J. Math. **70**, 433–446 (1948).
[16] REINER, M., and D. ABIR: Proc. IUTAM Symp. on Second-Order Effects, Haifa 1962. New York-London: Pergamon Press. 1964.
[17] LODGE, A. S.: Elastic Fluids. New York: Academic Press. 1964.
[18] RONAY, M.: British J. Appl. Phys. **16**, 727 (1965).
[19] PRAGER, W.: J. Appl. Phys. **16**, 837–840. (1945)
[20] HANDELMAN, G. H., C. C. LIN, and W. PRAGER: Q. Appl. Math. **4**, 397–407 (1947).

[21] RIVLIN, R. S., and J. L. ERICKSON: J. Ratl Mech. Anal. **4**, 323—425 (1955).
[22] ALBRECHT, B., and A. M. FREUDENTHAL: J. Solids and Struct. **1966**.
[23] PIPKIN, A. C., and R. S. RIVLIN: J. Appl. Math. Physics (ZAMP) **16**, 313—327 (1965).
[24] RAYMOND, M. H., and L. F. COFFIN: Trans. Am. Soc. Mech. Engrs **1964**.

Discussion

ONAT: Is the phenomenon you observed in room-temperature tests frequency dependent?

FREUDENTHAL: No.

ONAT: So the name creep is really a misnomer?

FREUDENTHAL: The name "cyclic creep" has been introduced by metallurgists. Since the form of the observed extension-time relation is similar to the form of a creep curve, they call it "creep".

KESTIN: It seems to me that the phenomena in the elastic-plastic domain fit rather well into the theory I described yesterday.

FREUDENTHAL: It is surprising that Prof. KESTIN should make a point that is usually brought up by metallurgist who are unwilling to accept the principle of tensorial invariance as a valid concept and are therefore looking for a geometric mechanism to explain the observed effect. I believe that in this case there is strong support for the assumption that the phenomenon is a second-order effect. We have found that the largest effect is in aluminum and that it is associated with very extensive subgrain formation, not with slip. The material essentially produces the isotropy which it needs in order to follow the principle of material invariance. In cases where subgrain formation is not possible for metallurgical reasons or inhibited, the axial extension is small. And we have also found that materials with high stacking fault energy produce high second-order effects; as the stacking fault energy goes down, the second-order effect decreases since the material approaches more closely the conditions Prof. KESTIN has in mind. In this case of anisotropic deformation the effect could be explained by pure geometry since it is well known that shearing of single crystals produces axial deformation.

ONAT: Would it make sense to make tests with single-crystals?

FREUDENTHAL: Torsion tests on single-crystals have been performed by WOOD and NINE. But in single-crystal tests yield is an anisotropic phenomenon. A shear stress in a general anisotropic medium produces elongations without the need of a second-order effect which is characteristic of the isotropic medium.

DRUCKER: Did you use solid bars and tubes and calculate the difference?

FREUDENTHAL: We have tested both solid bars and tubes. RIVLIN has calculated the difference in extension between elastic solid bars and tubes, and SWIFT has performed tests on both solid bars and tubes. All have shown that tubes produce larger extensions. The reason for not working with tubes consistently is the second-order buckling effect of a similar type as RIVLIN has predicted for elastic tubes and which stops tests on tubes very early (see Fig. 14).

A Thermodynamic Development of Elastic-Plastic Continua

By

A. E. Green and **P. M. Naghdi**

Newcastle-upon-Tyne (England) and Berkeley, Calif. (U.S.A.)

Summary

A general thermodynamical development of elastic-plastic continua for non-polar media is presented in which the plastic strain is not necessarily a symmetric tensor. Alternative, but entirely equivalent, developments of the theory are outlined and some special cases are discussed.

1. Introduction

The classical theory of elastic-plastic media is, for the most part, limited to infinitesimal deformation and is frequently developed without thermodynamical considerations. However, during the past decade, a number of publications have dealt with the thermodynamical aspects of the subject and these reflect the differences (which are partly conceptual) in the existing points of view of a thermodynamical approach to continuum mechanics.

It is the object of this paper to present a development of a nonlinear theory of elastic-plastic continua in which full use is made of the basic thermodynamical equations. This theory, which is exact and complete from both dynamical and thermodynamical points of view, is based on two recent papers by Green and Naghdi [1, 2] and may be referred to as a "rate-type" theory, in contrast to one whose constitutive equations may be obtained via a general functional theory. For background information and related contributions to the subject, we refer the reader to the references cited in [1, 2, 5].

While attention is confined here to non-polar media, the plastic strain tensor (which has the same invariance properties as the total strain tensor) need not necessarily be symmetric, even though the total strain is a symmetric tensor. However, upon removal of all stress components and temperature (in excess of the initial temperature of the medium) in the neighborhood of a point, the symmetric part of the plastic (or permanent) strain must become equal to the total strain. Moreover, as is well-known, only the total strain may be expressed in terms of displacements by the usual kinematical relations.

We use the principle of balance of energy and the Clausius-Duhem inequality. After stating certain properties of the plastic strain and its rate, we begin our analysis with constitutive assumptions for the stress, the Helmholtz free energy, the entropy and the heat flux vector, all of which depend on the deformation gradient (or a suitable invariant measure of strain), the plastic strain, as well as other relevant variables. The constitutive equations for the plastic strain rate (which is independent of the time scale used in computing the rate of change) are constructed with the help of loading criteria, using conventional terminology, and loading (or yield) surfaces, i.e., hypersurfaces in seven-dimensional Euclidean space of symmetric stress and temperature. Using a suitable form of an entropy production inequality, we then deduce explicit thermodynamical restrictions which must hold in the presence of the plastic strain.

Many of the results given here are slight generalizations of our previous work, but detailed proofs are omitted as these are similar to those in [1]. We conclude the present paper with some remarks concerning alternative, but entirely equivalent, developments of the theory of elastic-plastic continua and also discuss briefly some features of the theory in special cases, including the infinitesimal theory.

2. Preliminaries. Basic Equations

In this paper, we find it convenient to use the *direct*, in contrast to the *component* tensor notation. In general, we use lightface Greek lower case letters to designate scalar-valued functions, lightface Latin capital letters for tensors of order two, while vectors and points in Euclidean space will be denoted by boldface Latin lower case letters. Thus, e.g., the deformation gradient $\partial x_m / \partial X_M$ and the strain measure with components e_{KL} used previously [1, 2] will now be denoted by F and E, respectively. When the surface force acting across a surface of the continuum at time t is measured per unit area of the corresponding surface in the reference configuration, the force per unit area in [1] is denoted by π_{Kk}; here, this (non-symmetric) tensor will be designated as[1] T_R. Similarly, we employ T for the (symmetric) Cauchy stress tensor with components σ_{kl}.

---

[1] Often in the literature on continuum mechanics, this stress tensor is defined as the transpose of T_R with components π_{kK}.

We refer the motion of the continuum to a fixed system of rectangular Cartesian axes and designate the position of a typical particle at time τ by a column vector (or 3×1 matrix), $\boldsymbol{x}(\tau) = \{x_m(\tau)\}$, where

$$\boldsymbol{x}(\tau) = \boldsymbol{x}(\boldsymbol{X}, \tau) \qquad (-\infty < \tau \leq t), \tag{2.1}$$

$\boldsymbol{X} = \{X_M\}$ is a reference position of the particle which will be taken as the initial position, and all Latin subscripts have the range 1, 2, 3. We also use the notation

$$\boldsymbol{x} = \boldsymbol{x}(t) \tag{2.2}$$

and, in addition to the usual continuity and differentiability assumptions, we impose the restriction

$$\det [F(\tau)] > 0 \tag{2.3}$$

for physically possible motions. A second motion of the continuum which differs from that specified by (2.1) only by superposed rigid body motions, at different times, is

$$\boldsymbol{x}^*(\tau^*) = \boldsymbol{c}(\tau^*) + Q(\tau)\,[\boldsymbol{x}(\tau) - \boldsymbol{c}(\tau)], \tag{2.4}$$

where $\boldsymbol{c}(\tau)$, $\boldsymbol{c}^*(\tau^*)$ are vector functions of τ and $\tau^*\,(= \tau + a)$, respectively, a is an arbitrary constant, and Q is a proper orthogonal tensor function of τ so that

$$QQ^T = I, \qquad \det Q = 1, \tag{2.5}$$

I being the identity tensor, and Q^T stands for the transpose of Q.

The velocity vector at the point $\boldsymbol{x}(\tau)$ will be denoted by $\boldsymbol{v}(\tau)$, where

$$\boldsymbol{v}(\tau) = \dot{\boldsymbol{x}}(\tau), \qquad \boldsymbol{v} = \dot{\boldsymbol{x}}(t) \tag{2.6}$$

and where a superposed dot denotes differentiation with respect to τ holding $\boldsymbol{X}$ fixed.

In what follows, we reserve "grad" to denote spatial derivatives and use the symbol ∇ for $\nabla(//) \equiv \left\{\frac{\partial}{\partial X_M}(//)\right\}$. Let the displacement gradient at time t, with respect to the reference configuration $\boldsymbol{X}$, be denoted by

$$F = \nabla \boldsymbol{x}(\boldsymbol{X}, t) \tag{2.7}$$

and let

$$L = \dot{F}F^{-1} = \operatorname{grad} \boldsymbol{v}. \tag{2.8}$$

Then, for our present purpose, a suitable measure of deformation is

$$E = \frac{1}{2}(C - I) \tag{2.9}$$

and

$$\dot{E} = F^T D F, \tag{2.10}$$

where C is the right Cauchy-Green tensor and D is the rate of deformation tensor given by

$$C = F^T F, \quad D = \frac{1}{2}[L + L^T]. \tag{2.11}$$

Consider now an arbitrary volume V of the continuum at time t bounded by a surface A and let V_0 (the points of which have coordinates $\boldsymbol{X}$) be the corresponding volume in the reference state, bounded by a surface A_0. Let $\boldsymbol{n}_R$ be the outward unit normal to A_0, let h_0 be the flux of heat across A, measured per unit area of A_0, and let $\boldsymbol{t}_R$ be the surface force which acts across A and is measured per unit area of A_0. Then, we have the well-known results

$$\boldsymbol{t}_R = T_R^T \boldsymbol{n}_R, \qquad h_0 = \boldsymbol{q}_R \cdot \boldsymbol{n}_R, \tag{2.12}$$

where $\boldsymbol{q}_R$ is the heat flux vector measured per unit time and per unit area of A_0, $T_R = \{\pi_{Kk}\}$ is the (non-symmetric) Piola stress tensor and is related to the usual Cauchy stress tensor $T = T^T = \{\sigma_{kl}\}$ by

$$T\,(\det F) = F T_R = T_R^T F^T. \tag{2.13}$$

From the integral form of the principle of balance of energy and the Clausius-Duhem inequality, we can deduce the (local) rate of work equation (or energy equation)

$$\varrho_0 r - \operatorname{div} \boldsymbol{q}_R - \varrho_0 \dot{\varepsilon} + \operatorname{tr}\{T_R L F\} = 0 \tag{2.14}$$

and the inequality

$$\varrho_0 \dot{\eta} \geq -\operatorname{div}\left(\frac{\boldsymbol{q}_R}{\theta}\right) + \varrho_0 \frac{r}{\theta}, \tag{2.15}$$

where ϱ_0 is the mass density of the reference state, ε is the internal energy per unit mass, θ is the temperature (and will be assumed to be always positive), η is the entropy per unit mass, and r is the heat supply function per unit mass. Introducing the Helmholtz free energy through[1]

$$\psi = \varepsilon - \theta\eta, \tag{2.16}$$

[1] Previously, in [1, 2], the internal energy, the free energy, the temperature and entropy were denoted by U, A, T and S, respectively.

then with the use of (2.16) and (2.10), the rate of work Eq. (2.14) may be written as

$$\varrho_0 r - \operatorname{div} \boldsymbol{q}_R - \varrho_0 (\dot{\theta}\eta + \theta\dot{\eta} + \dot{\psi}) + \operatorname{tr}\{S\dot{E}\} = 0, \tag{2.17}$$

where $S = \{s_{KL}\}$ is the symmetric Piola stress tensor defined by[1]

$$\begin{aligned} S &= F^{-1} T_R^T = S^T \\ &= (\det F)\, F^{-1} T (F^T)^{-1} \\ &= T_R (F^T)^{-1}. \end{aligned} \tag{2.18}$$

If we use the rate of work Eq. (2.17) to eliminate r (and div $\boldsymbol{q}_R$) from (2.15), the resulting inequality may be written in the form

$$-\varrho_0 (\dot{\psi} + \eta\dot{\theta}) + \operatorname{tr}\{S\dot{E}\} - \frac{1}{\theta}\, \boldsymbol{q}_R \cdot \nabla\theta \geq 0. \tag{2.19}$$

The kinematical quantities E and $\dot{E}$ defined by (2.9) and (2.10) will be called the strain tensor and the strain rate tensor, respectively. The strain measure E remains unaltered when the continuum is subjected to superposed rigid body motions (2.4) at all times τ^*. We assume that r, ψ, θ, η, $\boldsymbol{q}_R$ and S are all unaltered under superposed rigid body motions so that the left-hand side of the rate of work Eq. (2.17) is then also unaltered. In the next Section, we shall be concerned with constitutive equations (and related thermodynamical restrictions) which involve mainly stress, stress rate, displacement gradients and velocity gradients, and such equations and related thermodynamical restrictions must be unaltered by superposed rigid body motions. The stress and the stress rate will be taken to be S and $\dot{S}$, both being invariant under superposed rigid body motions. We can then show that the displacement gradients and the velocity gradients must be replaced by E and $\dot{E}$, the constitutive equations (and related thermodynamical restrictions) then being relations between E, $\dot{E}$, S, $\dot{S}$, as well as other relevant variables.

3. Elastic-Plastic Continua

Before stating the necessary constitutive assumptions for a complete thermodynamical theory of elastic-plastic continua, it is desirable to discuss first some preliminary aspects of the constitutive assumption

---

[1] The stress tensor S may be regarded as the stress referred to a convected system of coordinates and measured per unit area of the coordinate planes in the reference state.

pertaining to the plastic strain. With reference to the configuration of the continuum at time t, we define at each point of the continuum (with coordinates $\boldsymbol{x}$) a strain tensor E by the relation (2.9). Next, we introduce a tensor $E'' = \{e''_{KL}\}$ through the following properties: (1) It is not necessarily symmetric; (2) it has the same invariance properties as E; (3) it is such that upon removal (not necessarily by unloading) of all stress components and the temperature difference from the initial temperature θ_0 in the neighborhood of a point, the symmetric part of E'' becomes identical with the strain tensor E; and (4) the constitutive equation for the rate of $\dot{E}''$ to be introduced presently [see Eq. (3.8)], is independent of the time scale used to compute the rate of change. We call E'', with the above properties, a "plastic" strain.

For convenience, we define a tensor $E' = \{e'_{KL}\}$ by the relation[1]

$$E' = E - E'', \tag{3.1}$$

and since E is symmetric, it follows that

$$E'_s = E - E''_s, \quad E'_a = -E''_a, \tag{3.2}$$

where

$$E''_s = \frac{1}{2}(E'' + E''^T), \quad E''_a = \frac{1}{2}(E'' - E''^T). \tag{3.3}$$

In view of the properties of the plastic strain stated above, it is clear that $E'_s = 0$ and $E''_s = E$ when $S = 0$ and $\theta = \theta_0$. We introduce a constitutive assumption for E in the form

$$E = E(S, E'', \varkappa, \theta), \tag{3.4}$$

and, in view of (3.2), also write

$$E'_s = E'_s(S, E'', \varkappa, \theta), \tag{3.5}$$

where the scalar function $\varkappa$ (to be specified presently) depends on the whole history of motion. We further assume that for fixed set of values of E'', $\varkappa$ and θ, $\dfrac{\partial E'_s}{\partial S} \neq 0$, i.e., we assume that E'_s is smoothly invertible in S so that there exists a unique inverse of (3.5) in the form

$$S = S(E'_s, E'', \varkappa, \theta). \tag{3.6}$$

[1] The tensor E' can be defined in a number of ways in terms of E and E'', but (3.1) is a convenient definition for our purpose especially in connection with thermodynamics.

Also, we admit the existence of a scalar function $f(S, E'', \theta)$ — called the *yield* or *loading* function — which is continuously differentiable with respect to its arguments and is such that the equation[1]

$$f(S, E'', \theta) = \varkappa \tag{3.7}$$

for a fixed value of $\varkappa$ and E'' represents a hypersurface in seven-dimensional Euclidean space — six components of stress S and one component of temperature[2].

We now postulate additional constitutive assumptions for $\dot{E}''$ and $\dot{\varkappa}$ as follows:

$$\dot{E}'' = \mathcal{G}(S, \dot{S}, E'', \varkappa, \theta, \dot{\theta}), \tag{3.8}$$

together with the requirements that

$$\begin{aligned} &\dot{E}'' = 0 \quad \text{when} \quad f < \varkappa \quad \text{with} \quad \dot{\varkappa} = 0, \\ &\dot{E}'' = 0 \quad \text{when} \quad f = \varkappa \quad \text{with} \quad \dot{\varkappa} = 0 \quad \text{and} \quad \hat{f} < 0, \\ &\dot{E}'' = 0 \quad \text{when} \quad f = \varkappa \quad \text{with} \quad \dot{\varkappa} = 0 \quad \text{and} \quad \hat{f} = 0, \\ &\dot{E}'' \neq 0 \quad \text{when} \quad f = \varkappa \quad \text{and} \quad \hat{f} > 0, \end{aligned} \tag{3.9}$$

and

$$\dot{\varkappa} = K(S, \dot{S}, E'', \dot{E}'', \varkappa, \theta, \dot{\theta}) \tag{3.10}$$

together with the restriction

$$\dot{\varkappa} = 0 \quad \text{when} \quad \dot{E}'' = 0, \tag{3.11}$$

where

$$\hat{f} = tr\left\{\frac{\partial f}{\partial S}\,\dot{S}\right\} + \frac{\partial f}{\partial \theta}\,\dot{\theta}. \tag{3.12}$$

We also add the condition that the constitutive Eqs. (3.8), as well as (3.9), are independent of which time scale is used to compute the rate of change, so that $\mathcal{G}$ (as well as K) is homogeneous of degree one in $\dot{S}$ and $\dot{\theta}$.

---

[1] The inclusion of a non-symmetric tensor in the yield function (effectively similar to E'' in f) has been proposed by Kröner [3]. Kröner's development of plasticity theory, with small deformations in [3], is motivated by consideration of plastic deformation in the context of dislocation in solid state physics.

[2] More fundamentally, instead of $f(S, E'', \theta)$ and (3.7), we may postulate the existence of a function of the form $f(S, F, \theta)$ and the Eq. $f(S, F, \theta) = \varkappa$ and then consider the continuum subjected to superposed rigid body motions. Under such motions, the point in stress and temperature space must still lie on this surface, so that f is invariant under such superposed rigid body motions. It then follows that f reduces to a function of S, E and θ and hence, by (3.4), to the form (3.7).

The reasons for the requirements (3.9) and the background information concerning the Eq. (3.7) are discussed by GREEN and NAGHDI [1]. Using conventional terminology, the family of surfaces (in stress and temperature space) represented by (3.7) are usually called *yield surfaces* or *loading surfaces* and the four conditions in (3.9), in the order listed, correspond to an elastic state, *unloading* from an elastic-plastic state, *neutral loading*, and *loading* from an elastic-plastic state.

In the rest of the paper, we restrict our discussion to the case in which K is a linear function of $\dot{S}$, $\dot{E}''$, $\dot{\theta}$ and $\mathscr{G}$ is a linear function of $\dot{S}$ and $\dot{\theta}$. Since (3.10) must be satisfied for all $\dot{S}$ and $\dot{\theta}$, we now assume that[1]

$$\dot{\varkappa} = tr\,\{H(S, E'', \varkappa, \theta)\,\dot{E}''\}, \tag{3.13}$$

and

$$\dot{E}'' = \mathscr{G} = \mathscr{A}\dot{S} + A\dot{\theta}, \tag{3.14}$$

where H and A are (non-symmetric) second order tensor functions of $S, E'', \varkappa, \theta$; $\mathscr{A}$ is a fourth order tensor function of $S, E'', \varkappa, \theta$ and where $\mathscr{A}\dot{S}$ represents $\{a_{KLMN}\,\dot{S}_{MN}\}$ with $a_{KLMN} = a_{KLNM}$.

We now complete our constitutive assumptions for the development of our thermodynamical theory. Thus, supplementary to the assumptions (3.4) to (3.7) and (3.13) and (3.14) subject to the conditions (3.9) and (3.11), we introduce the following assumptions for the free energy ψ the entropy η and the heat flux vector $\boldsymbol{q}_R$

$$\psi = \psi(E'_s, E'', \varkappa, \theta), \tag{3.15}$$

$$\eta = \eta(E'_s, E'', \varkappa, \theta), \tag{3.16}$$

$$\boldsymbol{q}_R = \boldsymbol{q}_R(E'_s, E'', \varkappa, \theta, \nabla\theta), \tag{3.17}$$

which will be assumed to hold throughout the motion during both loading and unloading.

Since the constitutive Eqs. $(3.9)_3$ and $(3.9)_4$ with $\dot{E}''$ given by (3.14) must be continuous throughout stress and temperature space, it follows that $\mathscr{G}$ must vanish during neutral loading, i.e. when $f = \varkappa$ with $\dot{\varkappa} = 0$ and $\hat{f} = 0$. From this fact we can deduce the result[2]

$$\dot{E}'' = \mathscr{G} = \lambda B\hat{f} \quad \text{when} \quad f = \varkappa, \quad \text{and} \quad \hat{f} > 0, \tag{3.18}$$

[1] In general, $\dot{\varkappa}$ cannot be integrated to give $\varkappa$ as a function of S, E'', θ. Anypart of (3.13) which is integrable as a function of E'' can be absorbed into f in (3.7) and may also be omitted from $\varkappa$ in the other constitutive equations, e.g., (3.4) to (3.6).

[2] For details see [1].

where $\lambda(S, E'', \varkappa, \theta)$ is a scalar and $B(S, E'', \varkappa, \theta)$ is a non-symmetric (second order) tensor function. Without loss of generality, we choose $\lambda > 0$ since at least one component of $\dot{E}''$ must be non-zero during loading. Also, from $\dot{f} = \dot{\varkappa}$ and (3.18), we obtain

$$\lambda\, tr\left\{B\left(H - \frac{\partial f}{\partial E''}\right)\right\} = 1 \tag{3.19}$$

during loading. Thus, the constitutive Eq. (3.18) replaces $(3.9)_4$ during loading and the remaining constitutive equations in (3.9) remain as before. The constitutive Eqs. (3.18) may also be put in the form

$$\dot{E}''_s = \lambda B_s \hat{f}, \qquad \dot{E}''_a = \lambda B_a \hat{f},$$

where $B_s = \frac{1}{2}(B + B^T)$ and $B_a = \frac{1}{2}(B - B^T)$. Similarly, with $H = H_s + H_a$, the left-hand side of (3.19) may be written as the sum of two terms representing its symmetric and antisymmetric parts.

We now proceed to discuss the restrictions which are imposed by the thermodynamical equations. With the help of (3.15), Eqs. (2.17) and (2.19) may be written as

$$\varrho_0 r - \varrho_0\left(\eta + \frac{\partial\psi}{\partial\theta}\right)\dot{\theta} + tr\left\{\left(S - \varrho_0 \frac{\partial\psi}{\partial E'_s}\right)\dot{E}'_s\right\}$$

$$+ tr\left\{\left[S - \varrho_0\left(\frac{\partial\psi}{\partial E''} + \frac{\partial\psi}{\partial\varkappa}H\right)\right]\dot{E}''\right\} - \varrho_0\dot{\eta}\theta - \operatorname{div} \boldsymbol{q}_R = 0, \tag{3.20}$$

and

$$-\varrho_0\left(\eta + \frac{\partial\psi}{\partial\theta}\right)\dot{\theta} + tr\left\{S - \varrho_0 \frac{\partial\psi}{\partial E'_s}\right)\dot{E}'_s\right\}$$

$$+ tr\left\{\left[S - \varrho_0\left(\frac{\partial\psi}{\partial E''} + \frac{\partial\psi}{\partial\varkappa}H\right)\dot{E}''\right\} - \frac{1}{\theta}\boldsymbol{q}_R \cdot \nabla\theta \geq 0 \tag{3.21}$$

both of which hold during loading, as well as unloading. In particular, for motions corresponding to unloading during which $\dot{E}'' = 0$ and $f < \varkappa$, the stress and temperature is any point within the loading surface and we can then deduce from (3.21) the relations[1]

$$\eta = -\frac{\partial\psi}{\partial\theta}, \quad S = \varrho_0 \frac{\partial\psi}{\partial E'_s}, \tag{3.22}$$

---

[1] For details see [1]. The argument leading to (3.22) and (3.23) is similar to that of Coleman and Noll [4]. The second of (3.22) can be used to express E'_s in terms of S, E'', $\varkappa$ and θ. In view of the property of the plastic strain when $S = 0$ and $\theta = \theta_0$, the right-hand side of $(3.22)_2$ must vanish when evaluated at $E'_s = 0$, $\theta = \theta_0$.

as well as the inequality

$$-\boldsymbol{q}_R \cdot \nabla\theta \geq 0. \tag{3.23}$$

Using (3.22), Eqs. (3.20) and (3.21) become

$$\varrho_0 r + tr\left\{\left[S - \varrho_0\left(\frac{\partial\psi}{\partial E''} + \frac{\partial\psi}{\partial\varkappa} H\right)\right]\dot{E}''\right\} - \varrho_0\dot{\eta}\theta - \operatorname{div}\boldsymbol{q}_R = 0, \tag{3.24}$$

and

$$tr\left\{\left[S - \varrho_0\left(\frac{\partial\psi}{\partial E''} + \frac{\partial\psi}{\partial\varkappa} H\right)\right]\dot{E}''\right\} - \frac{1}{\theta}\boldsymbol{q}_R \cdot \nabla\theta \geq 0, \tag{3.25}$$

respectively. Let σ denote the scalar function

$$\begin{aligned}\sigma &= tr\left\{\left[S - \varrho_0\left(\frac{\partial\psi}{\partial E''} + \frac{\partial\psi}{\partial\varkappa} H\right)\right] B\right\} \\ &= tr\left\{\left[S - \varrho_0\left(\frac{\partial\psi}{\partial E''_s} + \frac{\partial\psi}{\partial\varkappa} H_s\right)\right] B_s\right\} - tr\left\{\varrho_0\left(\frac{\partial\psi}{\partial E''_a} + \frac{\partial\psi}{\partial\varkappa} H_a\right) B_a\right\}\end{aligned} \tag{3.26}$$

which may be called the *dissipation* function. Then, by combining (3.18) and (3.25) and remembering that $\lambda > 0$, we may conclude that

$$\lambda\hat{f}\sigma - \frac{1}{\theta}\boldsymbol{q}_R \cdot \nabla\theta \geq 0 \quad \text{when} \quad f = \varkappa \quad \text{and} \quad \lambda\hat{f} > 0 \tag{3.27}$$

during loading, and

$$\sigma \geq 0, \tag{3.28}$$

where (3.28) is obtained from (3.27) by considering an arbitrary homogeneous temperature distribution for which $\nabla\theta = 0$.

The rate of work Eq. (3.24) holds during loading, and during unloading or neutral loading (since $\dot{E}'' = 0$) this reduces to

$$\varrho_0 r - \varrho_0\dot{\eta}\theta - \operatorname{div}\boldsymbol{q}_R = 0. \tag{3.29}$$

We observe that the inequalities (3.23) and (3.28) are independent of rates and are valid both during loading and unloading.

4. Further Remarks

The above development in Section 3, valid for anisotropic elastic-plastic continua, represents the material description of the theory. An alternative spatial description of the general theory, together with the special case of the theory for an elastic-perfectly plastic medium may be discussed in a similar manner and is included in [1]. The elastic-plastic theory of Section 3 slightly generalizes our previous results

[1], where the constitutive assumptions, corresponding to (3.4) to (3.6) and (3.15) to (3.17), were assumed not to depend explicitly on $\varkappa$. The effect of this generalization is reflected in the resulting equations and some of the thermodynamical restrictions. Questions may be raised as to whether or not other variables, in addition to $\varkappa$, should also be included in the constitutive equations. While, in principle, there is no difficulty in the inclusion of additional variables, at present it is difficult to see what these additional variables represent. Again, there is no difficulty in writing (3.10) i. e., the constitutive assumption for $\dot{\varkappa}$, as a functional of the entire history of deformation and temperature. But this would amount to ascribing to a generality far exceeding that of (3.8) for the plastic strain rate.

It is also possible to construct an equivalent theory without the explicit introduction of the difference strain E'. To see this, again let points of the continuum be referred to seven-dimensional Euclidean space of stress and temperature. We begin with the constitutive assumption (3.4) and assume that this is invertible in the form

$$S = S(E, E'', \varkappa, \theta). \tag{4.1}$$

The loading function still depends on S, E'', θ and all constitutive assumptions (3.7) to (3.19) remain the same, except that wherever $E_s{}'$ occurs this should be replaced with E and instead of Eqs. (3.20) and (3.21) we now have

$$\begin{aligned} &\varrho_0 r - \varrho_0 \left(\eta + \frac{\partial \psi}{\partial \theta}\right) \dot{\theta} + tr \left\{\left(S - \varrho_0 \frac{\partial \psi}{\partial E}\right) \dot{E}\right\} \\ &- tr \left\{\varrho_0 \left(\frac{\partial \psi}{\partial E''} + \frac{\partial \psi}{\partial \varkappa} H\right) \dot{E}''\right\} - \varrho_0 \dot{\eta} \theta - \operatorname{div} \boldsymbol{q}_R = 0, \end{aligned} \tag{4.2}$$

and

$$\begin{aligned} &- \varrho_0 \left(\eta + \frac{\partial \psi}{\partial \theta}\right) \dot{\theta} + tr \left\{\left(S - \varrho_0 \frac{\partial \psi}{\partial E}\right) \dot{E}\right\} \\ &- tr \left\{\varrho_0 \left(\frac{\partial \psi}{\partial E''} + \frac{\partial \psi}{\partial \varkappa} H\right) \dot{E}''\right\} - \frac{1}{\theta} \boldsymbol{q}_R \cdot \nabla \theta \geq 0. \end{aligned} \tag{4.3}$$

As in Section 3, we can then deduce $(3.22)_1$ and

$$S = \varrho_0 \frac{\partial \psi}{\partial E}, \tag{4.4}$$

in place of $(3.22)_2$ and finally, instead of the inequalities (3.27) and (3.28), we obtain

$$\begin{aligned} &\lambda \hat{f} tr \left\{-\varrho_0 \left(\frac{\partial \psi}{\partial E''} + \frac{\partial \psi}{\partial \varkappa} H\right) B\right\} - \frac{1}{\theta} \boldsymbol{q}_R \cdot \nabla \theta \geq 0 \\ &\qquad \text{when } f = \varkappa \text{ and } \lambda \hat{f} > 0, \end{aligned} \tag{4.5}$$

and

$$tr \left\{ \left(\frac{\partial \psi}{\partial E''} + \frac{\partial \psi}{\partial \varkappa} H \right) B \right\} \leq 0 . \tag{4.6}$$

Returning to the general theory of Section 3, if it is desired to identify E_s' as purely an "elastic" strain tensor in the usual context of elasticity, i.e., if during unloading, S and η are to be functions of E_s' and θ alone, then the right-hand side of each of the expressions (3.22) must be independent of E'' and $\varkappa$. It then follows that, if the behaviour of the medium during unloading is to be purely elastic, ψ must have the special form

$$\psi = \psi'(E_s', \theta) + \psi''(E'', \varkappa) , \tag{4.7}$$

together with[1]

$$\eta = - \frac{\partial \psi'}{\partial \theta} , \quad S = \varrho_0 \frac{\partial \psi'}{\partial E_s'} , \tag{4.8}$$

and ψ in (3.25) and (3.26) must be replaced with the function ψ''. In certain special cases or some particular application of the theory, the form (4.7) can result in some simplifications.

The infinitesimal theory in which the displacements and their space and time derivatives, as well as other relevant variables [including stress (in suitable non-dimensional form), temperature and their rates], are all "small" of order δ may be deduced as a special case of the general theory of Section 3. If the continuum with zero initial stress is initially homogeneous and in the state of rest at a constant temperature θ_0 and constant entropy η_0, then it is sufficient to express ψ (to the order of approximation considered) by

$$\psi = \psi_1 + \psi_2 + a_{KL} e''_{KL} \theta , \tag{4.9}$$

where

$$\psi_1 = \frac{1}{2} a'_{KLMN} \, e'_{(KL)} \, e'_{(MN)} + a'_{KL} \theta e'_{(KL)} + \frac{1}{2} \alpha' \theta^2 , \tag{4.10}$$

$$\psi_2 = \frac{1}{2} a''_{KLMN} e''_{KL} e''_{MN} + \alpha'' \varkappa .$$

In the above expression, e'_{KL} and e''_{KL} are infinitesimals of the first order so that the total strain tensor

$$e_{KL} = e'_{KL} + e''_{KL} = \frac{1}{2} (u_{K,L} + u_{K,L}) = 0(\delta) , \tag{4.11}$$

[1] During initial loading when $E'' = 0$, the results (4.7) and (4.8) still hold and ψ'' (0,0), a constant, can be absorbed into ψ'.

u_K is the infinitesimal displacement, $e'_{(KL)}$ stands for the symmetric components of e'_{KL}, θ is now the infinitesimal temperature difference from θ_0 and is of the order δ, $\varkappa$ is of the order δ^2, all coefficients a'_{KLMN}, a'_{KL}, α' etc, are constants and the tensor coefficients satisfy the symmetry relations

$$a'_{KLMN} = a'_{LKMN} = a'_{KLNM} = a'_{MNKL}, \tag{4.12}$$
$$a''_{KLMN} = a''_{MNKL}, \quad a'_{KL} = a'_{LK}.$$

As noted previously [1], if the yield surface is sufficiently smooth, then f may be represented by a polynomial of its arguments. Since the yield surface must be a closed surface, the function f cannot contain linear terms and must at least be a quadratic function in the variables S, E'' and θ, each of which is now of the order δ. The infinitesimal theory can be completed by consistent linearization of all equations of the general theory in a manner similar to that in [1], but we omit the details here. It should be noted, however, that the constitutive equation for S in the foregoing infinitesimal theory fulfills the requirement that the components of stress s_{KL} all vanish when $e'_{(KL)} = 0$ and $\theta = \theta_0$. If, in addition, we also require that $\eta = \eta_0$ (the initial value of entropy) when $e'_{(KL)} = 0$ and $\theta = \theta_0$, then the function ψ must have the special form (4.7). In this case, the last term in (4.9) is absent and, for the infinitesimal theory, ψ_1 and ψ_2 represent the counterparts of ψ' and ψ''.

In the rest of this Section, we restrict our attention to the isothermal theory of Section 3 and refer points of the continuum to a six-dimensional Euclidean space of stress S. If the plastic strain, and therefore B, is assumed to be a symmetric tensor, $B_a \equiv 0$ and then (3.28) may be written in the form

$$tr\left\{[S - \bar{S}]B_s\right\} \geq 0, \tag{4.13}$$

where for convenience we have set

$$\bar{S} = \varrho_0 \left(\frac{\partial \psi}{\partial E''_s} + \frac{\partial \psi}{\partial \varkappa} H_s\right). \tag{4.14}$$

Further, if $B = B_s$ in the constitutive Eq. (3.18) is such that

$$B_s = \frac{\partial f}{\partial S}, \tag{4.15}$$

i.e., if the plastic strain rate is assumed to be directed along the normal to the loading surface in stress space, then combination of (4.13) and (4.15) results in

$$tr\left\{[S - \bar{S}]\frac{\partial f}{\partial S}\right\} \geq 0. \tag{4.16}$$

For a given loading surface in stress space and given plastic strain and $\varkappa$, $\bar{S}$ may be regarded as a point in stress space and according to (4.16) the function f is star-shaped with respect to this point. Elsewhere [5], with the limitation to the infinitesimal theory and with reference to the specific example of uniform extension of an elastic-plastic bar, we have shown that if the yield surface is quadratic and independent of mean stress and the Helmholtz function depends only on elastic strain, then we can only have compressible plastic strain without the special assumption (4.15).

Acknowledgement

The results reported here were obtained in the course of research supported by the U. S. Office of Naval Research under Contract Nonr 222(69) with the University of California, Berkeley. One of us (P. M. N.) would like to acknowledge the hospitality of the Mathematics Department of the University of Newcastle upon Tyne, where he held a visiting appointment during May and June 1966.

References

[1] Green, A. E., and P. M. Naghdi: Arch. Ratl Mech. Anal. **18**, 251 and 408 (1965).
[2] Green, A. E., and P. M. Naghdi: Mathematika **12**, 21 (1965).
[3] Kröner, E.: J. Math. and Phys. **42**, 27 (1963).
[4] Coleman, B. D., and W. Noll: Arch. Ratl Mech. Anal. **13**, 167 (1963).
[5] Green, A. E., and P. M. Naghdi: Acta Mechanica **1**, 334 (1965).

Discussion

Kröner: Would you explain once more why you need E'' in the yield condition?

Naghdi: If E'' is not included, the yield function would correspond to that for an elastic-perfectly plastic solid.

Kröner: You have the other information in the $\varkappa$, as I understand it.

Naghdi: Yes, but E'' should also be included in f. I think you are also concerned about the presence of E'' in the free energy. In a general theory, there is no reason why E'' should be excluded from the yield function, free energy, entropy, etc. I really don't know why the remark has been made earlier that the free energy must be a function of the so-called state variables alone.

Perzyna: Would you explain the physical meaning of the assumption that the free energy depends on the inelastic part of the deformation, and would you explain the formal assumption that it is possible to split the free energy into two parts, elastic and plastic?

Naghdi: Your first question is related to Prof. Kröner's.

Perzyna: The second question: is it possible to split the free energy?

Naghdi: I said "if" — I do not suggest that we necessarily do this — if you wish to identify E' as purely an elastic strain in the usual context of nonlinear elasticity, then that statement is equivalent to the requirement that

$$\eta = -\frac{\partial \psi'}{\partial \theta}, \qquad S = \varrho_0 \frac{\partial \psi'}{\partial E'_s},$$

where ψ' is independent of E''.

DRUCKER: When you write E'' as a state variable, you have a two — element — model with various complications. Although admissible, I believe it to be a rather restricted theory.

NAGHDI: While I don't understand what is meant by a "two-element-model", I believe the theory that I have presented here to be less restrictive than most previous theories of "rate type" which are almost all linear theories. Formally there is no difficulty in making some of the constitutive assumptions a functional and therefore having a functional theory. But that would not be consistent with the aim of the present theory which is a "rate type" theory. Earlier this morning the question was raised: is a functional theory of plasticity desirable. Indeed it is. Also, the suggestion was made this morning — I believe Prof. TRUESDELL has made it at various times — that the yield function must drop out of the results. Well, this statement is not quite as startling as it may sound. Of course, if suitable smoothness assumptions are introduced into the constitutive functionals, it may then be possible to deduce the yield function.

BESSELING: I understand, this theory is valid also at large inelastic deformations and for large plastic deformations. In that case the initial reference geometry would not be of any value in determining the final state.

NAGHDI: Other descriptions of the theory are possible. We have done that elsewhere. Here, I have presented a material description of the theory. We have briefly discussed alternative developments of the theory in the text of the paper.

ONAT: It seems to me, that the stress tensor S employed in your theory does not provide complete information on the surface tractions acting on the unit area in the deformed state. To obtain these you must also specify the deformation of the element. Therefore, the form that you assume for your yield condition may not be ideal because S really does not reflect the sort of forces which control the physical phenomena at the instant of interest.

NAGHDI: The development that I have outlined is a material description of the theory in which the stress tensor which naturally occurs in the equations of motion is the non-symmetric Piola stress. But, this stress is related to the symmetric stress tensor S and the deformation gradient F. Since a function — here a yield function — of the non-symmetric Piola stress tensor may be expressed as another function of S and F and since the yield function must remain unaltered when the continuum is subjected to superposed rigid body motions, it follows that the yield function must be a function of S and E. In view of the constitutive assumption for E, the yield function can be reduced further to a function of S and E''.

KLUITENBERG: You have an equation for the rate of plastic flow $\dot{E}'' = \mathscr{A}\dot{S} + A\dot{\theta}$, you have also an expression for the entropy production. Now I want to substitute the rate equation into the entropy production equation. Then I get an expression which is linear in $\dot{S}$.

NAGHDI: That's right.

KLUITENBERG: If I now change the time which I have for a certain value of $\dot{S}$ found, then —

NAGHDI: But I have already assumed that the constitutive equation for $\dot{E}''$ is independent of which time scale is used to calculate the rate of change, so that the function $\dot{E}''$ is homogeneous of degree one in the variables $\dot{S}$ and $\dot{\theta}$.

Generalized Continuum Mechanics

By

A. E. Green and R. S. Rivlin

Newcastle-upon-Tyne (England) and Bethlehem, Pa. (U.S.A.)

Summary

A continuum-mechanical theory is presented in which the deformation is described by a number of generalized coordinate fields. The applied force system is described by conjugate generalized force fields distributed throughout the volume of the body and on the surface. These arise as the coefficients of the rates of change of the generalized forces in an expression for the rate at which work is done by the force system acting on the body. The differential equations of the theories are obtained by the systematic use of the first and second laws of thermodynamics, together with invariant-theoretical considerations. The theory is motivated by a particle model in which each particle consists of a number of mass-points. It is seen that starting with a given model there is a considerable ambiguity in the choice of the generalized coordinates describing the deformation and hence in the form of the resulting theory.

1. Introduction

In classical continuum mechanics, the deformation of a body is described by a single vector field — the displacement field. The manner in which a rigid motion changes the components, in a fixed rectangular cartesian coordinate system, of this vector field is, of course, implied by the geometrical meaning of displacement.

In the last few years there has been a flurry of interest in continuum-mechanical theories in which the deformation of a body is described not only by the vector displacement field but also by other vector or tensor fields. This interest was stimulated largely by the papers of Ericksen (1960, a, b, c, 1961), Mindlin and Tiersten (1962), Mindlin (1964), Toupin (1962) and others and by some remarks in the article by Truesdell and Toupin (1960). Green and Rivlin (1964) developed a rather general theory in which the deformation of a deformable body was described by a number of deformation fields[1]. One of these was given

[1] See also Eringen and Suhubi (1964a, b), Eringen (1964), and the comments by Green (1965) on these papers.

the properties of the usual displacement field while the others were tensor fields of various orders. This theory was called *multipolar theory.* In a more recent paper GREEN and RIVLIN (1967) have discussed two further theories — *director theory* and *multiple displacement theory.* In both of these the deformation is described by a number of vector fields, but these are assumed to behave differently under rigid body motion in the two theories. In this paper (GREEN and RIVLIN 1967) the director and multiple displacement theories are related to each other and it is shown that the earlier multipolar theory may be regarded as a special case of the director theory.

In developing the field equations of their theories, GREEN and RIVLIN rely mainly on the first and second laws of thermodynamics and on invariance conditions arising from the assumed behavior, under rigid body motion, of the vector or tensor fields describing the deformation.

In the present paper a somewhat more general point-of-view is adopted, although the approach taken is similar to that of the earlier papers (GREEN and RIVLIN, 1964, 1967). The deformation is described by a number of generalized coordinate fields. As special cases these may be the components of the vectors employed in the director or multiple displacement theories, or of the tensors employed in the multipolar theory.

In § 1 of the present paper, we consider a discrete system of particles, each of which may as a special case be a point-particle, but is here regarded more generally as a deformable element or a system of point-particles. The configuration of each of these particles is described by a finite[1] number of generalized coordinates. The rate at which the external forces acting on the system of particles do work is assumed to be a linear function of the rates of change of the generalized coordinates, the coefficients in which are the generalized forces. The kinetic energy of the system is assumed to be a quadratic function of the rates of change of the generalized coordinates, the coefficients in this expression being themselves functions of the generalized coordinates.

The first and second laws of thermodynamics are written down for the system of particles and the transition to the continuum is effected by replacing the summation over the particles by integration over the domain which they occupy. This procedure parallels, in a general sense, one procedure for passing from a system of point-particles to the usual continuum mechanics in which the deformation is described by the usual displacement field alone.

These expressions of the first and second laws of thermodynamics are applied to an elementary tetrahedron and in this way the generalized

[1] The extension to the case when the number is infinite is trivial and is omitted.

stresses and heat flux vector at a point are defined and related to the generalized force and heat flux respectively on a surface at the point of arbitrary orientation. We use these relations in the original integral expressions for the first and second laws of thermodynamics and apply the resulting equations to an arbitrary infinitesimal element. In this way we obtain relations which express in differential form the validity of the first and second laws.

We then write down formally the implications of the fact that the superposition on the assumed deformation of a rigid translational velocity, without changing instantaneously the values of the generalized coordinates, leaves the heat flow, internal energy and entropy unchanged. The relations so obtained cannot be developed further without making some assumption regarding the manner in which the rigid translational velocity affects the rates of change of the generalized coordinates. The way in which this is done in various previous theories is sketched.

Up to this point the argument applies equally to reversible and irreversible deformations. In the case when the material is elastic, we can proceed further. We do so here only for isothermal deformations.

It should be noted that in the replacement of summation over the particles by integration over the domain which they occupy, the generalized forces acting on the particles become a body-force field and a surface-force field. The presence of the latter implies that the particles in a thin layer near the boundary are replaced, in the continuum model, by the geometrical surface of the domain occupied by the continuum, the forces acting on the particles in this layer becoming the surface forces. These particles will also have associated with them kinetic and internal energies and entropy. Strictly, we should therefore introduce, into the expressions of the first and second laws of thermodynamics for the continuum, surface kinetic and internal energies and entropy. These terms are omitted in the present paper, but we hope to introduce them in a later work. It may be hoped that their introduction will help to classify the nature of the boundary conditions which are appropriate in continuum-mechanical theories involving generalized deformation and force fields.

It may be noted also that the passage from a system of particles to a continuum model can be effected only if the generalized displacements and applied forces vary but little as we pass from one particle to its neighbors. This implies that the resulting generalized continuum theory will be applicable to situations in which the generalized forces and displacements vary but slightly over distances comparable with those of the structural elements of the material, but will provide more detailed information regarding the deformation of these structural elements than does the usual continuum theory. This implication raises some question as to the applicability of theories of this type to such problems as the determination of stress distributions near a hole in a polycrystalline material of diameter comparable with the crystallite size, cf. MINDLIN and TIERSTEN (1962). It may, however, be that some other way of passing to the continuum theory will negate this comment.

2. The Mechanics of a System of Deformable Particles

We consider a system in motion consisting of N sub-systems which we shall call *particles*. We assume that the configuration of the Pth particle may be described by ν generalized coordinates $q_\alpha^{(P)} (\alpha = 1, \ldots, \nu)$ which are functions of time. We shall assume that the kinetic energy $\mathscr{T}^{(P)}$ of the Pth particle is given by a homogeneous quadratic expression in the time derivatives $\dot{q}_\alpha^{(P)}$ of the generalized coordinates, the coefficients in this expression being functions of the generalized coordinates themselves, but not depending explicitly on the time[1]. Thus,

$$\mathscr{T}^{(P)} = \frac{1}{2} f_{\alpha\beta}^{(P)} \dot{q}_\alpha^{(P)} \dot{q}_\beta^{(P)}, \quad f_{\alpha\beta}^{(P)} = f_{\alpha\beta}^{(P)} (q_\gamma^{(P)}). \tag{2.1}$$

Here and throughout the paper, the Einstein summation convention with range $1, \ldots, \nu$ applies to repeated Greek subscripts.

The kinetic energy $\mathscr{T}$ of the system of N particles is therefore given by

$$\mathscr{T} = \sum_{P=1}^{N} \mathscr{T}^{(P)} = \frac{1}{2} \sum_{P=1}^{N} f_{\alpha\beta}^{(P)} \dot{q}_\alpha^{(P)} \dot{q}_\beta^{(P)}. \tag{2.2}$$

We may, without loss of generality, take $f_{\alpha\beta} = f_{\beta\alpha}$.

---

[1] The motivation for this assumption can be seen if we consider each particle (i.e. sub-system) to consist of χ point-particles with masses $m^{(\lambda)}$ $(\lambda = 1, \ldots \chi)$ and vector positions $\mathbf{r}^{(\lambda)}$ with respect to a fixed origin. The kinetic energy of the particle is then $\frac{1}{2} \sum_{\lambda=1}^{\chi} m^{(\lambda)} \dot{\mathbf{r}}^{(\lambda)} \cdot \dot{\mathbf{r}}^{(\lambda)}$. The vectors $\mathbf{r}^{(\lambda)}$ are, in general, vector functions of the generalized coordinates $q_\alpha (\alpha = 1, \ldots \nu)$ and of time t. The kinetic energy of the particle is then

$$\frac{1}{2} \sum_{\lambda=1}^{\chi} m^{(\lambda)} \left(\frac{\partial \mathbf{r}^{(\lambda)}}{\partial q_\alpha} \dot{q}_\alpha + \frac{\partial \mathbf{r}^{(\lambda)}}{\partial t} \right) \cdot \left(\frac{\partial \mathbf{r}^{(\lambda)}}{\partial q_\beta} \dot{q}_\beta + \frac{\partial \mathbf{r}^{(\lambda)}}{\partial t} \right).$$

If the generalized coordinates are chosen in such a way that $\mathbf{r}^{(\lambda)}$ does not depend explicitly on t, the kinetic energy of the particle becomes

$$\frac{1}{2} \sum_{\lambda=1}^{\chi} m^{(\lambda)} \frac{\partial \mathbf{r}^{(\lambda)}}{\partial q_\alpha} \cdot \frac{\partial \mathbf{r}^{(\lambda)}}{\partial q_\beta} \dot{q}_\alpha \dot{q}_\beta.$$

We assume, in agreement with the usual assumptions of analytical mechanics, that the rate $\mathscr{R}$ at which work is done by the system of forces acting on the particle system is given by an expression of the form

$$\mathscr{R} = \sum_{P=1}^{N} F_\alpha^{(P)} \dot{q}_\alpha^{(P)}. \tag{2.3}$$

$F_\alpha^{(P)}$ $(\alpha = 1, \ldots, \nu)$ are called the components of the generalized force acting on the Pth particle.

Applying the first law of thermodynamics to the system, we have

$$\mathscr{R} = \dot{\mathscr{T}} + \dot{\mathscr{U}} - \mathscr{Q}, \tag{2.4}$$

where $\mathscr{Q}$ is the rate at which heat enters the system and $\mathscr{U}$ is the internal energy of the system.

Introducing (2.2) and (2.3) into (2.4), we obtain

$$\sum_{P=1}^{N} F_\alpha^{(P)} \dot{q}_\alpha^{(P)} = \frac{1}{2}\frac{d}{dt} \sum_{P=1}^{N} f_{\alpha\beta}^{(P)} \dot{q}_\alpha^{(P)} \dot{q}_\beta^{(P)} + \dot{\mathscr{U}} - \mathscr{Q}. \tag{2.5}$$

We consider now that, at some reference time t_0, the particles occupy a volume V bounded by a surface A. Then, instead of identifying the particles by labelling them $1, 2, \ldots, N$, we may identify them by their coordinates X_B $(B = 1, 2, 3)$ in some rectangular cartesian coordinate system X at time t_0. We assume that throughout V (except possibly in a thin layer at the surface of V), $F_\alpha^{(P)}$ and $q_\alpha^{(P)}$ vary sufficiently slowly, as we pass from one particle to its neighbors, that we may replace summations over these particles by volume integrations. We also assume that in a surface layer at A, $F_\alpha^{(P)}$ and $q_\alpha^{(P)}$ vary sufficiently slowly, as we pass from one particle in the layer to its neighbors in the layer, that we may replace summations over the particles in this layer by surface integrations.

With these assumptions, we make the replacements

$$\mathscr{R} = \sum_{P=1}^{N} F_\alpha^{(P)} \dot{q}_\alpha^{(P)} = \int_V F_\alpha \dot{q}_\alpha \, dV + \int_A F_\alpha^* \dot{q}_\alpha \, dA, \tag{2.6}$$

where F_α and F_α^* are functions of position in V and on A, respectively, and of time, and

$$\dot{\mathscr{T}} = \frac{1}{2}\frac{d}{dt} \int_V f_{\alpha\beta}(q_\gamma) \dot{q}_\alpha \dot{q}_\beta \, dV + \frac{1}{2}\frac{d}{dt} \int_A f_{\alpha\beta}^*(q_\gamma) \dot{q}_\alpha \dot{q}_\beta \, dA. \tag{2.7}$$

F_α $(\alpha = 1, \ldots, \nu)$ are the components of the *generalized body force field*, per unit volume measured at time t_0, and F_α^* $(\alpha = 1, \ldots, \nu)$ are the components of the *generalized surface force,* per unit area measured at time t_0. The first term on the right-hand side of Eq. (2.7) is the rate of change of the volume kinetic energy and the second term is the rate of change of the surface kinetic energy. We shall, in order to simplify matters, assume that the surface kinetic energy is zero, so that

$$\dot{\mathcal{T}} = \frac{1}{2} \frac{d}{dt} \int_V f_{\alpha\beta} (q_\gamma) \dot{q}_\alpha \dot{q}_\beta \, d V . \tag{2.8}$$

3. The First Law of Thermodynamics for a Body

Using the expressions (2.6) and (2.8) for the rate at which the forces do work and the kinetic energy, respectively, in the first law of thermodynamics (2.4), we obtain

$$\int_V F_\alpha \dot{q}_\alpha \, d V + \int_A F^*_\alpha \dot{q}_\alpha \, d A = \frac{1}{2} \frac{d}{dt} \int_V f_{\alpha\beta} \dot{q}_\alpha \dot{q}_\beta d V + \dot{\mathcal{U}} - \mathcal{Q}, \tag{3.1}$$

where $\mathcal{U}$ is the internal energy of the body and $\mathcal{Q}$ is the rate at which heat enters it.

Instead of writing down the first law of thermodynamics for the system of particles and then passing to the continuum, we may take the generalized coordinate fields q_α as primitive concepts and define generalized body force and surface force fields through the rate of work expression (2.6) and kinetic energy by an expression of the form (2.8). The particle model is introduced here solely to provide motivation.

We now consider that heat may enter the body in two ways — by generation throughout the volume[1] and by flow through the surface. Thus,

$$\mathcal{Q} = \int_V R d V - \int_A Q d A , \tag{3.2}$$

where R is the heat generation per unit volume and Q is the rate at which heat flows out of the surface, per unit area measured at time t_0.

We also consider that the internal energy $\mathcal{U}$ may be expressed as a volume integral, the surface internal energy being zero. We then have

$$\mathcal{U} = \int_V U d V, \tag{3.3}$$

---

[1] This process was introduced, in a less general context, by TRUESDELL and TOUPIN (1960) and employed by COLEMAN and NOLL (1963).

where U is the internal energy per unit volume measured at time t_0. Introducing (3.2) and (3.3) into (3.1), we obtain

$$\int_V F_\alpha \dot{q}_\alpha \, dV + \int_A F^*_\alpha \dot{q}_\alpha \, dA = \int_V \left[\frac{d}{dt}\left(\frac{1}{2} f_{\alpha\beta} \dot{q}_\alpha \dot{q}_\beta\right) + \dot{U} - R\right] dV + \int_A Q dA. \tag{3.4}$$

4. The Second Law of Thermodynamics for a Body

We may express the second law of thermodynamics for the body, regarded as a continuum, by means of the Clausius—Duhem inequality

$$\dot{\mathscr{S}} \geq \int_V \frac{R}{\theta} \, dV - \int_A \frac{Q}{\theta} \, dA, \tag{4.1}$$

where $\mathscr{S}$ is the total entropy of the body and $\theta (> 0)$ is the absolute temperature at a generic point of the body.

Again, as in the case of the internal energy, we assume that the entropy $\mathscr{S}$ may be expressed as a volume integral thus:

$$\mathscr{S} = \int_V S \, dV, \tag{4.2}$$

where S is the entropy per unit volume, measured at time t_0. With (4.2), the Clausius—Duhem inequality (4.1) becomes

$$\int_V \dot{S} \, dV \geq \int_V \frac{R}{\theta} \, dV - \int_A \frac{Q}{\theta} \, dA. \tag{4.3}$$

5. The First and Second Laws of Thermodynamics in Differential Form

We now apply the relations (3.4) and (4.3) to an elementary tetrahedron, which at time t_0 has three of its bounding planes parallel to the coordinate planes of a rectangular cartesian coordinate system X. The unit normal to its fourth bounding plane has components N_A in the system X at time t_0.

Let $\Pi_{A\alpha}$ be the generalized force at time t acting on the surface of the tetrahedron which is normal to the X_A-axis at time t_0, per unit area measured at time t_0. Let Q_A be the rate at which heat flows across this surface. Then, applying Eq. (3.4) to this elementary tetrahedron, we obtain

$$(F^*_\alpha - \Pi_{A\alpha} N_A) \, \dot{q}_\alpha - (Q - Q_A N_A) = 0. \tag{5.1}$$

Applying the inequality (4.3) to the same tetrahedron we obtain, bearing in mind that θ is positive,

$$Q - Q_A N_A \geq 0. \tag{5.2}$$

We have preserved the notation F^*_α and Q for the surface force acting on the slant surface of the tetrahedron and the heat flux through it. Since we must replace Q by $-Q$ if we replace N_A by $-N_A$, the equality sign in (5.2) is applicable and we have

$$Q - Q_A N_A = 0. \tag{5.3}$$

With (5.1), we obtain

$$(F^*_\alpha - \Pi_{A\alpha} N_A)\, \dot{q}_\alpha = 0. \tag{5.4}$$

Employing (5.3) and (5.4) to substitute for Q and $F^*_\alpha \dot{q}_\alpha$ in (3.4), we obtain, using the divergence theorem,

$$\int_V \left\{(F_\alpha + \Pi_{A\alpha,A})\dot{q}_\alpha - \frac{d}{dt}\left(\frac{1}{2} f_{\alpha\beta}\dot{q}_\alpha\dot{q}_\beta\right)\right\} dV$$
$$+ \int_V \Pi_{A\alpha}\dot{q}_{\alpha,A}\, dV = \int_V (\dot{U} - R + Q_{A,A})\, dV. \tag{5.5}$$

Equation (5.5) is valid whether or not V is taken as the volume of the whole body or part of the body. Accordingly, we may take V to be an infinitesimal volume in the body and obtain

$$(F_\alpha + \Pi_{A\alpha,A})\dot{q}_\alpha - \frac{d}{dt}\left(\frac{1}{2} f_{\alpha\beta}\dot{q}_\alpha\dot{q}_\beta\right) + \Pi_{A\alpha}\dot{q}_{\alpha,A}$$
$$= \dot{U} - R + Q_{A,A}. \tag{5.6}$$

We introduce the notation

$$T = \frac{1}{2} f_{\alpha\beta}\dot{q}_\alpha\dot{q}_\beta. \tag{5.7}$$

Then, it can easily be seen that

$$\dot{T} = \left[\frac{d}{dt}\left(\frac{\partial T}{\partial \dot{q}_\alpha}\right) - \frac{\partial T}{\partial q_\alpha}\right]\dot{q}_\alpha. \tag{5.8}$$

With (5.7) and (5.8), we may re-write (5.6) as

$$L_\alpha \dot{q}_\alpha + \Pi_{A\alpha}\dot{q}_{\alpha,A} = \dot{U} - R + Q_{A,A'} \tag{5.9}$$

where

$$L_\alpha = F_\alpha + \Pi_{A\alpha,A} - \left[\frac{d}{dt}\left(\frac{\partial T}{\partial \dot{q}_\alpha}\right) - \frac{\partial T}{\partial q_\alpha}\right]. \tag{5.10}$$

Again, introducing (5.3) into (4.3), using the divergence theorem and applying the resulting inequality to an infinitesimally small volume element, we obtain

$$\theta \dot{S} - R + Q_{A,A} - \frac{1}{\theta} Q_A \theta_{,A} \geq 0. \tag{5.11}$$

We now use Eq. (5.9) to eliminate $R - Q_{A,A}$ from (5.11) and obtain

$$-\dot{U} + \theta \dot{S} + L_\alpha \dot{q}_\alpha + \Pi_{A\alpha} \dot{q}_{\alpha,A} - \frac{1}{\theta} Q_A \theta_{,A} \geq 0. \tag{5.12}$$

6. Invariance under Rigid Translations

We now assume that the motion of the body is changed by the superposition of an arbitrary, constant, rigid-body, translational velocity. Let the generalized coordinates for the new deformation be $\bar{q}_\alpha$ and we suppose the superposed motion to be such that $\bar{q}_\alpha = q_\alpha$ at time t. We also consider that L_α, $\Pi_{A\alpha}$, Q_A, U, R are unchanged by the superposed constant rigid-body translational velocity. Then, we obtain, from (5.9),

$$L_\alpha(\dot{\bar{q}}_\alpha - \dot{q}_\alpha) + \Pi_{A\alpha}(\dot{\bar{q}}_{\alpha,A} - \dot{q}_{\alpha,A}) = 0. \tag{6.1}$$

Assuming further that F^*_α is unchanged by the superposed constant rigid-body translational velocity, we obtain, from (5.4),

$$(F^*_\alpha - \Pi_{A\alpha} N_A)(\dot{\bar{q}}_\alpha - \dot{q}_\alpha) = 0. \tag{6.2}$$

7. Some Special Cases

In order to proceed further, we must make some assumption regarding the relation between $\dot{\bar{q}}_\alpha$ and $\dot{q}_\alpha$ and the form of $f_{\alpha\beta}$ in the expression (2.8) for the rate of change of kinetic energy. This has been done previously in a number of ways.

(i) In classical continuum mechanics $\nu = 3$ and $q_\alpha (\alpha = 1, 2, 3)$ may be taken as the components of a vector field x_α in the rectangular cartesian coordinate system X. x_α are, of course, the coordinates of the various points of the body at time t. Then, writing

$$q_\alpha = x_\alpha \quad \text{and} \quad \bar{q}_\alpha = \bar{x}_\alpha, \tag{7.1}$$

we have

$$\dot{\bar{x}}_\alpha = \dot{x}_\alpha + c_\alpha, \tag{7.2}$$

where c_α is the superposed velocity. We assume also that

$$f_{\alpha\beta} = \varrho \delta_{\alpha\beta}, \tag{7.3}$$

where ϱ is the density at time t_0.

Introducing (7.1) and (7.2) into (6.2) and bearing in mind that c_α may be chosen arbitrarily, it is easy to see that

$$F^*_\alpha = \Pi_{A\alpha} N_A. \tag{7.4}$$

Introducing (7.2) and (7.3) into (6.1), we obtain in a similar manner, with (5.7) and (5.10),

$$L_\alpha = \Pi_{A\alpha,A} + F_\alpha - \varrho \ddot{x}_\alpha = 0 \quad (\alpha = 1, 2, 3). \tag{7.5}$$

Equation (7.5) is the usual Kirchoff—Piola equation of motion and (7.4) the usual force boundary condition of classical continuum mechanics.

Using (7.1), (7.2), (7.3) and (7.5) in (5.9), we obtain

$$\Pi_{A\alpha} \dot{x}_{\alpha,A} = \dot{U} - R + Q_{A,A}. \tag{7.6}$$

The inequality (5.12) becomes

$$(-\dot{U} + \theta \dot{S}) + \Pi_{A\alpha} \dot{x}_{\alpha,A} - \frac{1}{\theta} Q_A \theta_{,A} \geq 0. \tag{7.7}$$

(ii) Green and Rivlin (1967) have recently developed a theory in which the generalized coordinates q_α are taken to be the coordinates $x_i (i = 1, 2, 3)$, in the rectangular cartesian coordinate system X, employed in classical continuum mechanics and the components in this system of μ vectors $u_i^{(\alpha)} (\alpha = 1, \ldots, \mu)$.

Denoting the generalized body forces and surface forces corresponding to $u_i^{(\alpha)}$ by $F_i^{(\alpha)}$ and $F_i^{*(\alpha)}$ respectively and those corresponding to x_i by F_i and F_i^* respectively, with corresponding meanings for Π_{Ai} and $\Pi_{Ai}^{(\alpha)}$, we may re-write Eq. (6.2) as

$$(F_i - \Pi_{Ai} N_A)(\dot{\bar{x}}_i - \dot{x}_i) + \sum_{\alpha=1}^{\mu} (F_i^{*(\alpha)} - \Pi_{Ai}^{(\alpha)} N_A)(\dot{\bar{u}}_i^{(\alpha)} - \dot{u}_i^{(\alpha)}) = 0. \tag{7.8}$$

The kinetic energy T, per unit volume measured at time t_0, was taken in the form

$$T = \frac{1}{2} a_{ij} \dot{x}_i \dot{x}_j + \frac{1}{2} \sum_{\alpha=1}^{\mu} b_{ij}^{(\alpha)} \dot{x}_i \dot{u}_j^{(\alpha)} + \frac{1}{2} \sum_{\alpha,\beta=1}^{\mu} c_{ij}^{(\alpha\beta)} \dot{u}_i^{(\alpha)} \dot{u}_j^{(\beta)}, \tag{7.9}$$

where a_{ij}, $b_{ij}^{(\alpha)}$ and $c_{ij}^{(\alpha\beta)}$ are constants. It is assumed in the theory of Green and Rivlin (1967) that a rigid time-independent rotation Q_{ij} superposed on the assumed deformation changes the components $u_i^{(\alpha)}$ to $Q_{ij}\, u_j^{(\alpha)}$. Since the kinetic energy is unaltered by this rotation, it follows that a_{ij}, $b_{ij}^{(\alpha)}$ and $c_{ij}^{(\alpha\beta)}$ must have the forms

$$a_{ij} = a\,\delta_{ij}, \quad b_{ij}^{(\alpha)} = b^{(\alpha)}\,\delta_{ij} \quad \text{and} \quad c_{ij}^{(\alpha\beta)} = c^{(\alpha\beta)}\,\delta_{ij}, \tag{7.10}$$

where δ_{ij} is the Kronecker delta and a, $b^{(\alpha)}$ and $c^{(\alpha\beta)}$ are constants. The expression (7.9) for the kinetic energy per unit volume then becomes

$$\frac{1}{2}\,a\dot{x}_i\dot{x}_i + \frac{1}{2}\sum_{\alpha=1}^{\mu} b^{(\alpha)}\dot{x}_i\dot{u}_i^{(\alpha)} + \frac{1}{2}\sum_{\alpha,\beta=1}^{\mu} c^{(\alpha\beta)}\dot{u}_i^{(\alpha)}\dot{u}_i^{(\beta)}. \tag{7.11}$$

We can, of course, take $c^{(\alpha\beta)} = c^{(\beta\alpha)}$, without loss of generality. Introducing this form for the kinetic energy into (5.10), and making the replacements for $q_\alpha\,(\alpha = 1, \ldots, \nu)$ indicated, we obtain from the energy balance eq. (5.9)

$$\begin{aligned}&\left(F_i + \Pi_{Ai,A} - a\ddot{x}_i - \frac{1}{2}\sum_{\alpha=1}^{\mu} b^{(\alpha)}\ddot{u}_i^{(\alpha)}\right)\dot{x}_i \\ &+ \sum_{\alpha=1}^{\mu}\left(F_i^{(\alpha)} + \Pi_{Ai,A}^{(\alpha)} - \sum_{\beta=1}^{\mu} c^{(\alpha\beta)}\ddot{u}_i^{(\beta)} - \frac{1}{2}\,b^{(\alpha)}\ddot{x}_i\right)\dot{u}_i^{(\alpha)} \\ &+ \Pi_{Ai}\dot{x}_{i,A} + \sum_{\alpha=1}^{\mu}\Pi_{Ai}^{(\alpha)}\dot{u}_{i,A}^{(\alpha)} = \dot{U} - R + Q_{A,A}\,.\end{aligned} \tag{7.12}$$

In a similar manner, Eq. (6.1) becomes

$$\begin{aligned}&\left(F_i + \Pi_{Ai,A} - a\ddot{x}_i - \frac{1}{2}\sum_{\alpha=1}^{\mu} b^{(\alpha)}\ddot{u}^{(\alpha)}\right)(\dot{\bar{x}}_i - \dot{x}_i) \\ &+ \sum_{\alpha=1}^{\mu}\left(F_i^{(\alpha)} + \Pi_{Ai,A}^{(\alpha)} - \sum_{\beta=1}^{\mu} c^{(\alpha\beta)}\ddot{u}_i^{(\beta)} - \frac{1}{2}\,b^{(\alpha)}\ddot{x}_i\right)(\dot{\bar{u}}_i^{(\alpha)} - \dot{u}_i^{(\alpha)}) \\ &+ \Pi_{Ai}\,(\dot{\bar{x}}_{i,A} - \dot{x}_{i,A}) + \sum_{\alpha=1}^{\mu}\Pi_{Ai}^{(\alpha)}\,(\dot{\bar{u}}_{i,A}^{(\alpha)} - \dot{u}_{i,A}^{(\alpha)}) = 0\,.\end{aligned} \tag{7.13}$$

Further deductions from (7.12) and (7.13), as developed by Green and Rivlin (1967), depend on the manner in which it is assumed that a uniform, rigid, constant velocity c_i superposed on the assumed deformation affects $\dot{u}_i^{(\alpha)}$, i.e. on how $\dot{\bar{u}}_i^{(\alpha)}$ is related to $\dot{u}_i^{(\alpha)}$.

In the director theory it is assumed that

$$\dot{\bar{u}}_i^{(\alpha)} = \dot{u}_i^{(\alpha)}. \tag{7.14}$$

In the multiple displacement theory, on the other hand, it is assumed that the vectors $u_i^{(\alpha)}$ behave under translation in precisely the same way as does x_i; i.e.

$$\dot{\bar{u}}_i^{(\alpha)} = \dot{u}_i^{(\alpha)} + c_i . \tag{7.15}$$

It follows from (7.13) that in the director theory

$$F_i + \Pi_{Ai,A} - a\ddot{x}_i - \frac{1}{2} \sum_{\alpha=1}^{\mu} b^{(\alpha)} \ddot{u}_i^{(\alpha)} = 0 . \tag{7.16}$$

With (7.12), we then obtain

$$\sum_{\alpha=1}^{\mu} \left(F_i^{(\alpha)} + \Pi_{Ai,A}^{(\alpha)} - \sum_{\beta=1}^{\mu} c^{(\alpha\beta)} \ddot{u}_i^{(\beta)} - \frac{1}{2} b^{(\alpha)} \ddot{x}_i \right) \dot{u}_i^{(\alpha)} + \Pi_{Ai} \dot{x}_{i,A}$$
$$+ \sum_{\alpha=1}^{\mu} \Pi_{Ai}^{(\alpha)} \dot{u}_{i,A}^{(\alpha)} = \dot{U} - R + Q_{A,A} . \tag{7.17}$$

In the multiple displacement theory it follows from (7.13) that

$$\left(F_i + \Pi_{Ai,A} - a\ddot{x}_i - \frac{1}{2} \sum_{\alpha=1}^{\mu} b^{(\alpha)} \ddot{u}_i^{(\alpha)} \right)$$
$$+ \sum_{\alpha=1}^{\mu} \left(F_i^{(\alpha)} + \Pi_{Ai,A}^{(\alpha)} - \sum_{\beta=1}^{\mu} c^{(\alpha\beta)} \ddot{u}_i^{(\beta)} - \frac{1}{2} b^{(\alpha)} \ddot{x}_i \right) = 0 . \tag{7.18}$$

With (7.12), we then obtain

$$\sum_{\alpha=1}^{\mu} \left(F_i^{(\alpha)} + \Pi_{Ai,A}^{(\alpha)} - \sum_{\beta=1}^{\mu} c^{(\alpha\beta)} \ddot{u}_i^{(\beta)} - \frac{1}{2} b^{(\alpha)} \ddot{x}_i \right) (\dot{u}_i^{(\alpha)} - \dot{x}_i)$$
$$+ \Pi_{Ai} \dot{x}_{i,A} + \sum_{\alpha=1}^{\mu} \Pi_{Ai}^{(\alpha)} \dot{u}_{i,A}^{(\alpha)} = \dot{U} - R + Q_{A,A} \tag{7.19}$$

Returning to the system of particles discussed in § 2, we can see the motivation for the various choices of the generalized coordinates so far discussed. In the case of classical continuum mechanics, we consider the particles to be point-particles so that the configuration of the system is completely described by their coordinates in a three-dimensional rectangular cartesian coordinate system. The kinetic energy of the system of particles is then the sum of the usual kinetic energies for a number of point-particles.

In the director and multiple displacement theories each particle may be a system of point-particles. In the director theory the generalized coordinates for a particle may be taken, for example, as the coordinates, in a rectangular cartesian coordinate system, of one of the constituent point-particles and the components in this system of the vectors joining the remaining point-particles to this point-particle[1]. In the multiple displacement theory the generalized coordinates may be taken as the coordinates, in a rectangular cartesian system, of the constituent point-particles.

8. Elasticity Theory

We consider only the case when the deformation takes place isothermally, i.e. $\dot{\theta} = 0$, and the temperature is uniform throughout the body, i.e. $\theta_{,A} = 0$. Then, the inequality (5.12) can be re-written

$$-\dot{A} + \Pi_{A\alpha}\dot{q}_{\alpha,A} + L_\alpha \dot{q}_\alpha \geq 0, \tag{8.1}$$

where A is the Helmholtz free-energy, per unit volume measured at time t_0, defined by

$$A = U - \theta S. \tag{8.2}$$

We now make the assumption that A is a function of q_α and its spatial gradients only, thus:

$$A = A(q_\alpha, q_{\alpha,A}). \tag{8.3}$$

We obtain

$$\dot{A} = \frac{\partial A}{\partial q_\alpha}\dot{q}_\alpha + \frac{\partial A}{\partial q_{\alpha,A}}\dot{q}_{\alpha,A}. \tag{8.4}$$

The inequality (8.1) then becomes

$$\left(-\frac{\partial A}{\partial q_\alpha} + L_\alpha\right)\dot{q}_\alpha + \left(-\frac{\partial A}{\partial q_{\alpha,A}} + \Pi_{A\alpha}\right)\dot{q}_{\alpha,A} \geq 0. \tag{8.5}$$

We now make the constitutive assumption that L_α and $\Pi_{A\alpha}$ depend only on q_α and $q_{\alpha,A}$ and that $\dot{q}_\alpha$ and $\dot{q}_{\alpha,A}$ $(\alpha = 1, \ldots, \nu;\ \ A = 1, 2, 3)$ can be varied independently. We then obtain, from (8.5),

$$L_\alpha = \frac{\partial A}{\partial q_\alpha}, \quad \Pi_{A\alpha} = \frac{\partial A}{\partial q_{\alpha,A}}. \tag{8.6}$$

We note that if A is independent of one of the generalized coordinates, q_1 say, we obtain from (8.6),

$$L_1 = 0. \tag{8.7}$$

---

[1] Alternatively, the generalized coordinates for a particle may be taken as the cartesian coordinates of its center of mass and the components in the cartesian coordinate system of the joins of the constituent point-particles to this center of mass. We note that in this case the generalized coordinates for a particle are related.

In a similar manner, we obtain from eq. (5.4) the relation

$$F_\alpha^* = \Pi_{A\alpha} N_A, \tag{8.8}$$

with the assumption that F_α^* depends only on q_α, $q_{\alpha,A}$ and N_A.

Acknowledgement

This work was supported, as far as one of the authors (R. S. RIVLIN) was concerned, under Office of Naval Research, U. S. Navy, Contract Nonr 562(40) with Brown University.

References

[1] COLEMAN, B. D., and W. NOLL: Arch. Ratl Mech. Anal. **13**, 167 (1963).
[2] ERICKSEN, J. L.: Arch. Ratl Mech. Anal. **4**, 231 (1960a).
[3] — Kolloid-Z. **173**, 117 (1960b).
[4] — Trans. Soc. Rheol. **4**, 29 (1960c).
[5] — Trans. Soc. Rheol. **5**, 23 (1961).
[6] ERINGEN, A. C.: Int. J. Engng Sci. **2**, 205 (1964).
[7] — and E. S. SUHUBI: Int. J. Engng Sci. **2**, 189 (1964a).
[8] — — Int. J. Engng Sci. **2**, 389 (1964b).
[9] GREEN, A. E.: Int. J. Engng Sci. **3**, 533 (1965).
[10] — and R. S. RIVLIN: Arch. Ratl Mech. Anal. **17**, 113 (1964).
[11] — — ZAMP **18**, 208 (1967).
[12] MINDLIN, R.: Arch. Ratl Mech. Anal. **16**, 51 (1964).
[13] — and H. F. TIERSTEN: Arch. Ratl Mech. Anal. **11**, 415 (1962).
[14] TOUPIN, R.: Arch. Ratl Mech. Anal. **11**, 385 (1962).
[15] TRUESDELL, C. A., and R. TOUPIN: The Classical Field Theories. Handbuch der Physik. Vol. III. Berlin-Göttingen-Heidelberg: Springer. 1960.

Discussion

ONAT: I have a feeling that there may be some restrictions on the q_α in the case of n finite. What would be the nature of such restrictions if you go to a continuum.

RIVLIN: In general the only restriction which one must impose on the q_α is that implied by the consideration that two different material elements cannot be in the same place at the same time. I don't see that this would introduce any serious problem.

ONAT: It does! I feel that without restrictions, it would be too rich a theory.

Phenomenological Aspects of Creep Irreversibility

By

Jan Hult

Gothenburg (Sweden)

Summary

Creep irreversibility is discussed from a structural point of view. The partial strain recovery following upon unloading of a hyperstatic structure from a state of nonlinear creep deformation is here attributed to two causes: *isostatic recovery*, being an inherent material property and *hyperstatic recovery*, being a result of transient stress redistribution between various parts of the structure. The magnitude of hyperstatic recovery strain is studied for a few structural models. It is found to be at most of the order of the elastic strain, and may therefore be disregarded in most cases where isostatic recovery is present. In such cases recovery in various structures may profitably be described by the hereditary theory due to Rabotnov. Some aspects of this theory are finally discussed.

Introduction

The concept of irreversibility means different things to different people. Physicists in recent time have found the study of irreversible thermodynamic processes very rewarding, and the result of that activity is bound to have large influence on the future development of solid mechanics. Mechanical engineers, working with structural design problems, are likely to attach a more limited meaning to the word. It will mostly lead their thoughts to the deformation response of an engineering structure being exposed to a prescribed loading sequence.

Although seemingly different, these attitudes may often lead in the same direction. All solid materials can be looked upon as aggregates of small structural elements, such as molecules, crystal grains, etc. The behaviour of such materials *per se* may therefore be expected to show similarities to the behaviour of those aggregates of other, larger elements, which make up various engineering structures. Knowledge of structural irreversibility on a macroscopic scale may therefore aid the understanding of the phenomenon on a microscopic scale.

Engineering mechanics comprises the study of various structures built from various hypothetical materials. The elastic materials, linear or nonlinear, are fully reversible, whereas the rigid plastic material is fully irreversible. In elastic-plastic materials a certain fraction of the total strain is recovered instantaneously upon unloading. This causes the creation of residual stresses in hyperstatic structures; stresses which may disappear again in continued cyclic loading (shake down). This kind of irreversible strain response may be utilized to improve the functioning of engineering structures; examples of this are the autofrettage of gun barrels and the running of turbine disks at overspeed in order to create a state of residual stresses, counteracting the most dangerous stresses arising during service.

The viscoelastic materials are also characterized by irreversible strain response, but of a different nature. In contrast to plastic deformation, which is time independent and fully irreversible, viscoelastic deformation is time dependent and partly reversible. Hence in most real materials displaying creep the response to unloading is a gradual recovery of a certain fraction of the previously developed creep strain.

Creep recovery in a hyperstatic structure may be attributed to two causes: (1) the relaxation of residual (micro)-stresses being present in the microstructure after unloading; (2) the relaxation of residual (macro)-stresses arising upon unloading from a state of nonlinear creep. Since the first kind of recovery is present already in an isostatic structure, such as an ordinary creep tensile test specimen, it will be termed *isostatic recovery*. The second kind, being a characteristic of the hyperstatic structure itself, will be termed *hyperstatic recovery*.

In linearly viscoelastic structures, loaded by constant surface tractions, no hyperstatic recovery occurs upon unloading, provided the material is homogeneous and isotropic. The isostatic recovery in linearly viscoelastic materials is accurately described by the established theory.

Most structural metals employed in engineering designs, however, display nonlinear creep in a certain temperature range, and here the situation is different as regards recovery. Even though isostatic recovery is nearly always observed after unloading from a creep test, most of the phenomenological theories of metallic creep, which have come into practical use, disregard the isostatic recovery phenomenon altogether. One notable exception is the hereditary theory due to RABOTNOV (1948), which predicts isostatic recovery as a consequence of the non-stationary rate of creep. Since the amount of hyperstatic recovery is at most of the same order as the elastic strain, isostatic recovery as a rule is the dominating of the two.

In Fig. 1 is shown the deformation response for Maxwell and Norton type structures upon stepwise loading and unloading. In both these materials isostatic recovery is absent, whereas hyperstatic recovery (δ_r), as well as hyperstatic primary creep (δ_p) develops for the nonlinear Norton material. The magnitudes of δ_r and δ_p will now be examined for the case of a simple slightly hyperstatic structure, where the analysis may be performed without much difficulty.

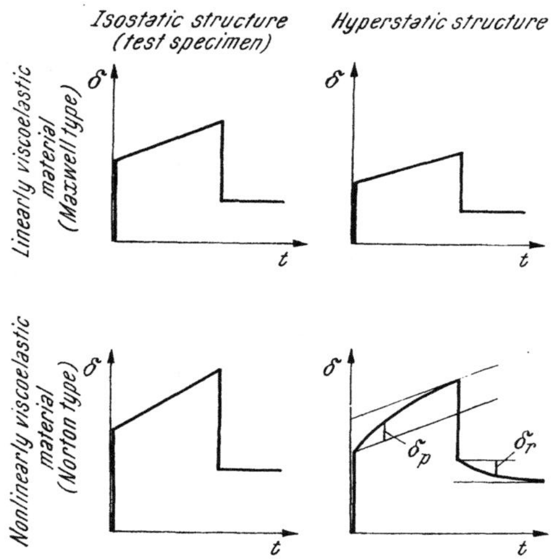

Fig. 1. Deformation response following stepwise loading and unloading on viscoelastic structures

Elementary Model

Consider the structure shown in Fig. 2. Two parallel straight bars, indexed 1 and 2, are joined to a rigid block, which is constrained to move parallel to the bars. The lengths and cross sectional areas of the bars are

$$\left.\begin{aligned} L_1 &= L(1-\vartheta) \qquad & L_2 &= L(1+\vartheta) \\ A_1 &= A(1-\Theta) \qquad & A_2 &= A(1+\Theta) \end{aligned}\right\} \tag{1}$$

where L denotes the average length and A the average cross sectional area.

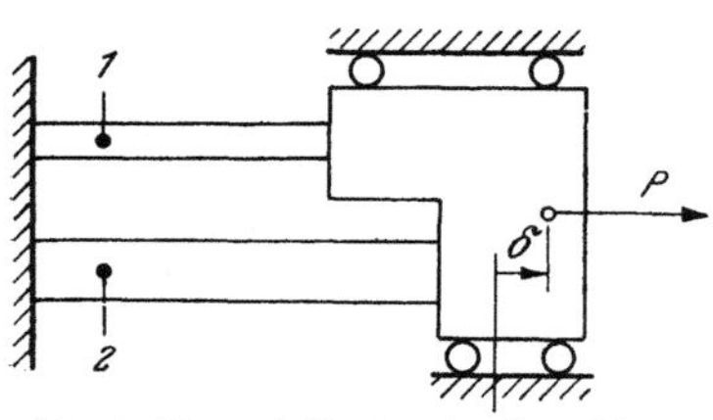

Fig. 2. Hyperstatic structural model

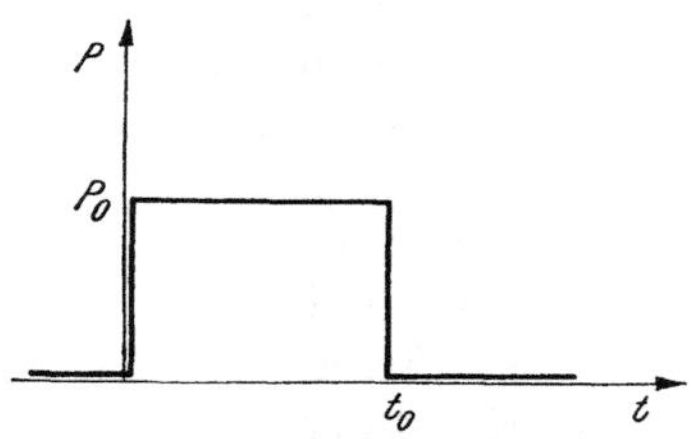

Fig. 3. Load history producing hyperstatic creep and recovery

The system will be subjected to the load history shown in Fig. 3, i.e.

$$\left.\begin{aligned} &t < 0; & &P = 0 \\ &0 \leq t < t_0; & &P = P_0 \text{ (creep period)} \\ &t_0 \leq t; & &P = 0 \text{ (recovery period)} \end{aligned}\right\}. \tag{2}$$

(a) Creep Period

Equilibrium requires

$$\sigma_1 A(1-\Theta)+\sigma_2 A(1+\Theta)=P_0. \tag{3}$$

Denoting by $\bar{\sigma}=P_0/2A$ the average stress in the structure, equilibrium is always maintained if the bar stresses are written as

$$\left.\begin{aligned}\sigma_1&=\bar{\sigma}(1+T)/(1-\Theta)\\ \sigma_2&=\bar{\sigma}(1-T)/(1+\Theta)\end{aligned}\right\} \tag{4}$$

where $T=T(t)$ remains to be determined.

Geometrical compatibility requires

$$\delta=\varepsilon_1 L(1-\vartheta)=\varepsilon_2 L(1+\vartheta), \tag{5}$$

where δ denotes the displacement of the block.

The bars are assumed to consist of a Hooke-Norton material, defined by the constitutive equation

$$\dot{\varepsilon}=\dot{\sigma}/E+k\sigma^n;\quad n\geq 1 \tag{6}$$

where the dot denotes differentiation with respect to time t, and where E denotes Young's modulus and k and n are material parameters characterizing the creep behaviour. Introduction of the transformed dimensionless time unit

$$t'=Ek\bar{\sigma}^{n-1}t \tag{7}$$

yields the following constitutive equations for the two bars

$$\left.\begin{aligned}\varepsilon_1' E/\bar{\sigma}&=(\sigma_1/\bar{\sigma})'+(\sigma_1/\bar{\sigma})^n\\ \varepsilon_2' E/\bar{\sigma}&=(\sigma_2/\bar{\sigma})'+(\sigma_2/\bar{\sigma})^n\end{aligned}\right\} \tag{8}$$

where the prime denotes differentiation with respect to time t'.

From eqs. (4), (5), and (8) follows the differential equation for $T=T(t')$

$$T'\left(\frac{1+\vartheta}{1+\Theta}+\frac{1-\vartheta}{1-\Theta}\right)=(1+\vartheta)\left(\frac{1-T}{1+\Theta}\right)^n-(1-\vartheta)\left(\frac{1+T}{1-\Theta}\right)^n. \tag{9}$$

The initial state, indexed 0, is obtained by replacing (6) by Hooke's law

$$\varepsilon=\sigma/E, \tag{10}$$

since creep strain requires time to develop. From eqs. (4), (5), and (10) follows

$$T_0=(\vartheta-\Theta)/(1-\Theta\vartheta) \tag{11}$$

and hence the initial condition supplementing the differential eq. (9) is $T(0) = T_0$.

From the nature of the problem follows that a stationary state of constant stress distribution and creep rate will be asymptotically approached. Denoting this state T_∞, eq. (9) yields with $T'_\infty = 0$

$$T_\infty = \left[1 - \frac{1+\Theta}{1-\Theta}\left(\frac{1-\vartheta}{1+\vartheta}\right)^n\right] \Big/ \left[1 + \frac{1+\Theta}{1-\Theta}\left(\frac{1-\vartheta}{1+\vartheta}\right)^n\right]. \tag{12}$$

From eqs. (11) and (12) follows that $T_0 \equiv T_\infty$ if $n = 1$ or $\vartheta = 0$. Hence the creep rate will be constant for all times (no hyperstatic primary creep) if the material is of the Maxwell type or if the two bars are of equal length. If $n > 1$ and $\vartheta \neq 0$, a transient state of creep will develop, as shown by Fig. 1. The study of this state is much simplified if $\vartheta \ll 1$ and $\Theta \ll 1$. The first of these conditions may be said to imply that the model is only slightly hyperstatic. Under these assumptions it follows from eqs. (11) and (12), that

$$|T_0| \ll 1, \quad |T_\infty| \ll 1,$$

and hence it is concluded that $|T| \ll 1$ throughout. The differential eq. (9) may then be written as

$$T' + \alpha T = \beta - \frac{1}{2}\beta n (n-1) T^2 + O(T^3), \tag{13}$$

where

$$\alpha = n(C + D), \quad \beta = C - D, \tag{14}$$

with

$$C = (1+\vartheta)(1-\Theta^2)/2(1-\vartheta\Theta)(1+\Theta)^n;$$
$$D = (1-\vartheta)(1-\Theta^2)/2(1-\vartheta\Theta)(1-\Theta)^n. \tag{15}$$

Neglecting first the T^2-term, the solution is found to be

$$T = \beta/\alpha + (T_0 - \beta/\alpha)\exp(-\alpha t'). \tag{16}$$

Insertion of this first approximation into the right hand member of the differential eq. (13) yields the second approximation

$$\begin{aligned} T = \beta/\alpha &- \frac{1}{2}(\beta^3/\alpha^3)\, n(n-1) - \\ &- (\beta^2/\alpha^2)\, n(n-1)(T_0 - \beta/\alpha)\,\alpha t' \exp(-\alpha t') + \\ &+ \frac{1}{2}(\beta/\alpha)\, n(n-1)(T_0 - \beta/\alpha)^2 \exp(-2\,\alpha t') + \\ &+ \vartheta\,[(n-1)/n]\exp(-\alpha t'). \end{aligned} \tag{17}$$

The corresponding strains may now be obtained from eqs. (4), (8), and (17). There results after some manipulation

$$\begin{aligned}\varepsilon_1 E/\bar{\sigma} = {} & (1+\vartheta)/(1-\vartheta\Theta) + \left[1 + nC + \frac{1}{2} n(n-1) C^2\right] t'/(1-\Theta)^n + \\ & + [nD/\alpha(1-\Theta)^n - D/(1-\Theta) + \\ & + n(n-1) CD/\alpha(1-\Theta)^n] [1 - \exp(-\alpha t')] + \\ & + [n(n-1) D^2/4\alpha(1-\Theta)^n] [1 - \exp(-2\alpha t')] \end{aligned} \tag{18}$$

where α, C, and D are as given by eqs. (14) and (15). The lapse of strain versus time is shown by Fig. 4. The development of hyperstatic primary

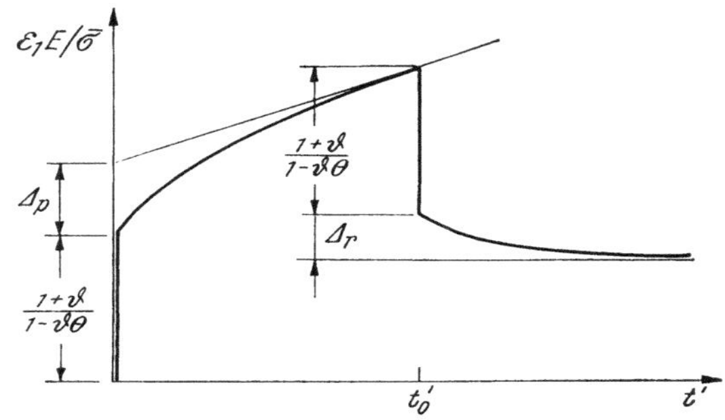

Fig. 4. Strain response of hyperstatic two bar model in Fig. 2 to load history of Fig. 3

creep in this type of structure was first indicated by Hoff (1957). The quantity Δ_p, being the asymptotic amount of hyperstatic primary creep is found from the last two terms in eq. (18)

$$\Delta_p = (n-1)^2 (n+3) \vartheta^2/4n^2. \tag{19}$$

The neglected terms in eqs. (17), (18) and (19) are all of third order in ϑ and Θ.

(b) Recovery Period

Equilibrium now requires, cf. eq. (3)

$$\sigma_1 A(1-\Theta) + \sigma_2 A(1+\Theta) = 0, \tag{20}$$

which is automatically satisfied by

$$\left.\begin{aligned}\sigma_1 &= -\bar{\sigma} R/(1-\Theta) \\ \sigma_2 &= \bar{\sigma} R/(1+\Theta).\end{aligned}\right\} \tag{21}$$

$\bar{\sigma}$ being as before $P_0/2A$, while $R = R(t)$ remains to be determined.

The geometrical compatibility requirement (5) still holds, whereas the constitutive relations (8) are somewhat modified due to the possible occurence of negative stresses

$$\left.\begin{aligned}\varepsilon_1' E/\bar{\sigma} &= (\sigma_1/\bar{\sigma})' + |\sigma_1/\bar{\sigma}|^n \operatorname{sgn}(\sigma_1/\bar{\sigma}) \\ \varepsilon_2' E/\bar{\sigma} &= (\sigma_2/\bar{\sigma})' + |\sigma_2/\bar{\sigma}|^n \operatorname{sgn}(\sigma_2/\bar{\sigma})\end{aligned}\right\}. \tag{22}$$

From eqs. (8), (21), and (22) follows the differential equation for $R = R(t')$, valid when $t' \geq t_0'$

$$R' = -\varkappa_n R^n \tag{23}$$

where

$$\varkappa_n = [(1-\vartheta)/(1-\Theta)^n + (1+\vartheta)/(1+\Theta)^n]/[(1-\vartheta)/(1-\Theta) + + (1+\vartheta)/(1+\Theta)]. \tag{24}$$

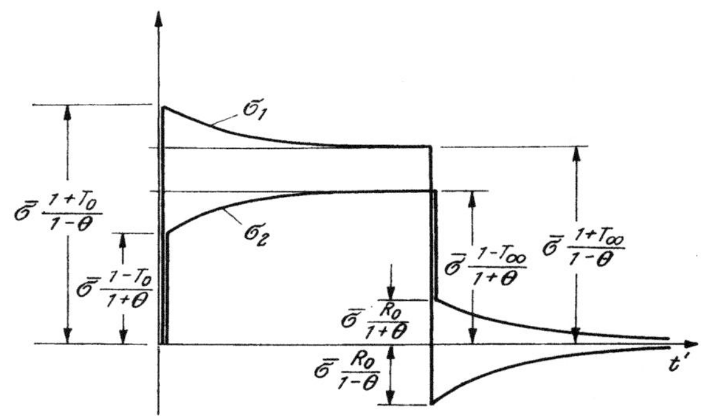

Fig. 5. Stress response of hyperstatic two bar model in Fig. 2 to load history of Fig. 3

The magnitude of R at t_0' may be found with sufficient accuracy from the stationary creep state T_∞ (Fig. 5). It follows, considering that the instantaneous change of state at t_0' is linearly elastic

$$R(t_0') = \vartheta(n-1)/n \equiv R_0. \tag{25}$$

Hence the solution to the differential eq. (23) may be written

$$R = [R_0^{1-n} + (n-1)\varkappa_n(t' - t_0')]^{1/(1-n)}. \tag{26}$$

The corresponding strains may now be obtained from eqs. (21), (22), and (26). The quantity Δ_r, being the asymptotic amount of hyperstatic recovery as defined by Fig. 4, is finally found to be

$$\Delta_r = (n-1)^2 \vartheta\Theta/n. \tag{27}$$

Comparison with eq. (19) shows, that hyperstatic primary creep and hyperstatic recovery are *not* identical, as has sometimes been suggested from a presumed analogy with the behaviour of linearly viscoelastic materials, where isostatic primary creep is identical to isostatic recovery. It should be noted that in this particular case, if $\Theta = 0$ i.e. the two bars have equal areas but different lengths, then $\Delta_p \neq 0$ while $\Delta_r = 0$. Hence the presence of hyperstatic primary creep does not even necessarily imply the presence of hyperstatic recovery. It should also be noted that $\Delta_p = \Delta_r = 0$ if $n = 1$, i.e. no hyperstatic effects appear in a Maxwell type structure.

Other Structures

Hyperstatic primary creep has been studied for various engineering structures by several authors. The analysis is greatly simplified when the structure is only slightly hyperstatic, such as in I-beams, moderately thickwalled pressure vessels, etc., cf. Hult (1962), Rosengren (1963). In all such cases the magnitude of hyperstatic primary creep strain has been found to be at most of the order of the elastic strains present.

Hyperstatic recovery, on the other hand, does not seem to have attracted as much attention. One reason for this might be that the mathematical difficulties involved are as a rule much larger. This is because stress reversals normally occur during the recovery period, cf. Sonnerup (1965). The particular problems involved will here be indicated for the case of an I-beam subject to recovery after a state of nonlinear bending creep deformation.

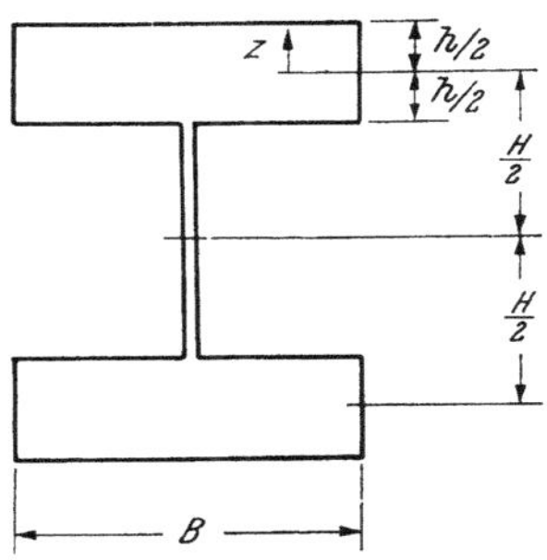

Fig. 6. Thin web I-beam

Consider a beam with cross section according to Fig. 6, Denoting by $\varkappa$ the curvature of the center line, the strain at fibre z in the upper flange equals

$$\varepsilon = \varkappa(H/2 + z). \tag{28}$$

With $\vartheta = h/H \ll 1$, $\zeta = 2z/h$, and $\varepsilon_0 = \varkappa H/2$ this transforms into

$$\varepsilon(\zeta, t) = \varepsilon_0(t) \cdot (1 + \vartheta\zeta) \tag{29}$$

whereas the following equation

$$\int_{-1}^{+1} \sigma(\zeta, t) \cdot (1 + \vartheta\zeta)\, d\zeta = 0 \tag{30}$$

expresses that the resulting bending moment in the unloaded state equals zero. Finally the constitutive eq. (6) is assumed to hold.

Immediately after unloading from a stationary state of creep the stress distribution is found to be

$$\sigma(\zeta, 0)/m = -(n-1)\,\zeta\vartheta/n - \\ -(n-1)\,(3\zeta^2 - 2n - 1)\,\vartheta^2/6n^2 + O(\vartheta^3\zeta^3) \tag{31}$$

which, of course, is self equilibrating according to eq. (30). Here $m = M/BHh$, where M denotes the bending moment causing the previous creep deformation. The stress destribution during the recovery period is therefore written as

$$\sigma(\zeta, t)/m = F(t)\vartheta^2 - 3G(t)\vartheta\zeta - 3H(t)\vartheta^2\zeta^2 + O(\vartheta^3\zeta^3) \tag{32}$$

where F, G, and H are to be determined. The equilibrium condition (30) first requires

$$F = G + H. \tag{33}$$

Finally the constitutive eq. (6) is satisfied at $\zeta = 0$ and $\zeta = \pm 1$, if

$$\left.\begin{aligned} &(G' + H')\vartheta^2 + (G + H)^n\vartheta^{2n} = \varepsilon_0' E/m, \\ &(G' + H')\vartheta^2 \mp 3G'\vartheta - 3H'\vartheta^2 + \\ &\quad + [(G + H)\vartheta^2 \mp 3G\vartheta - 3H\vartheta^2]^n = (1 \pm \vartheta)\,\varepsilon_0' E/m, \end{aligned}\right\} \tag{34}$$

where a prime denotes differentiation with respect to the new transformed time unit

$$t' = Ekm^{n-1}t. \tag{35}$$

It is found that $|H| \ll |F|, |G|$ if $n \gg 1$, which expresses that the stress distribution is nearly linear. In this case eqs. (34) degenerate into

$$G' + (3\vartheta)^{n-1}G^n = 0 \tag{36}$$

with the initial condition $G(0) = (n-1)/3n$. The strain ε_0 and hence the curvature $\varkappa$ may then be determined during the recovery period, and the result is as shown by Fig. 7. In particular

$$\varkappa_p/\varkappa_0 = (n-1)^2\,(n+3)\,\vartheta^2/12n^2 \tag{37}$$

$$\varkappa_r/\varkappa_0 \simeq (n-1)\,\vartheta^2/3n \tag{38}$$

where the last relation is valid only if $n \gg 1$. The formal similarity between eqs. (19) and (37) suggests similar results for other slightly hyperstatic structures. The corresponding expressions for hyperstatic recovery, eqs. (27) and (38), however, show formal differences. It is concluded again, that the hyperstatic recovery process is intrinsically more dependent on the structural geometry than the hyperstatic primary creep process.

It is also concluded that hyperstatic primary creep strain as well as hyperstatic recovery strain is small, not exceeding the order of elastic strain. Hence in all cases where the material itself displays measurable isostatic recovery, the hyperstatic recovery effect may be neglected. A study by LECKIE and MARTIN (1966) based on bounds for the energy dissipation seems to indicate that this conclusion has a rather general validity.

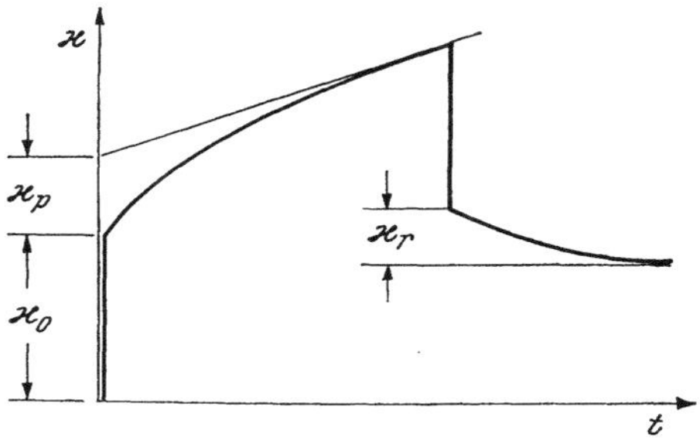

Fig. 7. Curvature response of I-beam in Fig. 6

The Rabotnov Theory of Isostatic Creep Recovery

On examining a large number of creep curves for various materials and constant stress levels RABOTNOV (1948) found that the corresponding isochronous stress-strain curves may be brought to coincide with one master curve, e.g. the rapid tensile test curve, by means of a simple transformation. Hence if the tensile test curve is described by

$$\varphi(\varepsilon) = \sigma \tag{39}$$

the required transformation may be written as

$$\varphi(\varepsilon) = [1 + G(t)]\sigma \tag{40}$$

where $G(0) = 0$. This representation is valid only when σ is constant throughout the creep process. By means of a Boltzmann type integral RABOTNOV generalized the transformation (40) to be valid also for arbitrarily varying stresses

$$\varphi(\varepsilon) = \sigma(t) + \int_0^t K(t-\tau)\,\sigma(\tau)\,d\tau. \tag{41}$$

From eqs. (40) and (41) follows the relation

$$G(t) = \int_0^t K(t-\tau)\,d\tau. \tag{42}$$

Creep problems may now be studied by examining the development in the φ-ε-plane, corresponding to the σ-ε-plane in the theories of elasticity and time independent plasticity, cf. Fig. 8.

One consequence of this representation is that isostatic creep recovery is predicted. For the loading sequence in Fig. 8 the amount of recovery strain at time $t = t_0 + t_s$ is found to be

$$\varepsilon_r = \sigma_0[G(t_0) + G(t_s) - G(t_0 + t_s)]/E. \tag{43}$$

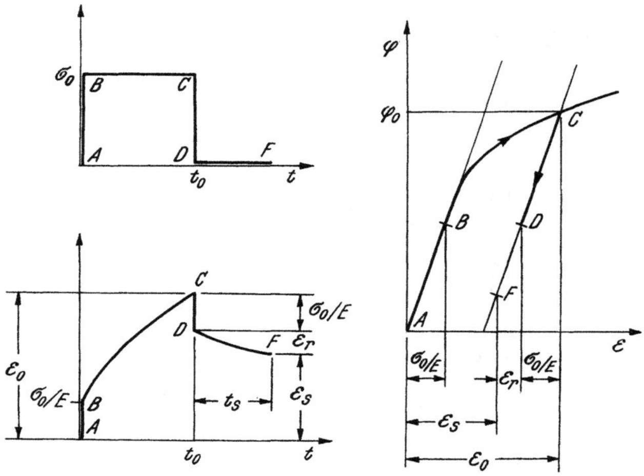

Fig. 8. Creep and recovery representation according to RABOTNOV

From this relation some conclusions may be derived regarding possible forms for the function $G(t)$. It follows that $\varepsilon_r \equiv 0$, if $G(t) = ct$ where c is a constant. Hence isostatic recovery is a consequence of $G(t)$ being a nonlinear function. Furthermore it follows from eq. (43)

$$d\varepsilon_r/dt_s = \sigma_0\,[dG(t_s)/dt_s - dG(t_0 + t_s)/dt_s]/E. \tag{44}$$

Hence for the recovery strain to be monotonically increasing, it is required that

$$dG(t)/dt > dG(t_0 + t)/dt \tag{45}$$

i.e. $dG(t)/dt$ must be monotonically decreasing. The conditions on $G(t)$ are therefore

$$\dot{G} > 0, \quad \ddot{G} < 0 \tag{46}$$

for all times t. The function

$$G(t) = \frac{\chi}{1 - \alpha}\, t^{1-\alpha}; \quad \alpha < 1 \tag{47}$$

found by RABOTNOV from test data by ROBINSON fulfills both these conditions.

Furthermore eq. (43) shows that no isostatic recovery will occur if $E = \infty$. Hence the function, suggested by RABOTNOV

$$\varphi(\varepsilon) = a\,\varepsilon^{\beta}, \quad \beta > 1 \tag{48}$$

although suitable for creep studies, where it implies stationary stress distributions, is not sufficient for cases where unloading occurs. In fact with $\varphi(\varepsilon)$ according to eq. (48) neither isostatic nor hyperstatic recovery will be predicted.

The RABOTNOV theory may also be used to predict the process following upon a partial unloading according to Fig. 9. Here the mechanisms of creep and recovery will be simultaneously at play. It is not *à priori* evident if the strain curve will follow path DE' or DE''. This may only be found by tracing the corresponding motion in the φ-ε-plane. For the particularly simple case where

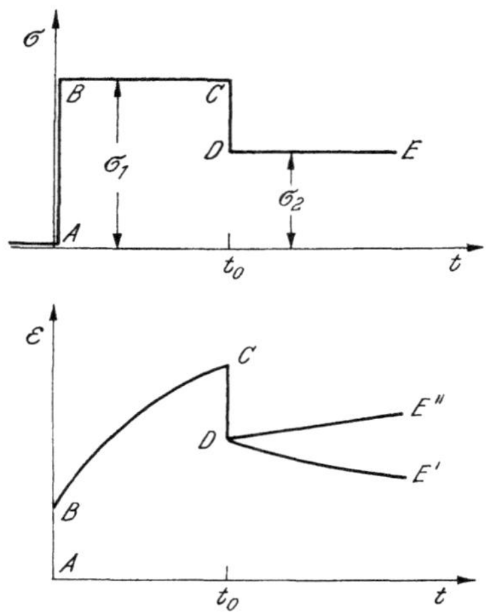

Fig. 9. Simultaneous creep and recovery after partial unloading

$$\varphi(\varepsilon) = \sigma/E \tag{49}$$

i.e. the master curve is a straight line, it follows that real recovery sets in at t_0 (path DE') if

$$\sigma_1/\sigma_2 > \dot{G}(0)/[\dot{G}(0) - \dot{G}(t_0)]. \tag{50}$$

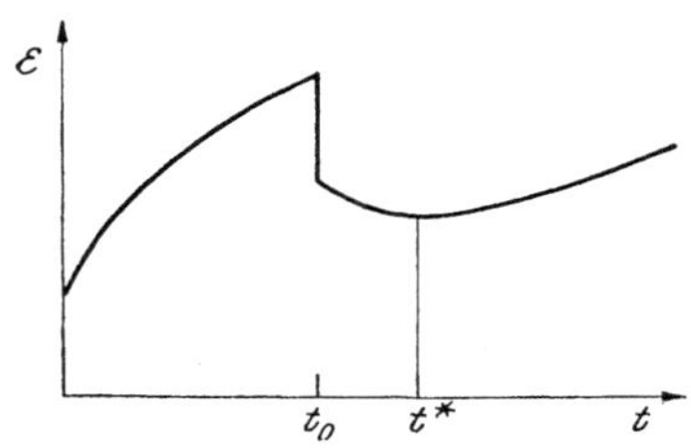

Fig. 10. Reversed recovery

Eventually, however, a state will be reached when $\dot{\varepsilon}_r$ changes from positive to negative, as shown by Fig. 10. The time t^* is found from the relation

$$\dot{G}(t^* - t_0)/[\dot{G}(t^* - t_0) - \dot{G}(t^*)] = \sigma_1/\sigma_2. \tag{51}$$

With $G(t)$ according to eq. (47) there results

$$t^* = t_0/[1 - (1 - \sigma_2/\sigma_1)^{1/\alpha}] \tag{52}$$

and hence a recovery rate reversal is always predicted if $\sigma_2 \neq 0$.

This particular kind of reversed recovery has been observed in experiments by NISHIHARA *et al.* (1959) and may be considered as a strong support for the RABOTNOV theory.

References

[1] HOFF, N. J.: On Primary Creep. J. Mech. Phys. Sol. **5** 150—151 (1957).
[2] HULT, J.: Primary Creep in Thickwalled Spherical Shells. Trans. Chalmers Univ. Techn. (Gothenburg) No. 264, 1963.
[3] LECKIE, F. A., and J. B. MARTIN: Deformation Bounds for Bodies in a State of Creep. Private Communication.
[4] NISHIHARA, T., S. TAIRA, K. TANAKA, and K. OHJI: Effect of Stress Reduction on Secondary Creep of Mild Steel. Bull. JSME **2** (5), 102—107 (1959).
[5] RABOTNOV, Y. N.: Some Problems on the Theory of Creep (1948). Transl. NACA (Washington D. C.) No. TM 1353, 1953.
[6] ROSENGREN, G.: Primary Creep in Cylindrical Pressure Vessels. IASS Symposium Non-Classical Shell Problems, Warsaw 1963. Amsterdam: North-Holland Publ. Co. 1964.
[7] SONNERUP, L.: Stress Reversals in Connection With Primary Creep. Trans. Chalmers Univ. Techn. (Gothenburg) No. 295, 1965.

Discussion

BESSELING: The little amount of recovery you get with this two-bar-model is closely connected with the type of creep-law you are using. If you use a different creep-law, for instance exponential, then you can get quite a lot of recovery. So even with models composed of subelements you may obtain an adequate description of the experimental behavior.

RABOTNOV: You mentioned my theory. The equation $\varphi[\varepsilon(t)]$ is a very particular case of a general non-linear functional equation due to VOLTERRA and FRÉCHET and used by some other authors for example Prof. COLEMAN. Compared with the experimental data we have found that the theory exaggerates recovery. I think that it is quite correct for the description of the behavior of high polymers, where most part of the deformation is reversible. For metals you can split the total creep-deformation into two parts: one of them is the purely irreversible creep-deformation, it might be described by the strain-hardening equation or steady-state equation. And the other, recoverable, part vanishes, when the temperature is raised. Perhaps at very high temperatures you have no recovery at all.

ONAT: We have studied recovery in one aluminum alloy at a high temperature, and we observed that recovery was almost nonexistent. As Prof. RABOTNOV remarked the description of such behavior by FRÉCHET's series of multiple integrals is very inconvenient, one needs a lot of terms. So for metals one has to use another form of representation.

HULT: If the function $G(t)$ is a linear one, then the RABOTNOV theory will predict no recovery, and if it is slightly non-linear the recovery will be quite small.

BRUN: The amount of recovery is very small. Doesn't that mean that there are other effects, like solid friction, to be taken into account, other than only viscous effects?

HULT: As I see it, the main use of the RABOTNOV theory, which of course is highly elaborate for a practical problem is in cases where we do have a large recovery, for materials like plastics.

Cerenkov Generation of Thermal Waves for the Wave Equations of Thermo—Electro—Magneto—Elasticity

By

S. Kaliski

Warsaw (Poland)

1. Introduction

In papers [1, 2, 3] phenomenological wave equations of heat conduction, thermoelasticity and thermoelectro-magneto-elasticity have been derived. A characteristic property of these equations is their hyperbolic character which means, that disturbations of any field, including the thermal propagate with finite velocity. These equations were obtained on the basis of the generalized Onsager relations by introducing some terms depending on the rate of variability of the stream, which permits the use of these equations not only for stationary dynamical processes but also for the processes of marked nonstationary character. This generalization and its consequences and thermodynamical modifications leads among other things to the elimination of the paradoxical parabolism of the heat equation and the fields coupled with the thermal field.

The question which arises is, how to design an experiment to prove the existence of that phenomenon.

One of the possible methods is by generation, by means of moving thermal sources, of Čerenkov thermal cones in a medium in which effects of coupled fields can be produced and by measuring the discontinuity of the electric field across the characteristic cone, coupled with the thermal cone. Unfortunately the complexity of measurement, short duration of the phenomenon, strong diffusivity and attenntaion of the discontinuity across the characteristic cone make the method of little practical value.

Another possibility, which will be discussed in this paper, is that of applying high frequency fields making possible the introduction of the Čerenkov generation of thermal waves of high frequency in piezocrystals by means of an electromagnetic wave moving in a slowing — down structure over the piezocrystal surface. It will be shown that with sufficiently high frequency, the thermal waves, calculated on the basis of the parabolic equations of thermo-piezo-electricity, are generated in quite a different way than the Čerenkov waves, obtained from the wave (hyperbolic) equations of thermo-piezo-electricity. In the first case the waves die out rapidly. In the second they die out (with sufficiently high frequency) very slowly. This feature suggests an experiment for proving the wave character of heat propagation. The entire analysis is performed with considerable simplifications and is based on asymptotic solutions. The fundamental wave equations of thermo-magneto-elasticity are given with reference to [3]. This is followed by the solution of the problem of Čerenkov generation of thermal waves in a piezocrystal by means of slowed-down electromagnetic surface waves for the wave equations and the classical equations of thermo-piezo-electricity. The characteristic properties of the solutions of the wave equations are shown. The properties can be used for the design of measurements of the effect in question.

2. Wave Equations of Thermo-Electro-Magneto-Elasticity

2.1. The wave (hyperbolic) equations of heat conduction, thermoelasticity, and thermo-electro-magneto-elasticity were derived in [1, 2, 3].

In [3] were obtained the wave equations of thermo-magneto-elasticity of real elastic conductors with thermoelectric effects and the wave equations of thermo-piezoelectricity. These equations were derived on the basis of the generalized Onsager relations by introducing some terms depending on the rate of variability of the stream which enables one to use these equations not only for dynamical stationary processes but also for processes of marked nonstationary character. In [1, 2, 3] were discussed the corresponding thermodynamic modifications connected with the generalization of the Onsager relations. The symmetry relations were obtained.

In the present section we shall discuss in brief the fundamental results obtained in [1, 2, 3] and used in the subsequent sections to study the problem of thermal Čerenkov radiation. In [3] a relativistic analysis is presented on the grounds of relativistic thermodynamics of irreversible processes [4, 5], to prove the possibility of obtaining wave equations (without generalizing the Onsager relations). A negative result was obtained similarly to [6]. In [3] was discussed the literature of the problem, in particular the papers by CATTANEO [7], VERNOTTE [8], GOLDSTEIN [9], SASAKI [10], FREUDENTHAL [11] and other works in which it was attempted to derive a wave type equation of heat conduction. In [2, 3] some critical remarks were made on the inconsistencies and the errors committed in some of the papers quoted.

2.2. Let us consider now the wave equation of heat conduction. The point of departure for the derivation of these equations are two postulates. The first is the postulate of finite propagation velocity of thermal disturbances and the other that of local character of the dynamic equations of heat conduction or, in other words, the possibility of describing the process by means of partial differential equations. The first postulate is physically obvious. The other, which is not, satisfies the requirement of maximum simplicity of description of the phenomenon, and also the requitement that the correction decisive for the wave character of the existing parabolic heat equation should not change the form of the classical equation which remains valid for stationary processes, or approximately for not too rapid processes. The derivation of a rigorous wave type heat equation is only possible on the ground of microscopic quantum processes of heat transport. The problem is very involved, however, due to, among other causes, the nonlinear unharmonic effects of crystal vibration which must be taken into account. In the phenomenological theory experimental parameters are used. It must be logically correct and adequate from the physical point of view. The generalization proposed will enable us to study comparably strong nonstationary processes and on the other hand leads to the removal of the paradox of infinitely high propagation velocity of thermal disturbances and those of the fields coupled with the thermal field. In other respects the phenomenological theory is the same for electron and phonon conductivity except that different propagation velocity of thermal disturbances is obtained in each case.

In our brief description of the derivation of the wave heat equation [1], we shall confine our considerations to linear processes and isotropic bodies.

We introduce the notations of [1], that is

$$\Theta = T_0 + T; \qquad \left(\frac{T}{T_0} \ll 1\right) \tag{2.1}$$

where Θ is the absolute temperature, T_0 is a reference temperature, T is the temperature of the disturbed field.

Further let us write

$$r_i = \frac{q_i}{\Theta}, \tag{2.2}$$

where q_i is the heat flux and r_i — the corresponding generalized coordinate discribing the flux ($\dot{r}_i$ — generalized velocity).

We assume the generalized Fourier law in the form

$$\frac{\tau}{k} (\Theta \dot{r}_i)_{,t} + \frac{\Theta}{k} \dot{r}_i = -\Theta_{,i} = -T_{,i} \tag{2.3}$$

or after linearization

$$\frac{\tau}{k} T_0 \ddot{r}_i + \frac{T_0}{k} \dot{r}_i = -T_{,i} \tag{2.4}$$

where τ will be called the relaxation constant of flux and k is the coefficient of heat conduction. Equation (2.4) reflects the fact that for a rapid change of the temperature gradient, the heat flux will follow this change with certain delay and the process of thermal flow will be characterized by a certain kinetic energy.

Equation (2.4), if we consider $\dot{r}_i$ as a generalized velocity, has formally the same character as that of mechanical motion of a body with energy dissipation, the quantity T corresponding to the mechanical stress tensor (pressure scalar).

It is clear that it is possible to generalize eq. (2.4) by introducing into this equation a generalized force of variable rate. This will not be done here.

We assume the total energy in the form

$$H = U + K, \tag{2.5}$$

where $U = U(s)$ is the internal energy,
$K = K(\dot{r}_i)$ — "kinetic" thermal energy,
s — entropy density.

Then in agreement with the first law of thermodynamics and bearing in mind the linearization we obtain (by analog to the mechanics of continua) the equation

$$\int_V (\dot{U} + \dot{K})\, dV = -\int_S \Theta \dot{r}_i \, dS_i . \tag{2.6}$$

Multiplying (2.4) by $\dot{r}_i$ we find

$$\frac{\tau}{k} T_0 \left(\frac{\dot{r}_i^2}{2}\right)_{,t} + \frac{T_0}{k} \dot{r}_i^2 = -T_{,i}\dot{r}_i . \tag{2.7}$$

Hence

$$\dot{K} + 2F = -T_{,i}\dot{r}_i^2, \tag{2.8}$$

where $K = \frac{\tau T_0}{k} \frac{\dot{r}_i^2}{2}$; $2F = \frac{T_0}{k} \dot{r}_i^2$ are quantities corresponding to the kinetic and the dissipation energy.

From (2.6) and (2.8) we find

$$\dot{K} + \dot{U} = -(\Theta\dot{r}_i)_{,i} = -(T_{,i}\dot{r}_i + \Theta\dot{r}_{i,i}) = \dot{K} + 2F - \Theta\dot{r}_{i,i} . \tag{2.9}$$

Hence

$$\dot{U} = 2F - \Theta\dot{r}_{i,i} = 2F - \Theta\left(\frac{q_i}{\Theta}\right)_{,i} = -q_{i,i} + \dot{K} . \tag{2.10}$$

or

$$\dot{U} + \dot{K} = -q_{i,i} \tag{2.11}$$

Bearing in mind that

$$\dot{U} = \Theta \dot{s} = \beta \dot{T} \tag{2.12}$$

we find from (2.10), (2.12)

$$\dot{s} = \frac{2F}{\Theta} - \left(\frac{q_i}{\Theta}\right)_{,i} \tag{2.13}$$

and hence the following expression for the entropy production rate

$$\dot{\sigma} = \dot{s} + \left(\frac{q_i}{\Theta}\right)_{,i} = \frac{2F}{\Theta}. \tag{2.14}$$

For anisotropic bodies we have

$$2F = T_0 k_{ij}^{-1} \dot{r}_i \dot{r}_j. \tag{2.15}$$

The tensor of the coefficients k_{ij}^{-1} must guarantee the positiveness of the quadratic form (2.15) and satisfy the familiar symmetry relations $k_{ij}^{-1} = k_{ji}^{-1}$.

Substituting now (2.12) into (2.11) we obtain after linearization

$$\beta \dot{T} = -q_{i,i} - \dot{K} \approx -q_{i,i} \tag{2.16}$$

and

$$\frac{q_i}{\Theta} \approx \frac{q_i}{T_0} \approx \dot{r}_i. \tag{2.17}$$

Differentiating (2.4) and substituting (2.16) we obtain

$$\tau \beta \ddot{T} + \beta \dot{T} = k T_{,ii} \tag{2.18}$$

which is the requiered hyperbolic (wave-type) equation of heat conduction. The propagation velocity of the thermal wave is

$$a_T = \sqrt{\frac{k}{\tau\beta}}. \tag{2.19}$$

2.3. The wave equations of thermoelasticity were derived in (2) in the following form

$$\left.\begin{aligned} &\varrho \dot{v}_i = E_{ikmn}^T \varepsilon_{mn,k} - \gamma_{ik} T_{,k} \\ &\tau \Theta_0 \gamma_{ij} \ddot{\varepsilon}_{ij} + \Theta_0 \gamma_{ij} \dot{\varepsilon}_{ij} + \tau \Theta_0 \beta_0 \ddot{T} + \Theta_0 \beta_0 \dot{T} = k_{ij} T_{,i})_{,j} \end{aligned}\right\} \tag{2.20}$$

where T_0 is now denoted by Θ_0.

It can be proved that the system of eqs. (2.20) is hyperbolic. If $\tau \to 0$ we get from (2.20) the classical parabolic system of equations of thermoelasticity and eq. (2.18) coincides with the classical equation of heat conduction.

The thermodynamical derivation of eqs. (2.20) is given in [2]. These equations will be obtained later as a particular case of the equations of thermo-piezoelectricity, which are more general.

2.4. We shall discuss now in brief the wave equations of thermomagnetoelasticity for real elastic conductors, the thermoelectric effects beeing taken into account.

This system of equations has after the linearization [3] the form

$$\left.\begin{aligned}
&\operatorname{rot}\vec{h} = \vec{j} + \dot{\vec{D}}; \quad \operatorname{rot}\vec{E} = -\dot{\vec{b}}; \quad \operatorname{div}\vec{b} = 0; \quad \operatorname{div}\vec{D} = 0,\\
&\varrho\ddot{u}_i = \sigma_{ik,k} + (\vec{j} \times \vec{B}_0)_i;\\
&T_0\beta_0(\tau\ddot{T} + \dot{T}) + T_0\alpha_{ij}(\tau\ddot{e}_{ij} + \dot{e}_{ij}) - (k_{ij}T_{,j})_i + (\pi_{ik}j_k)_{,i} = 0;\\
&b_i = \mu_{ik}h_k; \; D_i = \varepsilon_{ik}[E_k + (\dot{\vec{u}} \times \vec{B}_0)_k]; \; \sigma_{ik} = E_{ikmn}e_{mn} - \alpha_{ik}T;\\
&j_i + \tau\dot{j}_i + x_{ik}k_{kj}^{-1}\pi_{js}(j_s + \tau\dot{j}_s) =\\
&= \eta_{ik}\{E_k + \tau\dot{E}_k + [(\dot{\vec{u}} + \varepsilon\ddot{\vec{u}}) \times \dot{B}_0]_k\} + x_{ik}[k_{kj}^{-1}(\pi_{js}j_s - k_{js}T_{,s})];\\
&\text{or}\\
&\varepsilon\dot{q}_{i,i} = \tau T_0\beta_0\dot{T} + \tau T_0\alpha_{ij}\ddot{e}_{ij} =\\
&= (x_{ik}^{-1}k_{kj}j_i)_{,j} - \{\eta_{is}x_{ik}^{-1}k_{kj}[E_s + (\dot{\vec{u}} \times \vec{B}_0)_s]\}_{,j} - (k_{kj}T_{,k})_{,j}.
\end{aligned}\right\} \quad (2.21)$$

The thermodynamical derivation of these equations and the symmetry relations of the tensor coefficients are given in [3]. As was shown in [3], this system is hyperbolic (the notations used here are those of [3]).

In the case of isotropy the system of eqs. (2.21) takes on the following, much simpler, form:

$$\left.\begin{aligned}
&\operatorname{rot}\vec{h} = \vec{j} + \varepsilon[\vec{E} + (\ddot{\vec{u}} \times \vec{B}_0)]; \quad \operatorname{rot}\vec{E} = -\mu\dot{\vec{h}};\\
&\operatorname{div}\vec{h} = 0; \quad \operatorname{div}\vec{D} = 0;\\
&\varrho\ddot{\vec{u}} = G\nabla^2\vec{u} + (\lambda + G)\operatorname{grad}\operatorname{div}\vec{u} - \alpha\operatorname{grad}T + (\vec{j} \times \vec{B}_0);\\
&\tau\beta\varphi\ddot{T} + \beta\dot{T} + \tau\lambda\varphi\operatorname{div}\ddot{\vec{u}} + \lambda\operatorname{div}\dot{\vec{u}} - k\varphi\nabla^2T = 0;\\
&\vec{j} + \tau\varphi\dot{\vec{j}} = \eta\{\vec{E} + \tau\dot{\vec{E}} + [(\dot{\vec{u}} + \tau\ddot{\vec{u}}) \times \vec{B}_0]\} - x\operatorname{grad}T,
\end{aligned}\right\} \quad (2.22)$$

where

$$\lambda = T_0\alpha; \quad \beta = \beta_0T_0; \quad \varphi = 1 + \frac{\varkappa\pi}{\eta}.$$

If $\tau \to 0$ we obtain from these equations the classical equations of thermo-magnetoelasticity [12] and the symmetry relations [13].

2.5. To conclude this section we shall present the system of equations of thermo-piezoelectricity. In the next section we shall use this system of equations for further calculations. For this reason and since its thermodynamical derivation is relatively simple, we shall now sketch this derivation in brief.

The point of departure is the generalized Fourier law, the equations of motion and those of electrodynamics. Let us write them, with the notations introduced, in the form

$$\left.\begin{aligned} \tau\dot{q}_{i,i} + q_{i,i} &= -(k_{ij}T_{,j})_{,i}; \\ \varrho\ddot{u}_i &= \sigma_{ik,k}; \\ \operatorname{rot}\vec{H} = \dot{\vec{D}}; \quad \operatorname{rot}\vec{E} &= -\dot{\vec{B}}; \quad \operatorname{div}\vec{B} = 0; \quad \operatorname{div}\vec{D} = \varrho_e, \end{aligned}\right\} \tag{2.23}$$

where

$$\left.\begin{aligned} \sigma_{ik} &= E_{ikmn}^{ETH}\, e_{mn} - \alpha_{ik}^{EH} - r_{ikl}^{TH}\, E_l \\ D_i &= r_{ikl}^{TH}\, e_{kl} + \varepsilon_{ij}^{TeH}\, E_j + p_i^{eH}\, T \\ s &= \alpha_{ik}^{EH}\, e_{ik} + p_i^{eH}\, E_i + \beta_0^{eHE}\, T \\ B_i &= \mu_{ik}^{TeH}\, H_k \end{aligned}\right\} \tag{2.24}$$

and the superscripts of the coefficients determine their thermodynamical sense. These superscripts will be omitted in what follows. The quantity $q_{i,i}$ in the eqs. (2.23) must be expressed in terms of the variables E_i, u_i, T. The equations of state (2.24) must now be derived.

To this end let us consider the first and the second law of thermodynamics, confining ourselves to the linear approximation (Substantial derivatives are then replaced by ordinary ones).

On the basis of the first thermodynamical law, we obtain

$$\int_V (\dot{K} + \dot{U})\, dV = -\int_A N_i\, dA_i - \int_A \Theta\dot{r}_i\, dA_i + \int_A \sigma_{ij}\dot{u}_j\, dA_i, \tag{2.25}$$

where $N_i = \varepsilon_{ijk}E_jH_k$ — Poyntings vector,

$K = K_1 + K_2$,

$K_1 = \frac{1}{2}\varrho\dot{u}_i^2$; $\quad K_2 = \frac{1}{2}\tau T_0 k_{ij}^{-1}\dot{r}_i\dot{r}_j$,

U — density of internal energy,

k_{ij}^{-1} — symmetric tensor to be defined.

Making use of the Gauss formula and eq. (2.23), eq. (2.25) can be reduced to the form

$$\begin{aligned} \dot{U} &= \sigma_{ij}\dot{e}_{ij} + E_i\dot{D}_i + H_i\dot{B}_i + 2F - \Theta\dot{r}_{i,i} = \sigma_{ij}\dot{e}_{ij} + \\ &+ E_i\dot{D}_i + H_i\dot{B}_i + 2F - \Theta\left(\frac{q_i}{\Theta}\right)_{,i}, \end{aligned} \tag{2.26}$$

where

$$2F = \Theta k_{ij}^{-1}\dot{r}_i\dot{r}_j \approx T_0 k_{ij}^{-1}\dot{r}_i\dot{r}_j.$$

We have also

$$\dot{U} = \sigma_{ij}\dot{e}_{ij} + E_i\dot{D}_i + H_i\dot{B}_i + \Theta\dot{s}. \tag{2.27}$$

Then

$$\Theta\dot{s} = 2F - \Theta\left(\frac{q_i}{\Theta}\right)_{,i} \tag{2.28}$$

and the expression for the entropy production rate becomes

$$\dot{\sigma} = \dot{s} + \left(\frac{q_i}{\Theta}\right)_{,i} = \frac{2F}{\Theta} > 0. \tag{2.29}$$

This is ensured if the quadratic form $2F$ is positive, i.e., if the tensor k_{ij}^{-1} is symmetric.
If we introduce now, after [14], the potential

$$U_0 = U - T_0\dot{s}, \tag{2.30}$$

(2.27) can be reduced to the form

$$\dot{U}_0 = \sigma_{ij}\dot{e}_{ij} + E_i\dot{D}_i + H_i\dot{B}_i + T\dot{s}. \tag{2.31}$$

If the independent variables are chosen as shown in the expression

$$U_0 = U_0(e_{ij}, D_i, B_i, s), \tag{2.32}$$

we can calculate

$$\sigma_{ij} = \frac{\partial U_0}{\partial e_{ij}};\ H_i = \frac{\partial U_0}{\partial B_i};\ E_i = \frac{\partial U_0}{\partial D_i};\ T = \frac{\partial U_0}{\partial s}, \tag{2.33}$$

where

$$\begin{aligned} U_0 &= \frac{1}{2}\, E_{ikmn}^{DsB}\, e_{ij} e_{mn} + \frac{1}{2}\,\beta_{ij}^{seB}\, D_i D_j + \frac{1}{2}\, b^{eDB} s^2 - \\ &- h_{kij}^{sB}\, D_k e_{ij} - g_i^{eB}\, s D_i - \gamma_{ij}^{DB}\, s e_{ii} + \frac{1}{2}\,(\mu_{ik})^{-1} B_i B_k, \end{aligned} \tag{2.34}$$

where

$$\left.\begin{aligned}&E^{DsB}_{ikmn} = E^{DsB}_{kimn} = E^{DsB}_{iknm} = E^{DsB}_{mnik};\\&\beta^{seB}_{ij} = \beta^{seB}_{ji}; \quad h^{sB}_{kij} = h^{sB}_{kji}; \quad \gamma^{DB}_{ij} = \gamma^{DB}_{ji}; \quad (\mu_{ik})\,(\mu_{ik})^{-1} = 1.\end{aligned}\right\} \tag{2.35}$$

From (2.33) we obtain σ_{ij}, H_i, E_i, T and then, using the transformation formulae [3] we can easily obtain the relations of state (2.24).

Let us return now to the problem of expressing the quantity q_{ii} in terms of E_i, u_i, T. Equation (2.26) can be written in the following form

$$\dot{U} + \dot{K}_2 = \sigma_{ij}\dot{e}_{ij} + E_i\dot{D}_i + H_i\dot{B}_i - q_{i,i}. \tag{2.36}$$

Then we obtain, instead of (2.28),

$$\Theta\dot{s} \approx T_0\dot{s} = -q_{i,i} + \dot{K}_2 \tag{2.37}$$

or, confining ourselves to the linear terms only

$$T_0\dot{s} = q_{i,i}. \tag{2.38}$$

Substituting (2.38) and (2.24) in (2.23) we obtain finally

$$\left.\begin{aligned}&E_{ikmn}u_{m,nk} - r_{lik}E_{l,k} - \alpha_{ik}T_{,k} = \varrho\ddot{u}_i; \qquad (i = 1, 2, 3)\\&\varepsilon_{ijk}H_{k,j} = r_{ikl}\frac{1}{2}(\dot{u}_{k,l} + \dot{u}_{l,k}) + \varepsilon_{ij}\dot{E}_j + p_i\dot{T};\\&\varepsilon_{ijk}E_{j,k} = \mu_{ik}\dot{H}_k;\\&T_0\alpha_{ij}(\tau\ddot{u}_{i,j} + \dot{u}_{i,j}) + T_0p_i(\tau\ddot{E}_i + \dot{E}_i) + T_0\beta(\tau\ddot{T} + \dot{T}) -\\&\quad - (k_{ij}T_{,j})_{,i} = 0.\end{aligned}\right\} \tag{2.39}$$

If we neglect the magnetic induction and introduce the potential

$$E_i = \varphi_{,i} \tag{2.40}$$

the system of eqs. (2.39) can be reduced to

$$\left.\begin{aligned}&E_{ikmn}u_{m,nk} - r_{lik}\varphi_{,lk} - \alpha_{ik}T_{,k} = \varrho\ddot{u}_i; \qquad (i = 1, 2, 3)\\&[r_{ilk}u_{l,k} + \varepsilon_{ij}\varphi_{,j} + p_iT]_{,i} = 0;\\&(1 + \varepsilon\,\partial_t)(\alpha_{ij}\dot{u}_{ij} + p_i\dot{\varphi}_{,i} + \beta\dot{T}) - (x_{ij}T_{,j})_{,i} = 0,\\&\qquad\qquad\qquad\qquad\text{where } x_{ij} = \frac{k_{ij}}{T_0}.\end{aligned}\right\} \tag{2.41}$$

The set of eqs. (2.39) and (2.41) can be shown to be hyperbolic [3]. If $\tau \to 0$ these equations become the classical parabolic equations of thermo-piezoelectricity. If we neglect the piezoelectric effects, it can be shown that equations of thermoelasticity (2.20) are obtained from these equations.

3. The Čerenkov Generation of Thermal Waves by Means of Slowed-Down Electromagnetic Waves

In the phenomenological equations of the foregoing section the experimental constant characterizing the wave process was the constant τ. Its measurement in solids being very complicated, it is essential to invent a measurement method which could confirm in a qualitative manner at least the hyperbolic (wave) character of the propagation of thermal disturbances.

As already mentioned in the introduction one of such possibilities would be the generation, by means of moving heat sources, of thermal cones of Čerenkov radiation in the medium showing effects of coupled fields, and the measurement of the discontinuity of an electric field, for instance at the characteristic cone coupled with the thermal cone. As already metioned this method is, however, of little practical use in view of the complexity of the measurement, short duration of the phenomenon, strong diffusivity and rapid attenuation of the discontinuity across the characteristic cone. The other possibility which will now be discussed is connected with the application of high frequency fields enabling the consideration of processes non-stationary to a rather high degree.

The generation of high frequency thermal fields is possible only by electromagnetic means. Use should therefore be made of the effects of coupled fields. Similary to [15], where ultra and hypersounds were generated in piezo-crystals by means of an electromagnetic wave moving in a slowing-down structure over the piezoelectric body, we can generate in piezocrystals, making additional use of the thermo-electro-elastic coupling, a Čerenkov thermal wave of high frequency radiating from the surface towards the inside of the crystal. Such a method enables us to generate waves of frequency of the order ot 4.10^{10} cps at the temperature of liquid helium.

It will be shown that with such a frequency (or even higher if possible) the effect of the non-stationary character of the phenomenon produces a marked influence of the terms decisive for the wave character of the equations. It will be seen that for a sufficiently high frequency, the solutions obtained on the basis of the (classical) parabolic-hyperbolic set of equations represent a thermal wave radiating from the surface with such a decrement that it vanishes practically at once. On the other haud, the solution obtained on the grounds of the hyperbolic (wave) set of equations represent with the same frequency a thermal wave radiated with a relatively weak decrement, that is a wave of amplitude which is approximately constant for depths of the order of some dozens of wavelengths. The qualitative difference between the solutions above furnishes a qualitative experimental proof of the wave (hyperbolic) character of propagation of thermal disturbances. We shall confine ourselves to conduction phenomena of the non-electron type. Then the velocity of propagation of thermal perturbation is expected to be of the order approaching the velocity of sound. (with frequency in the acoustical branch)

Let us consider the following system. A piezoelectric (thermopiezoelectric) crystal is covered by a slowing-down structure in which a slowed-down electromagnetic wave moves along the surface of the crystal. Our assumptions will be those of [15]. In particular, by considering asymptotic solutions (for high frequencies) we shall reject the action of the electric field of the crystal on the electromagnetic wave in the slowing-down structure.

The velocity of the wave is assumed to be reduced by some ten times. This enables us to disregard the influence of the induction in the equations of the potential of the electric field in the crystal.

The electromagnetic wave moving in the slowing-down structure generates in the piezo-crystal an electric field, which in turn generates an elastic and a thermal field. In view of the fact that the phase velocity of the electromagnetic wave in the slowing-down structure (which is of the order of the velocity of light in the slowing-down structure) exceeds considerably the velocity of sound and that of the thermal wave an elastic and thermal Čerenkov wave radiates from the surface of the crystal. Making use of asymptotic methods we can obtain for high frequencies the amplitudes of the elastic and thermal waves and show the influence of the non-stationary character of the phenomenon resulting in a qualitatively different type of radiation in the case of hyperbolic (wave type) set of equations of the problem. The point of departure is eqs. (2.41).

For simplicity of calculations further simplifications will be introduced.

We shall assume, similarly to [15], that the elastic and thermal properties of the crystal are isotropic. This is of no essential influence on the qualitative character of the wave process studied but will substantially simplify the quantitative side of the problem. In particular, it will enable the limitation of the investigation of wave motion to a plane problem.

Assuming that the x_1 axis is directed along the surface of the crystal, the axis x_2 being normal to that surface, and that the solution is independent of x_3 (which is possible in the case of elastic and thermal isotropy), eqs. (2.41) can be written in the form

$$\left.\begin{aligned}
&(a_1^2 - a_2^2)\,(u_{1,11} + u_{2,12}) + a_2^2 \nabla^2 u_1 - \ddot{u}_1 - \frac{1}{\varrho} r_{l1k}\varphi_{l,k} - \frac{\alpha}{\varrho} T_{,1} = 0;\\
&(a_1^2 - a_2^2)\,(u_{1,12} + u_{2,22}) + a_2^2 \nabla^2 u_2 - \ddot{u}_2 - \frac{1}{\varrho} r_{l2k}\varphi_{l,k} - \frac{\alpha}{\varrho} T_{,2} = 0;\\
&\varkappa \nabla^2 T = (1 + \tau\,\partial_t)\,(\alpha \dot{u}_{i,i} + p_i \dot{\varphi}_{,i} + \beta \dot{T});\\
&\nabla^2 \varphi + \frac{p_i}{\varepsilon} T_{,i} = -\frac{1}{\varepsilon} r_{ilk} u_{l,ki},
\end{aligned}\right\} \tag{3.1}$$

where ε has been written in place of ε_{ik} (for quartz we have f. i. $\varepsilon_{11} = \varepsilon_{22} = \varepsilon_{33}$). The quantities a_1, a_2 denote the mean velocity of longitudinal and transversal waves, respectively. This set of equations must be completed by relevant boundary conditions, that is, the condition of zero stress at the surface of the crystal, zero heat flow in the direction normal to the surface and prescribed electric components (the action of the crystal field on the electromagnetic wave in the slowing-down structure being rejected in the asymptotic state).

We turn now to the derivation of asymptotic solutions.

4. Asymptotic Solutions

Let us consider the asymptotic case of high phase velocity of the electromagnetic wave in the slowing-down structure. The radiating elastic and thermal waves will have the character of a Čerenkov radiation. Let us introduce a small parameter which will be the ratio of the velocity of sound or that of the thermal wave to the velocity c_0 of the electromagnetic wave,

$$\mu = \frac{a_2}{c_0} \tag{4.1}$$

and construct solutions for certain approximations. Let us observe first, that in the first approximation we can reject as small in the last of eqs. (3.1) the action of the deformation and the temperature field. Then, for a given value of the electric field, at the surface of the crystal the solution for φ or directly for E_1, E_2 takes the form

$$\left.\begin{aligned}
E_1 &= E_{01} e^{i(\omega t - k_0 x_1) - h x_2}\\
E_2 &= E_{02} e^{i(\omega t - k_0 x_1) - h x_2}
\end{aligned}\right\} \tag{4.2}$$

where in an approximate manner

$$k_0 = h \tag{4.3}$$

and the values of E_{01}, E_{02} prescribed at the surface are interrelated by $E_{02} = iE_{01}$. On substituting (4.2) and (4.3) in the remaining of eqs. (3.1) we shall obtain particular integrals of the nonhomogeneous equations. They will be denoted by the superscript zero. From the third of eqs. (3.1) it follows that the principal terms retained after the rejection of the small terms are

$$i\omega\,(p_j\varphi_{,j} + \beta T^0) = 0. \tag{4.4}$$

Hence

$$T_0^0 = -\frac{p_i E_{0i}}{\beta} \tag{4.5}$$

From the two remaining eqs. (3.1) we obtain in an analogous manner, after rejecting lower-order terms

$$u_{0j}^0 = (-1)^j \frac{\mu^2}{\varrho a_2^2 k_0} R_j^0, \tag{4.6}$$

where

$$\mu = \frac{1}{\xi};\quad \xi = \frac{\omega}{a_2 k_0};\quad \mu \ll 1, \tag{4.7}$$

$$\left.\begin{aligned} R_1^0 &= (ir_{111} + r_{112})\,E_{01} + (ir_{211} + r_{212})\,E_{02} - i\alpha T_0^0 \\ R_2^0 &= (ir_{121} + r_{122})\,E_{01} + (ir_{221} + r_{222})\,E_{02} - \alpha T_0^0 \end{aligned}\right\} \tag{4.8}$$

Thus finally

$$\left.\begin{aligned} u_j^0 &= (-1)^j \frac{\mu^2}{\varrho a_2^2 k_0} R_j^0 F_0 \\ T_0 &= \frac{p_i E_{0i}}{\beta} F_0 \end{aligned}\right\} \tag{4.9}$$

where

$$F_0 = e^{i(\omega t - k_0 x_1) - h x_2}. \tag{4.10}$$

As is seen from (4.9) the displacement u_j^0 constitutes a small quantity of the second order in μ and can be rejected in further considerations. The particular solutions of the non-homogeneous equations being known, we proced to calculate the general solutions of eqs. (3.1). The solutions of the complete nonhomogeneous equations will be sought for in the form

$$\left.\begin{aligned} u_j &= A_{1j}F_1 + A_{2j}F_2 + u_j^0 \\ T &= T^0 + BF_3 \end{aligned}\right\} \tag{4.11}$$

where

$$F_j = e^{i(\omega t - k_0 x_1 - k_j x_2)};\qquad (j = 1, 2, 3). \tag{4.12}$$

The integrals (4.12) have the form of radiation integrals of the Čerenkov type except of F_3 which has a different structure for $\tau = 0$ (the classical parabolic-hyperbolic equations) and $\tau \neq 0$ (hyperbolic equations). Therefore these two cases will be considered separately in what follows. In eq. (4.12) we have

$$k_j^2 = \frac{\omega^2}{a_j^2} - k_0^2; \qquad (j = 1, 2) \tag{4.13}$$

or

$$\frac{k_j^2}{k_0^2} = \frac{\omega^2}{k_0^2 a_j^2} - 1 \approx \frac{\omega^2}{k_0^2 a_j^2}. \tag{4.14}$$

Hence

$$\frac{k_2}{k_0} = \xi; \quad \frac{k_1}{k_0} = \sqrt{m}\, \xi, \quad \text{where } m = \frac{a_2^2}{a_1^2}. \tag{4.15}$$

It can be shown, after rejecting u_j^0, by discussing the relations between integrals of the homogeneous equations that the constants A_{ki} in (4.11) are of different order in the asymptotic state. We have $A_{11} = O(\mu^2)$; $A_{22} = O(\mu^2)$; $A_{12}, A_{21} = O(\mu)$. Therefore the solution (4.11) can be expressed in the final form

$$u_1 = A_{21} F_2 = A_2 F_2; \quad u_2 = A_{12} F_1 = A_1 F_1; \quad T = T^0 + B F_3. \tag{4.16}$$

We proceed now to determine the wave vector k_3 from the third of eqs. (3.1) in its homogeneous form. As observed above two variants will be considered: 1) the classical case of $\tau = 0$, 2) the case of $\tau \neq 0$. To distinguish the first variant from the second, we shall assume $k_3 = k_3^*$. It will be determined by the eq.

$$\frac{k_3^{*2}}{k_0^2} = -\left(1 + i \frac{\omega}{k_0 a_2} \frac{a_2 \bar{\beta}}{k_0}\right) \approx -i \frac{a_2 \bar{\beta}}{k_0} \xi, \quad \text{where } \bar{\beta} = \frac{\beta}{\varkappa}; \tag{4.17}$$

hence

$$k_3^* = \sqrt{a_2 \bar{\beta} k_0} \left(\frac{\sqrt{2}}{2} - i \frac{\sqrt{2}}{2}\right) \sqrt{\xi}. \tag{4.18}$$

On substituting (4.18) in (4.16) we calculate A_1, B from the boundary conditions at the surface of the crystal

$$\sigma_{22} = 0; \quad \sigma_{21} = 0; \quad T_{,2} = 0, \tag{4.19}$$

or

$$\left.\begin{aligned} &\varrho a_1^2 u_{2,2} + \varrho (a_1^2 - 2a_2^2)\, u_{1,1} - r_{l22} E_l - \alpha T_{,2} = 0; \\ &\varrho a_2^2 (u_{1,2} + u_{2,1}) - r_{l21} E_l = 0; \\ &T_{,2} = 0. \end{aligned}\right\} \tag{4.20}$$

Hence we obtain, within terms of the first order,

$$u_{01} = A_2 = \frac{r_{l21} E_{0l}}{i \varrho a_2^2 k_0} \mu; \quad u_{02} = A_1 = \frac{r_{l22} E_{0l}}{i \varrho a_1^2 \sqrt{m}\, k_0} \mu \tag{4.21}$$

and

$$B = -\frac{T_0}{i} \sqrt{\mu} \sqrt{\frac{k_0}{a_2 \bar{\beta}}} \frac{1}{\frac{\sqrt{2}}{2} - i \frac{\sqrt{2}}{2}} = \frac{p_i E_{0i}}{\beta} \frac{1}{i \frac{\sqrt{2}}{2} + \frac{\sqrt{2}}{2}} \sqrt{\frac{k_0}{a_2 \bar{\beta}}} \sqrt{\mu} = -\frac{k_0}{i k_3^*} T_0. \tag{4.22}$$

Although it follows from (4.22) that B is of the first order in μ we should bear in mind that the representation with k_3^* in the denominator is not comparable with the previous small quantities in view of the fact that the order of magnitude of k_3^* is different from that of $k_i (= 1, 2)$. However a specific feature of this solution is the fact that it dies out very rapidly towards the interior of the crystal. This follows from (4.18) becouse the imaginary and the real part of k_3^* are identical. It can be stated therefore, that such a wave vanishes practically at the surface of the crystal. Let us consider now the second variant, $\tau \neq 0$.

From the third of eqs. (3.1) we have

$$\frac{k_3^2}{k_0^2} = \left[-1 + \frac{\omega^2}{k_0^2 a_T^2} - \frac{i \bar{\beta} a_2}{k_0} \frac{\omega}{a_2 k_0}\right] \approx n\, \xi^2 - \frac{i \bar{\beta} a_2}{k_0} \xi, \tag{4.23}$$

where

$$a_T = \sqrt{\frac{\varkappa}{\beta \tau}} = \sqrt{\frac{1}{\bar{\beta} \tau}}; \quad n = \frac{a_2^2}{a_T^2}. \tag{4.24}$$

Since, for average materials and for the frequency applied, the term $\frac{\bar{\beta} a_2}{k_0}$ represents a large quantity (of the order of 10^7) we cannot reject it as small in relation to $n \xi^2$ in (4.23).

We transform (4.23) to obtain

$$\frac{k_3^2}{k_0^2} = \frac{1}{k_0^2} \left(\frac{\omega^2}{a_T^2} - i \omega \bar{\beta}\right) = \frac{1}{k_0^2} \omega \bar{\beta} \tau \left(\omega - \frac{i}{\tau}\right). \tag{4.25}$$

Assuming that in our case the frequency is such that

$$\omega > \frac{1}{\tau} \tag{4.26}$$

we have

$$k_3 = \sqrt{n}\, \xi k_0. \tag{4.27}$$

The condition (4.26) requires application of a very high frequency (of the order of 10^{11} cps) which is at the limits of practical realizability.

Introducing (4.27) in (4.10) and then in the boundary conditions (4.20) the third of which should now be written in the accurate manner

$$q_2 = 0, \text{ hence } q_2 \approx \frac{\bar{\varkappa}}{i\omega\tau + 1} T_{,2} = 0 \quad \text{and} \quad T_{,2} = 0. \tag{4.28}$$

we find A_1, A_2 to be identical with (4.21) and

$$B = \frac{i k_0}{k_3} T_0 = i\mu T_0 = - i \frac{p_j E_{0j}}{\beta} \mu. \tag{4.29}$$

Although (4.29) is a small quantity of the first order in μ, the amplitude of B from (4.29) is comparable with (4.22) because $\sqrt{\frac{k_0}{a_2\beta}}$ in (4.22) is, in general, a very small quantity (k_0 is a small number in our conditions of the problem). The essential difference between the two solution lies, however, in the fact that although in the first variant a wave of high frequency vanishes practically at once, the solution of the wave equations for $\tau \neq 0$ shows an attenuation which is practically zero over depths equal to a considerable number of wave lengths.

The wave solution shows a smaller phase velocity of the wave radiated towards the interior of the crystal. This is an essential feature of the solution of the wave equation. Its experimental confirmation will constitute a qualitative proof of finite velocity of propagation of thermal disturbances. Let us observe finally, that the measurement of the effect mentioned is not a simple matter either. To avoid complicated calculations we have not taken into consideration the mechanical damping (depending on the first derivatives of strain with respect to time). This will not alter the qualititive result, but will influence in an essential manner the quantitative side of the relations and the experimental realization of the measurement. It appears, however the possibility of creating conditions for the occurence of a distinct wave effect of propagation of thermal disturbances may only be founded on processes of marked nonstationary character and of rather long duration. These processes are connected in turn with high frequency fields which require the application of effects of coupled fields. In the present paper we have introduced a number of far reaching simplifications (use of the asymptotic method etc.). Therefore, the results obtained should be regarded as furnishing merely qualitative information. This is connected also with the fact that we have used in the whole paper the quantity T, defined exactly for equilibrum processes only, to describe the nonstationary process.

It is clear that this can be done without deviating from the physical meaning of T for $\tau < \frac{2\pi}{\omega}$ only.

However, with this restriction the measuring of wave effects should become rather complicated becouse of the small difference between k^* and k_3. The question remains as to the practical limits of extrapolation of T for some higher ω. There is, however, some reason to expect the physical meaning of T, extrapolated in a linear formulation of the problem, to remain applicable to the cases of some higher ω allowing the use of relations (4.27) for k_3[1].

The final answer must be given by the experiment. The same holds, of course, for the problem of a temperature discontinuity on the characteristics which may be treated as an approximation only.

References

[1] KALISKI, S.: Wave Equation of Heat Conduction. Bull. Acad. Polon. Sci. Ser. Sci. Techn. **4**, **13** (1965).

[2] — Wave Equations of Thermoelasticity. Bull. Adac. Polon. Sci. Ser. Sci. Techn. **5**, 13 (1965).

[3) — Wave Equations of Thermo-electro-magneto-elasticity. Proc. Vibr. Probl. **3**, 6 (1965).

[4] ECKART, C.: The Thermodynamics of Irreversible Processes. Phys. Rev. 58 (1940).

[5] MEIXNER, J., and H. G. REIK: Thermodynamik der irreversiblen Prozesse. Handbuch der Physik, Vol. II/3. Berlin-Göttingen-Heidelberg: Springer. 1959.

[6] HUGHES, W. F.: Relativistic Magneto-hydrodynamics and Irreversible Thermodynamics. Proc. Cambr. Phil. Soc. **4**, 57 (1961).

[7] CATTANEO, C.: Sulla conduzione del calore. Atti de Seminario Math. u Fisico della Univ. di Modena **3** (1948).

[8] VERNOTTE, P.: Les paradoxes de la théorie continue de l'équation de la chaleur. Comptes Rendus **246**, 3154 (1958).
La véritable équation de la chaleur. Comptes Rendus **247**, 2103 (1958).

[9] GOLDSTEIN, S.: On the Diffusion by Discontinuous Movements and the Telegraph Equation. Q. J. Mech. Appl. Math. **2**, 4 (1954).

[10] SASAKI, M.: On the Relativistic Gases. Reprint from the Max-Planck-Festschrift. Berlin: 1958.

[11] FREUDENTHAL, A. M.: One Dimensional Response and Coefficients of Thermal Expansion in Timesensitive Materials. Acta. Techn. Acad. Sci. Hungaricae **3/4**, 41 (1962).

[12] KALISKI, S., and J. PETYKIEWICZ: Dynamical Equations of Motion Coupled with the Field of Temperatures and Resolving Functions for Elastic and Inelastic Anisotropic Bodies in the Magnetic Field. Proc. Vibr. Probl. **1**, 3 (1960).

[13] GROOT, S. R., and P. MAZUR: Non-Equilibrum Thermodynamics. Amsterdam: North-Holland Publ. Co. 1962.

[14] MINDLIN, R.: On the Equations of Motion of Piezoelectric Crystals. In: Problems of Continuum Mechanics. Philadelphia, Penn.: Soc. Ind. Appl. Math. 1961.

[15] KALISKI, S.: Čerenkov Generation of Ultra and Hypersounds. I — Cubic Crystals. II — Piezo-quartz. Proc. Vibr. Probl. **3**, 7 (1966).

---

[1] Private communication by I. MÜLLER concerning his dissertation on heat propagation problem.

Discussion

DRUCKER: How big is this effect going to be in a physical material? What amplitude will you get from an experiment?

KALISKI: The amplitude is not great but is real for physical measurement.

MÜLLER: I do not think, that the generalisation of the parabolic heat equation is so easy as you have assumed and only consists of a term which is the second derivative of temperature. You may get better results by consideration of additional terms, which follow from a rigorous thermodynamical treatment and affect the finite propagation velocity.

SEDOV: The parabolic equation follows only as result of the approximation?

KALISKI: No, I think, it is some different situation. If you consider as a statistical measure the Boltzmann transport equation and the equation of motion it would be extremely difficult to get the hyperbolic equation in solids. And therefore the real way is such a phenomenological way or the microscopic quantum mechanics way, but this, I suppose is not really possible in short time. It is a very interesting thing, that if you use the equations of the special theory of relativity, and if you do not generalize the Fourier relation then you will get parabolic equations.

MÜLLER: It is true, that the special theory of relativity does not eliminate this paradox. That is so, because it is not a relativistic effect, it is a purely classical effect. And the statement that you made just now in your answer to Prof. SEDOV's question, that the hyperbolic equation does not result from kinetic theory of gases is right only for the case of the Chapman-Enskog approximation, but not right for a more general solution of the Boltzmann equation. The situation is as follows: The usual thermodynamics of irreversible processes with Fourier's law is — in terms of the kinetic theory — the first approximation in deviations over mean free paths or over mean collision times, and the term which you introduced in your lecture is part of the second approximation.

On the Application of the Principles of Thermodynamics to Strained Solid Materials

By

J. Kestin

Providence, R.I. (U.S.A.)

"Physics, whether we choose to call it thermodynamics or not, should not be impotent, in the face of any situation which can be completely characterized by measurements with macroscopic instruments."

(P. W. Bridgman [1])

Summary

It is shown that a meaningful analysis of strained solid materials must begin with the identification of the independent thermodynamic parameters which determine their state. For example, in plasticity it is necessary to split the total strain into a part which is and a part which is not a parameter of state. In viscoelasticity, a similar splitting affects the stress, and in creep and relaxation it is necessary to introduce one or more additional variables of state. The importance of these observations lies in the fact that thermodynamic potentials can be functions of the parameters of state only.

An outline is given of a general method which allows us to extend our knowledge of uniform systems to include continuous systems. This method is based on the acceptance of the principle of local state which has been used successfully in the theories of elasticity and fluid mechanics or heat transfer and which forms the basis of the summary treatment due to de Groot and Mazur [4].

Particular attention is paid to plastic deformation and to strain hardening. First it is pointed out that the relation between the stress and the plastic strain rate does not seem to come within the scope of Onsager's relations whose validity is restricted to small departures from equilibrium. Secondly, a fully three-dimensional theory of elastic-plastic domains is proposed. Quoting some results obtained by J. R. Rice, it is made plausible that such a theory is likely to lead to a physically realistic description of the Bauschinger effect and of the variations in the yield surface under repeated loading cycles.

1. Introductory Statement

Many attempts have been made to fit the behavior of solid, strained materials into the framework of classical thermodynamics. One notable attempt was made by Bridgman [1] who came to the conclusion that such an aim is, indeed, possible, but that the foundations of thermodynamics required to be extended. This led him to formulate the concepts of generalized entropy and thermodynamic heat. We think that nothing so drastic is necessary and that all that is required is a careful interpretation of experimental results in terms of quite familiar ideas. In this way, we believe, it is possible to present a coherent and logically satisfactory framework. It is true, however, that use must be made of certain observations which are quite general but which are not normally in explicit use.

In order to concentrate exclusively on the physical aspects of the problem at hand we shall cast our reasoning in terms of the simplest possible solid system, namely a thin rod subjected to a uniform stress σ, positive for tension and negative for compression, and undergoing a single strain ε, also positive for extension and negative for compression. We shall completely desregard the possibility of the occurrence of instability through buckling or of the formation of a local "neck". Finally, we shall assume that our material obeys Hooke's law in the elastic range, and that its Young's modulus, E, its coefficient of linear thermal expansion, α, and its specific heat, c_ε, at constant strain are constant. Needless to add, the assumption of infinitely small strains rather than those of finite elasticity will also be implied.

The extension of the present interpretations to include elastic continua which sustain complete tensorial fields of stress and strain, to materials whose properties depend on temperature and whose stress-strain relations are nonlinear and, ultimately, to finite elasticity, requires a good deal of detailed work, though no difficulties of a fundamental nature occur if the principle of local state rather than the so-called "memory" approach is adopted. The same can be said about visco-elastic behavior, but the inclusion of plastic stress-strain relations, particularly in the presence of strain hardening, requires the adoption of a more detailed model. This model leads to the *theory of elastic-plastic domains* which is given in outline in Sections 13 and 14.

2. Elastic Range

We begin by reviewing the most elementary theory of elastic behavior in order to provide for easy reference later.

The first task which faces us before the principles of thermodynamics are applied is to establish the form of the equation of state of the system, and in order to achieve this, the number and admissible nature of the independent variables which describe the state of the system must be established. This can only be done by an appeal to experiment.

We shall accept that the state of an elastic system is determined by *two* independent variables of state. These can be chosen arbitrarily from among the three thermodynamic properties: thermodynamic temperature T, stress σ, and strain ε. Anticipating the need for an extension to continua, we may state here that in the more general case it is necessary to make a distinction between those components of the strain tensor ε_{ij} which describe the volume of the system, the trace ε_{ii}, and those which describe the shape, the strain deviator $\tilde{\varepsilon}_{ij}$ of trace zero ($\varepsilon_{ij} = \tilde{\varepsilon}_{ij} + \frac{1}{3}\,\delta_{ij}\varepsilon_{kk}$ with $\tilde{\varepsilon}_{kk} = 0$). The ability to make this distinction is lost in our one-dimensional system.

It follows that there exists a single-valued relation

$$f(\sigma, \varepsilon, T) = 0 \tag{1}$$

which we shall call the *thermal equation of state* of the system. For our simple system, this equation has the form

$$\varepsilon = \frac{\sigma}{E} + \alpha (T - T_0), \tag{2}$$

where α is the coefficient of linear thermal expansion and T_0 is an arbitrary reference temperature. This means that the system can be relaxed at $T = T_0$ so that ε and σ vanish simultaneously. The state described by $T = T_0$, $\sigma = 0$, and $\varepsilon = 0$ is conveniently called the *natural state*. For simplicity, it is assumed that the coefficient of thermal expansion, α, as well as Young's modulus, E, are independent of stress and temperature.

Equation (2) represents a plane in the three-dimensional stress-strain-temperature space of states, and isotherms in the stress-strain state diagram are represented by a family of parallel straight lines as shown in Fig. 1.

Processes of quasistatic loading and unloading are reversible, and the work required to perform them is

$$d\,W^0 = -\sigma\, d\varepsilon \tag{3}$$

per unit volume of system. Here, and later, the superscript o in $d\,W^0$ and dQ^0 will be used to emphasize the reversible nature of the process under consideration. The reversible work $d\,W^0$ is of the standard form of a generalized force (an intensive quantity), multiplied into the perfect differential of a generalized displacement (a specific quantity) and constitutes a Pfaffian expression. Since $d\,W^0$ relates to a *reversible* process, both quantities which appear in it constitute *parameters of state*. If the process were not reversible, still remaining quasistatic, the quantities need *not* represent a parameter of state[1].

[1] Since the preceding statement may not be immediately understandable to the reader, we shall illustrate it with examples in Section 4.

The thermal equation of state is not a fundamental one[1] and does not, therefore, contain a complete statement of the equilibrium properties of the system. This is achieved, as is well known, if the specific internal energy per unit volume u is given as a function of specific entropy s and strain ε,

$$u = u(s, \varepsilon). \tag{3a}$$

Alternatively, the enthalpy

$$u - \sigma\varepsilon = h(s, \sigma), \tag{3b}$$

the Helmholtz function (free energy)

$$u - Ts = f(T, \varepsilon), \tag{3c}$$

or the Gibbs function (free enthalpy)

$$u - \sigma\varepsilon - Ts = g(T, \sigma), \tag{3d}$$

each in terms of the variables indicated, can be used. The four fundamental forms are entirely equivalent and arise from each other through contact (Legendre) transformations.

We concentrate on eq. (3a) and proceed to calculate the internal energy and entropy, making full use of the thermal equation of state. The internal energy u is determined by the two partial derivatives $(\partial u/\partial \varepsilon)_T$ and $(\partial u/\partial T)_\varepsilon$. The former is determined from the combination of the First Law and the First Part of the Second Law of thermodynamics written for a reversible process. This leads to the eqs.

$$du = dQ^0 - dW^0 \quad \text{and} \quad du = T\,ds + \sigma\,d\varepsilon \tag{4a, b}$$

since

$$dQ^0 = T\,ds.$$

It follows that

$$\left(\frac{\partial u}{\partial \varepsilon}\right)_T = T\left(\frac{\partial s}{\partial \varepsilon}\right)_T + \sigma, \tag{4c}$$

and Maxwell's reciprocal relation

$$-\left(\frac{\partial s}{\partial \varepsilon}\right)_T = \left(\frac{\partial \sigma}{\partial T}\right)_\varepsilon \tag{4d}$$

leads to

$$\left(\frac{\partial u}{\partial \varepsilon}\right)_T = -T\left(\frac{\partial \sigma}{\partial T}\right)_\varepsilon + \sigma. \tag{4e}$$

---

[1] A fundamental equation of state allows us to compute all equilibrium properties of a system by differentiation alone; see, for example, [5] p. 533.

Maxwell's reciprocal relation is obtained by casting eq. (4b) in the form (3c) through the application of a Legendre transformation, that is from

$$df = d(u - Ts) = -s\,dT + \sigma\,d\varepsilon. \tag{5}$$

Substitution of eq. (2) into (4e) provides us with the partial derivative

$$\left(\frac{\partial u}{\partial \varepsilon}\right)_T = E\varepsilon + \alpha E T_0 \tag{5a}$$

of internal energy with respect to strain in terms of the thermal equation of state. It is well-known that the integral of this equation leads to that fundamental equation which is consistent with the adopted thermal equation of state. Integration yields

$$u = \frac{1}{2} E\varepsilon^2 + \alpha E \varepsilon T_0 + \psi(T),$$

where $\psi(T)$ is the constant of integration — a function of temperature alone.

The specific heat at constant strain, c_ε, is defined as $(\partial u/\partial T)_\varepsilon$ and

$$c_\varepsilon = \left(\frac{\partial u}{\partial T}\right)_\varepsilon = \psi'(T).$$

We assume the simplest form for $c_\varepsilon = \psi'(T)$ in line with our policy to reduce the mathematical complexity to a minimum, and postulate that c_ε is a constant. Thus

$$u(\varepsilon, T) = \frac{1}{2} E\varepsilon^2 + \alpha E \varepsilon T_0 + c_\varepsilon(T - T_0), \tag{6}$$

with $u = 0$ at $T = T_0$ and $\varepsilon = \sigma = 0$. At $T = T_0$

$$u = \frac{1}{2} E\varepsilon^2 + \alpha E\varepsilon T_0 = \frac{\sigma^2}{2E} + \alpha\sigma T_0. \quad (T = T_0) \tag{7}$$

It is worth noting here in passing that the term $\alpha\sigma T_0$ in eq. (7) is usually much larger than the term which represents strain energy, $\sigma^2/2E$.[1] Thus the internal energy, as opposed to strain energy, decreases on isothermal compression, reverting to its original value at the unrealistically high compressive stress $\sigma' = -2\alpha E T_0$.[2]

The entropy must be calculated from the heat dQ^0 exchanged along a *reversible* path according to the equation

$$ds = \frac{dQ^0}{T}. \tag{8}$$

[1] For aluminum $E = 7 \times 10^5$ kp cm/cm^3, $\alpha = 23.5 \times 10^{-6}$ 1/°K; for steel $E = 21 \times 10^5$ kp cm/cm^3, $\alpha = 13.3 \times 10^{-6}$ 1/°K; at $\sigma = 400$ kp/cm^2; $\alpha\sigma T_0 / \frac{1}{2}\sigma^2 E$ ≈ 20 for aluminum and ≈ 40 for steel.

[2] $\sigma' \approx -10{,}000$ kp/cm^2 for aluminum and $\sigma' \approx -17{,}000$ kp/cm^2 for steel.

The quantity of *reversible* heat dQ^0 exchanged during an elastic process is not known directly, but is computed from the First Law of thermodynamics applied to a *reversivible path*

$$dQ^0 = du + d\,W^0 = \alpha E T\, d\varepsilon + c_\varepsilon\, dT. \tag{9}$$

Hence

$$ds = \alpha E\, d\varepsilon + c_\varepsilon \frac{dT}{T} \tag{10}$$

and

$$s = \alpha E \varepsilon + c_\varepsilon \ln \frac{T}{T_0} \tag{10a}$$

with $s = 0$ at the natural state. At $T = T_0$

$$s = \alpha E \varepsilon = \alpha \sigma. \qquad (T = T_0) \tag{10b}$$

The elimination of temperature from eqs. (6) and (10a) leads to the fundamental equation of state

$$u(s, \varepsilon) = \frac{1}{2} E \varepsilon^2 + \alpha E \varepsilon T_0 + c_\varepsilon T_0 \left\{ \frac{\exp(s/c_\varepsilon)}{\exp(\alpha E \varepsilon / c_\varepsilon)} - 1 \right\} \tag{11}$$

$$\approx \frac{1}{2} E \varepsilon^2 + (c_\varepsilon - \alpha E \varepsilon)\, T_0 (e^{s/c_\varepsilon} - 1); \tag{11a}$$

the approximation (11a) may be used when

$$\frac{\alpha E \varepsilon}{c_\varepsilon} \ll 1.$$

The fundamental equation exhibits the constants E, α, c_ε of the material which are necessary and sufficient to determine *all* its *thermodynamic equilibrium* properties.

It may be worth noting that stressing a rod at constant temperature $T \neq T_0$ from $\sigma = 0$ to σ requires the following amounts of reversible work and heat:

$$W^0 = - \int_{\alpha(T - T_0)}^{\varepsilon} \sigma\, d\varepsilon = \frac{1}{2} E [\varepsilon - \alpha (T - T_0)]^2 = - \frac{\sigma^2}{2E} \tag{12}$$

$$Q^0 = \int_{\alpha(T - T_0)}^{\varepsilon} T\, ds = \alpha E T [\varepsilon - \alpha (T - T_0)] = \alpha \sigma T. \tag{13}$$

We note again, as already stated in connection with eq. (7), that the quantity of reversible heat during such an isothermal process, Q^0, is many times larger than the amount of reversible work, $-W^0$, performed on the system, contrary to what is asserted in some elementary books on elasticity. The negative of reversible work measured with respect to $\sigma = 0$ (but not $\varepsilon = 0$), W^0, represents the strain energy of the system, i. e. the change in the Helmholtz function at constant temperature.

The equations derived in this section allow us to supplement the state diagram of Fig. 1 with curves of constant internal energy and entropy. This has been done in Fig. 2 which also shows a u, ε state diagram. We omit the derivation of the respective equations for the sake of brevity, and confine the diagrams to the regions of interest in engineering elasticity.

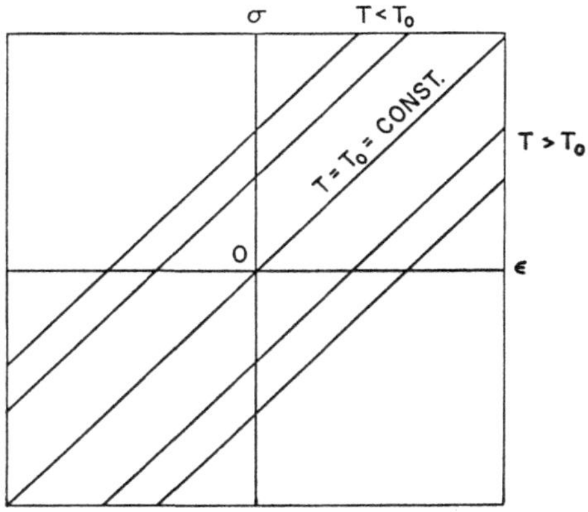

Fig. 1. Equation of state for elastic system

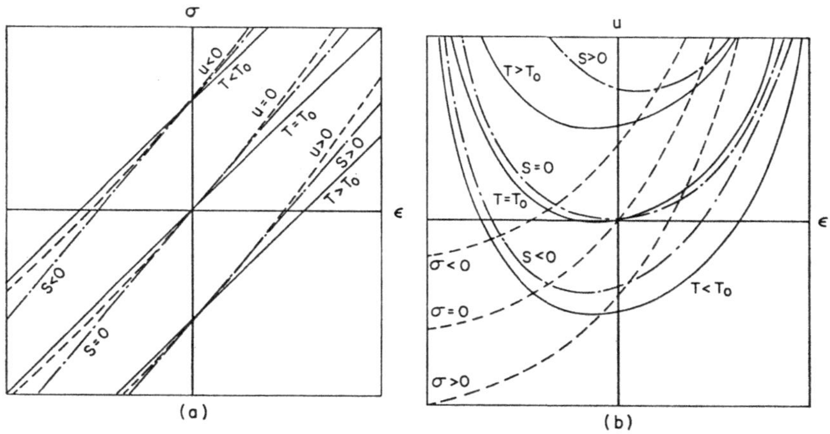

Fig. 2. State diagrams for a linear elastic rod

3. Elastic Continuum[1]

It might be useful to record here the equations of linear, infinitesimal elasticity employing thermodynamic terms. The thermal equation of state is

$$\varepsilon_{ij} = \frac{1+\nu}{E}\,\sigma_{ij} - \frac{\nu}{E}\,\partial_{ij}\sigma_{kk} + \delta_{ij}\alpha(T - T_0), \tag{14}$$

[1] The equations given in this Section have been derived by Professor J. R. Rice of Brown University.

or

$$\sigma_{ij} = \frac{E}{1+\nu}\,\varepsilon_{ij} + \frac{E\nu}{(1+\nu)(1-2\nu)}\,\delta_{ij}\varepsilon_{kk} - \frac{E}{1-2\nu}\,\delta_{ij}\alpha(T - T_0). \tag{15}$$

The equation is restricted to an isotropic continuum because it has been assumed that a change in temperature causes only a change in volume. In this case, the equation contains four independent variables: three components of stress or strain (or their invariants) and temperature, and the three material constants: E, ν, and α.

Maxwell's reciprocal relation (4d) is replaced by the equation

$$\left\{\frac{\partial \sigma_{ij}(\varepsilon_{kl}, T)}{\partial T}\right\}_{\varepsilon_{kl}} = -\left\{\frac{\partial s(\varepsilon_{kl}, T)}{\partial \varepsilon_{ij}}\right\}_T, \tag{16}$$

because the Helmholtz function $f = u - Ts$ satisfies the Gibbs equation

$$df = -\,s\,dT + \sigma_{ij}\,d\varepsilon_{ij}.$$

Hence the energy is

$$u(\varepsilon_{kl}, T) = \frac{1}{2}\,\sigma_{ij}\varepsilon_{ij} + \frac{1}{2}\,\frac{E}{1-2\nu}\,\alpha(T - T_0)\,\varepsilon_{kk} + \\ + \frac{E}{1-2\nu}\,\alpha T_0\varepsilon_{kk} + c_\varepsilon(T - T_0), \tag{17}$$

and the entropy becomes

$$s(\varepsilon_{kl}, T) = \frac{\alpha E}{1-2\nu}\,\varepsilon_{kk} + c_\varepsilon \ln\frac{T}{T_0}. \tag{18}$$

It is noteworthy that the entropy is not changed when a change in stress produces merely a change in shape. This is a direct consequence of the isotropy expressed by the assumption of a scalar coefficient of thermal expansion. For this reason, a reversible isothermal process at constant volume is also isentropic and occurs without an exchange of heat with the surroundings, since then

$$Q^0 = T\Delta s = 0.$$

The fundamental equation of state $u = u(s, e_{ij})$ can now be written in the form

$$u = \frac{1}{2}\,\frac{E}{1+\nu}\,\varepsilon_{ij}\varepsilon_{ij} + \frac{1}{2}\,\frac{E\nu}{(1+\nu)(1-2\nu)}\,(\varepsilon_{kk})^2 + \frac{E}{1-2\nu}\,\alpha T_0\varepsilon_{kk} + \\ + c_\varepsilon T_0\left\{\frac{\exp(s/c_\varepsilon)}{\exp[\alpha E\varepsilon_{kk}/c_\varepsilon(1-2\nu)} - 1\right\}. \tag{19}$$

4. A Remark on Quasistatic Work

The expression for work in thermodynamics is a natural generalization of the concept of work in mechanics. When the process is quasistatic, reversible or irreversible, the form of the expression for elementary work is that of some *external*, generalized force, multiplied into the differential of a generalized displacement of the system. When the process is reversible, i. e. when it consists of a continuous sequence of equilibrium states, the external generalized force is exactly balanced by an internal force. It follows that the generalized force enters into the description of the instantaneous state of the system and thus *constitutes a parameter of state*. The displacement of the external force in a reversible process is equal and opposite to that of the system, and it, too, *constitutes a parameter of state*. The most familiar example is the work of reversibly changing the volume of a fluid,

$$d\,W^0 = p\,d\,V, \tag{20}$$

in which both the pressure p and the volume V are parameters of state. The same is true of the stress and strain which appear in eq. (2).

The preceding assertion cannot be made about quasistatic *irreversible* processes because during such processes the system is *not* in an equilibrium state but traverses a continuous sequence of *near-equilibrium* states. We agree to describe such states by those values of the properties which would result if the system were allowed to reach equilibrium in adiabatic isolation. Hence at least one of the parameters cannot possibly constitute a property.

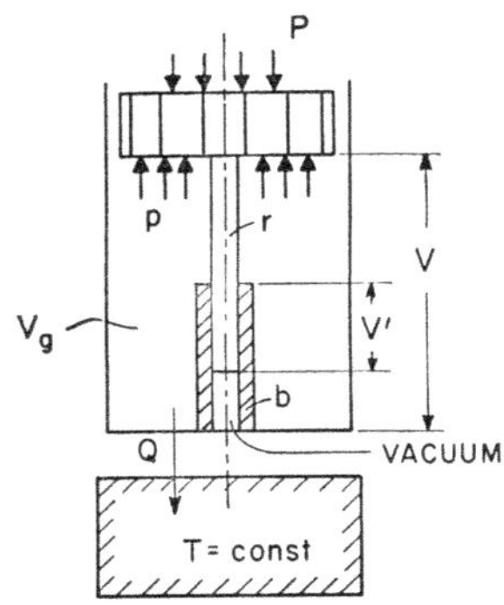

Fig. 3. Piston moving with friction

Several examples may serve to illustrate the point. The system in Fig. 3 is a simple variation on the most familiar thermodynamic system consisting of a gas in a cylinder whose volume is varied by a piston. In this case we shall assume that the piston moves with friction produced by the motion of rod r in bushing b (imagined evacuated). We assume that the process is isothermal (or nearly so) in that heat is exchanged with a reservoir of constant temperature. The process of compression or expansion is quasistatic, but clearly irreversible. The system is not in exact equilibrium, because the transfer of heat into the reservoir requires a temperature gradient. The work of the process is

$$dW = P\,dV, \tag{21}$$

where P denotes the external pressure. Clearly, P is not equal to the pressure p of the gas, but

$$P = p \pm p' \quad \text{with} \quad p' = |F/A|. \tag{22}$$

Here F is the friction force in the bushing-rod assembly and A is the area of the piston. Whereas p is a true parameter of state, P is not. Furthermore, V is not quite equal to the volume of the gas if the displacement of rod r is accounted for and can be represented as the sum

$$V = V_g + V',$$

of the volume of the gas V_g and the displaced (positive or negative) volume of the rod. Hence in eq. (18), V represents a parameter of state if the displacement of the rod is neglected, otherwise it does not. Thus in the expression

$$dW = (p \pm p')dV_g + (p \pm p')dV',$$

only the term

$$dW^0 = p\,dV_g,$$

represents reversible work, and at least one of the parameters P and V does not constitute a parameter of state.

The second example (Fig. 4) consists of two elastic rods which are free to move with respect to each other when a friction force $F_f = \mu N$ produced by the transverse compression forces N is overcome. We again assume isothermal operation and note that

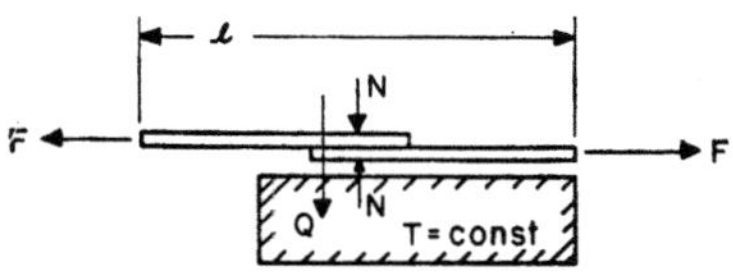

Fig. 4. Elastic rods with friction

$$dW = -F\,dl, \tag{23}$$

regardless whether the process is reversible or not. However as long as $F < F_f$, the process will be reversible, because only elastic straining will occur. Hence

$$dW^0 = -F\,dl \quad \text{if} \quad F < F_f. \tag{24}$$

As soon as the external force F becomes equal to the friction force, slippage will occur at the junction and heat will flow into the reservoir irreversibly; it flowed reversibly when F was smaller than F_f. Now the force transmitted by the elastically strained rods continues to be equal to the external force F but the sum of the strains of the elastic rods is no longer equal to the total displacement l measured from tip to tip. We continue to write the expression for work for this process in the form of eq. (23), but note that the length l ceases to be a parameter of state. This length is approximately equal to the sum of the strained lengths of the two rods, l_r, and the length of relative displacement l_d; the difference stems from the overlapping portions of the rod which are strained differently as l_d changes. Neglecting this small complication, we shall write

$$dW = -F\,d(l_r + l_d) = -(F\,dl_r + F\,dl_d). \tag{24a}$$

Here only the first term, $dW^0 = -F\,dl_r$, represents reversible work.

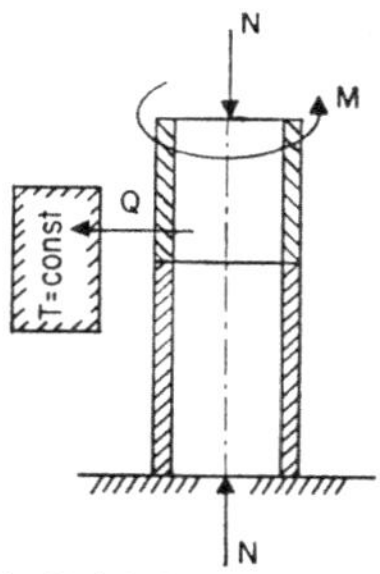

Fig. 5. Twisted cylindrical shells

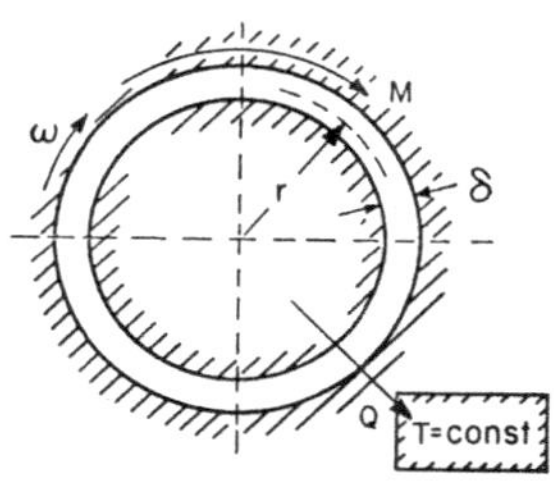

Fig. 6. Couette flow

An analogous example is represented in Fig. 5 where two very thin cylindrical shells are pressed together and twisted with torque M. As long as there is no slippage

$$dW^0 = M\,d\gamma, \tag{25}$$

where γ denotes the angle of twist, and both M and γ constitute parameters of state. With slip

$$dW = M\,d\gamma, \tag{26}$$

but γ ceases to be a parameter of state.

As our final example, we shall consider a thin, cylindrical, infinitely long lamina of a viscous fluid whose shape is distorted at an angular velocity ω by a torque M, while the volume remains constant (Fig. 6). An exchange of heat assures (nearly) isothermal operation. Work is performed at a rate

$$\frac{dW}{dt} = M\,\frac{d\gamma}{dt} \tag{27}$$

which can also be written

$$dW = M\,d\gamma. \tag{27a}$$

Here γ denotes the angle of twist.

It is well-known that the state of a fluid is described by two parameters of state, say its temperature T and volume V. In the process under consideration both quantities remain constant, and we must conclude that the state of the system remains unaltered. In other words, $dW^0 = 0$, and the parameters which appear in eq. (27a) are irrelevant to the description of the state. The entropy of the system remains unchanged (because the state is fixed), and since

$$ds = \frac{dQ^0}{T} = 0,$$

we must conclude that $dQ^0 = 0$, though heat is being exchanged. This is a process of energy dissipation in steady-state.

It is noted that in this example, the torque M is directly related to the rate of strain $d\gamma/dt$, and that

$$M = 0 \quad \text{when} \quad \frac{d\gamma}{dt} = 0.$$

No such relation between the generalized force and generalized displacement existed in the previous examples where the relationship between the generalized force and generalized displacement could be altered at will by suitably manipulating the forces of friction. Furthermore, in Fig. 3, the irreversible process occurred at a variable state, whereas in Figs. 4, 5 and 6 the irreversible process occurred at a fixed (or nearly fixed) state of the system.

The preceding examples lead to the valuable conclusion that the expression for reversible work dW^0 of a system can be used to identify true parameters of state. The corresponding expression dW for an irreversible process can be used to check whether a particular quantity does or does not constitute a parameter of state.

5. Ideal Plastic Deformation

We now proceed to analyze the case of ideal plastic deformation of an elastic rod. To this end, we shall consider a hysteresis loop performed with this system at constant temperature. Such a loop is represented in the stress-strain diagram of Fig. 7a by the process 0 1 2 3 4 5 0. We suppose that the natural state 0 is completely restored after one cycle and that unloading is reversible and proceeds with the original value of the elastic modulus. The former need not necessarily be the case, and whether the natural state is restored completely or not depends on the properties of the bar. For example, the material of the bar may be in a state of constrained chemical equilibrium which is displaced quasistatically during the process of deformation as a result of a chemical reaction. Thus the initial extent of reaction, ξ_1 at 0, may be different from the final extent ξ_2, also at 0. In such cases, two properties are inadequate for the description of the state, and the variable extent of reaction, ξ, must also be included, one such additional parameter appearing for every independent reaction in progress. The same can be said about the elastic modulus for unloading. We shall exclude such possibilities for the time being and will show in Section 12 that the restoration of the natural state is not generally possible in metallic bars.

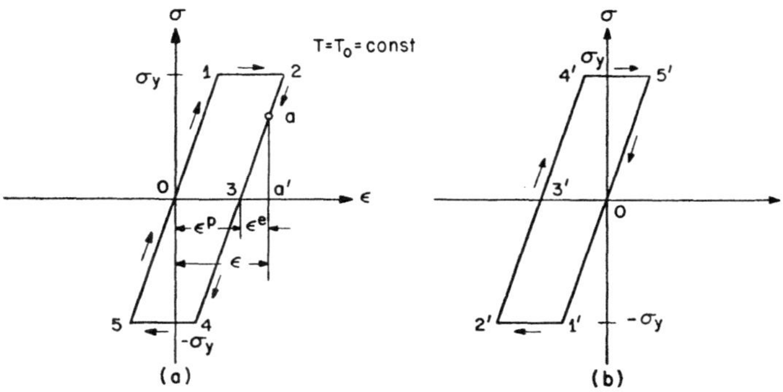

Fig. 7. Hysteresis loop of an elastic-ideally plastic rod

It must be clearly understood that the hysteresis loop 0 1 2 3 4 5 0 does not represent a reversible cycle, because reversing the direction of loading from 0 to 1′, as shown in Fig. 7b, produces a different loop, namely 01′2′3′4′5′0. This indicates that the processes of plastic deformation, 12 and 45 in Fig. 7a and 1′2′ and 4′5′ in Fig. 7b, are irreversible, because the remaining processes are reversible.

Regardless of this, it may be asserted that the total amount of work

$$W = -\oint \sigma d\varepsilon,$$

exactly balances the total amount of heat

$$Q = \oint dQ,$$

as long as the initial state is restored at 0. This is a consequence of the fact that the First Law applies equally to reversible and irreversible processes. Conversely, if the equality of heat and work is confirmed by experiment, it must necessarily be concluded that one complete hysteresis loop restores the original value of internal energy.

Experiment further suggests that the work of plastic deformation tends to balance the heat in the limit of ideal plastic behavior. Thus for any ideally plastic path, say 12 in Fig. 7a, we may write

$$\int_1^2 dQ = \int_1^2 dW,$$

so that

$$dQ = dW = -\sigma d\varepsilon \quad \text{(ideal plastic path).} \tag{28}$$

Appeal to the First Law, $du = dQ - dW$, shows that

$$du = 0 \quad \text{(along plastic path).} \tag{28a}$$

We must therefore conclude that the internal energy of a bar deformed plastically and isothermally remains constant. This means that the internal energy of a bar deformed plastically at the yield stress σ_y is independent of the amount of plastic deformation imposed upon it.

The preceding argument has shown that along the lines of plastic deformation 1 2, 4 5, 1′2′ and 4′5′ the following parameters remain constant:

$$\sigma = \sigma_y; \quad T; \quad u.$$

On the other hand, the strain ε can attain an infinity of values. It is usually stated that this circumstance removes plastic deformation from the province of classical thermodynamics, because the equation of state

$$f(\sigma, \varepsilon, T) = 0,$$

must be single-valued. The conclusion is reached that a thermal equation of state for this system "does not exist", because the preceding equation is obviously multi-valued, and that "it is necessary to find a suitable generalization of thermodynamics" [1] to meet this contingency. Conclusions of this type are inescapable, *if and only if* the assumption is made that *the total strain ε continues to be a true parameter of state.*

In order to restore the situation, we must abandon the preceding, unjustified assumption, no matter how compelling it appears to be on intuitive grounds. The fact that $dW = -\sigma d\varepsilon$ along the irreversible, plastic paths 1 2 and 4 5 should be an indication that the total strain ε is not a true parameter of state.

We interpret experiments to mean that the state of a bar in plastic deformation continues to be described by *two* independent properties. We have shown earlier that along any path of isothermal plastic deformation as many as *three* parameters remain constant. The conclusion is inescapable: *all parameters of state remain constant.* In other words, the process of isothermal plastic deformation occurs at a *constant state.* Thus, all states along a horizontal line $\sigma = \text{const}$ $(\sigma \leq \sigma_y)$ in an isothermal stress-strain diagram *are identical.*

The preceding conclusion has (almost) been reached by BRIDGMAN [1] who ascribed to all such points the same entropy. He thought, however, that he introduced thereby a "generalized entropy", it being clear that he implicitly assumed that the total strain, ε, continued to be a parameter of state. If this is recognized, no generalization is required.

We must, therefore, make a clear distinction between a stress-strain diagram and a state diagram. The state diagram for Fig. 7 is

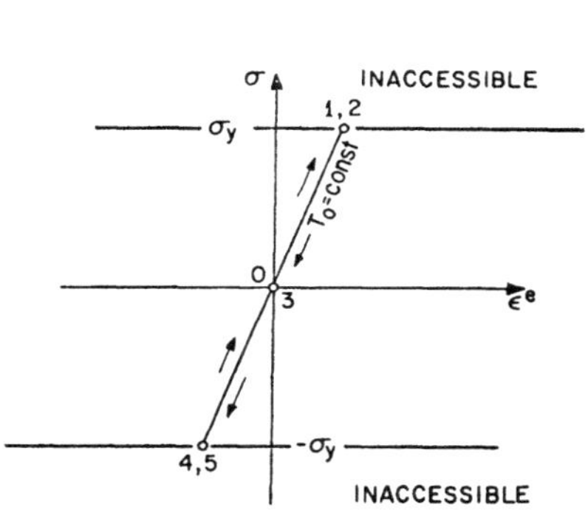

Fig. 8. State diagram for Fig. 7

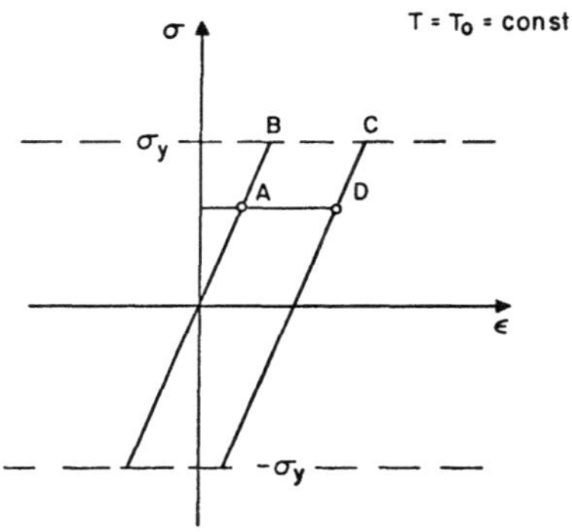

Fig. 9. The computation of entropy

shown represented in Fig. 8. No such distinction was necessary in the case of an elastic system.

In order to identify the proper strain parameter, it is sufficient to allow the system to relax from a given state of plastic deformation, say from state 2 in Fig. 7 along 2 3. The resulting process is reversible, and the work is

$$d W^0 = -\sigma d \varepsilon^e,$$

where ε^e must now be measured with respect to point 3 which is in the natural state. Thus, at any intermediate point, say a in Fig. 7a, we must distinguish between the elastic strain, ε^e, which constitutes a parameter of state, and the plastic strain (or permanent set), ε^p, which does not. It is clear that

$$\varepsilon = \varepsilon^e + \varepsilon^p, \tag{29}$$

and the preceding argument shows that the total strain must be separated into an elastic part and a plastic part (permanent set).

We must now devote a few remarks to entropy. Since isothermal plastic deformation does not change the *state* of the bar, it cannot change its entropy. It is often argued as follows [1]: Consider two arbitrary states A and D as shown in Fig. 9. Any attempt to compute the difference in their entropies must fail, because in order to do so it is necessary to span a *reversible* path between them. Now, clearly, no reversible path joining such two "states" can be found. Consequently eq. (8) cannot be applied, because dQ^0 cannot be determined even *in principle*. Thoughts of this kind compelled BRIDGMAN [1] to write: "There is a twofold problem here — the problem of extending the conceptual machinery so as to be capable of handling the new situations, and the experimental problem of finding what the facts are in the new domain". As already indicated, the only conceptual change required is to recognize that states A and D are identical. It is clear, as shown above, that states B and C are identical. Furthermore, since $s_B = s_C$, and

$$s_B - s_A = s_C - s_D,$$

we also have

$$s_A = s_D.$$

There is no need to find a reversible path, because the difference in total strain between "state" A and D *does not amount to a difference in state*. Indeed, inverting the argument, we can assert that

$$dQ^0 = 0 \quad \text{along} \quad \sigma = \text{const}, \quad T = \text{const}.$$

This is the quantity which BRIDGMAN called the "thermodynamical heat" as opposed to "experimental heat" — heat in an irreversible process in our present terminology. BRIDGMAN concluded correctly that the "thermodynamical" heat for a fictitious path of $\sigma = \text{const}$, $T = \text{const}$ is zero. In this manner the difficulties connected with P. W. BRIDGMAN's "sea of irreversibility" also disappear; they will re-appear in a different guise in Section 12. The impression of a "sea of irreversibility" is created when an irreversible process (such as rubbing) occurs nearly isothermally at an essentially constant state.

Accepting the preceding conclusions, we recognize that the fundamental equation of state for our system is identical with eq. (11) on condition that the permanent set ε^p is subtracted from the total strain ε, or that ε is replaced by ε^e. Hence, for systems in plastic deformation

$$u(s, \varepsilon) = \frac{1}{2} E(\varepsilon^e)^2 + \alpha E \varepsilon^e T_0 + c_\varepsilon T_0 \left\{ \frac{\exp(s/c_\varepsilon)}{\exp(\alpha E \varepsilon^e / c_\varepsilon)} - 1 \right\}. \qquad (30)$$

Since all other properties are derivable from this equation, all equations of Sections 2 and 3 remain valid subject to the substitution of ε^e for ε. More generally, if φ denotes any one of the thermodynamic potentials u, s, h, f, g, we can assert that

$$\frac{\partial \varphi}{\partial \varepsilon^p} \equiv 0 \quad \text{but, generally} \quad \frac{\partial \varphi}{\partial \varepsilon^e} = \frac{\partial \varphi}{\partial \varepsilon} \neq 0, \qquad (30\text{a})$$

with any one other parameter kept constant. We also note that

$$d\,W^0 = -\sigma\,d\,\varepsilon^e \quad \text{(elastic)} \tag{30b}$$

whereas

$$d\,W = -(\sigma\,d\,\varepsilon^p + \sigma\,d\,\varepsilon^e) \quad \text{(plastic)} \tag{30c}$$

with

$$dQ = -\sigma\,d\,\varepsilon^p, \tag{30d}$$

denoting the heat dissipated.

The elastic strain ε^e can always be determined by allowing the system to relax, and it must not be forgotten that any theories which assume that the energy u, entropy s (or the Helmholtz function f, or any thermodynamic potential) depend on the total strain ε or on the plastic strain ε^p cannot possibly lead to results which agree with experiments.

6. Strain Hardening

The extension of our ideas to include strain-hardening materials can proceed on two levels of approximation. In the first approximation we continue to ignore changes in cross-sectional area and other three-dimensional effects. In the second approximation, to be presented in the succeeding section, we propose a theory of elastic-plastic domains and take the view that three-dimensional effects are important even in the case of slender rods subjected to uniaxial loading.

The diagram in Fig. 10 represents an idealized hysteresis loop with the reversible portions 4 1 and 2 3, and the irreversible, strainhardening portions 1 2 and 3 4. We suppose that the hysteresis loop can be closed and that the bar can be restored to its natural state at 0. It is realized that many experiments can be quoted which suggest that the initial state is not restored at 0, and we shall deal with this point later. As a first approximation we assume that this is possible and so establish the validity of the unstated principle which pervades thermodynamics and which asserts that any given equilibrium state of a system *can* be restored by a sequence of processes, reversible or irreversible.

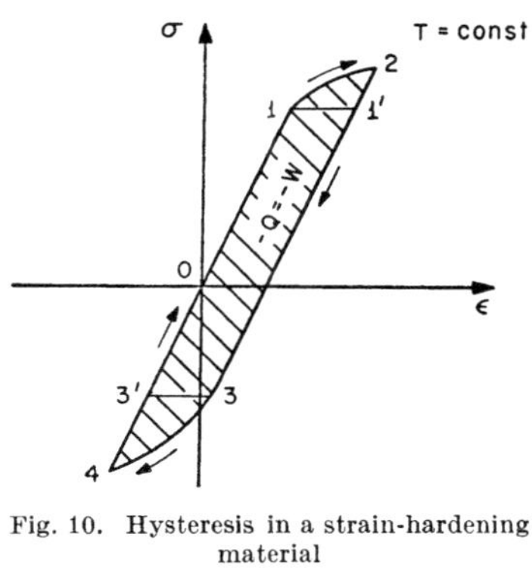

Fig. 10. Hysteresis in a strain-hardening material

In order to identify the stress parameter which represents a thermodynamic property, we use the device of determining the expression for reversible work in unloading, assuming that the material constants, including Young's modulus, have retained their values. We conclude that the elastic strain

$$\varepsilon^e = \frac{\sigma}{E} \tag{31}$$

with

$$\varepsilon = \varepsilon^p + \varepsilon^e, \tag{32}$$

continues to play that part. The method of determining ε^e is suggested by Fig. 11 which is in this respect similar to Fig. 7, except for the fact that the yield stress is now variable, being given by an empirical relation.

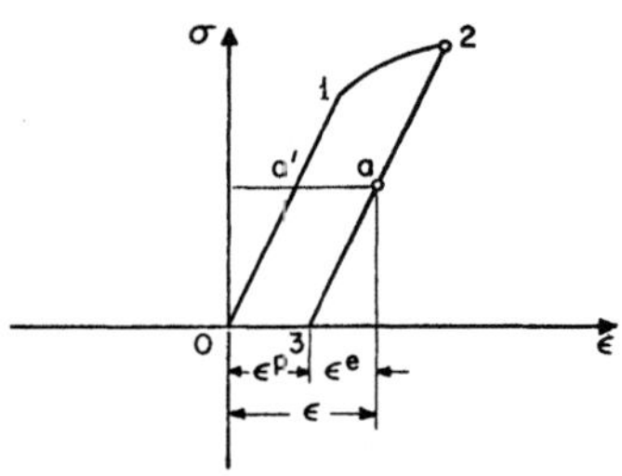

Fig. 11. The elastic stress

Since at two states of equal stress (a and a' in Fig. 11) at least three state variables, σ, ε^e and T, have the same values, we conclude that such two states are identical. This demonstrates that the sequence of the irreversible process 1 2 followed by the reversible process 2 1′ constitutes an irreversible, isothermal cycle.

The thermal equations of state, the expressions for internal energy, entropy, reversible work and heat of an elastic bar, retain their validity with the substitution of ε^e for ε. Similarly eqs. (30a, b, c) remain applicable. The state diagram is identical with that shown previously in Fig. 8 (or Figs. 1 and 2), except that the isotherms are now bounded by the rupture stress σ_r instead of the unique yield stress σ_y for an elastic ideally-plastic system.

The expression for work and heat dissipated, eqs. (30b, c, d), can be represented by areas in the isothermal stress-strain diagram; they are displayed in Fig. 12. The area 0 1 2 b represents the actual work performed on the system

$$\int_0^2 \sigma d\varepsilon = -W,$$

whereas area a 2 b represents the *reversible* work performed on the system. The difference, area 0 1 2 a, is proportional to the energy dissipated during the process (see eq. (50) of Section 10); it is not equal to the reversible heat Q^0 or the actual heat Q exchanged but to their *difference*. When the initial state is restored by unloading to a, area 0 1 2 a becomes equal to the energy dissipated during the cycle, that is to the heat rejected to the surroundings. This heat is then equal to the total work done.

Figure 5 contains a simple mechanical analog of a rod deformed plastically. The case when $N = \text{const}$ corresponds to ideal plastic behavior, whereas an increasing N represents the preceding simplified case of strain hardening.

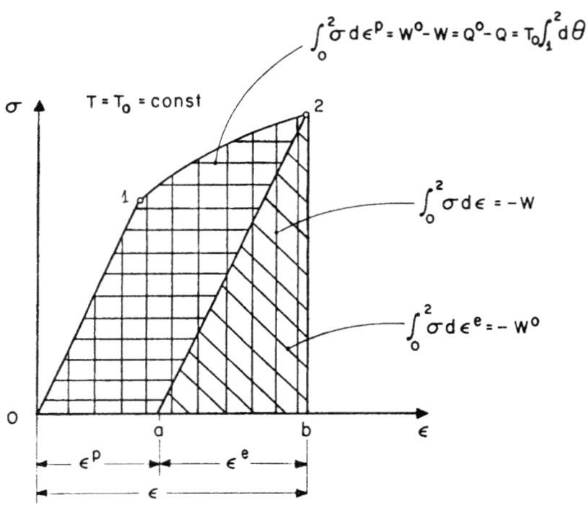

Fig. 12. Areas in an isothermal stress-strain diagram

7. Viscoelastic Material

We now turn our attention to the simplest possible viscoelastic material, namely one which obeys Hooke's law at any given strain rate $\dot{\varepsilon}$. Although the stress-strain relation is affected by the strain rate, it is supposed that the effects of inertia are negligible. The isothermal stress-strain diagram of such a material is shown in Fig. 13 in which the strain rate $\dot{\varepsilon}$ appears as an additional parameter. The first impulse is to say that the state of the system, for example at point A, is completely described by any three of the four variables, σ, ε, T, and $\dot{\varepsilon}$. In other words, it seems "natural" to include the strain rate as an additional property and to theorize that three rather than two independent properties determine the state. Whether such a conclusion is warranted or not cannot be decided by an appeal to any principles, but must be based on the interpretation of experimental results. We interpret these to indicate that only processes at $\dot{\varepsilon} = 0$ are reversible, all others being

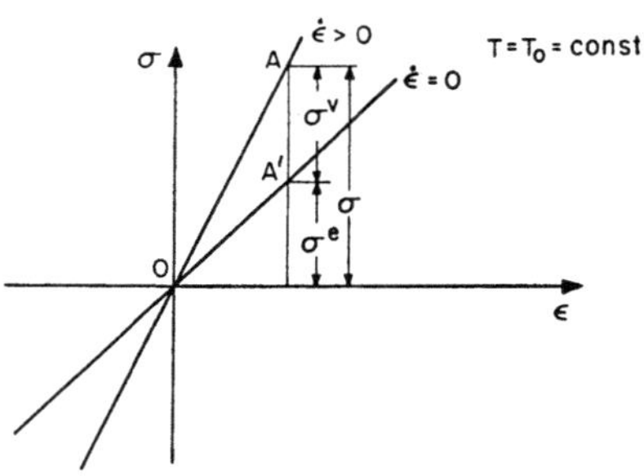

Fig. 13. Viscoelastic material

irreversible. This, as we shall see presently, excludes the possibility that $\dot{\varepsilon}$ represents a true parameter of state and leads to the conclusion that such properties as entropy and internal energy, or that any one of the thermodynamic potentials, must be independent of $\dot{\varepsilon}$.

If $\dot{\varepsilon}$ were a true parameter of state, and if the thermal equation of state were of the form

$$f(\sigma, \varepsilon, \dot{\varepsilon}, T) = 0, \tag{33}$$

it would be possible to devise a *reversible* process, i. e. a continous sequence of equilibrium states characterized by the constraint that $\dot{\varepsilon} = \text{const}$. We have asserted that this is possible only for $\dot{\varepsilon} = 0$. This excludes $\dot{\varepsilon}$ as a variable from eq. (33).

The rate of performing (irreversible) work at $\dot{\varepsilon} > 0$ is

$$\frac{dW}{dt} = -\sigma\dot{\varepsilon}, \tag{34}$$

which suggests that the stress σ *may* have ceased to be a parameter of state. Equation (34) can also be written

$$dW = -\sigma d\varepsilon, \tag{34a}$$

and for $\dot{\varepsilon} = 0$, the *reversible* work is

$$dW = -\sigma^e d\varepsilon, \tag{34b}$$

where $\sigma^e < \sigma$. This identifies σ^e as the true parameter of state. It follows that now the stress must be subdivided into an elastic, "reversible" portion σ^e and a viscous portion σ^v, with

$$\sigma = \sigma^e + \sigma^v, \tag{35}$$

as suggested in Fig. 13. Consequently, from the point of view of thermodynamics, the states at A and A' in the diagram are identical. In order to calculate the internal energy u or the entropy s at point A it is not necessary to span a reversible process between A and, say, the natural state. This may be done for state A' instead.

It has now become clear that all equations for an elastic system remain valid. In particular, the fundamental equation (11) remains unaltered, but in all other equations, σ^e must be substituted for σ. For example, instead of the well-known equation

$$\left(\frac{\partial u}{\partial \varepsilon}\right)_s = \sigma, \tag{36}$$

for an elastic system, which follows directly from

$$du = T ds + \sigma d\varepsilon,$$

we must now write

$$\left(\frac{\partial u}{\partial \varepsilon}\right)_s = \sigma^e. \tag{36a}$$

Similarly, when Legendre transformations of the fundamental equation (11) are performed, it must not be forgotten that for a viscoelastic material

$$\begin{aligned} h(s, \sigma^e) &= u - \sigma^e \varepsilon, \\ f(T, \varepsilon) &= u - T s, \\ g(T, \sigma^e) &= u - \sigma^e \varepsilon - T s, \end{aligned} \tag{37a, b, c}$$

and that

$$\frac{\partial \varphi}{\partial \sigma^v} = 0 \quad \text{but, generally} \quad \frac{\partial \varphi}{\partial \sigma} = \frac{\partial \varphi}{\partial \sigma^e} \neq 0, \tag{38}$$

for any thermodynamic potential φ with any other variable held constant.

The present situation is analogous to that existing in a viscous fluid and discussed briefly in connection with Fig. 6. In the preceding case it was not suggested that the torque, M, or the angular velocity ω should enter the equation of state. In other words, since the torque M is an integral of the shearing stresses τ in the viscous fluid, and the angular velocity ω is proportional to the velocity gradient in the lamina, it has not been suggested that either of these two quantities should enter the equation of state of a viscous fluid. Instead, only the pressure, i. e. one third of the trace of the complete stress tensor and the volume, one third of the time-integral of the trace of the rate-of-strain tensor, were retained in the equation of state.

In the viscoelastic case, the process at $\dot{\varepsilon} > 0$ is irreversible and entropy is produced at a rate

$$T \frac{d\theta}{dt} = \sigma^v \dot{\varepsilon} \tag{39}$$

in complete analogy with a viscous fluid. During isothermal operation, a compensating heat flux maintains the entropy at the same level as that which would exist at the same strain but at zero strain rate.

The further difference between a viscoelastic material and a plastic material resides in the fact that now processes which occur at finite rates are discussed instead of the quasistatic ones which were the only ones to occur before. The ability to continue to use the equations of classical thermodynamics rests on an implied hypothesis which asserts that instantaneously, finite-rate processes can be treated as being quasi-static. It is worth noting that all of fluid mechanics rests on the acceptance of this hypothesis.

8. Creep and Relaxation

The considerations of the preceding sections may suggest that the thermal equation of state of a material capable of creep and relaxation continues to be of the form

$$f(\varepsilon, \sigma, T) = 0. \tag{40}$$

This, however, is obviously impossible, because in creep the strain varies when $\sigma = \text{const}$ and $T = \text{const}$, whereas during relaxation σ varies while $\varepsilon = \text{const}$ and $T = \text{const}$. We must, inevitably, come to the conclusion that eq. (40) should contain at least one more variable, ξ, whose variation compensates for the variation of σ or ε, as the case may be. Consequently, the thermal equation of state must be of the form

$$f(\varepsilon, \sigma, T, \xi_i) = 0, \tag{40a}$$

where the symbol ξ_i stands for as many variables ξ as experiment reveals in each case.

We interpret the results of experiments in a manner to suggest that the variable or variables ξ_i are of the same nature as the extent of a chemical reaction. The inclusion of such variables in the equations represents a continuum approximation which makes allowances for the processes of chemical reaction or diffusion as they occur in systems in retarded chemical equilibrium.

The preceding view is implied in MEIXNER's papers on relaxation phenomena [2,3] and in the account of elastic relaxation given by DE GROOT and MAZUR [3]. The paper by MEIXNER [2] contains a particularly simple treatment of relaxation in a rod subjected to uniaxial stressing. The theory is based on a Taylor series expansion of the Helmholtz function (assumed to contain a single internal parameter, ξ) near the point of chemical equilibrium for which $\xi = 0$ is stipulated.

Restricting oneself to isothermal processes, one may write the fundamental equation in the form

$$f(\varepsilon, \xi) = f(0, 0) + \frac{1}{2} E_\infty \varepsilon^2 + b \varepsilon \xi + \frac{1}{2} c \xi^2, \tag{41}$$

its differential being

$$df = \sigma d\varepsilon - A d\xi, \tag{41a}$$

where A is the affinity conjugate to ξ.

The relation assumed for the Helmholtz function shows that

$$\sigma = \left(\frac{\partial f}{\partial \varepsilon}\right)_\xi = E_\infty \varepsilon + b\,\xi, \tag{42a}$$

and that

$$-A = \left(\frac{\partial f}{\partial \xi}\right)_\varepsilon = b\,\varepsilon + c\,\xi. \tag{42b}$$

During a reversible process $A = 0$, and

$$\sigma = \left(E_\infty - \frac{b^2}{c}\right)\varepsilon,$$

showing that Young's modulus is given by

$$E = E_\infty - \frac{b^2}{c} < 0. \tag{43}$$

Postulating a linear relation between the affinity A and the generalized flux $d\xi/dt$, i.e. assuming

$$A = C \cdot \frac{d\xi}{dt}, \tag{44}$$

it is possible to derive the so-called dynamic equation of state[1]

$$\sigma + \tau_\varepsilon \dot{\sigma} = E(\varepsilon + T_\sigma \dot{\varepsilon}), \tag{44a}$$

where

$$\tau_\varepsilon = \frac{C}{c} \quad \text{and} \quad \tau_\sigma = \frac{C}{c}\,\frac{E_\infty}{E}. \tag{44b, c}$$

The general integral of eq. (44a) can be written in the form

$$\sigma(t) = E_\infty \varepsilon(t) - \frac{E_\infty - E}{\tau_\varepsilon} \int_0^\infty e^{-u/\tau_\varepsilon} \varepsilon(t-u)\,du, \tag{44d}$$

which neglects the transient const $\times \exp(-t/\tau_\varepsilon)$.

The interpretation of this equation has given rise to some misunderstandings. As it stands, the equation describes a *thermodynamic process* which results when a given variable strain $\varepsilon(t)$ has been applied to the ends of the rod. If we eliminated the parameter t from $\varepsilon(t)$ and $\sigma(t)$, we would obtain the representation of the process in the stress-strain plane, realizing, however, that the internal parameter ξ would change

[1] The use of the term "equation of state" in this context is not justified because the equation contains quantities which are not true parameters of state.

from point to point as demanded by eq. (42a). Thus a relation of this type would not constitute an equation of state, all equations of state for isothermal constrained internal equilibrium being implied in eq. (41). It is, therefore, not surprising that the details of the *process* depend on all "present" and "past" values of the straining program ("history" of the strain *at the boundary*) embodied in the function $\varepsilon(t)$. This does not justify the statement, often interpreted quite widely, that the system under consideration is equipped with a "memory" (fading or otherwise), and that its present *state* depends on the system's "history". Whenever a thermodynamic process is considered it is found that all parameters of state become functions of time and that their values at a particular instant of time depend on all the values of all the parameters at the given and at past instants. This does not mean, however, that time must appear explicitly in the equations of state which constitute relations between the thermodynamic parameters *at a given instant*. On the contrary, if the correct independent parameters of state of a system have been identified, time must disappear from any true equation of state.

The presence of the internal parameter, ξ, in the preceding treatment suggests strongly that the theory of viscoelastic materials, treated descriptively in the preceding section, can be built on the assumption of the existence of internal variables in systems which exhibit such behavior. This point of view is strongly represented in the numerous papers due to MEIXNER [4].

9. General Pattern

The extension to continuous systems in the case of elastic materials appears to be straightforward and is not subject to controversy. When more complex systems are studied, the approaches differ, and it may be worthwhile to restate the relevant principles before proceeding to the study of strain hardening in terms of the theory of elastic-plastic domains.

We believe that the thermodynamic description of continuous systems requires the acceptance of the validity of the *principle of local state* [5]. This principle asserts that all relations between thermodynamic properties which are valid for uniform systems continue to be valid locally at a point and instantaneously, even if processes occur at finite rates. The sciences of elasticity and fluid mechanics are built on the implied validity of this principle. The agreement between these theories and experiment suggests that the principle of local state may be accepted for the study of inelastic behavior without reservations. Whether it can be raised to the rank of a universally valid principle of continuum thermodynamics need not be discussed here.

At any instant t the continuous system exists in a definite state characterized by the instantaneous distribution of its (true) thermodynamic properties φ_i over space. Hence the state is fixed by n fields $\varphi_i(\boldsymbol{x}, t)$ where $\boldsymbol{x}$ is a position vector and where n denotes the number of independent, thermodynamic properties of the corresponding uniform system. The time t plays the part of a parameter. When time varies, we speak of a *process*. If ψ is an arbitrary dependent property, then there exists a single-valued equation of state

$$\psi(\boldsymbol{x}, t) = F\{\varphi_1(\boldsymbol{x}, t), \ldots, \varphi_n(\boldsymbol{x}, t)\}, \tag{45}$$

which is satisfied at any point $\boldsymbol{x}$ and at any instant t. The function F cannot contain the position vector $\boldsymbol{x}$ or the time t *explicitly*.

It is convenient to review the equations of thermodynamics which must be generalized to three dimensions in a compact manner[1]. In all preceding examples, experiment suggests expressions for total work dW and for reversible work dW^0. This allows us to write the First Law in two forms. Considering an elementary quasistatic change of state between two identical neighboring states, we can write for the *reversible* process

$$dU = dQ^0 - dW^0, \tag{45a}$$

and, separately, for the *irreversible* process

$$dU = dQ - dW. \tag{45b}$$

In these equations, U constitutes an integral taken over the volume V of the system at $t = \text{const}$; dU then represents the change in this quantity in time dt. The quantities dQ, dW, dQ^0 and dW^0 are computed as integrals of the local rates of heat or work flux taken over the surface Σ of the system if the presence of electric, magnetic or gravitational fields and of radiation is excluded; these fluxes can be described in terms of suitable vector fields.

Equation (45b) is the energy equation for the process in the system. In it, U is a function of the true thermodynamic properties of the system only[2]. The expressions for dQ and dW, however, may contain quantities which are not necessarily thermodynamic properties. By contrast, eq. (45a) can contain only thermodynamic properties. In various modified forms, eq. (45a) is referred to as the Gibbs equation, particularly when the substitution $dQ^0 = TdS$ is made in it, and when the appro-

[1] For a more extensive discussion see Chapter 13 in [5].

[2] In some papers, the internal energy U is frequently and erroneously assumed to be a function of all parameters in the problem.

priate expression for dW^0 is also inserted. For example, for a system in retarded chemical equilibrium undergoing a single chemical reaction, eq. (45a) assumes the well-known form:

$$dU = TdS - PdV - Ad\xi, \tag{46}$$

where A is the chemical affinity, and ξ is the extent of reaction. Here ξ constitutes a parameter of state, and $U = U(S, V, \xi)$. Evidently, in this case

$$dW^0 = PdV + Ad\xi,$$

and it is easy to see that PdV is the total, and not the reversible work, except when $\xi = \text{const}$ or $A = 0$. In the case of a continuous system, S must be calculated as a volume integral, and dS is its change in time dt.

Upon comparing eqs. (45a) and (45b), we obtain

$$dQ^0 - dQ = dW^0 - dW, \tag{47}$$

realizing that the change in internal energy is independent of the process. This enables us to write the expression for entropy

$$dS = \frac{dQ^0}{T}$$

in the form

$$dS - \frac{dQ}{T} = \frac{dW^0 - dW}{T}. \tag{48}$$

This equation expresses the fact that the difference between the change in the entropy of the system, dS, and that produced by the exchange of heat in an irreversible process, the entropy flux, is equal to a quantity

$$d\theta = \frac{dW^0 - dW}{T} \tag{49}$$

which can be interpreted as the amount of entropy *generated* by the irreversible process during time dt. Clausius' inequality shows that

$$d\theta > 0 \quad \text{and} \quad \frac{dW^0 - dW}{T} > 0$$

must be positive definite for an irreversible process. Equation (48) can be transformed to

$$dQ = TdS - Td\theta. \tag{49a}$$

In view of (45b), this is equivalent to

$$dU = TdS - dW - Td\theta, \tag{50}$$

which is another form of Gibbs' equation. Comparison with eq. (46) shows that for a chemical system,

$$dW = PdV; \quad d\theta = \frac{A}{T}\,d\xi \quad \text{and} \quad dW^0 - dW = A\,d\xi.$$

When expressions for reversible and irreversible work are known, there is no need to generalize Gibbs' equation to three dimensions[1], because the expression for entropy production can be written at once, eq. (49). The strength of the entropy source $d\theta/dt$ per unit volume (or mass) and time will be of a particular mathematical form. It has been shown in the preceding examples, that both expressions of work appear in the forms

$$dW^0 = X^0 dy^0 \quad \text{and} \quad dW = X dy,$$

where X and X^0 are intensive quantities, and y and y^0 are specific quantities[2]. In these forms, only X^0 and y^0 are *known* to be parameters of state. Thus

$$\frac{d\theta}{dt} = \frac{X^0 dy^0/dt - X dy/dt}{T},$$

and the strength of the entropy source appears in the form of products of generalized forces X or X^0 into generalized fluxes $\dot{y}$ or $\dot{y}^0$. The results of experiments on uniform systems embodied in their equations of state, for example in the fundamental equation of state, provide information *only* about the relation between X^0 and y^0. The relation between X and y *must be determined by separate experiments.*

10. Constitutive Equations

The preceding remarks lead to an important interpretation of constitutive equations. Since stress and strain may appear in the form of thermodynamic properties or quantities which occur in the expression for the total work dW in an irreversible process, the relation between them may stem from two unrelated sets of experiments: those which lead to an equation of state and experiments which link the forces and fluxes in an irreversible process. It is best to keep these two statements

[1] In view of the appearance of T, this equation must *first* be written for an elementary volume and *then* integrated.

[2] In the more general case sums of such expressions may occur; we omit the summation sign for simplicity.

separate in order clearly to distinguish between those parameters that do and those that do not enter the expressions for energy, entropy and the remaining thermodynamic potentials. The complete constitutive relation enters the equations of motion (the Cauchy equations), but in the energy equation care must be exercised to insert into the term dU only that portion which is determined by the equation of state, while the expression for total work dW again contains the complete expressions.

For example, in a viscoelastic system

$$\frac{d\theta}{dt} = \frac{\sigma^v \dot{\varepsilon}}{T} \tag{51}$$

and we may assume a linear relation between σ^v and ε of the form

$$\sigma^v = \mu \dot{\varepsilon}, \tag{51a}$$

with μ playing the part of a viscosity. The system does, necessarily, sustain an additional distribution of entropy sources caused by a conduction of heat. Curie's principle[1] allows us to recognize that these two sets of entropy sources do not interact in an isotropic system, and we may disregard it for simplicity. It appears, therefore, that the complete constitutive equation for our system is

$$\sigma = \varepsilon E - \alpha E (T - T_0) + \mu \dot{\varepsilon}. \tag{52}$$

However,

$$\begin{aligned} u &= \frac{1}{2} E \varepsilon^2 + \alpha E \varepsilon T_0 + c_\varepsilon (T - T_0) \\ &= \frac{1}{2} E \left\{ \frac{\sigma^e}{E} + \alpha (T - T_0) \right\}^2 + \alpha \sigma^e T_0 + \alpha^2 E T_0 (T - T_0) + \\ &\qquad + c_\varepsilon (T - T_0) \end{aligned} \tag{52a}$$

and

$$\begin{aligned} s &= \alpha E \varepsilon + c_\varepsilon \ln T/T_0 \\ &= \alpha \sigma^e + \alpha^2 E (T - T_0) + c_\varepsilon \ln T/T_0. \end{aligned} \tag{52b}$$

11. Onsager's Relations

It is recalled that a matrix of phenomenological coefficients is obtained when the relation between the generalized forces and generalized fluxes is investigated with the aid of the entropy production term. In

[1] See [4] p. 57.

its simplest form[1] Onsager's theorem asserts that this matrix must be symmetrical. In this treatment[2], small deviations from equilibrium are considered, and it is clear that we are justified in saying that a given flux is "caused" by a given force if both vanish simultaneously. It is also clear in this treatment that the fluxes which appear in the general expression constitute divergences of specific, thermodynamic properties.

In our previous examples, this is the case with the viscous stress σ^v and the associated flux $\dot{\varepsilon}$. The case of plastic flow, with or without strain hardening, constitutes an exception. Now the force does not vanish when the flux vanishes, and the system is disturbed from equilibrium only by virtue of the temperature field, being so to say in equilibrium with respect to the stress field. It is difficult to see how a mere re-arrangement of the system can be fitted into the normal scheme which leads to the matrix of phenomenological coefficients, and it seems that we have no option but to conclude that plastic flow is *not* governed by Onsager's relations.

12. The Theory of Elastic-plastic Domains

A careful examination of experimental evidence shows that metallic rods are not truly homogeneous. Viewed under a microscope, they present a picture of randomly shaped, randomly oriented *domains* which slip in favorable circumstances. The slip occurs along clearly visible planes and is produced by the passage of dislocations.

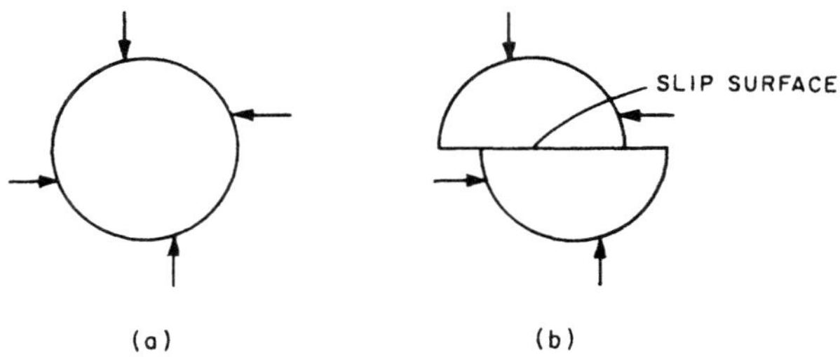

Fig. 14. Deformation of an elastic-plastic domain

We propose to explain the process of strain hardening by assuming that each domain taken separately behaves elastically up to a certain limit; when this limit is exceeded, the domain deforms plastically without a change in volume in a manner suggested by the sketch in Fig. 14. Plastic deformation is supposed to occur along a slip surface. The

[1] For simplicity we disregard the complications arising from the existence of two types of generalized forces ([2], Encyclopedia), because this circumstance has no bearing on our argument.

[2] [4] pp. 36, 40 and 57.

thermodynamic state of an elastic-plastic domain is thus described by the parameters identified in Section 5 for the ideal plastic deformation of a rod.

At first, the system of domains retains contiguity, and when this is lost, conditions for fracture are created. The requirement of contiguity means that domains which adjoin a domain undergoing plastic deformation will be subjected to internal elastic stresses and may themselves undergo plastic deformation when conditions for it become favorable. The thermodynamic state of the whole system is described by a field of stresses $\sigma_{ij}(\boldsymbol{x})$, by a field of *elastic* strains $\varepsilon_{ij}^{e}(\boldsymbol{x})$, by a temperature field $T(\boldsymbol{x})$ etc., but exclusive of the field of plastic strain $\varepsilon_{ij}^{p}(\boldsymbol{x})$. A process occurs when the fields vary with time but the principle of local state asserts that the equations of Section 3 retain their validity. The occurrence of slip by itself does not cause the state to change; this occurs indirectly through the effect of slip on the stresses and through the effect of the entropy sources on the distribution of temperature.

The differences in the behavior of different materials is explained by the differences in the average size of their domains and by the criterion of slip. It is assumed that slip occurs when the shearing stress acting in the slip surface attains a critical value τ_{cr}. The exact criteria for the orientation of the slip surface and of the number of possible slip surfaces in a domain in a particular material cannot be stated at this point, a good deal of prior detailed experimental research being required for this purpose. However, two limiting cases can be discerned in a general way.

We consider an annealed material completely free of internal stresses. When the domains possess no preferred direction of the slip surface, we can foresee the following sequence of events, say, when tension at the ends is applied. At first the system will behave completely elastically and reversibly. When the external load reaches a value which creates a shearing stress in excess of τ_{cr} the whole system will begin to yield. Thus the first yield surface of the system will correspond exactly to Tresca's yield condition. The second limiting condition will exist when the domains are small and numerous and possess a single preferred direction of slip each, the directions being arranged in a completely random manner throughout the system. Such a system will also at first behave completely elastically. However, yield will not occur when the limiting shear stress τ_{cr} is created at some point because at this point the preferred direction may differ from the direction of maximum shear produced by the external loads. Owing to the assumed randomness of orientation and to the assumed absence of volume change in plastic deformation, a general criterion for the system might express the condition that the external forces produce only an additional change in shape at the moment of macroscopically observable yielding. Assuming a very large coefficient of thermal conductivity and, therefore, a process at constant temperature, the resulting yield criterion is likely to appear in the form postulated independently by HUBER [6], MISES [7] and HAIGH [8]. In other words, we would expect that the yield criterion relates to a state reached by the system as a whole and characterized by the amount of reversible isothermal work required to produce a critical change in shape from a completely annealed state. The thermodynamic origin of this criterion was recognized by KLUITENBERG [9].

It is well known that most annealed materials exhibit a first isothermal yield surface which lies between these two extremes. This demonstrates that a good deal of information regarding microstructure must be available before the first yield surface can be computed analytically. It is also noteworthy that the preceding picture suggests that the plastic strain tensor field $\varepsilon^p(\boldsymbol{x}, t)$ during a process is essentially discontinuous because the assumption of the absence of a preferred direction of slip is not tenable.

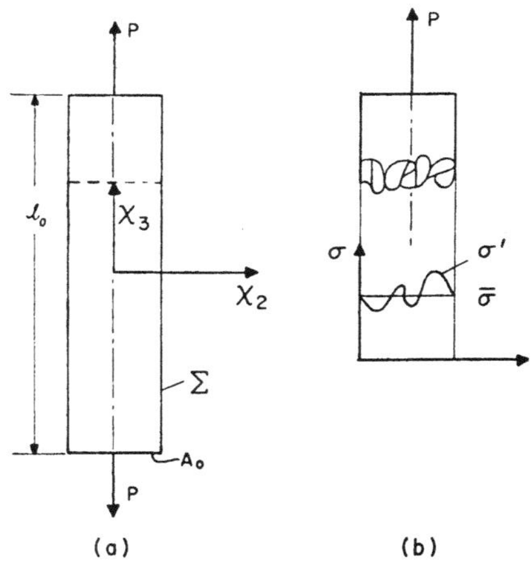

Fig. 15. Origin of internal stresses

To obtain a somewhat more visual appreciation of the behavior of such a composite system, we may again consider a bar under uniform tension as shown in Fig. 15. We consider its upper part after yield had occurred, and recognize that an originally flat surface will not remain flat owing to the random distribution of plastically deformed domains. This is illustrated in Fig. 15b. It is clear that the distribution of normal stresses which would restore contiguity cannot be uniform but must consist of an average stress

$$\bar{\sigma}_{ij} = \frac{P}{A_0} \delta_{3i} \delta_{3j} \tag{53}$$

where P is the uniform load and A_0 is the original cross-sectional area. In addition there will exist a non-uniform stress distribution σ'_{ij} of zero average value. The stress field $\sigma'_{ij}(\boldsymbol{x}, t)$ constitutes the internal field which vanishes at the boundary.

On unloading, the average stress $\bar{\sigma}_{ij}$ will decrease uniformly to zero, but the same cannot be said about the internal stress field σ'_{ij} which will remain locked in. The internal stress field will eventually contain compressive components and this circumstance will affect the yield surface on compression leading to the Bauschinger effect.

According to this theory, a cycle of loading and unloading cannot restore the original state. Thus in Fig. 11 states 0 and 3 are no longer identical as assumed in the elementary theory of Section 6. This circumstance is implied in BRIDGMAN's "sea of irreversibility", It is quite true that the original state of a composite system cannot, generally speaking, be restored by forces applied on its boundary. However, the original state of every domain can be restored, at least conceptually, and so the obstacles to the application of the equations of classical thermodynamics have been removed.

13. Some General Results Due to J. R. Rice

Professor J. R. RICE has derived several interesting results based on the preceding ideas. They are quoted here with his permission.

Reverting to the case of the bar depicted in Fig. 15, we find that

$$n_j \sigma'_{ij} = 0$$

on the surface Σ, if n_j denotes the outward normal. The equations of equilibrium are

$$\frac{\partial \sigma'_{ij}}{\partial x_j} = 0.$$

Hence

$$\int_V \frac{\partial}{\partial x_k} (x_j \sigma'_{ik}) \, dV = \int_V \left(\delta_{jk} \sigma'_{ik} + x_j \frac{\partial \sigma'_{ik}}{\partial x_k} \right) dV = \int_V \sigma'_{ij} \, dV,$$

but

$$\int_V \frac{\partial}{\partial x_k} (x_j \sigma'_{ik}) \, dV = \int_\Sigma x_j n_k \sigma'_{ik} \, d\Sigma$$

where Σ is the surface area. This proves that

$$\int_V \sigma'_{ij} \, dV = 0. \tag{54}$$

The equation of state (19) together with the principle of local state allow us to write

$$U = \int_V \left\{ \frac{1}{2} (\bar{\sigma}_{ij} + \sigma'_{ij}) (\bar{\varepsilon}_{ij} + \varepsilon'_{ij}) + \frac{E \alpha T_0}{1 - 2\nu} (\bar{\varepsilon}_{kk} + \varepsilon'_{kk}) \right\} dV,$$

where $\bar{\varepsilon}_{ij}$ is the average strain produced by the average stress $\bar{\sigma}_{ij}$. Hence

$$U = \left(\frac{1}{2}\,\bar{\sigma}_{ij}\bar{\varepsilon}_{ij} + \frac{E\alpha T_0}{1-2\nu}\,\bar{\varepsilon}_{kk}\right)V + \frac{1}{2}\int_V \sigma'_{ij}\varepsilon'_{ij}\,dV;$$

here $V = A_0 l_0$ denotes the volume of the bar. The equation separates the internal energy into two terms, the *macroenergy*

$$U_M = \left(\frac{1}{2}\,\bar{\sigma}_{ij}\bar{\varepsilon}_{ij} + \frac{E\alpha T_0}{1-2\nu}\,\bar{\varepsilon}_{kk}\right)V \tag{55a}$$

which would exist in the absence of internal stresses, and the *microenergy*

$$U_m = \frac{1}{2}\int_V \sigma'_{ij}\varepsilon'_{ij}\,dV \tag{55b}$$

locked in the system by the variable or microstresses.

Performing the same calculation for entropy from eq. (18), we find that

$$S = \int_V \frac{\alpha E}{1-2\nu}\,(\bar{\varepsilon}_{kk} + \varepsilon'_{kk})\,dV = \frac{\alpha E V}{1-2\nu}\,\bar{\varepsilon}_{kk}. \tag{56}$$

This equation demonstrates that the system possesses only macroentropy S_M, its microentropy S_m being equal to zero identically.

We have now prepared the ground for the application of Gibbs' eq. (50). The total work is

$$dW = -P\,dl$$

with

$$dW^0 = -P\,d\bar{l}$$

where $\bar{l}$ denotes the elastic elongation of the bar. Hence

$$dU_M + dU_m = T\,dS_M + P\,dl - T\,d\theta.$$

Since the microentropy vanishes, we may also write

$$dU_M = T\,dS_M + P\,d\bar{l},$$

and obtain

$$P\,d(l-\bar{l}) - dU_m = T\,d\theta \geq 0. \tag{57}$$

This equation proves that the plastic work $P\,dl^p$ with

$$l^p = l - \bar{l}$$

exceeds the amount of locked-in microenergy stored in the bar.

These very perliminary, and so far quite general, results lead us to express the hope that the theory of elastic-plastic domains is capable of further development and refinement. It is clear that more detailed statement about the nature of the slip planes must be incorporated, and that the nature of the yield surface of the system must depend not only on such details but also on the loading program prescribed for the surface of the system. This explains why the existing experimental results on yield surfaces are so difficult to classify and to interpret.

14. A Stress Cycle

When a system is subjected to an isothermal stress cycle from an initial fully annealed elastic state to a final elastic state, its volume is restored, and so is its entropy, but its shape must change if the cycle included plastic deformation. We may also assert that its internal energy has increased. Thus in eq. (50),

$$\oint T dS = 0, \quad \oint dU > 0, \quad T\oint d\theta > 0.$$

Hence

$$-W = -\oint dW = \oint dU + T\oint d\Phi - T\oint dS$$

and

$$-W = \oint dU + T\oint d\theta \geq 0.$$

The total work is given by a line integral of tractions t_i into displacements ξ_i measured on the surface Σ. Consequently we obtain

$$\oint \int_{\Sigma} t_i \, d\xi_i \, d\Sigma \geq 0. \tag{58}$$

This equation is reminiscent of, but not equivalent to, DRUCKER's stability criterion.

Acknowledgements

The author wishes to express his gratitude to his colleagues at Brown University for many exhausting discussions on the subject of this paper. Particular thanks are due to Professors D. C. DRUCKER, J. MEIXNER (who was Visiting Professor in the fall of 1965), E. T. ONAT and J. R. RICE. Professor J. R. RICE derived a number of specific equations and established some confidence in the viability of the theory of elastic-plastic domains. I am grateful for his permission to quote samples of such results hoping that his full contribution will become available without much delay.

In thanking my colleagues I do not wish to imply that they necessarily agree with the contents of this paper or that they are in any way responsible for such errors as it may contain. These must be blamed entirely on me.

Mr. R. T. WOOD prepared the state diagrams.

References

[1] BRIDGMAN, P. W.: Rev. Mod. Phys. **22**, 56 (1950).
[2] MEIXNER, J.: Natuurk. **26**, 259 (1960). See also article by MEIXNER, J., and REIK, H. G., in: "Encyclopedia of Physics", Vol. III/12, p. 413. Berlin–Göttingen–Heidelberg: Springer 1965.
[3] MEIXNER, J.: Consequences of an Inequality in Nonequilibrium Thermodynamics. Trans. ASME (Ser. E. J. Appl. Mech.) **88**, 481 (1966).
[4] DE GROOT, S. R., and P. MAZUR: Non-Equilibrium Thermodynamics. Amsterdam: North-Holland Publ. Co. 1962.
[5] KESTIN, J.: A Course in Thermodynamics. Blaisdell 1966.
[6] HUBER, M. T.: Czasopismo techniczne. Lwow: 1904.
[7] MISES, R. VON: Gött. Nachr. (Math. phys. Kl.) **1913**, 582.
[8] HAIGH, B. P.: Rep. Brit. Association **1919**, 486.
[9] KLUITENBERG, G. A.: Physica **28**, 217 (1962).

Discussion

LEE: In regard to your simpler theory of plasticity it seems to me that one should consider the experimental results of TAYLOR and FARREN[1] with regard to energy production, in which they found that in plastic flow something like 0.9, or actually a proportion which increases with increasing workhardening, of the mechanical work appeared as heat and the rest appeared as distortion in the crystal lattice through dislocations. Attempts have been made to associate this with an increase of entropy due to the increased randomness in the distribution of dislocations in the crystal. This stored energy can in fact be recovered on annealing so that it seems to me one would need such an energy storage mechanism and I presume this really comes in your third theory.

KESTIN: This fits exactly in the second theory. The statement is that each domain performs an ideal plastic deformation due to the passage of dislocations. If we perform a large amount of strain hardening and release the strain, the energy at the end is higher than that at the beginning due to the existence of a field of micro-stresses.

LEE: This maybe is an idealization which is not sufficiently justified, because presumably dislocations were needed there also, and presumably then some energy must be lost in producing this state. Maybe one can argue that this is a physically realizable idealization and therefore one need not worry about it?

KESTIN: No, this is included! Because we say that a domain when dislocations pass through it, left on its own and by itself, will perform an isothermal process. This occurs in such a way, that all the work that is done on it is taken away in the form of heat in order to conserve its temperature. I think, this is an assumption that is reasonable and that is sufficiently well confirmed by experiments.

LEE: This would presumably mean there are no dislocations left in the body?

KESTIN: No, there are dislocations in the body.

LEE: Yet there is no energy associated with them?

KESTIN: There is no energy associated with the passage of dislocations in this theory.

LEE: You are considering a continuous constant distribution?

KESTIN: That's right, but at this point the theory is fully capable of being refined exactly with respect to such details.

---

[1] Proc. Roy. Soc. London A **107**, 422 (1925).

LEE: You mentioned that in order to get the general energy relation one must load slowly, but in the limit of very fast loading one essentially gets elastic behavior, so that at this limit one can also work on an energy basis and on simple thermodynamic relations.

KESTIN: You can work on an energy basis and with thermodynamic relations at all levels and this behavior is reproduced by the theory. All you have to do is to admit the possibility of the existence of an additional set of internal variables.

KLUITENBERG: You mentioned that in the case of an ideal plastic yield you cannot apply Onsager relations. Onsager relations are symmetry relations concerning cross effects. Which cross effects do you have in mind?

KESTIN: Any cross effects that may occur in the system. If the system is ideally elastic-plastic and also viscoelastic, you have cross effects due to the transfer of mass, due to the transfer of heat and due to the transfer of work.

KLUITENBERG: From which arguments do you conclude that you cannot apply the Onsager relations?

KESTIN: Because the stress would not vanish simultaneously with the rate of strain in ideal plastic deformation.

KLUITENBERG: But the possibility to apply non-equilibrium thermodynamics mainly depends on the question whether or not a system is far outside of thermodynamic equilibrium during a process. Do you think that during the process of plastic flow materials are always far outside of thermodynamic equilibrium?

KESTIN, No, on the contrary!

KLUITENBERG: Why can you then not apply the Onsager relations?

KESTIN: Because entropy production due to a departure from equilibrium is zero.

KOITER: In Fig. 11 you gave a stress — strain diagram of a strain hardening material and pointed out two points at the same stress level but at different strain due to previous plastic strain in one of them, and then you said that the states in these two points were identical. This puzzles me in some way because the behavior of the system in these two points is entirely different.

KESTIN: I said it in two approximations. In the first approximation, when we admit the possibility that the system returns to its original state after the cycle, the statement is correct. In the second approximation when we try to understand the behavior of the system in terms of a distributed stress with a micro-stress field these states will not be identical.

Note Added in Proof

Since the above paper has been presented at the Symposium my attention was drawn to the fact that there exist calculations of the strain-hardening curve by the consideration of internal stresses caused by slip in different crystalline structures. The following three papers (which contain additional references) may be consulted:

[a] BUDIANSKY, B., Z. HASHIN and J. L. SANDERS, Jr.: The Stress Field of a Slipped Crystal and the Early Plastic Behavior of Polycrystalline Materials. In: Plasticity. p. 239. Ed. E. H. LEE and P. S. SYMONDS. New York: Pergamon Press. 1960.

[b] BUDIANSKY, B., and T. T. WU: Theoretical Prediction of Plastic Strains in Polycrystals. Proc. 4th US National Congr. Appl. Mech., p. 1175, 1962.

[c] LIN, T. H., and M. ITO: Latent Elastic Strain Energy Due to Residual Stresses in a Plastically Deformed Polycrystal. ASME Paper 67-APM-16 (to be published in J. Appl. Mech.).

Finite Strain Elastic-Plastic Theory

By

E. H. Lee and **D. T. Liu**

Stanford, Calif., and Palo Alto, Calif. (U.S.A.)

Summary

In the generation of plane waves in metal plates by detonation of contact explosives, the pressures reached can be sufficiently high to produce finite elastic and plastic strain components, and appreciable changes in temperature due to thermo-mechanical coupling effects. A theory of elastic-plastic deformation is developed to include these conditions. Finite elastic and plastic deformations are represented in terms of the initial and final configurations of the body, and an unstressed configuration at the constant base temperature subject only to plastic flow. In general this configuration cannot be represented by a continuous displacement function. The analysis of the kinematics of these three states is carried out for general deformation history. The elastic strain component is related to the stress through thermo-elastic theory for finite strains. A plasticity law is developed which includes the influence of temperature change and the finite elastic strain existing during the duration of plastic deformation. Thermo-mechanical coupling is included in the plasticity theory according to experiments on the dissipation of mechanical work into heat energy during plastic flow. Following plastic flow the internal energy and the entropy are greater than in a body of the same material under the same stress and at the same temperature, but not subject to plastic flow. However, these contributions are not considered to modify the elastic law appreciably, as has been found experimentally.

Application of this law to the theory of the propagation of plane waves of one-dimensional strain is discussed.

Introduction

This contribution differs from most of those presented at the Symposium in that it is concerned with the solution of a particular physical problem. However, analysis of the problem brings up basic questions of the development of the theory of plasticity, including aspects of it relevant to the subject of the Symposium.

We are concerned with the analysis of elastic-plastic waves of one-dimensional strain, that is plane waves with only one non-zero displacement component normal to the plane of the wave front. The waves are considered to be uniform over planes parallel to the wave-front, so that the single non-zero displacement component is a function only of time and of the coordinate with axis normal to the wave front. Such waves can be generated [1] by a slab of explosive placed in contact with the surface of a metal plate and detonated so that a plane detonation wave generates a stress wave travelling through the plate. Such a configuration can be accurately realized over most of the plate, apart from a region of the order of the plate thickness from the edge, where conditions are modified by reflections from the free edges.

Most of the experimental investigations of plastic waves have been concerned with propagation along rods or wires, for which the free lateral surface generates waves which are essentially of one-dimensional stress, with the longitudinal component being virtually the only non-zero one. But because of the lateral motion, and associated inertia, generated by the longitudinal strain, lateral stresses arise, and the usual one-dimensional stress theory is only an approximation [2]. This can prevent precise interpretation of experimental findings, and thus invalidate deductions concerning, for example, the magnitude of a strain-rate influence. It has proved virtually impossible to analyze satisfactorily the axially symmetrical motion including the influence of the lateral components. The investigation of waves of one-dimensional strain, in which lateral motion is prevented, eliminates this difficulty, and hence may provide a better vehicle for measurements in dynamic plasticity. In this case only a single equation of motion arises, for components normal to the wave front, and the ease of analytical interpretation is therefore greatly enhanced. However, other difficulties arise to offset this advantage. Because no lateral motion occurs, the longitudinal strain also represents the dilatation, and compressibility thus dominates the wave motion. For appreciable plastic flow, appreciable changes in volume occur, and this introduces a dominant influence of thermo-mechanical coupling. Resulting temperature changes must therefore be included in the analysis of the motion. The constraint of the compressibility on the generation of plastic flow causes the speeds of propagation of plastic waves to be of the order of elastic waves speeds, in contrast to the much smaller values associated with plastic waves in rods. This reduces the time scale for measurement. Since plastic flow is only a secondary effect on the resistance to volume compression, satisfactory analysis can be achieved with only the simplest versions of plasticity theory. In fact, much analytical work on the propagation of waves of one-dimensional strain has completely neglected plasticity effects by assuming the stress tensor to represent pure hydrostatic pressure yielding the common hydrodynamic theory [1]. While this circumstance offers the possibility of accurate analysis on the basis of simple plasticity laws, it also implies that extremely accurate measurements are needed to deduce plasticity characteristics, since the dominant influence of compressibility must first be subtracted. In spite of this difficulty, current techniques permit measurement of work hardening and unloading characteristics and rate effects [3, 4] from measurements on waves of one-dimensional strain.

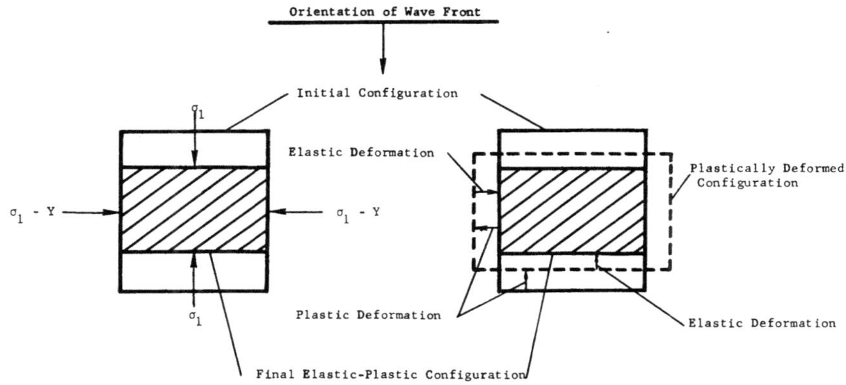

Fig. 1. Stress and strain configurations in one-dimensional strain

Fig. 1 illustrates the stress and strain configurations in a material element deformed by a plastic wave of one dimensional strain. The sketch on the left shows the stress system acting on the deformed element. For plastic flow generated by a stress σ_1 on planes parallel to the wave front, the Mises or Tresca yield conditions both prescribe the lateral stress to be of magnitude $(\sigma_1 - Y)$, where Y is the yield stress in compression. For explosive loading on an aluminum plate, σ_1 may be of the order 100 kilobars, with Y of the order 2.5 kilobars. Thus the stress system is almost hydrostatic. The corresponding elastic strains are therefore almost isotropic compression since we are concerned with an essentially isotropic material. The sketch on the right illustrates the form the plastic deformation must take, so that combined with the elastic deformation described above, zero lateral strain results. It is clear that for finite total strain, which can occur in explosive loading, both the elastic and plastic components must be finite and of the same order. This eliminates direct adoption of the usual relation, based on infinitesimal theory, which expresses the total strain as the algebraic sum of elastic and plastic components. The plastic strain represents the change in the unstressed shape of the body, which can then be considered to be deformed by the elastic strains. An increment of plastic strain is thus modified by the finite elastic strain which could remain constant while the plastic increment occurs, and this rules out the simple addition law, for the increment of total strain is the modified increment of plastic strain. An elastic-plastic theory which encompasses these conditions is presented below. It includes the kinematics of both finite elastic and plastic strains, and thermo-mechanical coupling. The latter must include the irreversible nature of plastic flow. The present discussion provides a condensed version of a more complete description of the theory [5].

Kinematics

As shown in Fig. 2 a single set of Cartesian axes (x_1, x_2, x_3) is used to define the positions of particles of a body in its initial configuration $\boldsymbol{x}^1$

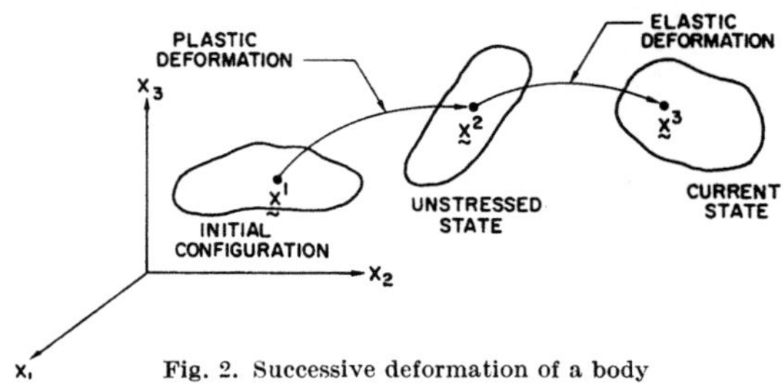

Fig. 2. Successive deformation of a body

(bold face type denoting a vector or column matrix with components x_1^1, x_2^1, x_3^1) and also after elastic-plastic deformation: $\boldsymbol{x}^3$. The finite total strain can be expressed in terms of the deformation gradient matrix:

$$\boldsymbol{D} = \frac{\partial x_i^3}{\partial x_j^1} \tag{1}$$

The plastic deformation is given by the change in permanent shape of the body from the initial configuration, considered to have uniform temperature T^1, to the current state modified by removal of the elastic strains and the thermal expansion by reducing the entire body again to the temperature T^1. The configuration devoid of elastic and thermal strains is termed the unstressed state and is expressed by the particle positions $\boldsymbol{x}^2$. Thus the plastic strain is related to the plastic deformation gradient:

$$\boldsymbol{D}^p = \frac{\partial x_i^2}{\partial x_j^1} \tag{2}$$

and the thermo-elastic strain to the corresponding deformation gradient:

$$\boldsymbol{D}^e = \frac{\partial x_i^3}{\partial x_j^2} \tag{3}$$

The chain rule for partial derivatives gives the matrix product relation:

$$\boldsymbol{D} = \boldsymbol{D}^e \boldsymbol{D}^p \tag{4}$$

This replaces the usual expression that the total strain is the algebraic sum of the elastic and plastic components. The pre-multiplication of $\boldsymbol{D}^p$ by $\boldsymbol{D}^e$ expresses the influence of the finite elastic strain on the plastic strain in determining the total strain. If it is desired to work with actual strain components, the Lagrange strain associated with $\boldsymbol{D}$ is:

$$\boldsymbol{E} = (\boldsymbol{D}^T \boldsymbol{D} - \boldsymbol{I})/2 \tag{5}$$

where the superscript T denotes the transpose of a matrix, and $\boldsymbol{I}$ is the unit matrix.

In speaking of deformation gradients above, it is assumed that the three configurations depicted in Fig. 2 are continuous, and that the transformations between them are differentiable. In general this will not be so since unloading a plastically deformed body will usually leave residual stresses, and to unstress all elements it would be necessary to dissect the body into infinitesimal volume elements, which will not fit together without the elastic deformations associated with the residual stresses. Although configuration $\boldsymbol{x}^2$ will thus not usually be continuous, linear transformations for volume elements will exist:

$$d\boldsymbol{x}^2 = \boldsymbol{D}^p d\boldsymbol{x}^1, \qquad d\boldsymbol{x}^3 = \boldsymbol{D}^e d\boldsymbol{x}^2, \qquad d\boldsymbol{x}^3 = \boldsymbol{D}\, d\boldsymbol{x}^1 \tag{6}$$

where the $\boldsymbol{D}$'s are functions of position and time, since for sufficiently small elements the linear parts of the transformations will dominate. $\boldsymbol{D}^p$ and $\boldsymbol{D}^e$ will not satisfy the compatibility relations required for partial derivatives, though $\boldsymbol{D}$ will do so since the initial and final configurations represent actual positions of the body. The relation (4) still applies according to the algebra of successive linear transformations.

A series of mappings as shown in Fig. 2 has already been suggested by Backman [6] and Sedov [7], although without the uniform temperature requirement in the unstressed state. However, the separation into elastic and plastic contributions as in (4) appears to have been handled in a more complicated fashion in [6] by means of a second order expansion in strains, and in [7] by use of convected coordinates. The separation of the elastic and plastic laws is thus achieved in [7] only for small strains and rotations, in contrast to the treatment described in the next section.

For waves of one-dimensional strain with the coordinate axes principal axes by symmetry, the deformation gradient matrices take simple diagonal forms in terms of the stretch ratios λ

$$\boldsymbol{D} = \begin{pmatrix} \lambda_1 & 0 & 0 \\ 0 & 1 & 0 \\ 0 & 0 & 1 \end{pmatrix}, \quad \boldsymbol{D}^e = \begin{pmatrix} \lambda_1^e & 0 & 0 \\ 0 & \lambda_2^e & 0 \\ 0 & 0 & \lambda_2^e \end{pmatrix}, \quad \boldsymbol{D}^p = \begin{pmatrix} \lambda_1^p & 0 & 0 \\ 0 & \lambda_2^p & 0 \\ 0 & 0 & \lambda_2^p \end{pmatrix} \tag{7}$$

and (4) reduces to:

$$\lambda_1^e \lambda_1^p = \lambda_1, \qquad \lambda_2^e \lambda_2^p = 1 \tag{8}$$

in view of the zero lateral motion.

Constitutive Relations

The kinematic analysis of the previous section separates the elastic and plastic deformations. It is known that following plastic flow elastic constants based on true stress and natural strain are effectively unchanged [8, 9]. For the usual circumstance of small elastic strains and finite plastic strains, the use of natural strain for elastic unloading and loading following plastic flow is equivalent to defining infinitesimal strain from the unstressed state, since the unstressed state differs only infinitesimally from the current state. For the situation covered in the present paper the unstressed state can differ by finite deformation from the current state, so that Hooke's law elastic constants in terms of natural strain cannot adequately represent the elastic behavior. However, the fact that the Hooke's law elastic constants are effectively unchanged by plastic flow suggests that a nonlinear finite strain elastic law based on strains corresponding to $\boldsymbol{D}^e$ will be effectively independent of previous plastic flow. Since finite strains occur, thermo-mechanical coupling will be important and the general isotropic nonlinear thermo-elastic theory will be needed (see for example PEARSON [10], Chap. 10 or TRUESDELL and NOLL [11], Chap. D 11). In terms of the Helmholtz elastic free energy function per unit mass

$$A^e = A^e(\lambda_1^e, \lambda_2^e, \lambda_3^e, T) \tag{9}$$

the stress and the elastic component of the entropy, S^e, are given by [5]:

$$\sigma_i = \varrho \lambda_i^e \frac{\partial A^e}{\partial \lambda_i^e}, \quad i = 1, 2, 3; \quad S^e = -\frac{\partial A^e}{\partial T} \tag{10}$$

These relations can be specialized for isotropic materials [5]. Appropriate forms for metals in compression have been given by MURNAGHAN [12] and the Mie-Grüneisen equation of state used in the hydrodynamic theory [1] can readily be expressed in the form (9).

Although the elastic law can be formally separated from the influence of plasticity, the irreversible nature of plastic flow introduces a thermo-mechanical coupling term which modifies the temperature and hence the elastic response. The experiments of FARREN and TAYLOR [13] showed that during plastic flow the temperature increases above that corresponding to thermo-elastic theory, because part of the irrecoverable plastic work is dissipated into heat. The rest is stored in the distortion of the crystal lattice due to the passage of dislocations, and is recoverable only by annealing [14]. The fact, discussed above, that the thermo-elastic law is not appreciably influenced by plastic flow indicates that in the neighborhood of a large proportion of the atoms, the crystal structure is not modified by plastic flow. However, the additional internal energy stored due to plastic flow must be considered, and an associated component of entropy arises due to the added randomness corresponding to the distribution of dislocations [15].

It was mentioned in the introduction, that since plasticity is only a secondary influence in one-dimensional strain, a simple version of the theory is adequate. However the appreciable temperature changes which take place due to thermo-mechanical coupling necessitate consideration of a temperature dependent yield criterion, and a theory based on PRAGER's analysis [16] is therefore adopted. We shall assume no change of volume due to plastic flow:

$$\lambda_1^p (\lambda_2^p)^2 = 1 \tag{11}$$

The rate of working per unit mass, $\dot{W}$, can be represented in the form [5]

$$\varrho_0 \dot{W} = \lambda_1^e (\lambda_2^e)^2 \left\{ \left[\sigma_1 \frac{\dot{\lambda}_1^p}{\lambda_1^p} + 2\sigma_2 \frac{\dot{\lambda}_2^p}{\lambda_2^p} \right] + \left[\sigma_1 \frac{\dot{\lambda}_1^e}{\lambda_1^e} + 2\sigma_2 \frac{\dot{\lambda}_2^e}{\lambda_2^e} \right] \right\} \tag{12}$$

This permits separation of the elastic and plastic contributions, and making the hypothesis that the amount of plastic work required to produce given work hardening depends only on the temperature and not on the elastic strains occurring, the yield condition takes the form:

$$|\sigma_1 - \sigma_2| = Y / \lambda_1^e (\lambda_2^e)^2 \tag{13}$$

where Y is the yield stress in simple compression. Since, as detailed below, the plastic work determines the energy stored in the dislocation pattern, dependence of the yield on this seems appropriate.

For isotropic work hardening the yield stress is a function of plastic work and temperature, and (13) takes the form:

$$\lambda_1 |\sigma_1 - \sigma_2| = Y(W^p, T) \tag{14}$$

where W^p is determined from a simple compression test:

$$\varrho_0 W^p = \int_0^t Y \left| \frac{\dot{\lambda}_1^p}{\lambda_1^p} \right| dt \tag{15}$$

Plastic loading or unloading occurs according to the inequalities:

$$\frac{d}{dt} (\lambda_1 |\sigma_1 - \sigma_2|) \gtrless \frac{\partial Y}{\partial T} \dot{T} \tag{16}$$

and for loading the rate of change of plastic strain is given by:

$$\frac{d}{dt}(\lambda_1|\sigma_1 - \sigma_2|) = \frac{\partial Y}{\partial W^p}\frac{Y}{\varrho_0}\left|\frac{\dot{\lambda}_1^p}{\lambda_1^p}\right| + \frac{\partial Y}{\partial T}\dot{T} \tag{17}$$

There will, of course, be a corresponding change of elastic strain according to (10).

FARREN and TAYLOR [13] observed that during plastic flow the temperature rise exceeded that associated with the thermo-elastic law by an amount equivalent to a heat source $\gamma \dot{W}^p$ per unit mass, where γ is initially about 0.9 and increases towards unity with increasing plastic flow. The remainder of the rate of plastic working: $(1 - \gamma)\dot{W}^p$ is stored as internal energy in the crystal lattice independently of the elastic internal energy associated with the free energy function (9). The fall off in storage of energy seems to be associated with the growth of dislocation density to a limiting value and the associated reduction of the work hardening gradient. We shall assume γ to maintain a constant value since this contribution to the internal energy is minor compared to that associated with the work of volume compression. In the absence of an external heat source the elastic and plastic components of internal energy are therefore given by:

$$\dot{E}^e = \dot{W}^e + \gamma \dot{W}^p \tag{18}$$

$$\dot{E}^p = (1 - \gamma)\dot{W}^p \tag{19}$$

and the growth of the elastic component of entropy by:

$$T\dot{S}^e = \gamma \dot{W}^p \tag{20}$$

It is assumed in the above that the temperature does not increase sufficiently to anneal out the stored plastic work. Also only continuous changes in strain have been considered. If shock waves occur an additional irreversibility arises leading to an entropy increase over and above that represented by (20). This is determined from the Hugoniot energy balance equation of shock wave theory.

The above theory is suggested as an adequate basis for the analysis of the propagation of waves of one-dimensional strain through metals. It provides a generalization of the theories applied to this problem so far, but simpler versions of it may be adequate for particular circumstances. For example, in the lower range of stresses, finite strain effects and the associated temperature rise may be sufficiently small to be introduced in a more approximate way. For high pressures the increase in temperature could so reduce the yield stress that the limit comprising the hydrodynamic theory may prove adequate, or some approximation to this limit.

A word should perhaps be said concerning the relation of the theory presented above with the general thermo-plastic theory presented by Professors GREEN and NAGHDI at this Symposium. Whereas we were concerned with a particular physical situation which called for an elementary version of plasticity theory, they developed the most general theory on the basis of certain initial postulates. They imposed the constraints of only general laws, such as the Clausius-Duhem inequality of entropy production, without prescribing any particular form of irreversibility and associated quantitative entropy production expression. Although the philosophies behind the contributions are thus contrasting, some aspects of the theories overlap leading to different representations of the same effects as discussed in [5].

References

[1] RICE, M. H., R. G. MCQUEEN, and J. M. WALSH: In: Solid State Physics. SEITZ, F., and D. TURNBULL (Eds.) Vol. 6, p. 1. New York: Academic Press. 1958.
[2] LEE, E. H.: Proc. 5th U. S. Natl Congr. Appl. Mech. p. 405, 1966.
[3] BARKER, L. M., C. D. LUNDERGAN, and W. HERRMANN: J. Appl. Phys. **35**, 1203 (1964).
[4] BUTCHER, B. M., and C. H. KARNES: J. Appl. Phys. **37**, 402 (1966).
[5] LEE, E. H., and D. T. LIU: J. Appl. Phys. **38**, 19 (1967).
[6] BACKMAN, M. E.: J. Appl. Phys. **35**, 2524 (1964).
[7] SEDOV, L. I.: Introduction to the Mechanics of a Continuous Medium. Reading, Mass.: Addison-Wesley Publ. Co. 1965. [English translation of Vvedenie v Mekhaniku Sploshnoi Snedy. Moscow: State Physics and Mathematics Press. 1962.]
[8] DALBY, W. E.: Phil. Trans. Roy. Soc. (London) **A-221**, 117 (1921).
[9] HOWARD, J. V., and S. L. SMITH: Proc. Roy. Soc. (London) **A-107**, 113 (1925).
[10] PEARSON, C. E.: Theoretical Elasticity. Cambridge, Mass.: Harvard University Press. 1959.
[11] TRUESDELL, C., and W. NOLL: Handbuch der Physik, Vol. III/3. Ed. S. FLÜGGE. Berlin—Göttingen—Heidelberg: Springer. 1965.
[12] MURNAGHAN, F. D.: Amer. J. Math. **59**, 235 (1937).
[13] FARREN, W. S., and G. I. TAYLOR: Proc. Roy. Soc. (London) **A-107**, 422 (1925).
[14] QUINNEY, H., and G. I. TAYLOR: Proc. Roy. Soc. (London) **A-143**, 307 (1934).
[15] TITCHENER, A. L., and M. B. BEVER: In: Progress in Metal Physics. CHALMERS, B., and R. KING (Eds.), vol. 7, p. 247. New York—London: Pergamon Press. 1958.
[16] PRAGER, W.: K. Nederl. Ak. Wetersch. Proc. **61**, 176 (1958).

Discussion

RIMROTT: You have mentioned a change of modulus of elasticity as a function of strain. In stress waves, isn't there also a noticeable change of the modulus of elasticity as a function of the strain-rate and should that not be considered?

LEE: As far as the elastic part is concerned for metals, I believe that no appreciable strain-rate effect has been observed, so that from the elastic point of view such a strain-rate effect is minimal. Now, even in plasticity it tends to be a minor effect, and work has been done from shock wave measurements to check this [FOWLES, G. R.: J. Appl. Phys. **32**, 1475 (1961).]

NIORDSON: What about the stored energy that is part of the total energy in plastic deformation? Isn't that strongly dependent on the strain-rate?

LEE: I don't think so. One thing we know about plasticity is that it really is essentially rate independent for many metals. There is a strain-rate term, but you have to go to order 10^3 times in strain-rate, before you change the stress by about 10% for aluminum. I think — although it is certainly true for accurate analysis that one has to include rate effects — this is a very slow influence compared with the effect of strain. And so I would have thought that the stored energy would also be virtually independent of the strain-rate.

KESTIN: From the experimental arrangement which you show, I would guess that the duration of the experiment was very short.

LEE: Right!

KESTIN: This means that you have a rapid build-up of entropy sources which modify the distributed temperature field. This in turn, affects the local properties including the elastic modulus. Therefore, it seems to me that there is one quantity which is missing: the thermal conductivity. The local value of thermal conductivity determines the rate at which temperature changes locally and therefore also the rate, at which the local elastic properties change throughout the system.

LEE: I forgot to mention that we did carry out the wave analysis neglecting conductivity simply on the basis that waves move fast. I know for example that in the paper by TRUESDELL and NOLL [11] [The Nonlinear Field Theories of Mechanics. In: Handbuch der Physik, Vol. III/3. Berlin: Springer 1965] a statement is made that when one is considering dissipation, the dissipative effects are roughly of the same order as the similar influence from conduction. However, although this is probably true in, say, the propagation of sound in gases or maybe in polymers, I would not think that it is true in plasticity, because in plasticity the irreversibility is an enormous effect that comes in instantaneously as soon as one exceeds the yield limit. But with regard to the influence of temperature on the elastic law, this has been thought about, and most of the work on plane waves has considered this and has been done by physicists. The physicists by and large analyse this on the basis of hydrodynamic theory, but they have worried a good deal about the thermodynamics. And it is considered that even in shock-waves, the time available is essentially adequate for equilibrium. Under certain circumstances with some substances, I think with polymers, there is a relaxation influence measurable after the wave has passed. But with most metals one does reach a steady state behind the shock-wave.

PERZYNA: I think, that relation $\boldsymbol{D} = \boldsymbol{D}^e \cdot \boldsymbol{D}^p$ is made under very strong conditions which must be satisfied from a physical point of view, because there is a superposition of elastic deformation gradient $\boldsymbol{D}^e$ and plastic deformation gradient $\boldsymbol{D}^p$. When you introduce thermoelastic-plastic effect, it will be disturbed by this effect and I think there may be some influence on this relation.

LEE: I don't agree — this is a purely geometrical result that simply says that the plastic flow begins to change the permanent shape and this, if you like, new body is then stressed elastically.

Application of Variational Methods in the Non-Linear Theory of Dispersive Wave Propagation

By

M. J. Lighthill
London (England)

Summary

A brief account is given of recent work, making use of variational methods to obtain simplifying approximations in the non-linear theory of wave propagation in dispersive systems. Some experiments, verifying the conclusions of the approximate theory in particular cases, are briefly described.

Professor SEDOV suggested that I might like to mention some work, which Professor G. B. WHITHAM and I have both been engaged on, which is rather similar to Professor SEDOV's work, because it also uses a variational principle of a basically Hamiltonian character. I will speak only very briefly about this. An account of the work has already been published in the Soviet Union by the Siberian department of the Academy of Sciences [1], and it will also be published in England shortly by the Royal Society [2]. Other papers [3, 4, 5, 6] describe earlier phases of the work.

The problem is to consider the extension of "geometrical optics" to waves which satisfy non-linear equations. The idea of geometrical optics is to obtain a great simplification of the dynamics of waves when the wave length is very small compared with other length scales in the problem. But geometrical optics applies only when the waves satisfy linear equations. The difficulties in the non-linear case are not too formidable if we have a non-dispersive system, that is, if the wave velocity depends on amplitude, perhaps, but not on frequency. This is the case with sound waves, and the non-linear theory of sound waves is quite well known. They are non-dispersive waves because the wave velocity does not depend on frequency.

But if we have dispersive waves, there are greater complications when the equations are non-linear. Roughly speaking, we have dependence on frequency, because the waves are dispersive, *and* dependence on amplitude, because the waves are non-linear. Until, perhaps, five or six years ago, I think, there was no one at all who worked on this rather difficult case. Recently a number of papers on it have appeared; but WHITHAM's method has one advantage over the rest, in that it has very great generality.

As I said before, it is a theory which applies when the wave length is much smaller than other lengths in the problem, exactly as in geometrical optics. The theory of geometrical optics for a dispersive system which is linear, although the wave velocity depends on frequency, is quite well known. The extension of the ideas of this theory to *anisotropic* dispersive media was first made by HAMILTON, and his results are essentially equivalent to the correspondence principle of quantum mechanics.

But to do it in the non-linear case (when there is also dependence on amplitude) is harder. The idea of WHITHAM's method is this: that if we have a complex wave system and the wave length is small (compared with the distances in which things are changing), then we will assume that *locally* we have approximately a plane periodic wave. Thus, at each point it is approximately plane and periodic, but the parameters such as amplitude and frequency vary in space.

Now, if we have some complicated non-linear partial differential equation representing waves, which I shall not write down, it is not so difficult to determine plane periodic solutions of it, because they will simply depend on one variable. All physical quantities are functions of a single independent variable, and so it is only an ordinary differential equation that we have to solve to determine these plane periodic waves. This is easily done.

So for them we can work out their detailed dependence on wave number and on frequency, and, of course, on the amplitude. Now we know that for infinitesimal amplitude the wave satisfies a relationship between wave number and frequency which is called the dispersion relationship, so we can say that the frequency ω is a function of the wave number k_i when the amplitude is zero. However, this relationship, as we know, will change when the amplitude increases and so it will become a relation like this between the amplitude a and the wave number and the frequency:

$$\omega = \omega(k_i, a). \tag{1}$$

Other quantities will also depend on these variables; if you want to work out the energy density in the wave, that will also depend on the wave number and the amplitude.

Well, the idea of the method is quite simple. We know that any conservative system (and if it is possible to have periodic waves, the system will be conservative) must satisfy HAMILTON's Principle, and this states that the variation of the integrated Lagrangian will be zero:

$$\delta \int L\, dt = 0. \tag{2}$$

Of course, the Lagrangian may not be a classical Lagrangian; it may be a relativistic Lagrangian. Additions to the classical Lagrangian would be needed also to deal with electromagnetic fields, Coriolis forces, etc. But this will still be the equation defining the system in every case. Where we are dealing with a continuous medium, we have to use the Lagrangian density, Lagrangian per unit volume, so that the equation becomes

$$\delta \iint L dx dt = 0. \tag{3}$$

Well, now, the idea of WHITHAM's method for small wave lengths is expressible in a few lines only. It starts by saying that, although the Lagrangian density will vary up and down between the crests and the hollows of the wave, we could replace this in the integral by an average over a wave length.

You think: I must integrate over a large space (this is an integral over the whole space, this integral with respect to x); the integral is over a large space and the integrand varies up and down between the crests and the hollows of the waves. So we smooth this variation and replace L by an average Lagrangian $\boldsymbol{L}$:

$$\delta \iint \boldsymbol{L} dx dt = 0. \tag{4}$$

This average is simply the average over one wave, and we have said that *locally* the wave is a plane periodic wave. So in (4) we replace the average Lagrangian $\boldsymbol{L}$ by the value which it has for a plane periodic wave. And I said that we can calculate that plane periodic wave by solving an ordinary differential equation and, having calculated it, we can easily determine the Lagrangian and find its average over one wave length.

So the average Lagrangian density $\boldsymbol{L}$ over one wave length will be known as a function of the wave number and the amplitude a, and I call it the function

$$\boldsymbol{L} = \boldsymbol{L}(k_i, a) \tag{5}$$

of the wave number and the amplitude. Well, actually, it is convenient to eliminate the amplitude between eqs. (1) and (5), which we can obviously do, and write the average Lagrangian as a function of the wave number and the frequency:

$$\boldsymbol{L} = \boldsymbol{L}(k_i, \omega). \tag{6}$$

Now, we can regard the wave number and the frequency as derivatives of something which we call "phase". The phase will be some complicated function. If we call it θ, then it will take integer values on each of the wave crests, say $\theta = 1, \quad \theta = 2, \quad \theta = 3, \ldots$ Evidently, if that is so, the wave number is the gradient of θ with respect to the space variables and the frequency is minus the gradient of θ with respect to the time variable:

$$k_i = \frac{\partial \theta}{\partial x_i}, \qquad \omega = -\frac{\partial \theta}{\partial t}. \tag{7}$$

This is because as the wave passes the point the rate at which the phase increases is the frequency; and, if you keep the time constant, then the rate at which you cross wave crests when going in any direction will be the component of wave number in that direction.

So, eq. (4) equates to zero the variation of the integral of a function $\boldsymbol{L}$, which is itself a function of wave number and frequency, and these are the derivatives $\partial\theta/\partial x_i$ and $-\partial\theta/\partial t$:

$$\delta \iint \boldsymbol{L}\left(\frac{\partial \theta}{\partial x_i}, -\frac{\partial \theta}{\partial t}\right) dx dt = 0. \tag{8}$$

It is a function of four variables — of three components of the wave number, and of the frequency; and these are the derivatives of θ with respect to x_1, x_2, x_3 and t.

And from this we immediately obtain the non-linear analogue of geometrical optics. We simply write down the Euler equation of this variational principle (8), and the Euler equation, if we allow θ to vary, is

$$\frac{\partial}{\partial x_i}\left(\frac{\partial \boldsymbol{L}}{\partial k_i}\right) = \frac{\partial}{\partial t}\left(\frac{\partial \boldsymbol{L}}{\partial \omega}\right), \tag{9}$$

and this is the basic equation from which one can work to obtain solutions.

You see, because we know the function (6), we know everything that enters into the eq. (9) which determines the variation of the phase θ in space, and hence by eq. (7) the variation of wave number and frequency in space; and what type of equation is this? The answer is that it is a second-order quasi-linear equation, because, you see, $\partial\boldsymbol{L}/\partial k_i$ and $\partial\boldsymbol{L}/\partial\omega$ in (9) are functions of k_i and ω, which in turn are the first derivatives of θ. Hence, differentiating again, eq. (9) becomes

$$\frac{\partial^2 \boldsymbol{L}}{\partial k_i \, \partial k_j} \frac{\partial^2 \theta}{\partial x_i \, \partial x_j} - 2 \frac{\partial^2 \boldsymbol{L}}{\partial k_i \, \partial \omega} \frac{\partial^2 \theta}{\partial x_i \, \partial t} + \frac{\partial^2 \boldsymbol{L}}{\partial \omega^2} \frac{\partial^2 \theta}{\partial t^2} = 0. \tag{10}$$

This is a partial differential equation of second order which is quasi-linear; in other words, it is linear in the highest derivatives of θ. It can be either hyperbolic or elliptic, and we find quite different behaviour in the two cases. There is also one intermediate case: in fact, in the limit when the amplitude is zero, this equation is parabolic. Because it is parabolic, it has just one system of characteristics, and they are the rays of ordinary geometrical optics. But the rays split in the hyperbolic case, so that you have then a division of the group velocity into two group velocities.

And in the elliptic case you have no real characteristics. Actually, we find that the elliptic case is one in which the waves of finite amplitude are unstable. And BROOKE BENJAMIN and FEIR have done an experimental check of the theory, using waves on deep water, because the deep-water case is one in which the equation is elliptic. Actually, there is a very simple criterion for one-dimensional, that is, uni-directional wave motions. This very simple criterion for ellipticity is

$$[\omega - f(k)]\, f''(k) < 0, \tag{11}$$

where $\omega = f(k)$ is the dispersion relation for infinitesimal amplitude. Equation (11) is satisfied for surface gravity waves on deep water in the case when the amplitude is not zero.

This theory, then, has demonstrated the rather unexpected fact that a plane periodic wave on deep water of finite amplitude is an unstable state, and this fact has been verified experimentally by putting a modulation into the waves, a very small modulation, and it has been observed to grow exponentially in time at a rate close to that predicted by the theory.

I would emphasise, however, that the theory does not merely tell you the effect of small departures from an initial state; it will predict how such a wave will change over a long period, and it is quite interesting, for example, that we find that, if you make a wave group of constant frequency whose amplitude grows in time and then becomes small again, then during the passage down a water tank of such a wave group the energy becomes more concentrated in the centre. That is a rather surprising fact, which is not predicted by the linear theory at all, and which has also been verified experimentally.

So, we believe, as a result of such simple experiments in deep water using uni-directional waves, that the general theory has some validity, but we also believe that the important applications of the theory will, probably, be to other systems, including three-dimensional waves on water and, certainly, including various kinds of magnetohydrodynamic waves.

References

[1] Лайтхилл, Дж.: Теория волн при дисперсии и нелинейности. В сборнике "Некоторые вопросы вычислительной и прикладной математики". Новосибирск: Изд. "Наука", Сибирское отд. 1966.

[2] Lighthill, M. J. (Ed.): A Discussion on Non-linear Theory of Wave Propagation in Dispersive Systems. Contributions by M. J. Lighthill, G. B. Whitham, C. J. R. Garrett, J. E. Feir, T. Brooke Benjamin, K. Hasselmann, P. G. Saffman, O. M. Phillips, M. S. Longuet-Higgins, A. E. Gill and K. Kenyon. Proc. Royal Soc. (London) **A-299**. 1 (1967).

[3] — Group Velocity. J. Inst. Math. Appl. **1**, 1 (1965).

[4] — Contributions to the Theory of Waves in Non-linear Dispersive Systems. J. Inst. Math. Appl. **1**, 269 (1965).

[5] Whitham, G. B.: Non-linear Dispersive Waves. Proc. Royal Soc. (London) **A-283**, 238 (1965).

[6] — A General Approach to Linear and Non-linear Dispersive Waves Using a Lagrangian. J. Fluid Mech. **22**, 273 (1965).

Vibrational Dissociation Relaxation in a Multicomponent Mixture of Viscous Heat-Conducting Gases

By

Yu. P. Lunkin

Leningrad (U.S.S.R.)

1. Introduction

To take account of relaxation effects in problems of gasdynamics it is necessary to determine the thermal flow vector in the energy equation and to obtain additional diffusion-relaxation equations. In the present paper equations are derived which describe simultaneous process of vibrational relaxation and non-equilibrium dissociation in a multicomponent mixture of viscous and heat-conducting gases. These equations take into account two factors: a) the influence of non equilibrium excitation of molecule vibrational degrees of freedom on chemical reaction rate and total heat flow, b) the influence of chemical reaction rate on the change of average vibrational energy of each dissociating component.

On the basis of these equations the author investigated the influence of vibrational-dissociation relaxation on (1) the structure of a plane shock wave, (2) supersonic flow around blunt bodies, (3) laminar boundary layer on a plate.

2. Derivation of Relaxation Diffusion Equations

The equation describing the change of particle number n_i of the i-th component is

$$\frac{\partial n_i}{\partial t} + \operatorname{div} n_i\,(\boldsymbol{U} + \boldsymbol{W}_i) = \delta n_i,$$

where $\boldsymbol{U}$ and $\boldsymbol{W}_i$ are hydrodynamic and diffusion velocities, δn_i-rate of change of particle number, resulting from chemical reactions. Introducing mass concentrations $C_i = \frac{n_i m_i}{\Sigma n_i m_i}$ and transforming we obtain

$$\frac{dC_i}{dt} + \frac{1}{\varrho}\,\mathrm{div}\,(\varrho C_i \boldsymbol{W}_i) = \delta C_i,$$

where

$$\frac{d}{dt} = \frac{\partial}{\partial t} + \boldsymbol{U}\nabla, \quad \delta C_i = \frac{m_i \delta n_i}{\varrho},$$

W_i is given by J. O. HIRSCHFELDER, CH. F. CURTISS, R. B. BIRD [1].

To obtain explicit expressions for δC_i it is necessary to know the reaction kinetics and influence of the thermodynamic state of vibrational degree of freedom on reaction rates. Evidently, this influence is small as regards the recombination rate k_r and is considerable for the dissociation rate k_d.

We shall define this influence by the function $V = \frac{k_d}{k_{deq}}$, where k_{deq} is the value of k_d for the case of vibrational equilibrium. Derivation of the function V for diatomic gas is given C. E. TREANOR, P. V. MARRONE [3] and YU. P. LUNKIN, F. D. POPOV [2]. In a similar way expressions for V are obtained in the case of multicomponent mixture.

The equation which describes the change of i-th component molecule number on the vibrational level "k" (considering only quantum transitions to neighbour levels) is

$$\frac{\partial n_{ik}}{\partial t} + \mathrm{div}\, n_{ik}(\boldsymbol{U} + \boldsymbol{W}_{ik}) = \sum_{l=1}^{N_1+N_2} z_{il}\,\{n_{i,k+1}\,(P_{k+1,k})_i^{(l)} +$$
$$+ n_{i,k-1}(P_{k-1,k})_i^{(l)} - n_{ik}\,[(P_{k,\,k+1})_i^{(l)} + (P_{k,\,k-1})]\} +$$
$$+ \sum_{l=1}^{N_2} z_{il}\left\{n_{i,\,k+1}\sum_j \frac{n_{ej}}{n_e}\,(Q_{k+1,k}^{j,j+1})_i^{(l)} + n_{i,k-1}\sum_j \frac{n_{ej}}{n_e}\,(Q_{k-1,k}^{j,j-1})_i^{(l)} -\right.$$
$$\left. - n_{ik}\sum_j \frac{n_{ej}}{n_e}\,[(Q_{k,k+1}^{j,j-1})_i^{(l)} + (Q_{k,k-1}^{j,j+1})_i^{(l)}]\right\} + \sum_s [\delta n_{ik}]_s.$$

Here N_1 and N_2 are the numbers of monoatomic and diatomic components, $\boldsymbol{W}_{ik}$ is diffusion rate of the particles of the i-th component on vibrational level k; Z_{il} is the number of collisions per second of the i-th component molecule with the particles of the l-th kind, $(P_{k+1,\,k})_i^{(l)}$ is the probability of the transition of the i-th component molecule from the level $k + 1$ to the level k at collision with a particle of the l-th kind which is accompanied by the transition of vibrational energy into translational energy, $(Q_{k+1,k}^{j,j+1})_i^{(l)}$ is the probability of transfer of a vibrational quantum from a molecule of the i-th component with the level $k + 1$

to a molecule of the l-th component with the level j during their collision.

The first sum corresponds to the transition of translational energy into vibrational energy of the i-th component, the second sum corresponds to intermolecular exchange of vibrational quanta and the third to the change of n_{ik} resulting from chemical reactions

Multiplying both parts of the equation by $\varepsilon_{ik} = k h \nu_i$, adding up and introducing $\varepsilon_i = \frac{1}{\varrho} \sum_{\varkappa} n_{ik} \varepsilon_{ik}$ we obtain

$$\frac{d\varepsilon_i}{dt} + \frac{1}{\varrho} \operatorname{div} \left[\sum_{\varkappa} n_{ik} \varepsilon_{ik} \boldsymbol{W}_{ik} \right] \frac{\varepsilon_{ieq} - \varepsilon_i}{\tau_{ii}} +$$

$$+ \sum_{\substack{l=1 \\ l \neq i}}^{N_2} \frac{1}{\tau_{il}} \frac{m_e}{C_e} \frac{1}{h\nu_e} \left[\varepsilon_e \left(\varepsilon_i + C_i \frac{h\nu_i}{m_i} \right) \exp \frac{h(\nu_e - \nu_i)}{kT} - \right.$$

$$\left. - \varepsilon_i \left(\varepsilon_e + C_e \frac{h\nu_e}{m_e} \right) \right] + \frac{1}{m_i} \sum_s (\bar{\varepsilon}_i)_s^{(l)} (\delta C_i)_s$$

where

$$\tau_{il}^{-1} = \sum_{l=1}^{N} Z_{il} (P_{1,0})_i^{(l)} \left[1 - \exp\left(- \frac{h\nu_i}{kT} \right) \right],$$

$$\tau_{ii}^{-1} = Z_{il} (Q_{1,0}^{0,1})_i^{(l)} \qquad l \neq i,$$

$$(\bar{\varepsilon}_i)_s^{(l)} = \sum_k k h \nu_i (g_{ik})_s^{(l)}, \quad (g_{ik})_s^{(l)} = \frac{(\delta n_{ik})_s^{(l)}}{\sum_k (\delta n_{ik})_s^{(l)}}.$$

Expressions for $(g_{ik})_s^{(l)}$ are obtained for diatomic gas by Yu. P. Lunkin, F. D. Popov [2]. It is easy to generalize these expressions for the case of multicomponent mixtures.

Expressions for $\boldsymbol{W}_{ik}$ may be obtained in the following manner: let us consider molecules of the i-th component with different vibrational levels as particles differing in kind. We will get, then, a more complex mixture each component of which has no inner degrees of freedom. For such a mixture we may use the equations from J. O. Hirschfelder, Ch. F. Curtiss, R. B. Bird [1]. Let us consider now concetration diffusion, pressure and thermal diffusion; after necessary transformations we obtain

$$W_{ik} = - b_i D_{ii} \frac{n_i}{n_{ik}} \nabla \frac{n_{ik}}{n_i} + W_i = - \left[b_i D_{ii} \frac{n_i}{n_{ik}} \frac{d}{dT_{vi}} \frac{n_{ik}}{n_i} \right] \nabla T_{vi} + W_i$$

$$b_i = \frac{n}{\varrho} \left(\sum_{l=1}^{N} \frac{C_i}{m_i} \frac{D_{ii}}{D_{il}} \right)^{-1} \qquad n_i = \sum_k n_{ik}.$$

Hence

$$\sum_k n_{ik} \varepsilon_{ik} \boldsymbol{W}_{ik} = -\lambda_{vi} \nabla T_{vi} + \varrho \varepsilon_i \boldsymbol{W}_i$$

$$\lambda_{vi} = \varrho C_i b_i D_{ii} C_{vi}$$

where D_{ii} is the coefficient of self-diffusion and C_{vi} is vibrational specific heat.

The thermal flow vector in the energy equation may be determined using a method analogous to the method for determining $\boldsymbol{W}_{ik}$. Finally we obtain

$$\boldsymbol{q} = -\lambda_0 \nabla T - \sum_{i=N_1+1}^{N} \lambda_{vi} \nabla T_{vi} + \sum_{i=1}^{N} \varrho C_i h_i \boldsymbol{W}_i + \\ + \frac{n}{\varrho} k T \sum_{i,j} \frac{c_j D_i^T}{m_i m_j D_{ij}} (\boldsymbol{W}_i - \boldsymbol{W}_j)$$

$$\lambda_0 = \lambda_{tr} + \sum_{i=N_1+1}^{N} \lambda_{i\,\mathrm{rot}}$$

where D_i^T is the coefficient of thermal diffusion and λ_{tr}, $\lambda_{i\,\mathrm{rot}}$ denote coefficients of thermal conduction of translational and rotatory degrees of freedom.

3. Plane Shock Wave

For the case of a plane shock wave an appropriate system of one-dimensional equations was solved neglecting viscosity and thermal conduction of gases. Concentration profiles, profiles of translational and vibrational temperatures T and T_v were obtained for a whole range of initial conditions and reaction rates.

As an example profiles of T_v are given in Fig. 1 for oxygen at different M_∞ ($p_\infty = 0{,}01$ atm, $T_\infty = 290\,^0$K). Fig. 2 shows profiles of T, T_v and C for $M_\infty = 10$ at dissociation from any vibrational level (fully drawn curves) and from the upper vibrational level (dotted curves).

4. Supersonic Flow around Blunt Bodies

In our investigation we considered non-viscous and non-thermal conducting gas flow around blunt bodies. Calculations were carried out using the method of integral relations (Dorodnitsin's method).

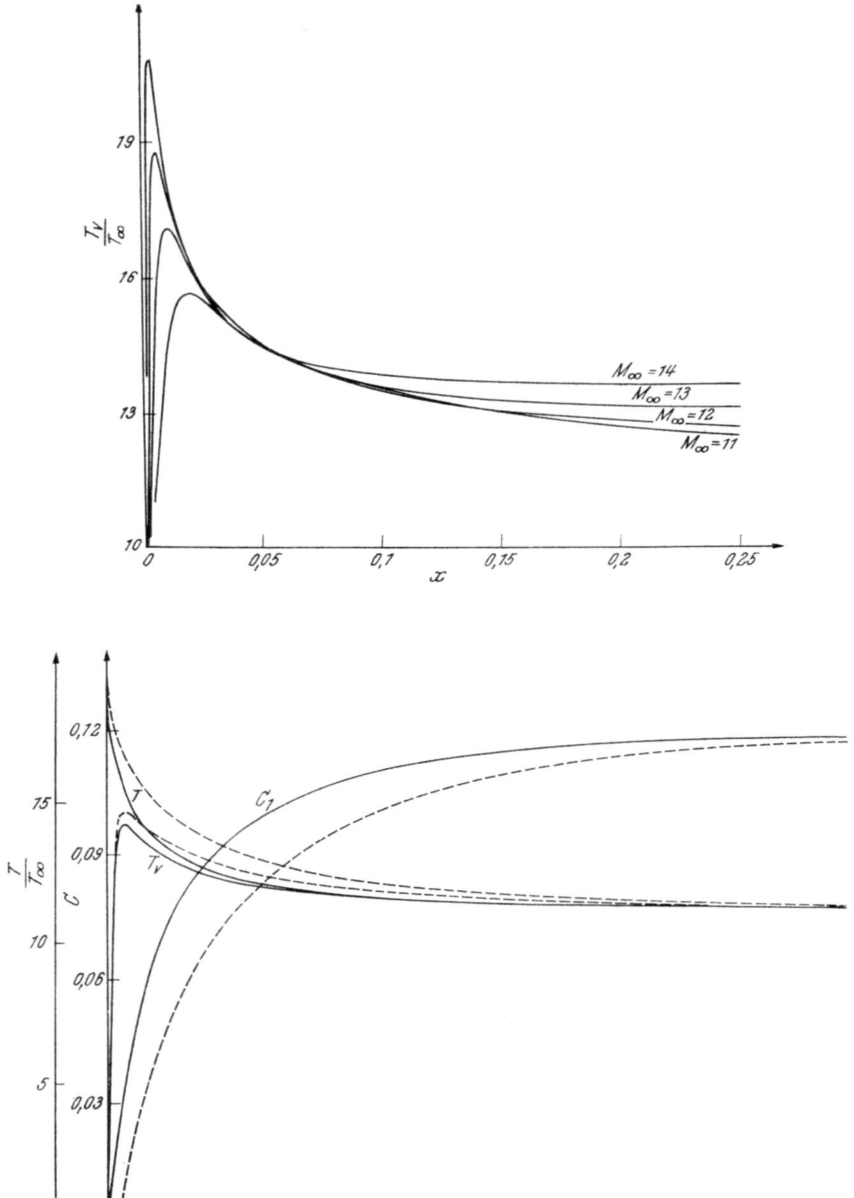

Fig. 2. Gas — O_2, $P_\infty = 0{,}01$ atm, $T_\infty = 290\,°K$

The calculations are illustrated by Fig. 3 showing shapes of shock waves and sound lines for oxygen flow around a sphere ($R_0 = 1$ cm, $p_\infty = 0{,}01$ atm, $T_\infty = 290$°K, $M_\infty = 10$); curves 1 — equilibrium flow, 2 — nonequilibrium dissociation at equilibrium vibrations, 3 — nonequilibrium vibrations and nonequilibrium dissociation, 4 — "frozen" flow corresponding to $\gamma = \frac{Cp}{Cv} = 1{,}4$.

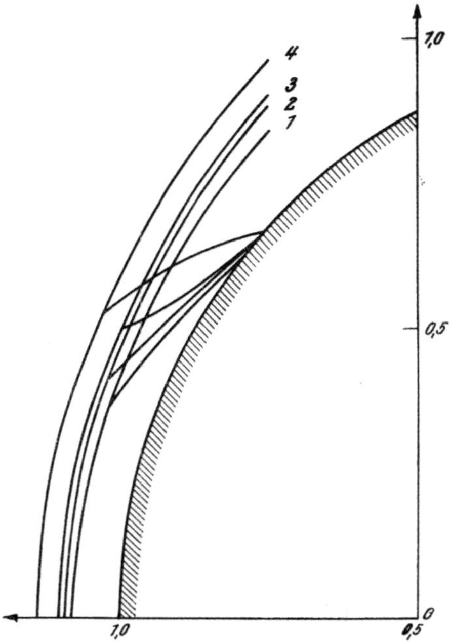

Fig. 3. Gas — O_2, $P_\infty = 0{,}01$ atm, $T_\infty = 290$°K, $M_\infty = 10$, $R_0 = 1$ cm

Fig. 4 shows temperature profiles along zero line of flow. It is seen that for a flow with nonequilibrium dissociation and nonequilibrium vibrations there exist two regions of temperature drop, namely, a region of an abrupt drop near the shock wave and a region near the body. The first region is due to vibrational relaxation, the second is due to dissociation relaxation.

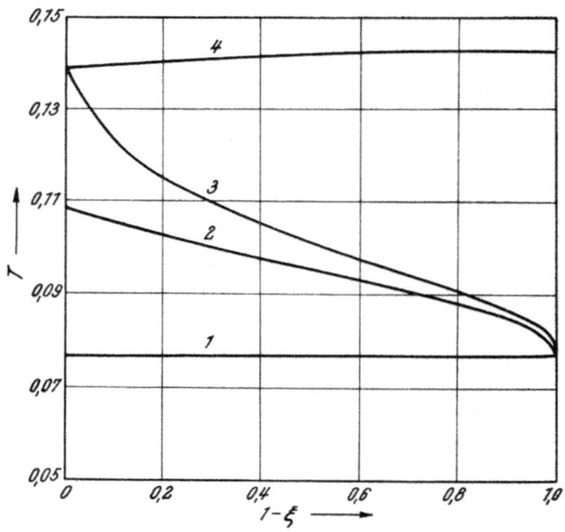

Fig. 4. Gas — O_2, $P_\infty = 0{,}01$ atm, $T_\infty = 290$°K, $M_\infty = 10$, $R_0 = 1$ cm

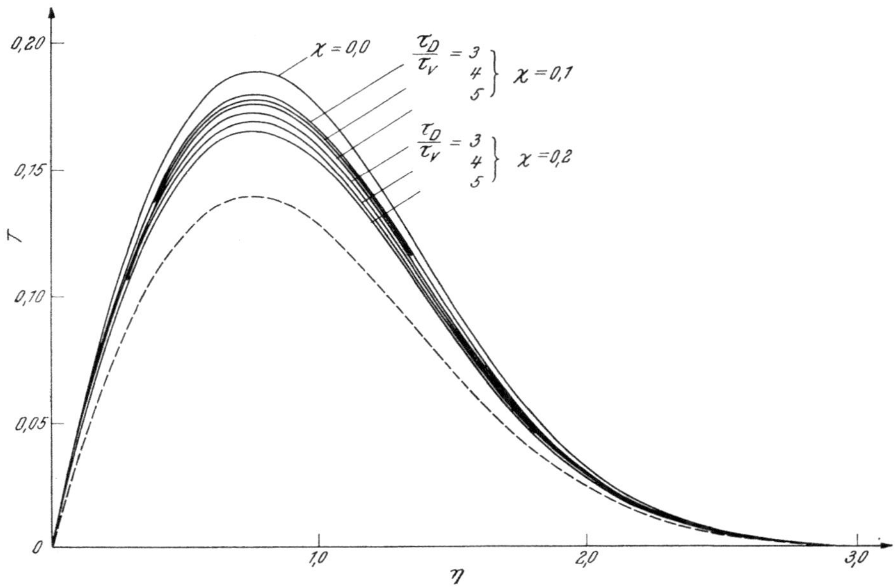

Fig. 5. Gas — O_2, $P_\infty = 1$ atm, $T_\infty = T_w = 2000\,°K$, $M_\infty = 2$

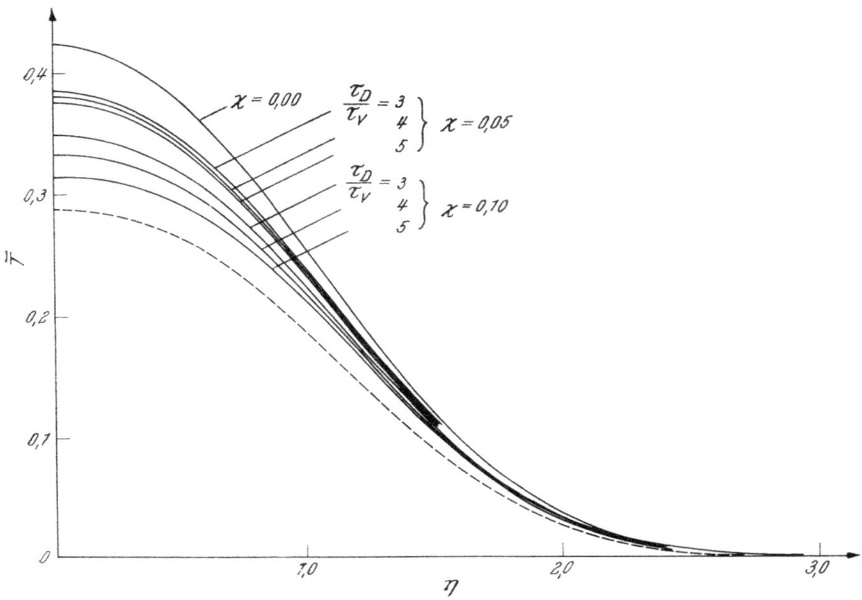

Fig. 6. Gas — O_2, $P_\infty = 1$ atm, $T_\infty = 2000\,°K$, $M_\infty = 1{,}5$

5. Laminary Boundary Layer on a Plate

Oxygen flow around a flat plate was considered as an example of laminary boundary layer. The Chapman-Rubezin parameter was taken constant. In such a case the velocity equation is the equation given by BLASIUS.

Solutions of three equations for T, ε, $C_1 = \alpha$ were arranged in series according to degrees of small parameter related with relaxation time.

$$T = \sum T_n(\eta)\chi^n \quad \varepsilon = \sum \varepsilon_n(\eta)\chi^n \quad \alpha = \sum \alpha_n(\eta)\chi^n$$

where $\chi = \left(\frac{x}{u_\infty \tau_v}\right)^{-1}$ for flows close to equilibrium flows, and $\chi = \left(\frac{x}{u_\infty \tau_v}\right)$ for flows close to "frozen" flows. In such a case the initial system of equations reduces to the system of ordinary differential equations.

Fig. 5 shows temperature profiles for a plate with a given temperature $\left(M_\infty = 2,\ A = \frac{\tau_d}{\tau_v} = 3, 4, 5;\ T_w = T_\infty = 2000\,^\circ\mathrm{K},\ p_\infty = 1\ \mathrm{atm}\right)$.

Fig. 6 shows temperature profiles for a heat insulated plate ($M_\infty = 1{,}5$; $A = 3, 4, 5$; $p_\infty = 1$ atm, $T_\infty = 2000\,^\circ$K).

Dotted curves correspond to the case of equilibrium flow.

It is seen that relaxation processes influence the temperature profile in a marked degree.

Acknowledgement

The author is grateful to S. B. KOLESHKO and F. D. POPOV who assisted in carrying out the present work.

References

[1] HIRSCHFELDER, J. O., Ch. F. CURTISS and R. B. BIRD: Molecular Theory of Gases and Liquids. New York—London: 1954.
[2] LUNKIN, Yu. P., and F. D. POPOV: J. techn. Phys. **36**, 4, 661 (1966).
[3] TREANOR, C. E., and P. V. MARRONE: Phys. Fluids **5**, 1022 (1962).

TIP Has Many Faces

By

J. Meixner
Aachen (Germany)

Summary

There are several approaches to the theory of continuous materials in non-equilibrium. They are the classical thermodynamics of irreversible processes, the so-called general non-linear continuum theory based on the Clausius-Duhem inequality and a recent approach which uses instead a fundamental inequality which follows from a thrifty application of the Second Law. Merits and shortcomings of these theories are discussed. It is also explained that the concept of entropy in non-equilibrium has no foundation in a consequent continuum theory; no unique entropy can be defined nor evaluated from measured dynamic and thermodynamic data. Only if other information on the atomistic structure and on the molecular mechanisms is available, may the entropy be given a meaning outside equilibrium. We would therefore assert that the concept of entropy in non-equilibrium should be completely discarded when building up a general continuum theory. Only the third of the approaches mentioned above meets this requirement.

It is regrettable that the field of continuum physics, the thermal aspects being included, does not have a simple and clear name. There is one name which would really be perfect if it were not used in another and too special connotation. This name is "thermodynamics" and it expresses the dynamic and the thermal aspects in an excellent way. But the word "thermodynamics" is unfortunately used for what should properly be called "thermostatics".

In order to avoid confusion I shall use the word "thermostatics" for equilibrium thermodynamics and the abbreviation "TIP" for the thermodynamic theory of irreversible processes in its various forms. There are several approaches and attitudes, special theories and more general ones, emphasis on physical aspects or mathematical developments, and there is, not surprisingly, a difference in the postulates on which the theories are based.

All these theories have in common that they aim at significant results without having recourse to statistical or kinetic theories. The limitations of such continuum theories can, of course, only be found by taking the atomistic structure into consideration.

While the concepts and ideas in thermostatics are conclusive and very well confirmed by many experiments, any form of TIP has to cope, right at the beginning, with the following problems.

1. How is temperature to be defined in a non-equilibrium state? The measurement of a local temperature by a thermometer is somewhat questionable in an irreversible process. (1) The thermometer cannot be made arbitrarily small and consequently changes the field of the irreversible process in its neighbourhood. (2) Due to its heat capacity and volume it will, at most, indicate an average of the temperature over a certain time interval and a certain volume. (3) Only for processes which are sufficiently slow and practically homogeneous over a volume which is large compared to the size of the thermometer can we expect to measure something like a true temperature provided the thermometer moves along with the material.

From a kinetic point of view one might define a temperature of the material at a point from the velocity distribution of the atoms or molecules. If it is nearly Maxwellian, then such a definition would be reasonable and would provide at least the existence of a temperature field on the thermodynamic scale. If, however, the velocity distribution is far away from a Maxwell distribution, then it is very doubtful whether one can define a meaningful temperature at all. We would even contend that under such conditions continuum physics breaks down completely.

2. What is to be understood by entropy in a non-equilibrium state? Thermostatics develops the entropy concept for equilibrium states only. Its only statement about non-equilibrium situations is the second part of the Second Law. If a process leads from one equilibrium state to another equilibrium state without transfer of components to the system, then the difference of the entropies in the final state and in the original state is always larger than the sum of the reduced imparted heats during the process. There is, as far as we can see, no way to define entropies in non-equilibrium states as long as we stay in the domain of continuum physics and consequently permit only such operations as are in conformity with continuum physics, e. g. application of heat baths, surface tractions, chemical baths (i. e. chemical potentials). Therefore we might ask: Is the concept of entropy not only impossible to define but even superfluous in the foundation of TIP? We know that kinetic theories lead to the existence of entropies and definite values of them in non-equilibrium states as well as in equilibrium. This leads to a further question: Can we infer some statement about this "kinetic" entropy from the operations of continuum physics?

3. Does there exist a state in every sufficiently small part of the material? By state we mean the existence of a set of parameters α_i such that the knowledge of their numerical values at a given time and of the external action $A(t)$ after this time permits us to predict the future behavior of the state, in other words, that the $\alpha_i(t + \tau)$ for $\tau > 0$ are functionals of the $\alpha_i(t)$ and of $A(t + t')$ with $0 \leqq t' \leqq \tau$. We speak of the principle of local state if such a functional relation exists and involves only quantities at the same material point, and in addition, if the α_i can be explained in terms of continuum physics and if the set α_i is complete.

Whether a principle of local state holds or not, we can, of course, characterize a small part of the material at a time t_0 by the treatment it has undergone for $t \leqq t_0$ provided it was in a well defined state at $t = -\infty$ or if the treatment in the very distant past is not relevant to the future behavior of the material under external actions.

4. What is entropy flow? If we confine our attention to pure heat conduction, then the entropy balance equation can be written — provided we ignore the difficulties pointed out before — as

$$\frac{ds}{dt} + \nabla \cdot \left(\frac{1}{T}\,\boldsymbol{q} + \mathbf{S}\right) = \boldsymbol{q} \cdot \nabla \frac{1}{T} + \nabla \cdot \mathbf{S}. \tag{1}$$

Here s denotes the entropy density, $\boldsymbol{q}$ the heat flow, T the thermodynamic temperature and $\boldsymbol{S}$ is an arbitrary vector field. One can interprete $\frac{1}{T}\,\boldsymbol{q} + \boldsymbol{S}$ as the entropy flow and the right member of this balance equation as the entropy production which is then required to be non-negative. It is a commonly accepted procedure to set $\boldsymbol{S} = 0$ and to require that $\boldsymbol{q} \cdot \nabla \frac{1}{T}$ is never negative. A justification for this choice may be possible in special cases, either on intuitive grounds or by its success. But it has recently been shown by MÜLLER [1] that this choice is not always compelling and he has also given a corroboration of this result from the kinetic theory of gases. The use of the Clausius-Duhem inequality

$$\frac{ds}{dt} + \nabla \cdot \frac{1}{T}\,\boldsymbol{q} \geqq 0, \tag{2}$$

meets therefore the same objection quite apart from the question what the entropy in non-equilibrium is. This inequality is, in a way, a guessed differential expression for the second part of the Second Law, but not derivable from it.

The classical theory of TIP will in the following be called Onsagerism[1]. Its basic assumptions are:

1. The principle of local state holds.
2. Gibbs' differential relation is valid (which implies the existence of temperature, entropy etc.).
3. The entropy balance equation has an entropy production term which is a sum $\sum X_i Y_i$ of products of "fluxes" and "forces", so that in a product $X_i Y_i$ one factor is even and the other factor is odd with respect to time reversal with both factors vanishing in equilibrium and the entropy production is never negative.
4. The fluxes are functions of the forces and of those parameters of state which remain independent in thermostatics.
5. For small departures from equilibrium these functions reduce to homogeneous linear functions whose coefficients satisfy the Onsager-Casimir reciprocal relations.

During the last decade another approach to TIP has developed. It will be called Truesdellism[2]. The existence of temperature and entropy in non-equilibrium is postulated as well as the Clausius-Duhem inequality. The concept of state is abandoned; instead the "state" of a "material point" at time t is characterized by the past treatment and it is assumed that, with increasing time interval, the influence of the past treatment at time t decays in a well defined way (materials with so-called fading memory). As well as in Onsagerism, the conservation laws are taken for granted. No restrictions on the size of the departure from equilibrium are made and the goal is a non-linear field theory of materials with fading memory. It turns out that the Clausius-Duhem inequality is particularly powerful in the development of the thermodynamic aspects of the theory[3].

A third approach to TIP was proposed quite recently [3, 4]. It will be called Passivism here because it is based on an inequality which, in the theory of electrical networks, is called the passivity property. Passivism is still in an early stage, but seems to have some merits. It avoids the concept of entropy in non-equilibrium, does not need the principle of local state, replaces the Clausius-Duhem inequality by the passivity inequality which can be derived from the second part of the Second Law. Its results give, among other things, a justification of Onsagerism and a new insight into the entropy concept in non-equilib-

---

[1] For this notation compare [2], p. 49.

[2] See footnote 1.

[3] This has been demonstrated by COLEMAN and MIZEL [2a]. An account of the subsequent development has been given by TRUESDELL and NOLL [2b], section 96 bis.

rium. Its only disadvantage — and here we should say, at the present time — is its restriction to small departures from an equilibrium state.

Table 1 gives (indicated by + signs) a survey of some of the basic assumptions made in the three approaches to TIP as discussed previously.

Table 1

	O	T	P
Existence of temperature in non-equilibrium	+	+	+
Existence of entropy in non-equilibrium	+	+	
Principle of local state	+		
Clausius-Duhem inequality or positive entropy production	+	+	
Entropy flow is q/T	+	+	
Linear rate equations			+

It should be stressed that in Onsagerism the rate equations must not necessarily be linearized. But the Onsager-Casimir reciprocal relations are statements about the coefficients of the linear terms only, in an expansion of the rate functions, and results on higher order coefficients have not yet been established.

We expound now the third approach in some detail.

We act in the true spirit of Thermodynamics if we take seriously the definition of a thermodynamic system, as it is given, e.g., in the classical treatise of SCHOTTKY [5]. It reads:

"In nature there exist systems whose interaction with their surroundings consists only in the exchange of work and heat. Such systems we shall designate as thermodynamic systems."

We exclude in the following electrical and magnetic phenomena also exchange of components of the system with its surroundings. That is we confine our attention to the exchange of mechanical work and heat.

We also restrict ourselves — for the sake of simplicity — to materials with a scalar pressure in equilibrium.

The system to be considered is explained by assigning to it the following properties:

1. The system has a smooth surface.

2. For given surface conditions (T and P = normal pressure, constant over the surface and in time) there exists a unique equlibrium state.

3. If the system is initially in an equilibrium state and time dependent temperatures and tractions are applied over its surface — they may vary from point to point — then a unique velocity field and heat flow field in the whole system is produced.

4. The materials contained in the system are thermodynamically stable and during the whole process the inequality

$$\int \varrho dV \left[u - u^+ + P^+ \left(\frac{1}{\varrho} - \frac{1}{\varrho^+}\right) - T^+ (s(u, \varrho) - s^+)\right] > 0, \tag{3}$$

holds. The integration goes over the whole system, ϱ is the density, u the specific internal energy, s the value of the equilibrium entropy function per mass unit, evaluated for the values of u and ϱ existing at the considered time.

5. It is possible, after any process of the system up to any time t_0, to manipulate the temperature baths and the surface tractions in such a manner that equilibrium is attained for $t \to \infty$ under the conditions of adiabatic insulation and of vanishing total work, i. e.

$$\boldsymbol{q} = 0, \quad \oint \varrho dV\, t_\alpha v_\alpha = 0 \quad \text{for} \quad t \geqq t_0, \tag{4}$$

with t_α being the surface tractions and v_α the velocity vector on the surface.

6. The conservation laws of mass, of linear and angular momentum and of energy hold.

7. In equlibrium states there exists a unique entropy of the system and in the passage from one equilibrium state to another the second part of the second law holds.

8. Temperatures and tractions can be controlled as functions of time and over the whole surface. That means we can prescribe the heat baths attached to the various parts of the surface and the surface tractions which are produced by appropriate mechanical devices quite arbitrarily as functions of time and over the surface. In order to define the temperature of a heat bath we must, however, make the assumption, that the heat baths are continuously in thermodynamic equilibrium states, i. e. that all processes in the heat baths have zero relaxation time, including heat conduction. By T we understand then the thermodynamic temperature of the heat bath.

9. The materials contained in the system are thermodynamically stable.

For a system with these properties which is, for $t \to -\infty$, in an equilibrium state with the temperature T^+ and the mechanical pressure P^+, one can derive the inequality

$$\int_{-\infty}^{t'} dt \oint \left[\left(\frac{T^+}{T} - 1\right) n_\alpha q_\alpha + (P^+ n_\alpha - t_\alpha)\, v_\alpha\right] dA \geqq 0, \tag{5}$$

valid for all values t'. The temperature T and the traction components t_α depend on time and on the surface point quite arbitrarily. The unit normal vector to the surface is n_α, dA is the area element of the surface. The subscript denotes the components and the summation convention is adopted.

The inequality derives from the inequality (3), from the second part of the Second Law and from the fact that the kinetic energy is non-negative.

A closely related inequality has been proposed by DRUCKER [6].

We call (5) the fundamental inequality for the following reasons:

1. Only a minimum of quantities is contained in it and it is a statement concerning only their surface values.
2. The temperatures and the tractions pertain only to the heat baths and to the mechanical devices that produce the tractions, i. e. to quantities which we can control from outside the system.
3. Neither the equilibrium entropy nor a questionable non-equilibrium entropy or entropy production enter this inequality.
4. It holds for arbitrary departures from the initial equilibrium state within the domain expressed by (3).
5. It has a better foundation than the Clausius-Duhem inequality.
6. One can retrieve the laws of equilibrium thermodynamics including the existence of an equilibrium entropy from this inequality and the conservation laws.
7. It has far reaching consequences for non-equilibrium situations, i. e. for processes.

For an external action with $\left(\frac{T^+}{T} - 1\right) q_\alpha n_\alpha = 0$ over the surface, the fundamental inequality (5) can be simply expressed in words in the following manner. The mechanical work done on the system during a process which starts from an equilibrium state is, at any moment, larger than the work done under dead loading (which is the work under a virtual process with the same velocity field over the surface but with the initial surface pressure).

From the third property of the system we infer that the heat flow and the velocity over the surface at time t are functionals of the present values of the temperatures and surface tractions and of their past histories. These functionals are, in their structure, restricted by the fundamental inequality.

There is a special case when these functionals can be immediately written down in an explicit form. If the departures of the surface temperature and of the surface tractions from their initial values are always small, then we may expect that the functionals are linear and then we have in the functional relation

$$\tau(\boldsymbol{x}, t), t_\alpha(\boldsymbol{x}, t) - P^+ n_\alpha(\boldsymbol{x}, t) \to \frac{dA}{dA^+} q_\alpha(\boldsymbol{x}, t)\, n_\alpha(\boldsymbol{x}, t), \frac{dA}{dA^+} v_\alpha(\boldsymbol{x}, t), \tag{6}$$

where $\tau = \frac{T^+}{T} - 1$, a linear passive system with an appropriate definition of the scalar product. The functional relation can then be described in terms of positive real admittance operators.

It is, of course, of extreme interest how these operators depend on the materials contained in the system and on their distribution over the system, that is, on the structure of the system. In order to solve this problem one has to make a few further assumptions.

1. The distribution of the materials can be expressed by density fields.

2. There exist fields $T(\boldsymbol{x},t)$, $P_{\alpha\beta}(\boldsymbol{x},t)$, $v_\alpha(\boldsymbol{x},t)$, $q_\alpha(\boldsymbol{x},t)$ inside the system such that the surface values of $t_\alpha = P_{\alpha\beta}(\boldsymbol{x},t)\, n_\beta$, $v_\alpha(\boldsymbol{x},t)$, $q_\alpha(\boldsymbol{x},t)\, n_\alpha$ agree with the respective quantities in the fundamental inequality (5).

3. The fundamental inequality can be applied to any part of the system with the understanding that the temperature baths and the tractions on its surface are represented by the temperature $T(\boldsymbol{x},t)$ on its surface and by the product of the pressure tensor $P_{\alpha\beta}(\boldsymbol{x},t)$ with the normal unit vector n_β on its surface, respectively.

4. As a part of the system we can choose an infinitesimal volume element.

The last assumption is perhaps the one most open to criticism. An infinitesimal element of matter is not very meaningful. But even an element of very small size, be it of the dimension of one free path in a gas or of an atomic diameter in the case of liquids or solids, is certainly no longer a thermodynamic system. Therefore the fundamental inequality should be inapplicable.

At present we shall just pass over this difficulty and shall comment on it and try to give a justification for this assumption somewhat later.

While the fundamental inequality (5) remains correct, even if diffusion occurs in the interior of the system, as long as there is no exchange of components through its surface, we have to exclude diffusion processes when applying the fundamental inequality to parts of the whole system.

When going to the limit of an infinitesimally small subsystem, we first transform the surface integral in the fundamental inequality into a volume integral by applying Gauß theorem. Then we obtain for all t'

$$\int_{-\infty}^{t'} dt \left[-\tau \frac{\partial q_\alpha}{\partial x_\alpha} - \Pi_{\alpha\beta} \frac{\partial v_\beta}{\partial x_\alpha} - q_\alpha \frac{\partial \tau}{\partial x_\alpha} - v_\beta \frac{\partial \Pi_{\alpha\beta}}{\partial x_\alpha} \right] \frac{dV}{dV^+} \geqq 0 \,. \tag{7}$$

where, excluding spin fields,

$$\Pi_{\alpha\beta} = \Pi_{\beta\alpha} = P_{\alpha\beta} - P^+ \delta_{\alpha\beta} \tag{8}$$

and dV/dV^+ is the size of the volume element relative to its size in the original equilibrium state. This quantity is proportional to ϱ^{-1}, the reciprocal of the density.

We may designate (7) as the fundamental inequality in local form and expect that there exists a functional relation

$$\tau,\ \Pi_{\alpha\beta},\ -\frac{\partial \tau}{\partial x_\alpha},\ -\frac{\partial \Pi_{\alpha\beta}}{\partial x_\alpha} \rightarrow -\frac{1}{\varrho}\frac{\partial q_\alpha}{\partial x_\alpha},\ -\frac{1}{2\varrho}\left(\frac{\partial v_\alpha}{\partial x_\beta}+\frac{\partial v_\beta}{\partial x_\alpha}\right),\ \frac{1}{\varrho}q_\alpha,\ \frac{1}{\varrho}v_\alpha \tag{9}$$

with the understanding that the terms to the right are, at time t, functionals of the present and previous values of the terms to the left. These functionals depend, moreover, only on the material and on the initial equilibrium state.

These functionals are, in their structure, now restricted by the fundamental inequality (7). This raises an important point. Are we permitted to admit quite arbitrary functions for the variables in the left member of (9)? In other words, which is the class of admissible thermodynamic processes for which the fundamental inequality (9) is required to hold? We certainly cannot control or prescribe temperature and tractions arbitrarily over the surface of an infinitesimal element of matter. The most we can assume is that we can control such quantities over the surface of a very small element of matter which, however, is still so large that it is a thermodynamic system. Furthermore the surface actions must certainly be sufficiently slow on some scale.

It is therefore at most an approximation if we require that the fundamental unequality (7) is to hold for quite arbitrary functions in the left member of (9) and one has to investigate the consequences of an assumption of that kind, to how good an approximation it leads unter wellstated conditions, or what its domain of validity is.

If we make no restriction on the class of admissible processes, and if we confine our attention to small departures from the initial equilibrium state, then the functional relation (9) becomes a linear passive system and in the case of isotropic fluids it can be expressed explicitly by

$$-\frac{\partial q_\alpha}{\partial x_\alpha} = \zeta_{11}(p)\tau + \zeta_{12}(p)\cdot\frac{1}{3}\Pi_{\alpha\alpha}, \tag{10}$$

$$-\frac{\partial v_\alpha}{\partial x_\alpha} = \zeta_{21}(p)\tau + \zeta_{22}(p)\cdot\frac{1}{3}\Pi_{\alpha\alpha}, \tag{11}$$

$$\frac{\partial v_\alpha}{\partial x_\beta}+\frac{\partial v_\beta}{\partial x_\alpha}-\frac{2}{3}\frac{\partial v_\gamma}{\partial x_\gamma}\delta_{\alpha\beta} = -\zeta(p)\left[\Pi_{\alpha\beta}-\frac{1}{3}\Pi_{\gamma\gamma}\,\delta_{\alpha\beta}\right], \tag{12}$$

$$-q_\alpha = \eta_{11}(p)\frac{\partial \tau}{\partial x_\alpha} + \eta_{12}(p)\frac{\partial \Pi_{\alpha\beta}}{\partial x_\beta}, \tag{13}$$

$$-v_\alpha = \eta_{21}(p)\frac{\partial \tau}{\partial x_\alpha} + \eta_{22}(p)\frac{\partial \Pi_{\alpha\beta}}{\partial x_\beta}. \tag{14}$$

These equations are to be understood as equations in the Laplace transforms of the various quantities, for instance $\tau(p)$ should be read as

$$\tau(p) = \int_0^\infty e^{-pt}\,\tau(t)\,dt. \tag{15}$$

It is thereby assumed that the equilibrium state is perturbed only after $t = 0$. These equations (10) to (14) can be converted into after-effect equations and therefore they describe in general materials with memory. But for this we refer to the literature [7, 8].

The matrices $\zeta_{ik}(p)$ and $\eta_{ik}(p)$ are positive matrices and $\zeta(p)$ is a positive function of the complex variable p in $\mathrm{Re}\, p > 0$. All these functions have real values for positive p. In particular, the matrices $p\,\zeta_{ik}(p)$ and $p\,\eta_{ik}(p)$ are symmetric for $p \to 0$ and the matrices $\frac{1}{p}\,\zeta_{ik}(p)$ and $\frac{1}{p}\,\eta_{ik}(p)$ are symmetric for $p \to \infty$. For all other values of p the symmetry of the matrices $\zeta_{ik}(p)$ and $\eta_{ik}(p)$ can be deduced by applying Onsager's ideas.

In addition to the equations (10) to (14) there are of course the conservation laws for mass, linear momentum and energy. Compatibility of all these equations requires for instance that $\eta_{21} = 0$, $\eta_{22} = (p\varrho)^{-1}$. Then (14) is the equation of motion. Moreover $\eta_{12} = 0$ and (13) is Fourier's law of heat conduction with a frequency dependent heat conductivity. For more details we refer to [3, 4].

The properties of the material are therefore reflected in the positive matrix $\zeta_{ik}(p)$ and in the positive functions $\zeta(p)$ and $\eta_{11}(p)$. They can, in principle, be measured by investigating harmonic variations in the whole frequency domain $-\infty < \omega < \infty$.

How can we now assign an entropy and an entropy production to the elements of the material in a process as functions of time?

A possible way to approach this problem is indicated by the fact that the matrix $\zeta_{ik}(p)$ is a positive matrix and that the functions $\zeta(p)$ and $\eta_{11}(p)$ are positive functions, all of them having real values for positive p. This fact permits one to model the relations (10) to (14) by electrical networks, at least if the $\zeta_{ik}(p)$, $\zeta(p)$, $\eta_{11}(p)$ are rational functions or if the matrix $\zeta_{ik}(p)$ can be approximated by a rational positive matrix and the functions $\zeta(p)$, $\eta_{11}(p)$ can be approximated by rational positive functions within the desired accuracy. Obviously there exists a state of the network. It can be explained in terms of the past histories of the applied voltages, but also by the charges on the capacitances and by the currents through the inductances in the network, all of them being taken at the same time t. In the second case there is no need to consult the past history. For such a network a complete thermodynamic formalism in the sense of

Onsagerism can be developed [3, 4] and an entropy of the system as well as an entropy production can be defined in terms of the external and internal variables of the network model.

So far, so good. But there is not only one electrical network for given functions $\zeta_{ik}(p), \eta_{11}(p), \zeta(p)$; there is an infinity of them. And not only one entropy can be defined in this way for a given state, but an infinity of entropy values exists. There seems to be no justification in preferring one of these entropy values over the others. Only the entropy flow proves to be $\boldsymbol{q}/T$, independently of the chosen network model.

One can, however, conclude that the infinity of entropy values lies between finite bounds, which can be evaluated in terms of the past history of the material. This is done by using the concept of the lost energy of an electrical network which has been introduced by BREUER and ONAT [9] and independently by KÖNIG and TOBERGTE [10, 11]. The lost energy at time t is a functional of the applied voltages $u(s)$ for s $= t$, defined by

$$W_{lost} = \int_{-\infty}^{t} (\boldsymbol{u}(s), \boldsymbol{i}(s))\, ds + \inf \int_{t}^{\infty} (\boldsymbol{u}_1(s), \boldsymbol{i}_1(s))\, ds, \tag{16}$$

the infimum being taken over all continuations $\boldsymbol{u}_1(s)$ of the voltage $\boldsymbol{u}(t)$ after time t. In words, the lost energy at time t is that part of the energy put into the network up to time t which cannot be recovered however the voltage be manipulated after time t. From the lost energy one can derive an upper bound for the entropy production and therefore also for the entropy change from the equilibrium state at $t = -\infty$.

But now I will return to a previous remark. Are we really permitted to make mass elements infinitely small, rates of change of the temperature and of the pressure tensor arbitrarily large, at least in a thought experiment?

Success seems to justify such a prodecure. But I would contend that behind it there is a conjecture which should be explicitly stated and which may even prove to be wrong in some cases. It is as follows:

1. The laws of continuum physics can be derived by ignoring the atomistic structure of the materials. Infinitely small mass elements may still be considered as thermodynamic system and processes are admissible which are produced by external actions with arbitrarily large rates of change. What one arrives at in this manner is not yet continuum physics but continuum mathematics.

2. Continuum mathematics is a good approximation to continuum physics in a limited domain of validity. This domain can be marked out in a second step by appealing to kinetic theories or, if such are not available, to intuition.

A more satisfactory approach to continuum physics would be to restrict oneself from the outset to mass elements which are small, but still so large that they can be considered as thermodynamic systems and to admit only external actions which vary sufficiently slowly so that one can imagine them to be producible by real operations. But nothing has been done along this line, so far.

References

[1] MÜLLER, I.: Zur Ausbreitungsgeschwindigkeit von Störungen in kontinuierlichen Medien. Aachen: 1966. (Dissertation.)

[2] TRUESDELL, C.: Six Lectures on Modern Natural Philosophy. Berlin—Heidelberg—New York: Springer. 1966.

[2a] COLEMAN, B. D., and V. J. MIZEL: J. Chem. Phys. **40**, **1116** (1964).

[2b] TRUESDELL, C., and W. NOLL: The Non-linear Field Theories of Mechanics. Encyclopedia of Physics, Vol. III/3. Ed. S. FLÜGGE. Berlin—Heidelberg—New York: Springer. 1965.

[3] MEIXNER, J.: J. Appl. Mech. **1966**.

[4] — Z. Physik **193**, **366—383** (1966).

[5] SCHOTTKY, W.: Thermodynamik. Berlin: Springer. 1929.

[6] DRUCKER, D. C.: J. Appl. Mech. **26**; Trans. ASME, Ser. E, **81**, 101—106 (1959), and "Plasticity", Second Symposium on Naval Structural Mechanics, p. 170. Providence, R. I.: Brown Univ. 1960.

[7] KÖNIG, H., and J. MEIXNER: Math. Nachrichten **19**, 265 (1958).

[8] MEIXNER, J.: Arch. Ratl. Mech. Anal. **17**, **278** (1964).

[9] BREUER, S., and E. T. ONAT: J. Appl. Math. Phys. **15**, 12 (1964).

[10] KONIG, H., and J. TOBERGTE: J. reine angew. Math. **212**, 104 (1963).

[11] TOBERGTE, J.: Invariante Teilräume und die verlorene Energie linearer passiver Transformationen. Köln: 1965. (Dissertation.)

Discussion

ROY: Prof. MEIXNER considers that the concept of entropy in non-equilibrium state should be completely discarded. Let me explain why in accordance with thoughts of DUHEM and JOUGUET I could not subscribe to such an idea. I share in fact the opinion that continuous media can be macroscopically defined by two kinds of informations, i. e. (1) functions of internal state only and (2) dissipative actions, these depending on the internal state and their time rate of variation. Entropy, as all functions (1), must be then defined independently of equilibrium or not.

Evidently the main difficulty of this physical and not mathematical concept lies in these basic questions: how to define the "internal state", i.e., what are the variables to be chosen for this definition? Which variables can be independent or non independent? Which variables can influence the functions (1) and/or the dissipative actions (2)? For brevity I would conclude by a general remark: a comprehensive theory as those of Macroscopic Thermodynamics can be presented in different manners according to a certain arbitrary choice of the division of a global subject in component parts, this division bringing of course no alteration to the completeness of the said subject. May be the apparent difference of opinion which I have just commented is connected with a difference of presentation.

MEIXNER: I think, there is no difference of opinion between you and me. If I know the state in terms of the real molecular variables and mechanisms, chemical reactions or whatever it is, then I know the entropy, then there is a uniqueness. If I don't have so much information but only the information I get from the outside — I'm acting on the surface of the system — I cannot define the entropy. And there may be two systems — I don't know, if one can construct such — which are exactly alike in their response to applied temperature baths and to applied tractions and are very different in their molecular structure. But the more I know about what's going on inside the system the more I can contract the bounds in the entropy, and if I know all molecular mechanism then I can give it very precisely.

Transfer Phenomena in Fluid-Saturated Porous Media

By

V. N. Nikolaevskii
Moscow (U.S.S.R.)

Summary

A phenomenological approach to the derivation of the equations of mass, energy and impulse continuity for multicomponent media. A case of heterogeneous (multiphase) continuum, an example of which is a fluid-saturated porous medium. The constitutive average laws of interphase mass, impulse and energy exchange. The theory of mass transfer in fluid flow through porous media accompanied by phase transfer and the effects of the convective diffusion. Statistical methods. Some applications of the theory to a tracer injection into a fluid flow through porous media (taking into account the adsorption effects) and to calculations of the recovery of hydrocarbon mixtures from gas-condensate reservoirs.

The mechanics of interacting continua is an important part of modern mechanics. The current investigations in this field are related to the problems of multi-component and multi-phase flows of liquid and gas complicated by simultaneous chemical reactions. They are also devoted to the problems of flow with suspended solid particles.

The principles of this branch of mechanics are analogous to those used in the theory of flow through porous media. Until recently the theory mainly treats either the quasi-stationery flow of a homogeneous fluid in a rigid or elastic body, or the slow flow of a two-phase mixture of unaltered component composition. Therefore, the equations of mass balance for liquid phases and solid matrix of a medium were sufficient to solve the problems. However, the study of wave propagation in saturated porous rocks has required the analysis of mass, impulse and energy transfer. The study of fluid dispersion and also of phase interchanges in the multiphase flow have required a deeper investigation of particle transfer in the flow through porous channels.

According to the mathematical model of interacting and interchanging continua the mass of each continuum is homogeneously spread over the differential volume of the medium considered [1, 2, 3, 4].

In reality the masses of these continua are presented as elementary particles and the total medium is a mixture of these particles. Two characteristic cases should be distinguished:

1. The elementary particles are molecules or molecular compounds of an unaltered chemical composition. Then the continua are called components. Their specific mass content is characterized by concentration.

2. The elementary particles are composed of a great number of molecules. Therefore they can be regarded as continuous and characterized by density, pressure and other parameters of continuum mechanics. Such continua are called phases. Their specific volume content in a total medium is characterized by saturation.

1. Though the formal principles for the construction of a phenomenological theory of multi-component and multi-phase flows are the same (the equations of mass, energy and momentum continuity are derived for each continuum the interaction being supposed to be of a volume nature), the interaction forces are related differently with the parameters of the continua. This is due to the fact [5] that in the first case (solutions, gas mixtures) the volumes of each component are equal to each other and to the whole volume

$$V_1 = V_2 = \dots = V \tag{1.1}$$

While in the second case the whole volume is the sum of the volumes of each phase

$$V_1 + V_2 + \dots = V \tag{1.2}$$

In turn, each phase may involve the particles of different components. Such complicated continuum models are necessary to describe the flow with phase transitions.

2. Fluid-saturated porous media consist of two phases (liquid and solid). Let us suppose each phase to be homogeneous and in the state of the local thermodynamic equilibrium characterized by the phase-averaged parameters: the phase temperature T_j the specific energy ε_j etc. ($j = 1, 2$). The saturation of a medium with a liquid is usually called porosity.

Having averaged within the volume of the whole medium one arrives at the equations of mass balance for each phase;

$$\frac{\partial}{\partial t}(1-m)\varrho_1 + \frac{\partial}{\partial x_i}(1-m)\varrho_1 u_i = 0, \quad \frac{\partial}{\partial t} m\varrho_2 + \frac{\partial}{\partial x_i} m\varrho_2 w_i = 0 \tag{2.1}$$

the equation of momentum balance for the whole medium,

$$\varrho_1(1 - m)\frac{d_1 u_i}{dt} + \varrho_2 m \frac{d_2 w_i}{dt} - \frac{\partial \Gamma_{ij}}{\partial x_j} = 0 \tag{2.2}$$

$$\frac{d_1}{dt} = \frac{\partial}{\partial t} + u_i \frac{\partial}{\partial x_i}, \qquad \frac{d_2}{dt} = \frac{\partial}{\partial t} + w_i \frac{\partial}{\partial x_i}$$

and for the liquid phase

$$\varrho_2 m \frac{d_2 w_i}{dt} - \frac{\partial P_{ij}}{\partial x_j} = \psi_i \tag{2.3}$$

where ϱ_1, ϱ_2 — densities, u_i, w_i — phase velocities, T_{ij} — total stress tensor, P_{ij} — the part of the stresses applied to a liquid phase; ψ_i — the force of an interphase interaction. According to (2.2) and (2.3) a solid phase is subjected to a surface force

$$\Pi_{ij} = \Gamma_{ij} - P_{ij} \tag{2.4}$$

The following relations are supposed to be valid

$$P_{ij} = -mp\delta_{ij}, \qquad \Pi_{ij} = (1 - m)\sigma_{ij} \tag{2.5}$$

$$\varrho_1 = \varrho_1(\sigma_{11} + \sigma_{22} + \sigma_{33}, T_1), \qquad \varrho_2 = \varrho_2(p, T_2) \tag{2.6}$$

and it is supposed that the viscous forces in the liquid are negligible as compared with the viscous force of the interphase interaction R_i,

$$\psi_i = p\frac{\partial m}{\partial x_i} + R_i \tag{2.7}$$

Relations (2.5) are related to the conditions of the total stress distribution between the volumes V_1 and V_2 (see (1.2), and the definition (2.7) reflects the occurence of solid wall reactions in the channels of variable cross-section. With $T_1 = T_2 =$ const., the eqs. (2.1—6) govern isothermal processes in saturated porous media [6, 7].

For comparison let us note that in the case of a twocomponent gas mixture these relations will be replaced by the Dalton law:

$$\Pi_{ij} = -P_1\delta_{ij}, \quad P_{ij} = -P_2\delta_{ij}, \quad \varrho_1 = \varrho_1(P_1), \quad \varrho_2 = \varrho_2(P_2) \tag{2.8}$$

and correspondingly [8, 9] eq. (2.7) will be replaced by

$$\psi_i \equiv R_i \tag{2.9}$$

Having averaged within the volumes V_1 and V_2 the energy balance equations for each phase can be derived:

$$\frac{\partial}{\partial t}\varrho_1(1-m)\left(\varepsilon_1+\frac{U_i^2}{2}\right)+\frac{\partial}{\partial x_i}\left\{\varrho_1(1-m)\left(\varepsilon_1+\frac{U_i^2}{2}\right)u_i-\Pi_{ij}u_j+j_i^{(1)}\right\}$$
$$-\frac{\delta W}{dt}+q=0 \tag{2.10}$$

$$\frac{\partial}{\partial t}\varrho_2\, m\left(\varepsilon_2+\frac{W_i^2}{2}\right)+\frac{\partial}{\partial x_i}\left\{\varrho_2 m\left(\varepsilon_2+\frac{W_i^2}{2}\right)w_i-P_{ij}w_j+j_i^{(2)}\right\}$$
$$+\frac{\delta W}{dt}-q=0 \tag{2.11}$$

where $j_i^{(k)}$ — components of heat inflow within k-phase, q — intensity of heat interchange between phases, $\delta W/dt$ — work of the forces on inner (as related to the volume V) interfaces of V_1 and V_2. Let us note that $\delta W/dt=0$ if the interfaces do not move. Here the difference between the models for multi-phase media and multi-component mixtures can easily be seen. The latter is characterized by the condition $\delta W/dt\equiv 0$.

Starting with the equations of motion one can derive the averaged equations of kinetic energy for each phase. Subtracting them from the equations for the whole phase energies (2.10) and (2.11) the equations of heat flow for each phase will be obtained. For example, for fluids the corresponding equation will be

$$m\,\varrho_2\frac{d_2 s_2}{dt}+pm\frac{\partial w_i}{\partial x_i}+pw_i\frac{\partial m}{\partial x_i}+\frac{\delta W}{dt}=\frac{\delta Q_2}{dt}+R_i w_i \tag{2.12}$$

$$\frac{\delta Q_2}{dt}=-\frac{\partial}{\partial x_i}j_i^{(2)}+q$$

Let us assume

$$\frac{\delta W}{dt}=p\frac{\partial m}{\partial t}+R_i u_i \tag{2.13}$$

This means that work is dissipated within a fluid and the specific averaged energy of the fluid particles is the same as if the whole volume were occupied by the fluid only. Actually in this case eq. (2.12) will be as follows:

$$\varrho_2 m\left(\frac{d_2\varepsilon_2}{dt}+p\frac{d_2}{dt}\frac{1}{\varrho_2}\right)=R_i(w_i-u_i)+\frac{\delta Q_2}{dt} \tag{2.14}$$

Correspondingly, the equation of heat flow to the solid phase will have the form:

$$\varrho_1(1-m)\frac{d_1\varepsilon}{dt}-(1-m)\,\sigma_{ij}\frac{1}{2}\left(\frac{\partial U_i}{\partial x_j}+\frac{\partial U_j}{\partial x_i}\right)+p\frac{d_1(1-m)}{dt}=\frac{\delta Q_1}{dt} \tag{2.15}$$

If we introduce the average phase entropies

$$\varrho_1(1-m)\,T_1\frac{d_1 s_1}{dt}=\frac{\delta Q_1}{dt}, \quad \varrho_2 m T_2\frac{d_2 s_2}{dt}=\frac{\delta Q_2}{dt}+R_i(w_i-u_i) \quad (2.16)$$

and then write out the expression for the rate of the local growth of entropy within the volume V and apply the Onsager principle, then in the simplest case we can obtain Fick's law for the heat flows $j_i^{(k)} = -\lambda_k(\partial T_k/\partial x_i)$ within each phase, Darcy's law $R_i = a(w_i - u_i)$ and an averaged law of an interphase heat exchange $q = b(T_1 - T_2)$. These laws are valid only for micro-stationary conditions for each phase and for the case of small deviations of the whole volume from equilibrium.

One must know the specific energy of each physe as a function of the state parameters in order to complete the set of equations for the chosen model of interacting continua. According to (2.14—16).

$$E_2 = E_2(p, T_2), \qquad E_1 = E_1(p, \sigma_{ij}; T_1)$$

Therefore, the solid phase strains are determined by the two systems of stresses [7, 10]. It should be emphasized that these deformations are reversible — in the case of plastic strains one must take into account the energy dissipation within the solid phase of a porous medium [11].

3. In case of multi-component and multi-phase mixtures the equations of mass balance are derived for each component in each phase taking mass transfer within a phase into account.

However, for mass transfer within a phase, the considerable fluctuations of velocity, concentration, temperature etc. from their mean values are of great importance. These fluctuations govern the process of a pulsation transfer. To analyse it one can use a statistical approach bearing in mind the random character of the relative motion of the elementary particles of every phase. It is assumed that the random fields (independent of time in case of micro-stationary conditions) of the above parameters are realized in the volumes V_1 and V_2.

Let us consider the simpliest example: fluid dispersion in the flow through porous media. We shall call one of the fluid components a tracer. The solid phase (matrix) is supposed to be rigid. The random character of a porous space produces a random field of local velocities v_i. The mean value of these velocities within the volume V_2 is w_i, and the difference $v_i - w_i \neq 0$ represents a velocity pulsation. The tracer fluctuates within the flow about its average transfer. A formal analogy with the Brownian motion shows [12] that the probability for a particle to reach a certain point within a medium at time t (that is an average concentration of a marked component) satisfies a common diffusion equation with convection. For the effective diffusion coefficient which must be determined by the mean velocity vector w_i, geometrical considerations [13] give the following formula valid for an isotropic medium,

$$D_{ij} = A \frac{w_i w_j}{w_i^2} + B\delta_{ij} \tag{3.1}$$

where A and B are scalars.

If the dimensionality requirements are satisfied relation (3.1) gives [13]:

$$D_{ij} = (\lambda_1 - \lambda_2) \frac{w_i w_j}{|\vec{w}|} + \lambda_2 |\vec{w}| \delta_{ij} \tag{3.2}$$

Here λ_1, λ_2 are longitudinal and transverse dispersion coefficients which are parameters of a medium for flows similar with respect to the Reynolds number. An experiment [14] shows that

$$\lambda_1 = 0{,}127 \text{ cm}, \qquad \lambda_2 = 0{,}0089 \text{ cm}, \qquad \text{i. e. } \lambda_1/\lambda_2 = 14{,}3,$$

for flow through river sand at $w = 4{,}5 \cdot 10^{-2}$ cm/sec. The longitudinal diffusion coefficient is $D_1 = 5.10^{-3}$ sqcm/sec which is much greater than the usual value of the molecular diffussion coefficient $D_0 = 10^{-5} - 10^{-6}$ sqcm/sec. Actually, a porous medium adsorbs the tracers which leads to the delay of transport [15].

The considered effect of mixing intensification is essential for heat transfer. However, at the mentioned velocities this effect does not influence the order of the heat conductivity coefficient. An attempt made by BIOT [16, 17] to introduce an additional force proportional to inertia forces and connected with pulsating transport of momentum should be noted.

4. If the particles of a component may be present at both phases and the rate of mass transfer between the phases is not infinite then the equations of mass conservation for the components should be constructed for each phase separately. In addition the volume character of interphase mass exchange is assumed and the intensity of this exchange is connected with the phase parameters (chemical potentials or concentrations by relations analogous to those for chemical reaction rates).

The case of an instant exchange between phases is simpler if the equations of mass conservation of the components are constructed for the entire medium but the different character of mass transfer in each phase is taken into consideration [18]. For example, in case of isothermal two-phase flow of multi-component mixtures fixed porous body these equations have the form:

$$\frac{\partial}{\partial t} \left\{ C_{\mathrm{I}}^k \varrho_{\mathrm{I}} s + C_{\mathrm{II}}^k \varrho_{\mathrm{II}} (1 - s) \right\} + \operatorname{div} \left\{ C_{\mathrm{I}}^k \varrho_{\mathrm{I}} \overrightarrow{w_{\mathrm{I}}} + C_{\mathrm{II}}^k \varrho_{\mathrm{II}} \overrightarrow{w_{\mathrm{II}}} \right\} = 0 \tag{4.1}$$

where s — the saturation of a porous space with the first phase, $\varrho_{\rm I}$, $\varrho_{\rm II}$ — phase densities, $C_{\rm I}^k$, $C_{\rm II}^k$ — phase concentrations of a k-th component, $\vec{w}_{\rm I}$, $\vec{w}_{\rm II}$ — mean phase velocities.

It is the system (4.1) which is used for the calculations of gas-condensate mixture flow through porous media. The velocities $w_{\rm I}$ and $w_{\rm II}$ ar determined by the equations of phase flows. If the inertial forces are neglected these equations take the form of the generalized Darcy law:

$$m\vec{w}_{\rm I} = -\frac{k}{\mu_{\rm I}} f_{\rm I}(s) \operatorname{grad} p, \qquad m\vec{w}_{\rm II} = -\frac{k}{\mu_{\rm II}} f_{\rm II}(s) \operatorname{grad} p \tag{4.2}$$

where k — permeability of the medium, $\mu_{\rm I}$, $\mu_{\rm II}$ — phase viscosities, $f_{\rm I}(s)$, $f_{\rm II}(s)$ — relative phase permeabilities. It is generally assumed that gas-condensate flows in the conditions of the local thermodynamic equilibrium of the phases. On account of this the phase concentrations are bound by the equality of the chemical potentials of the components in gas and liquid phases. If, as it is assumed in (4.2), phase pressures are equal, the number γ of independent parameters is determined by the Gibbs rule $\gamma = K - J + 2$, where K — number of components, J — number of phases. The difference of phase pressures is equal to the capillary pressure P_c, the latter being dependent on the saturation $P_c = P_c(S)$. For small values of fluid saturation this difference is essential. The conditions of chemical equilibrium begin then to depend on saturation and the number of degrees of freedom is equal to $\gamma_+ = \gamma + J - 1 = K + 1$ and coincides (at T = constant) with the number of unknown variables in the equation of motion (4.1).

To solve the set of eqs. (4.1—2) one must use numerical methods because of the non-linear experimental functions involved.

An essential simplification can be achieved by means of the following method: an actual multi-component mixture is divided into some groups which are then considered as "fictitious components". Thus one can introduce binary, ternary and other models of the mixture in question, but one should take into account that for binary models the conditions of thermodynamic equilibrium depend on pressure and temperature only while for ternary ones they depend on these and, in addition, on one of the concentrations. Therefore, if phase transitions due to a change of the chemical composition of a phase are considered one must use a ternary (or more complicated) model. In particular, such transitions take place in gas-condensate reservoirs while the cycling process is applied.

References

[1] РАХМАТУЛИН, Х. А.: Прикл. матем. и мех. **20**, 2 (1956).
[2] TRUESDELL, C., and R. A. TOUPIN: Handbuch der Physik, vol. III/1. Berlin—Göttingen—Heidelberg: Springer. 1960.
[3] ERINGEN, A. C., and I. D. INGRAM: Int. J. Engng Sci. **2**, 197—212 (1965).

[4] GREEN, A. E., and P. M. NAGHDI: Int. J. Engng Sci. **3** (1965).
[5] СЕДОВ, Л. И.: Проблемы науки. Изд. «Знание», 1966.
[6] ФРЕНКЕЛЬ, Я. И.: Известия АН СССР, сер. географ., геофиз. **4**, **1944**.
[7] НИКОЛАЕВСКИЙ, В. Н.: Добыча нефти. Теория и практика. Ежегодник **1963**. Изд-во «Недра», **1964**.
[8] TRUESDELL, C.: J. Chem. Phys. **37** (10), 2336—2344 (1962).
[9] ADKINS, J. E.: Philos. Trans. Roy. Soc. (London) **A-255**, 607, 635 (1963); **A-256**, 301 (1964).
[10] ЗОЛОТАРЕВ, П. П., В. Н. НИКОЛАЕВСКИЙ: Теория и практика добычи нефти. Ежегодник, Изд. «Недра», 1966.
[11] НИКОЛАЕВСКИЙ, В. Н.: Теория и практика добычи нефти. Ежегодник, Изд. «Недра», 1966.
[12] SCHEIDDEGER, A. E.: J. Appl. Phys. **25** (8), (1954).
[13] НИКОЛАЕВСКИЙ, В. Н.: Прикл. мат. и мех. **23**, **6** (1959).
[14] БОНДАРЕВ, Э. А., и А. Р. ШКИРИЧ: Известия АН СССР. Механика **1965** (6).
[15] БОНДАРЕВ, Э. А., и В. Н. НИКОЛАЕВСКИЙ: Прикл. мех. и техн. физ. **1962** (5).
[16] BIOT, M. A.: J. Acoust. Soc. Amer. **28**, 168—178 (1956).
[17] BIOT, M. A.: J. Appl. Phys. **33** (4), 1482—1498 (1962).
[18] НИКОЛАЕВСКИЙ, В. Н.: Известия АН Азерб. ССР, сер. геолого-географ., **1965** (1).

Discussion

FRENKIEL: Diffusion in this case is larger than the molecular diffusion. Is this also true if we consider diffusion after very small intervals of time, small compared with the scale of the process?

NIKOLAEVSKII: Convective dispersion is connected with the deviations from the velocity during a great number of random steps of a particle. Therefore, if the interval of time is small compared to the characteristic duration of one step, diffusion is much smaller and of the order of molecular diffusion.

MANDL: What is the definition of the averaging procedure? Have you assumed that the operations of spatial averaging and spatial differentiation commute? In the case of porous media this assumption does not hold in general.

NIKOLAEVSKII: The averaging procedure comes first. It is necessary to write out the equations of the microscopic motion, the Navier-Stokes equations for example. Then, for instance, a mean value of the pressure over the fluid part of a face is taken as macroscopic pressure.

LIGHTHILL: Did you consider the case of water and air in the soil? Did you consider any unsteady case?

NIKOLAEVSKII: In the case of partially saturated soils in general it is necessary to consider the motion of three phases and to introduce the equations of energy. We solved one problem of an acoustic wave propagation in partially saturated soils, when the gas phase does not move relatively to solid matter. The results are published in the book "Theory and practice of oil exploration" (Moscow, 1966).

LIGHTHILL: Did you obtain solutions in the quasistationary case?

NIKOLAEVSKII: We found some solutions by numerical methods, for instance the solution corresponding to the quasistationary gas-condensate flow where, besides the usual conditions, the pressure and concentrations were functions of the coordinates, but saturation was a function of coordinates and time.

FRÖSSLING: Have you had any troubles with the numerical side of this problem? You may possibly have instability in some solution and you may obtain different results if you take different intervals for the independent variables.

NIKOLAEVSKII: We computed solutions for the most simple models only and we did not find any troubles in the numerical procedure.

LEE: What is the meaning of quasistationary solutions?

NIKOLAEVSKII: We don't take into account the dynamical terms in the equations of momentum, but retain time-derivatives in the equations of mass balance.

MANDL: When pressure oscillations are superimposed on the driving pressure difference of a Poiseuille flow, the velocity field will change, but not uniformly. Therefore, the ratio between instantaneous mean flow rate and instantaneous pressure gradient will change. Likewise, in quasistationary flow through a porous medium the permeability will no longer remain a constant. Did you take this into account?

NIKOLAEVSKII: This theory is much simpler than nature. We tried to restrict our study to the boundary of the microstationary condition. In the opposite case it is necessary to change Darcy's law. Biot introduced some visco-dynamical operator to account for such effects, but we don't have a practical definition of this operator for the particular media. Besides, we know that in the case of the first longitudinal wave the boundary where microstationary conditions fail is very near the boundary where the length of such a wave has an order of diameters of solid particles and the methods of continuum mechanics are inapplicable.

Couple-Stresses in the Theory of Thermoelasticity

By

W. Nowacki

Warsaw (Poland)

Summary

The aim of the present paper is to generalize some theorems on the coupled thermoelasticity of a medium characterized by two vectors independent from each other: the displacement vector $\boldsymbol{u}$ and the rotation vector $\boldsymbol{\omega}$.

Basing on the thermodynamics of irreversible processes the constitutive equations and the expanded equation of heat conductivity for an isotropic medium are derived.

The author succeeded in obtaining a basic system of differential equations of coupled thermoelasticity. The propagation of thermoelastic waves in an unbounded medium is discussed.

Moreover, a generalization of the virtual work principle to dynamic problem of coupled thermoelasticity is advanced.

Finally, the reciprocity theorem is derived and some conclusions resulting from this theorem are discussed.

1. Introduction

The asymmetric theory of elasticity was first advanced in 1887 by Voigt [1], and then developed in 1909 by the brothers E. and F. Cosserat [2]. It is assumed in this theory that the entire action upon a material volume bounded by a surface is described completely in terms of the field of stress vectors and "couple-stress" vectors.

Recently we witness a further development of this theory, particularly since it proved useful in explaining some regularities of propagation of short acoustic waves in crystals, in polycrystalline structures as well as in high polymers.

Let us mention here the papers by Truesdell and Toupin [3], and by Aero and Kuvshinskii [4]. Grioli [5] and Toupin [6] succeeded in generalizing the theory of the Cosserat medium to finite deformations. Finally, Mindlin and Tiersten [7] presented an exhaustive discussion of the linear theory of a homogeneous, isotropic and centrosymmetric medium.

In subsequent years KUVSHINSKII and AERO (1963, [8]), PALMOV [9] and ERINGEN and SUHUBI [10] developed the theory of asymmetric elasticity describing the deformation of a body by the vectors of displacement $\boldsymbol{u}$ and rotation ω, mutually independent thus departing from the kinematic assumption $\omega = \frac{1}{2}$ curl $\boldsymbol{u}$, which is the very basis of the theory of the Cosserat medium.

The purpose of the present work is to consider, within the framework of the theory of asymmetric elasticity, the interaction between the fields of displacements $\boldsymbol{u}$, rotation ω and temperature θ.

Confining our considerations to the elastic, homogeneous, isotropic and centrosymmetric medium we derive constitutive equations based on the thermodynamics of irreversible processes. A complete set of differential equations of asymmetric thermoelasticity is given as well as equations of motion and an generalized equation of heat conductivity.

Finally, a variational theorem and a theorem on reciprocity are given, and some conclusions are derived from these theorems.

2. Equations of Motion. Energy Equation and Entropy Balance

The system of equations of motion consists of two equations: the equation of balance of linear momentum and the equation of angular momentum [4], [7]:

$$\sigma_{ji,j} + X_i - \varrho \ddot{u}_i = 0, \tag{2.1}$$

$$\varepsilon_{ijk}\sigma_{jk} + \mu_{ji,j} + Y_i - J\ddot{\omega}_i = 0, \qquad i, j, k = 1, 2, 3. \tag{2.2}$$

In these equations the symbol σ_{ij} denotes the asymmetric force-stress tensor, μ_{ji} the asymmetric couple-stress tensor. X_i and Y_i designate the volume density of the body forces and body couples; ε_{ijk} stands for the usual alternator. u_i are the components of the displacement vector $\boldsymbol{u}$, ω_i the components of the rotation vector.

The principle of conservation of energy written for an arbitrary volume V of a body, bounded by a surface A, has the form

$$\frac{d}{dt}\int_V \left[\frac{1}{2}(\varrho v_i v_i + J w_i w_i) + U\right] dV = \int_V (X_i v_i + Y_i \omega_i)\, dV + \int_A (p_i v_i + m_i w_i)\, dA - \int_A q_i n_i\, dA. \tag{2.3}$$

Here $v_i = \dot{u}_i$, $w_i = \dot{\omega}_i$. By U we denote the internal energy, and by q_i the components of the heat flux vector $\boldsymbol{q}$.

The quantities p_i and m_i are connected with tensors σ_{ij} and μ_{ji} by the following relations

$$p_i = \sigma_{ji} n_j, \qquad m_i = \mu_{ji} n_j, \tag{2.4}$$

where n_i are the components of the unit vector $\boldsymbol{n}$ on A.

The terms on the left hand side of (2.3) represent the rate of increase of the kinetic and internal energy, respectively, of the volume. The first term on the right-hand side represents the rate of work of the body forces and body couples, the second term-the rate of work of the surface tractions and couples. Finally, the last integral on the right-hand side of eq. (2.3) denotes the heat transferred to the volume V by heat conduction. Taking into consideration (2.1), (2.2) and (2.4) and making use of the divergence theorem, we obtain the following equation

$$\int_V \{\dot{U} - [\sigma_{ij}(v_{i,j} - \varepsilon_{kji} w_k) + \mu_{ji} w_{i,j}] + q_{i,i}\} \, dV = 0. \tag{2.5}$$

This expression holds for any arbitrary volume V. If the integrand is continuous, then the relation

$$\dot{U} = \sigma_{ji} \dot{\gamma}_{ji} + \mu_{ji} \dot{\varkappa}_{ji} - q_{i,i} \tag{2.6}$$

holds locally.

The following notations were introduced

$$\gamma_{ji} = u_{i,j} - \varepsilon_{kji} \omega_k, \qquad \varkappa_{ji} = \omega_{i,j}. \tag{2.7}$$

Here γ_{ji} is the asymmetric strain tensor, $\varkappa_{ji}$ the curvature-twist tensor.

The equation of entropy balance can be written in the form ([11], p. 29).

$$\int_V \dot{S} \, dV = -\int_A \frac{q_i n_i}{T} \, dA + \int_V \Theta \, dV. \tag{2.8}$$

The left-hand side of this equation represents the rate of increase of entropy. The first term on the right-hand side is the rate at which entropy is supplied to the volume across the surface. The second term on the right-hand side of (2.8) denotes the rate of entropy production, due to heat conduction. Using the divergence theorem we have

$$\int_V \left(\dot{S} - \Theta - \left(\frac{q_i}{T}\right)_{,i}\right) dV = 0, \tag{2.9}$$

and hence, since V is arbitrary we get

$$\dot{S} = \Theta - \frac{q_{i,i}}{T} + \frac{q_i T_{,i}}{T^2}, \tag{2.10}$$

which holds at each point of the body. In accordance with the postulate of thermodynamics of irreversible processes we have $\Theta \geqq 0$.

Eliminating $q_{i,i}$ from (2.6) and (2.10), and introducing the expression for the Helmholtz free energy $F = U - ST$, we obtain

$$\dot{F} = \sigma_{ji}\dot{\gamma}_{ji} + \mu_{ji}\dot{\varkappa}_{ji} - \dot{T}S - T\left(\Theta + \frac{q_i T_{,i}}{T^2}\right). \tag{2.11}$$

Since the free energy is a function of the independent variables γ_{ji}, $\varkappa_{ji}$, T, there is

$$\dot{F} = \frac{\partial F}{\partial \gamma_{ji}}\dot{\gamma}_{ji} + \frac{\partial F}{\partial \varkappa_{ji}}\dot{\varkappa}_{ji} + \frac{\partial F}{\partial F}\dot{T}. \tag{2.12}$$

Assuming that the functions Θ, q_i, . . σ_{ji}, μ_{ji} do not explicitly depend on the time derivatives of the functions γ_{ji}, $\varkappa_{ji}$, T and defining the entropy as $S = -\dfrac{\partial F}{\partial T}$, we obtain — after comparing (2.11) with (2.12) — the following relations

$$\sigma_{ji} = \frac{\partial F}{\partial \gamma_{ji}}, \quad \mu_{ji} = \frac{\partial F}{\varkappa_{ji}}, \quad S = -\frac{\partial F}{\partial T}, \quad \Theta + \frac{q_i T_{,i}}{T^2} = 0. \tag{2.13}$$

The second law of thermodynamics will be satisfied if $\Theta \geqq 0$, or if

$$-\frac{T_{,i} q_i}{T^2} \geqq 0. \tag{2.14}$$

The above inequality satisfies the Fourier law of thermal conductivity

$$-q_i = k_{ij} T_{,j} \qquad \text{or} \qquad -q_i = k_{ij}\theta_{,j} \qquad T = T_0 + \theta \tag{2.15}$$

Here T_0 denotes the temperature of the body in its natural state, in which stresses and deformations are equal to zero (i. e., for $\gamma_{ji} = 0$, $\varkappa_{ji} = 0$, $T = T_0$). The quantities k_{ij} are the coefficients of thermal conductivity and form a symetric tensor.

From eq. (2.10) — taking into account the last relation of the group (2.13) — we have

$$T\dot{S} = -q_{i,i} = k_{ij}\theta_{,ij}. \tag{2.16}$$

For a homogeneous and isotropic body, we get

$$T\dot{S} = k\theta_{,jj} \tag{2.17}$$

where k is constant.

3. Constitutive Equations

Let us expand the expression for the free energy $F(\gamma_{ji}, \varkappa_{ji}, T)$ in the neighborhood of a natural state ($\gamma_{ji} = 0$, $\varkappa_{ji} = 0$, $T = T_0$) into a Taylor series, omitting powers higher than 2. This expansion has, for a centrosymmetric body, the form

$$F = \frac{\mu + \alpha}{2} \gamma_{ji}\gamma_{ji} + \frac{\mu - \alpha}{2} \gamma_{ji}\gamma_{ij} + \frac{\lambda}{2} \gamma_{kk}\gamma_{nn}$$
$$+ \frac{\gamma + \varepsilon}{2} \varkappa_{ji}\varkappa_{ji} + \frac{\gamma - \varepsilon}{2} \varkappa_{ji}\varkappa_{ij} + \frac{\beta}{2} \varkappa_{kk}\varkappa_{nn}$$
$$- \nu\gamma_{kk}\theta - \chi\varkappa_{kk}\theta - \frac{m}{2}\theta^2. \tag{3.1}$$

This form results from the following considerations. Since the free energy is a scalar, each term on the right-hand side of expression (3.1) must be a scalar. Now, with the components of the asymmetric tensor γ_{ji} we can construct three independent quadratic invariants, namely $\gamma_{ji}\gamma_{ji}$, $\gamma_{ji}\gamma_{ij}$ and $\gamma_{kk}\gamma_{nn}$. The same holds for the tensor $\varkappa_{ji}$. However, the terms $\gamma_{ji}\varkappa_{ji}$, $\gamma_{ji}\varkappa_{ij}$ and $\gamma_{kk}\varkappa_{nn}$ cannot appear in eq. (3.1), since this would contradict the postulate of centrosymmetric isotropy. In the seventh and eighth term of relation (3.1) the invariants γ_{kk} and $\varkappa_{kk}$ appear. This results from the fact that with the components of the tensor γ_{ji} and $\varkappa_{ji}$ one can form invariants of the first kind only, namely γ_{kk} and $\varkappa_{kk}$.

Making use of relations (3.13) we have

$$\sigma_{ji} = (\mu + \alpha)\gamma_{ji} + (\mu - \alpha)\gamma_{ij} + (\lambda\gamma_{kk} - \nu\theta)\delta_{ji}, \tag{3.2}$$

$$\mu_{ji} = (\gamma + \varepsilon)\varkappa_{ji} + (\gamma - \varepsilon)\varkappa_{ij} + (\beta\varkappa_{kk} - \chi\theta)\delta_{ij} \tag{3.3}$$

$$S = \nu\gamma_{kk} + \chi\varkappa_{kk} + m\theta. \tag{3.4}$$

Relations (3.2), (3.3) may be rewritten in an equivalent form

$$\sigma_{ji} = 2\mu\gamma_{(ij)} + 2\alpha\gamma_{\langle ij\rangle} + (\lambda\gamma_{kk} - \chi\nu\theta)\delta_{ij}, \tag{3.5}$$

$$\mu_{ij} = 2\gamma\varkappa_{(ij)} + 2\varepsilon\varkappa_{\langle ij\rangle} + (\beta\varkappa_{kk} - \theta)\delta_{ij}. \tag{3.6}$$

Here μ, λ are Lame constants and α, γ, ε, β are new constants of elasticity referring to the isothermal state. The constants ν, χ depend on the mechanical as well as on the thermal properties of the body. The symbols ( ) and $\langle\rangle$ denote symmetric and skew-symmetric part of a tensor, respectively.

Solving eqs. (3.5) and (3.6) for γ_{ij} and $\varkappa_{ij}$ we obtain

$$\gamma_{ij} = \alpha_t \delta_{ij} \theta + 2\mu' \sigma_{(ij)} + 2\alpha' \sigma_{\langle ij \rangle} + \lambda' \delta_{ij} \sigma_{kk}, \tag{3.7}$$

$$\varkappa_{ij} = \beta_t \delta_{ij} \theta + 2\gamma' \mu_{(ij)} + 2\varepsilon' \mu_{\langle ij \rangle} + \beta' \delta_{ij} \mu_{kk}, \tag{3.8}$$

where

$$\mu' = \frac{1}{4\mu}, \quad \alpha' = \frac{1}{4\alpha}, \quad \gamma' = \frac{1}{4\gamma}, \quad \varepsilon' = \frac{1}{4\varepsilon}, \quad \lambda' = -\frac{\lambda}{6\mu K},$$

$$\beta' = -\frac{\beta}{6\gamma\Omega}, \quad \alpha_t = \frac{\nu}{3K}, \quad \beta_t = \frac{\chi}{3\Omega}, \quad K = 2\mu + 3\lambda, \quad \Omega = 2\gamma + 3\beta.$$

The symbol α_t appearing in relation (3.7) is the coefficient of thermal dilatation.

If the volume element, free from force-stresses and couple-stresses on its surface, produced temperature voluminal changes only, we obtain

$$\gamma_{ij}^{\circ} = \alpha_t \delta_{ij} \theta, \qquad \varkappa_{ij}^{\circ} = 0.$$

The coefficient β_t and, consequently, the quantity χ should be equal to zero. This assertion is also supported by another argument. Let us consider the problem of classical elastokinetics where an adiabatic process is assumed ($\dot{S} = 0$). In this case the temperature is given by

$$\theta = -\frac{1}{m} (\nu \gamma_{kk} + \chi \varkappa_{kk}). \tag{3.9}$$

Introducing (3.9) into relations (3.5) and (3.6) we obtain[1]

$$\sigma_{ij} = 2\mu \gamma_{(ij)} + 2\alpha \gamma_{\langle ij \rangle} + (\lambda^* \gamma_{kk} + \tau^* \varkappa_{kk}) \delta_{ij}, \tag{3.10}$$

$$\mu_{ij} = 2\mu \varkappa_{(ij)} + 2\varepsilon \varkappa_{\langle ij \rangle} + (\beta^* \varkappa_{kk} + \delta^* \gamma_{kk}) \delta_{ij} \tag{3.11}$$

where

$$\lambda^* = \lambda + \frac{\nu^2}{m}, \quad \beta^* = \beta + \frac{\chi^2}{m}, \quad \tau^* = \delta^* = \frac{\nu\chi}{m}.$$

It is easily seen that relations (3.10) and (3.11) are in contradiction with the formulae for an isotropic, centrosymmetric medium, where force stresses σ_{ij} are independent of $\varkappa_{ij}$ and couple-stresses are independent of γ_{ij} (cf., e. g., [7], [8]). Thus, we should have $\chi = 0$. Consequently, we have to put for the free energy $\chi = 0$ in (3.1).

[1] Observe that for a Cosserat medium, where $\omega_i = \frac{1}{2} \varepsilon_{ijk} u_{k,j}$, we have $\varkappa_{kk} = 0$.

In this way, the final forms of relations (3.2) (3.4) will be:

$$\sigma_{ij} = 2\mu\gamma_{(ij)} + 2\alpha\gamma_{\langle ij\rangle} + (\lambda\gamma_{kk} - \nu\theta)\,\delta_{ij}, \tag{3.12}$$

$$\mu_{ij} = 2\mu\varkappa_{(ij)} + 2\varepsilon\varkappa_{\langle ij\rangle} + \beta\varkappa_{kk}\delta_{ij}, \tag{3.13}$$

$$S = \nu\gamma_{kk} + m\theta. \tag{3.14}$$

Let us remark that in the last of the above relations the quantity m remains undefined. It will be determined from the following considerations.

To begin with, we shall consider the differential relation

$$dU = \sigma_{ji} d\gamma_{ji} + \mu_{ji} d\varkappa_{ji} + T dS. \tag{3.15}$$

Introducing into (3.15) the relation

$$dS = \left(\frac{\partial S}{\partial\gamma_{ji}}\right)_{\varkappa,T} d\gamma_{ji} + \left(\frac{\partial S}{\partial x_{ji}}\right)_{\gamma,T} d\varkappa_{ji} + \left(\frac{\partial S}{\partial T}\right)_{\varkappa,\gamma} dT, \tag{3.16}$$

and taking into account completeness conditions of dU, we arrive at the dependence

$$\left(\frac{\partial S}{\partial\gamma_{ji}}\right)_{\varkappa,T} - \nu\delta_{ij} = 0, \quad \left(\frac{\partial S}{\partial\varkappa_{ji}}\right)_{\gamma,T} = 0. \tag{3.17}$$

Substituting now the above formulae into (3.15) and (3.16) and taking into account $\left(\frac{\partial S}{\partial T}\right)_{\gamma,\varkappa} = \frac{c_\varepsilon}{T}$, where c_ε denotes the specific heat at constant deformation, we get

$$dU = \sigma_{ji} d\gamma_{ji} + \mu_{ji} d\varkappa_{ji} + \nu T d\gamma\varepsilon_{kk} + c_\varepsilon dT, \tag{3.18}$$

$$dS = \nu d\gamma_{kk} + c_\varepsilon \frac{dT}{T}. \tag{3.19}$$

Integrating (3.19) under the assumption that $S = 0$ for the natural state, we have

$$S = \nu\gamma_{kk} + c_\varepsilon \log\frac{T}{T_0}. \tag{3.20}$$

Assuming $\left|\frac{\theta}{T_0}\right| \ll 1$, expanding the logarithm into a series and retaining one term of this series only, we have

$$S = \gamma_{kk}\nu + c_\varepsilon \frac{\theta}{T_0}. \tag{3.21}$$

From a comparison of (3.14), and (3.21) it results that $m = \frac{c_\varepsilon}{T_0}$.

4. Differential Equations of Thermoelasticity

The constitutive relations (3.12), (3.13) enable us to express the equations of motion (2.1), (2.2) in terms of the displacement vector $\boldsymbol{u}$, the rotation vector $\boldsymbol{\omega}$ and temperature. We obtain the following differential equations

$$(\lambda + 2\mu) \operatorname{grad} \operatorname{div} \boldsymbol{u} - (\mu + \alpha) \operatorname{curl} \operatorname{curl} \boldsymbol{u} + 2\alpha \operatorname{curl} \boldsymbol{\omega} + \boldsymbol{X} = \varrho \ddot{\boldsymbol{u}} + \nu \operatorname{grad} \theta, \tag{4.1}$$

$$(\beta + 2\gamma) \operatorname{grad} \operatorname{div} \boldsymbol{\omega} - (\gamma + \varepsilon) \operatorname{curl} \operatorname{curl} \boldsymbol{\omega} + 2\alpha \operatorname{curl} \boldsymbol{u} - 4\alpha\boldsymbol{\omega} + \boldsymbol{Y} = J\ddot{\boldsymbol{\omega}}. \tag{4.2}$$

To eqs. (4.1), (4.2) we have to add the equation of heat conduction. For this purpose we consider the relations (2.17) and (3.20)

$$T\dot{S} = k\theta_{,jj} \tag{4.3}$$

$$T\dot{S} = \nu T \dot{\gamma}_{kk} + c_\varepsilon \dot{T}. \tag{4.4}$$

It results from a comparison of these relations that

$$\theta_{,jj} - \frac{1}{\varkappa}\dot{\theta} - \eta_0\left(1 + \frac{\theta}{T_0}\right) \operatorname{div} \dot{\boldsymbol{u}} = 0, \quad \varkappa = \frac{k}{c_\varepsilon}, \quad \eta_0 = \frac{\nu T_0}{k}. \tag{4.5}$$

When linearizing this equation, we assume that $\left|\frac{\theta}{T_0}\right| \ll 1$. Moreover, taking heat sources within the body into account and denoting by W the quantity of heat generated par unit volume and unit time, we obtain the following extended equation of thermal conduction

$$\theta_{,jj} - \frac{1}{\varkappa}\dot{\theta} - \eta_0 \operatorname{div} \dot{\boldsymbol{u}} = -\frac{Q}{\varkappa}, \qquad Q = \frac{W}{k}. \tag{4.6}$$

It is interesting to note that only the term derived from dilatation appears in eq. (4.6), consequently connected with the first deformation invariant γ_{kk}.

Equations (4.1), (4.2) and (4.6) constitute a set of equations of linear coupled thermoelasticity in the theory of asymmetric elasticity.

We have to supplement the differential equations of thermoelasticity with the boundary and initial conditions. If on the surface A, bounding the body of volume V, the force-stresses p_i, m_i and temperature θ are prescribed, boundary conditions will have the form

$$p_i(\boldsymbol{x}, t) = \sigma_{ji}(\boldsymbol{x}, t)\, n_j(\boldsymbol{x}), \quad m_i(\boldsymbol{x}, t) = \mu_{ji}(\boldsymbol{x}, t)\, n_j(\boldsymbol{x}),$$
$$\theta(\boldsymbol{x}, t) = h(\boldsymbol{x}, t), \qquad \boldsymbol{x} \in A, \qquad t > 0, \tag{4.7}$$

while the initial conditions will be expressed by the formulae

$$u_i(\boldsymbol{x}, 0) = f_i(\boldsymbol{x}), \qquad \dot{u}_i(\boldsymbol{x}, 0) = g_i(\boldsymbol{x}),$$
$$\omega_i(\boldsymbol{x}, 0) = l_i(\boldsymbol{x}), \qquad \dot{\omega}_i(\boldsymbol{x}, 0) = k_i(\boldsymbol{x}),$$
$$\theta(\boldsymbol{x}, 0) = r(\boldsymbol{x}), \qquad \boldsymbol{x} \varepsilon V, \qquad t = 0. \tag{4.8}$$

The dynamic equations of thermoelasticity (4.1), (4.2) can be separated by decomposing the vectors $\boldsymbol{u}$ and $\boldsymbol{\omega}$ into their potential and solenoidal parts

$$\boldsymbol{u} = \operatorname{grad} \varphi + \operatorname{curl} \boldsymbol{\Psi}, \qquad \boldsymbol{\omega} = \operatorname{grad} \zeta + \boldsymbol{\Phi},$$
$$\operatorname{div} \boldsymbol{\Psi} = 0, \qquad \operatorname{div} \boldsymbol{\Phi} = 0. \tag{4.9}$$

Decomposing in a similar way the expressions for the body forces and the couple-body forces

$$\boldsymbol{X} = \operatorname{grad} \vartheta + \operatorname{curl} \boldsymbol{\chi}, \qquad \boldsymbol{Y} = \operatorname{grad} \sigma + \boldsymbol{\Omega}, \tag{4.10}$$

we obtain the following system of equations

$$\Box_1^2 \varphi = n\theta - \frac{1}{c_1^2} \vartheta, \tag{4.11}$$

$$\Box_3^2 \zeta - s\zeta = -\frac{\sigma}{c_3^2}, \tag{4.12}$$

$$\Box_2^2 \boldsymbol{\Psi} - r\boldsymbol{\Phi} = -\frac{1}{c_2^2} \boldsymbol{\chi}, \tag{4.13}$$

$$\Box_4^2 \boldsymbol{\Phi} - p\nabla^2 \boldsymbol{\Psi} - 2p\boldsymbol{\Phi} = -\frac{\boldsymbol{\Omega}}{c_1^2}. \tag{4.14}$$

The notations introduced in (4.11) + (4.14) are as below

$$\Box_\alpha^2 = \nabla^2 - \frac{1}{c_\alpha^2} \partial_t^2, \quad \alpha = 1, 2, 3, 4, \quad c_1 = \left(\frac{\lambda + 2\mu}{\varrho}\right)^{1/2}, \quad c_2 = \left(\frac{\mu + a}{\varrho}\right)^{1/2},$$
$$c_3 = \left(\frac{\beta + 2\gamma}{J}\right)^{1/2}, \quad c_4 = \left(\frac{\gamma + \varepsilon}{J}\right)^{1/2}, \quad r = \frac{2\alpha}{\varrho c_2^2}, \quad s = \frac{4\alpha}{J c_3^2}, \quad p = \frac{2\alpha}{J c_4^2}.$$

To eq. (4.11) and eq. (4.14) we have to add the equation of heat conduction

$$D\theta - \eta_0 \nabla^2 \dot{\varphi} = -\frac{Q}{\varkappa}, \qquad D = \nabla^2 - \frac{1}{\varkappa} \partial_t. \tag{4.15}$$

We shall consider first the propagation of thermoelastic waves in an infinite space. Observe that eqs. (4.11) and (4.15) are coupled. Eliminating the temperature from these equations we obtain the equation for the longitudinal wave

$$(\square_1^2 D - \eta_0 \partial_t \nabla^2)\varphi = -\frac{nQ}{\varkappa} - \frac{1}{c_1^2} D\vartheta, \qquad \partial_t = \frac{\partial}{\partial t}. \tag{4.16}$$

If the quantities χ, $\boldsymbol{\Omega}$, σ and the initial conditions of the functions ζ, $\boldsymbol{\Psi}$, $\boldsymbol{\Phi}$ are equal to zero, then in an unbounded elastic space only dilatational waves will propagate. Equation (4.16) describing these waves is identical with that obtained for the elastic classical medium (with no couple-stresses). It is known [12] that these waves are attenuated and dispersed. Since

$$u_i = \varphi_{,i}, \quad \omega_i = 0, \quad \gamma_{ji} = \varphi_{,ij}, \quad \varkappa_{ji} = 0,$$

we have

$$\sigma_{(ij)} = 2\mu(\varphi_{,ij} - \delta_{ij}\varphi_{,kk}) + \varrho(\ddot{\Phi} - \vartheta)\delta_{ij},$$
$$\sigma_{\langle ij\rangle} = 0, \quad \mu_{ji} = 0. \tag{4.17}$$

If Q, ϑ, χ, $\boldsymbol{\Omega}$ are equal to zero and the initial conditions of the functions θ, φ, $\boldsymbol{\Psi}$, $\boldsymbol{\Phi}$ are homogeneous, then in an infinite medium only torsional waves, described by eq. (4.12), propagate. We have namely

$$u_i = 0, \quad \omega_i = \zeta_{,i} \quad \varkappa_{ji} = \omega_{i,j} = \zeta_{,ij}, \quad \gamma_{ji} = 0.$$

Only couple-stresses will appear in the medium forming the symmetric tensor

$$\mu_{(ij)} = 2\gamma\zeta_{,ij} + \beta\delta_{ij}\nabla^2\zeta, \quad \mu_{\langle ij\rangle} = 0, \quad \sigma_{ij} = 0. \tag{4.18}$$

The propagation of these waves is not accompanied by a temperature field.

Finally, in the case when the quantities Q, ϑ, σ are equal to zero and the initial conditions of the functions φ, ζ, θ are homogeneous, only transverse waves propagate in an infinite space (described by the set of eqs. (4.13) and (4.14)). Eliminating from these equations separately the function $\boldsymbol{\Psi}$ and then $\boldsymbol{\Phi}$ we arrive at the following equations

$$[(\square_4^2 - 2p)\,\square_2^2 - pr\nabla^2]\boldsymbol{\Phi} = -\frac{p}{c_2^2}\nabla^2\chi - \frac{1}{c_4^2}\square_2^2\boldsymbol{\Omega}, \tag{4.19}$$

$$[(\square_4^2 - 2p)\,\square_2^2 - pr\nabla^2]\boldsymbol{\Psi} = -\frac{r}{c_4^2}\boldsymbol{\Omega} - \frac{1}{c_2^2}(\square_4^2 - 2p)\chi. \tag{4.20}$$

In an infinite medium these waves are not accompanied by a temperature field. Since $\operatorname{div} \boldsymbol{u} = 0$ they do not induce any changes in the volume of the body. The force-stresses σ_{ji} and couple-stresses μ_{ji} form an asymmetric tensor.

In a finite medium all three kinds of waves discussed here appear. Equations (4.11) $\div$ (4.15) are coupled by means of the boundary conditions.

5. Virtual Work Principle

It may be easily shown that the following equation holds

$$\int_V [(X_i - \varrho \ddot{u}_i)\delta u_i + (Y_i - J\ddot{\omega}_i)\delta \omega_i] dV + \int_A (p_i \delta u_i + m_i \delta \omega_i) dA = \\ = \int_V (\sigma_{ij}\delta\gamma_{ji} + \mu_{ji}\delta\varkappa_{ji}) dV. \quad (5.1)$$

The left-hand side of this equation represents the variation of work of external forces, while the right-hand side equals that of internal forces.

In the above equation symbols δu_i and $\delta \omega_i$ stand for the virtual changes of the components of displacement vector $\boldsymbol{u}$ and rotation vector $\boldsymbol{\omega}$.

Introducing into the right-hand side of eq. (5.1) the relations (3.12) and (3.13) we reduce eq. (5.1) to the form

$$\int_V [(X_i - \varrho \ddot{u}_i)\delta u_i + (Y_i - J\ddot{\omega}_i)\delta \omega_i] dV + \int_A (p_i \delta u_i + m_i \delta \omega_i) dA = \\ = \delta W - \nu \int_V \theta \gamma_{kk} dV, \quad (5.2)$$

where

$$\delta W = \int_V (2\mu\gamma_{(ij)}\delta\gamma_{(ij)} + 2\alpha\gamma_{\langle ij\rangle}\delta\gamma_{\langle ij\rangle} + 2\gamma\varkappa_{(ij)}\delta\varkappa_{(ij)} + 2\varepsilon\varkappa_{\langle ij\rangle}\delta\varkappa_{\langle ij\rangle} + \\ + \lambda\gamma_{kk}\delta\gamma_{kk} + \beta\varkappa_{kk}\delta\varkappa_{kk}) dV.$$

We have to supplement eq. (5.2) by a further equation, since only four causes, namely X_i, Y_i, p_i, m_i, appear in this equation in explicit form. Thus, we adjoin to eq. (5.2) the relation

$$-\nu \int_V \theta \delta\gamma_{kk} dV = \int_A \theta n_i \delta H_i dA + \frac{c_\varepsilon}{T_0} \int_V \theta \delta\theta dV + \\ + \frac{T_0}{k} \int_V \dot{H}_i \delta H_i dV, \quad (5.3)$$

derived from the heat equation by Biot [13]. The vector $\boldsymbol{H}$ in (5.3) is connected with the vector of heat flow $\boldsymbol{q}$ and the entropy S by the following relations

$$\boldsymbol{q} = T_0 \dot{\boldsymbol{H}}, \quad S = -\operatorname{div}(\boldsymbol{H}). \quad (5.4)$$

Introducing (5.3) into eq. (5.2) we have

$$\delta(W+P+D)=\int_V[(X_i-\varrho\ddot{u}_i)\delta u_i+(Y_i-J\ddot{\omega}_i)\delta\omega_i]dV+$$
$$+\int_A(p_i\delta u_i+m_i\delta\omega_i)dA-\int_A\theta n_i\delta H_i dA. \qquad (5.5)$$

Here the heat potential $P=\frac{c_\varepsilon}{2T_0}\int_V\theta^2 dV$ and the dissipation function D were applied, where $\delta D=\frac{T_0}{k}\int_V\dot{H}_i\delta H_i dV$.

For $Y_i=0$, $m_i=0$, $\varkappa_{ji}=0$, $\alpha=\gamma=\varepsilon=\beta=0$, eq. (5.5) reduces to the variational equation of the coupled thermoelastic medium without couple-stresses.

The variational principle eq. (5.5) may be used for the derivation of the energy theorem, if we compare the functions u_i, ω_i, θ at point $\boldsymbol{x}$ and time t with those actually occurring in the same point after a time interval dt. Thus, introducing into eq. (5.5)

$$\delta u_i=v_i dt,\quad \delta\omega_i=w_i dt,\quad \delta\theta=\dot{\theta}dt,\quad \delta H_i=\dot{H}_i dt=-\frac{k}{T_0}\theta_{,i}dt,$$

and so on, we obtain the following formula

$$\frac{d}{dt}(K+W+P)+\chi_\theta=\int_V(X_i v_i+Y_i w_i)dV+\int_A(p_i v_i+$$
$$+m_i w_i)dA+\frac{k}{T_0}\int_A\theta n_i\theta_{,i}dA, \qquad (5.6)$$

where

$$K=\frac{\varrho}{2}\int_V v_i v_i dV+\frac{J}{2}\int_V w_i w_i dV,\quad \chi_\theta=\frac{dD}{dt}=\frac{k}{T_0}\int_V\theta_{,i}\theta_{,i}dV>0.$$

Here K denotes kinetic energy, and χ_θ is proportional to the source of entropy which is always a positive quantity.

The energy theorem (5.6) may be exploited to demonstrate the uniqueness theorem for a simply connected body. Such a demonstration may be carried out in a way similar to that indicated in [14] or [15].

6. The Reciprocity Theorem

Let us consider two systems of causes and effects acting on an elastic body of volume V bounded by the surface A.

We assign to the first group of causes the body forces X_i, body couples Y_i, heat sources Q, loadings p_i and m_i on the surface and the heating of this surface (the prescribed temperature or the heat flux on the surface A). The effects are: the components of the displacement vector $\boldsymbol{u}$, of the rotation vector $\boldsymbol{\omega}$ and temperature θ.

The second set of causes and effects will be distinguished from the first one by a prime. In the sequel we assume the initial conditions to be homogeneous.

Let us apply the one-sided Laplace transformation to the constitutive equations. We obtain the following relations

$$\bar{\sigma}_{ji} = (\mu + \alpha)\bar{\gamma}_{ji} + (\mu - \alpha)\bar{\gamma}_{ij} + (\lambda\bar{\gamma}_{kk} - \nu\bar{\theta})\delta_{ij}, \tag{6.1}$$

$$\bar{\mu}_{ji} = (\gamma + \varepsilon)\bar{\varkappa}_{ji} + (\gamma - \varepsilon)\bar{\varkappa}_{ij} + \beta\bar{\varkappa}_{kk}\delta_{ij}, \tag{6.2}$$

where

$$\bar{\sigma}_{ji}(\boldsymbol{x}, p) = L[\sigma_{ji}(\boldsymbol{x}, t)] = \int_0^\infty \sigma_{ji}(\boldsymbol{x}, t)e^{-pt}dt, \quad \text{e. c. t.}$$

and similar relations for $\bar{\sigma}'_{ji}$ and $\bar{\mu}'_{ji}$. It may easily be seen that the following identity holds

$$\bar{\sigma}_{ji}\bar{\gamma}'_{ji} + \bar{\mu}_{ji}\bar{\varkappa}'_{ji} - \bar{\sigma}'_{ji}\bar{\gamma}_{ji} - \bar{\mu}'_{ji}\bar{\varkappa}_{ji} = \nu(\bar{\theta}'\bar{\gamma}_{kk} - \bar{\theta}\bar{\gamma}'_{kk}). \tag{6.3}$$

Integrating relation (6.3) over the volume V we get

$$\int_V (\bar{\sigma}_{ji}\bar{\gamma}'_{ji} - \bar{\sigma}'_{ji}\bar{\gamma}_{ji} + \bar{\mu}_{ji}\bar{\varkappa}'_{ji} - \bar{\mu}'_{ji}\bar{\varkappa}_{ji})dV = \nu\int_V (\bar{\theta}'\bar{\gamma}_{kk} - \bar{\theta}\bar{\gamma}'_{kk})dV. \tag{6.4}$$

Now let us perform the Laplace transformation on the equations of motion

$$\bar{\sigma}_{ji,j} + \bar{X}_i = p^2\varrho\bar{u}_i, \quad \varepsilon_{ijk}\bar{\sigma}_{jk} + \bar{\mu}_{ji,j} + \bar{Y}_i = p^2J\bar{\omega}_i, \tag{6.5}$$

as well as on the equations of motion for the state marked with primes. Taking advantage of relations (6.5) we reduce eq. (6.4) to the form

$$\begin{aligned}\int_V(\bar{X}_i\bar{u}'_i + \bar{Y}_i\bar{\omega}'_i)dV + \int_A(\bar{p}_i\bar{u}'_i + \bar{m}_i\bar{\omega}'_i)dA &= \\ = \int_V(\bar{X}'_i\bar{u}_i + \bar{Y}'_i\bar{\omega}_i)dV + \int_A(\bar{p}'_i\bar{u}_i + \bar{m}'_i\bar{\omega}_i)dA &+ \\ + \nu\int_V(\bar{\theta}'\bar{\gamma}_{kk} - \bar{\theta}\bar{\gamma}'_{kk})dV.\end{aligned} \tag{6.6}$$

This is the first part of the reciprocity theorem. The second part is obtained by taking account of the heat conduction equations for both sets.

We apply the Laplace transformation to these equations assuming that the initial conditions are homogeneous

$$\bar{\theta}_{,jj} - \frac{p}{\varkappa}\bar{\theta} - \eta_0 p \bar{\gamma}_{kk} = -\frac{\bar{Q}}{\varkappa},$$

$$\bar{\theta}'_{,jj} - \frac{p}{\varkappa}\bar{\theta}' - \eta_0 p \bar{\gamma}'_{kk} = -\frac{\bar{Q}'}{\varkappa}. \tag{6.7}$$

Multiplying the first of eqs. (6.7) by $\bar{\theta}'$, the second by $\bar{\theta}$, subtracting one from another, integrating over the region V and making use of Green's transformation theorem, we get

$$p\eta_0 \int_V (\bar{\gamma}_{kk}\bar{\theta}' - \bar{\gamma}'_{kk}\bar{\theta})\,dV + \frac{1}{\varkappa}\int_V (\bar{Q}'\bar{\theta} - \bar{Q}\bar{\theta}')\,dV - \int_A (\bar{\theta}'\bar{\theta}_{,n} - \bar{\theta}'_{,n}\bar{\theta})\,dA = 0. \tag{6.8}$$

Eliminating the common term from (6.6) and (6.8), we obtain the final form of the theorem on reciprocity

$$\frac{\eta_0 \varkappa p}{\nu}\left[\int_V (\bar{X}_i\bar{u}'_i - \bar{X}'_i\bar{u}_i + \bar{Y}_i\bar{\omega}'_i - \bar{Y}'_i\bar{\omega}_i)\,dV + \int_A (\bar{p}_i\bar{u}'_i - \bar{p}'_i\bar{u}_i + \bar{m}_i\bar{\omega}'_i - \bar{m}'_i\bar{\omega}_i)\,dA\right] + \varkappa\int_A (\bar{\theta}\bar{\theta}'_{,n} - \bar{\theta}'\bar{\theta}_{,n})\,dA + \int_V (\bar{Q}'\bar{\theta} - \bar{Q}\bar{\theta}')\,dV = 0. \tag{6.9}$$

It is obvious that in this relation all causes and effects appear. Applying the inverse Laplace transformation on eq. (6.9) we have

$$\begin{aligned}
&\frac{\eta_0 \varkappa}{\nu}\Bigg\{\int_V dV(x)\int_0^t \Bigg[X_i(x,\tau)\frac{\partial u'_i(x,t-\tau)}{\partial\tau} - X'_i(x,t-\tau)\frac{\partial u_i(x,\tau)}{\partial\tau} + \\
&+ Y_i(x,\tau)\frac{\partial \omega'_i(x,t-\tau)}{\partial\tau} - Y'_i(x,t-\tau)\frac{\partial \omega_i(x,\tau)}{\partial\tau}\Bigg]d\tau + \\
&+ \int_A dA(x)\int_0^t \Bigg[p_i(x,\tau)\frac{\partial u'_i(x,t-\tau)}{\partial\tau} - p'_i(x,t-\tau)\frac{\partial u_i(x,\tau)}{\partial\tau} + \\
&+ m_i(x,\tau)\frac{\partial \omega'_i(x,t-\tau)}{\partial\tau} - m'_i(x,t-\tau)\frac{\partial \omega_i(x,\tau)}{\partial\tau}\Bigg]d\tau\Bigg\} + \\
&+ \varkappa\int_A dA(x)\int_0^t [\theta(x,\tau)\theta'_{,n}(x,t-\tau) - \theta'(x,t-\tau)\theta_{,n}(x,\tau)]\,d\tau + \\
&+ \int_V dV(x)\int_0^t [\theta(x,\tau)\,Q'(x,t-\tau) - \theta'(x,t-\tau)\,Q(x,\tau)]\,d\tau = 0.
\end{aligned} \tag{6.10}$$

With $Y_i = Y'_i = 0$, $m_i = m'_i = 0$ the theorem on reciprocity (6.10) goes over into the theorem on reciprocity for an elastic medium without couple-stresses, given by CAZIMIR-IONESCU [16]. For static loads and for stationary heat flow we get the system of equations

$$\int_V (X_i u'_i - X'_i u_i + Y_i \omega'_i - Y'_i \omega_i) d V + \int_A (p_i u'_i - p'_i u_i +$$

$$+ m_i \omega'_i - m'_i \omega_i) d A + \nu \int_V (\theta \gamma'_{kk} - \theta' \gamma_{kk}) d V = 0, \qquad (6.11)$$

$$\int_V (Q' \theta - Q \theta') d V + \varkappa \int_A (\theta'_{,n} \theta - \theta_{,n} \theta') d A = 0. \qquad (6.12)$$

In the reciprocity equation (6.11) tempeatures θ and θ' are treated as known functions, obtained from the solution of the heat conductivity equations

$$\theta_{,jj} = -\frac{Q}{\varkappa}, \qquad \theta'_{,jj} = -\frac{Q'}{\varkappa}. \qquad (6.13)$$

Equation (6.12) may be treated as a theorem of reciprocity for the problem of heat conduction.

7. Conclusions from the Theorem of Reciprocity

Let us consider first an infinite medium. Let a concentrated and instantaneous force $X_i = \delta(x - \xi) \delta(t) \delta_{ik}$, directed along the axis x_k, be acting at the point ξ of the medium. Denote by $U_i^{(k)}(x, \xi, t)$ the displacement caused by this force. Furthermore, let a concentrated and instantaneous force $X'_i = \delta(x - \eta) \delta(t) \delta_{ij}$, directed along the axis x_j, be acting at the point η. Denote the displacement caused by this force by $U_i^{(j)}(x, \eta, t)$. From the theorem of reciprocity (6.10), formulated for an infinite region, we have

$$\int_V d V(x) \int_0^t d\tau \left[[\delta(x - \xi) \delta(\tau) \delta_{ik} \frac{\partial U_i^{(j)}(x, \eta, t - \tau)}{\partial \tau} - \right.$$

$$\left. - \delta(x - \eta) \delta(t - \tau) \delta_{ij} \frac{\partial U_i^{(k)}(x, \xi, \tau)}{\partial \tau} \right] = 0, \qquad (7.1)$$

and hence

$$\dot{U}_k^{(j)}(\xi, \eta, t) = \dot{U}_j^{(k)}(\eta, \xi, t).$$

After integration with respect to time, we finally get

$$U_k^{(j)}(\xi, \eta, t) = U_j^{(k)}(\eta, \xi, t). \qquad (7.2)$$

Let a concentrated and instantaneous force $X_i = \delta(x - \xi)\delta(t)\delta_{ik}$ act at the point ξ of an infinite medium, and a concentrated and instantaneous source of heat $Q' = \delta(x - \eta)\delta(t)$ act at the point η. Denote by $\Theta^{(k)}(x, \xi, t)$ the temperature caused by the action of force X_i, and by $U_i(x, \eta, t)$ the displacement caused by the action of the source Q'. From eq. (6.10) we obtain the following relation

$$\int_V dV(x) \int_0^t d\tau \left[\delta(x - \eta)\delta(t - \tau) \frac{\partial \Theta(x, \xi, \tau)}{\partial \tau} + \right.$$
$$\left. + \frac{\eta_0 \varkappa}{\nu} \delta(x - \xi)\delta(\tau)\delta_{ik} \frac{\partial U_i(x, \eta, t - \tau)}{\partial \tau}\right] = 0, \tag{7.3}$$

wherefrom

$$\Theta^{(k)}(\eta, \xi, t) = -\frac{\eta_0 \varkappa}{\nu} \frac{\partial U_k(\xi, \eta, t)}{\partial t}. \tag{7.4}$$

Let a concentrated and instantaneous force $X_i = \delta(x - \xi)\delta(t)\delta_{ik}$ act at the point ξ of infinite medium, and the concentrated and instantaneous body couple $Y_i' = \delta(x - \eta)\delta(t)\delta_{ij}$ act at the point η. Denote by $\Omega_i^{(k)}(x, \xi, t)$ the rotation vector caused by the action of force X_i, and by $V_i^{(j)}(x, \eta, t)$ the displacement caused by the body couple Y_i'. From eq. (6.10) we get

$$V_k^{(j)}(\xi, \eta, \mathrm{t}) = \Omega^{(k)}(\eta, \xi, t). \tag{7.5}$$

Finally, let a body couple $Y_i = \delta(x - \xi)\delta(t)\delta_{ik}$ act at the point ξ, and a source of heat $Q' = \delta(x - \eta)\delta(t)$ act at the point η. Denote the temperature caused by the action of body couple by $\vartheta^{(k)}(x, \xi, t)$, and the rotation vector caused by the action of the source Q' by $\Omega_i(x, \eta, t)$. From the theorem of reciprocity (6.10) we obtain the following relation

$$\vartheta^{(k)}(\eta, \xi, t) = -\frac{\eta_0 \varkappa}{\nu} \frac{\partial \Omega_k(\xi, \eta, t)}{\partial t}. \tag{7.6}$$

It can be shown that the relations (7.2), (7.4) — (7.6) hold for a finite body at homogeneous boundary conditions

Let us consider a finite body V and assume that the causes which set the medium in motion are defined by the boundary conditions. We wish to find expressions for the displacements u_i, rotation vectors ω_i and temperature θ at an internal point $x \varepsilon V$ by means of integrals on the surface A bounding the region V. These functions should satisfy the equations of motion, the extended equation of heat conduction and the boundary conditions. When deriving the formulae for the functions $u_i(x, t)$, $\omega_i(x, t)$, $\theta(x, t)$ we shall use the theorem of reciprocity (6.10). Assume, first, that quantities marked with primes refer to displacements $u_i' = U_i^{(k)}(x, \xi, t)$, rotation vector $\omega_i' = \Omega_i^{(k)}(x, \xi, t)$ and temperature $\theta' = \Theta^{(k)}(x, \xi, t)$, caused in an infinite medium by a concentrated and instanteneous force $X_i' = \delta(x - \xi)\,\delta(t)\,\delta_{ik}$, applied at the point ξ and directed along the axis x_k. Assuming absence of body forces ($X_i = 0$), body couples ($Y_i = Y_i' = 0$) and heat sources ($Q = Q' = 0$) we obtain from (6.10) the following expression

$$\begin{aligned}\dot{u}_k(x, t) = \int_A dA(\xi) \int_0^t d\tau \Big\{ p_i(\xi, \tau) \frac{\partial U_i^{(k)}(\xi, x, t-\tau)}{\partial \tau} - \\ - p_i^{(k)}(\xi, x, t-\tau) \frac{\partial u_i(\xi, \tau)}{\partial \tau} + m_i(\xi, \tau) \frac{\partial \Omega_i^{(k)}(\xi, x, t-\tau)}{\partial \tau} - \\ - m_i^{(k)}(\xi, x, t-\tau) \frac{\partial \omega_i(\xi, \tau)}{\partial \tau} + \frac{\nu}{\eta_0} [\theta(\xi, \tau)\, \Theta_{,n}^{(k)}(\xi, x, t-\tau) - \\ - \Theta^{(k)}(\xi, x, t-\tau)\, \theta_{,n}(\xi, \tau)] \Big\}, \quad x \varepsilon V,\ \xi \varepsilon A. \qquad (7.7)\end{aligned}$$

Here we have introduced the notations

$$p_i^{(k)} = \sigma_{ji}^{(k)}(x, \xi, t)\, n_j(x), \qquad m_i^{(k)}(x, t) = \mu_{ji}^{(k)}(x, \xi, t)\, n_j(x),$$

where $\sigma_{ji}^{(k)}$ we understand stresses and by $\mu_{ji}^{(k)}$ couple-stresses caused by a concentrated force $X_i' = \delta(x - \xi)\,\delta(t)\,\delta_{ik}$.

Integration operations in the surface integrals are to be carried out with respect to the variable ξ. Formula (7.7) gives us the relation between the function $\dot{u}_k(x, t)$, $x \varepsilon V$, $t > 0$ and functions u_i, p_i, m_i, ω_i, θ, $\theta_{,n}$ on the surface A.

Now let us assume, in the system with "primes", a concentrated and instantaneous body couple $Y_i' = \delta(x - \xi)\,\delta(t)\,\delta_{ik}$ acting along the axis x_k. In an infinite medium the body couple will cause the displacement $u_i' = V_i^{(k)}(x, \xi, t)$, vector $\omega_i' = \Lambda_i^{(k)}(x, \xi, t)$, and temperature $\theta' = \vartheta^{(k)}(x, \xi, t)$. From the theorem of reciprocity (6.10), at $X_i = X_i' = 0$, $Y_i = 0$, $Q = Q' = 0$ we obtain the following formula:

$$\begin{aligned}\dot{\omega}_k(x, t) = \int_A dA(\xi) \int_0^t d\tau \Big\{ p_i(\xi, \tau) \frac{\partial V_i^{(k)}(\xi, x, t-\tau)}{\partial \tau} - \\ - \hat{p}_i^{(k)}(\xi, x, t-\tau) \frac{\partial u_i(\xi, \tau)}{\partial \tau} + m_i(\xi, \tau) \frac{\partial \Lambda_i^{(k)}(\xi, x, t-\tau)}{\partial \tau} - \\ - \hat{m}_i^{(k)}(\xi, x, t-\tau) \frac{\partial \omega_i(\xi, \tau)}{\partial \tau} + \frac{\nu}{\eta_0} [\theta(\xi, \tau)\, \vartheta_{,n}^{(k)}(\xi, x, t-\tau) - \\ - \vartheta^{(k)}(\xi, x, t-\tau)\, \theta_{,n}(\xi, \tau)] \Big\}, \quad x \varepsilon V,\ \xi \varepsilon A \qquad (7.8)\end{aligned}$$

Here

$$\hat{p}_i^{(k)}(x,t) = \sigma_{ji}^{(k)}(x,\xi,t)\, n_j(x), \quad \hat{m}_i^{(k)} = \mu_{ji}^{(k)}(x,\xi,t)\, n_j(x).$$

We denote $\hat{\sigma}_{ji}^{(k)}$ and $\hat{\mu}_{ji}^{(k)}$ the force-stress tensor and the couple-stress tensor, respectively. Also the function $\dot{\omega}_k(x,t)$, $x \varepsilon V$, $t > 0$, is expressed here by the functions u_i, p_i, ω_i, m_i, θ, $\theta_{,n}$ on the surface A. Now let the system with "primes" in an infinite medium be limited to the action of a concentrated and instantaneous heat source $Q' = \delta(x-\xi)\,\delta(t)$ causing displacements $u_i' = U_i(x,\xi,t)$, rotation vector $\omega_i' = \Omega_i(x,\xi,t)$ and temperature $\theta' = \Theta(x,\xi,t)$.

From (6.10), assuming that $X_i = X_i' = 0$, $Y_i = Y_i' = 0$, $Q = 0$ we get the temperature at the point

$$\begin{aligned}\theta(x,t) = \varkappa \int_A dA(\xi) \int_0^t d\tau \Big\{ &\theta_{,n}(\xi,\tau)\,\Theta(\xi,x,t-\tau) - \\ &- \theta(\xi,\tau)\,\Theta_{,n}(\xi,x,t-\tau) - \frac{\eta_0}{\nu}\Big[p_i(\xi,\tau)\frac{\partial U_i(\xi,x,t-\tau)}{\partial\tau} - \\ &- p_i^*(\xi,x,t-\tau)\frac{\partial u_i(\xi,\tau)}{\partial\tau} + m_i(\xi,\tau)\frac{\partial \Omega_i(\xi,x,t-\tau)}{\partial\tau} - \\ &- m_i^*(\xi,x,t-\tau),\frac{\partial\omega_i(\xi,\tau)}{\partial\tau}\Big]\Big\}, \quad x \varepsilon V, \quad \xi \varepsilon A. \end{aligned} \tag{7.9}$$

Here

$$p_i^*(x,t) = \sigma_{ji}^*(x,\xi,t)\, n_j(x), \quad m_i^*(x,t) = \mu_{ji}^*(x,\xi,t)\, n_j(x).$$

We denote by σ_{ji}^* and μ_{ji}^* the force-stress tensor and couple-stress tensor caused by the action of an instantaneous and concentrated heat source Q'.

Formulae (7.7)—(7.9) may be treated as an extension of Somigliana's formulae [17], to the problems of thermoelasticity. Some simplifications of these formulae can be obtained by taking into account the reciprocity relations (7.4)—(7.6).

If Green's functions $U_i^{(k)}$, $\Omega_i^{(k)}$, $\Theta^{(k)}$ etc. are selected in such a way as to satisfy on the surface A the homogeneous boundary conditions for displacements, rotation vector and temperature, then the formulae (7.7)—(7.9) will yield the solution of the first boundary problem, if functions u_i, ω_i and θ are given at the boundary.

Similarly, if we select Green's functions $U_i^{(k)}$, $\Omega_i^{(k)}$, $\Theta^{(k)}$ etc., in such a way that the boundary is free from loads and temperature, then the formulae (7.7)—(7.9) will yield the solution of the second boundary value problem, when on A loads p_i, m_i and temperature θ are given.

Let us still consider the stationary problems. Let the body contained within the region V and bounded by the surface A be subjected to heating. Let on the part A_u of the surface A, equal to zero, appear displacements u_i and vector of rotation ω_i, and on the part A_σ of the surface A, equal to zero, appear functions p_i and m_i. Moreover, let $X_i = 0$.

For determining the displacement $u_i(x)$, $x \varepsilon V$ consider a body of the same shape and the same boundary conditions. Let in this body $\theta' = 0$ and let a concentrated force $X_i' = \delta(x - \xi)\, \delta_{ik}$ be acting at the point ξ which is, consequently, directed along the axis x_k. This force will cause displacements $U_i^{(k)}(x, \xi)$ assuming that the functions $U_i^{(k)}(x, \xi)$ are selected such as to satisfy homogeneous boundary conditions on A_σ and A_u.

Making use of formula (6.11), we obtain

$$u_k(x) = \nu \int_V \theta(\xi)\, U_{i,i}^{(k)}(\xi, x)\, d\, V(\xi), \quad x \varepsilon V, \quad k = 1, 2, 3. \qquad (7.10)$$

Here $U_{i,i}^{(k)}(\xi, x)$ should be treated as a dilatation caused at the point ξ by a concentrated force X_i applied at the point x. The formula (7.10) may be treated as a generalization of MAYSEL's formula [18], for the problem of thermoelasticity with couple-stresses.

References

[1] VOIGT, W.: Theoretische Studien über die Elastizitätsverhältnisse der Kristalle. Abh. Ges. Wiss. Göttingen **34** (1887).

[2] COSSERAT, E., and F. COSSERAT: Théorie des corps deformables. Paris: A. Hermann et Fils. 1909.

[3] TRUESDELL, C., and R. A. TOUPIN: The Classical Field Theories. Encyclopedia of Physics, vol. III/1, Sec. 200, 203, 205. Berlin-Göttingen-Heidelberg: Springer. 1960.

[4] AERO, E. L., and E. V. KUVSHINSKII: Fundamental Equations of the Theory of Elastic Media with Rotationally Interacting Particles. Fizika Tverdogo Tela, **2**, 1399 (1960). [In Russian]

[5) GRIOLI, G.: Elasticita asimetrica. Ann. Mat. pura ed appl., Ser. IV **50**, 389 (1960)

[6] TOUPIN, R. A.: Elastic Materials with Couple Stresses. Arch. Ratl Mech. Anal. **11**, 385 (1962).

[7] MINDLIN, R. D., and H. F. TIERSTEN: Effects of Couple-stresses in Linear Elasticity. Arch. Ratl Mech. Anal. **11**, 415 (1962).

[8] KUVSHINSKII, E. V., and E. L. AERO: Continuum Theory of Asymmetric Elasticity. Fizika Tverdogo Tela **5**, 2592 (1963).

[9] PALMOV, N. A.: Fundamental Equations of the Theory of Asymmetric Elasticity. Prikl. Mat. Mekh. **28**, 401 (1964).

[10] ERINGEN, A. C., and E. S. SUHUBI: Nonlinear Theory of Micro-Elastic Solids. Int. J. Engng Sci. **2**, 189 (1964); **2**, 389 (1964).

[11] GREEN, A. E., and I. E. ADKINS: Large Elastic Deformations. Oxford: 1960.

[12] CHADWICK, P.: Thermoelasticity. The Dynamical Theory "Progress in Solid Mechanics", vol. **1**, Amsterdam: 1960.

[13] BIOT, M. A.: Thermoelastictiy and Irreversible Thermodynamics. J. Appl. Phys. **27** (1956).

[14] WEINER, J. H.: A Uniqueness Theorem for the Coupled Thermoelastic Problem. J. Appl. Math. **15** (1958).

[15] Ionescu-Cazimir, V.: Problem of Linear Coupled Thermoelasticity. Uniqueness Theorems. Bull. Acad. Polon. Sci., Ser. Sci. Techn. **12**, 12 (1964).
[16] Ionescu-Cazimir, V., Problem of Linear Coupled Thermoelastictiy. Reciprocal Theorems for the Dynamic Problem of Thermoelasticity. Bull. Acad. Polon. Sci., Ser. Sci. Techn. **12**, 9 (1964).
[17] Trefftz, E.: Mathematische Elastizitätstheorie. In: Handbuch der Physik, vol. VI. Berlin: Springer. 1926.
[18] Maysel, V. M.: Thermoelastic Problems. Kiev: 1951. [In Russian]

Discussion

Grioli: How do you determine the coefficient J in the kinetic energy of rotations?

Nowacki: J — ist eine spezielle dynamische Charakteristik des Mediums. Sie ist gleich dem Trägheitsmoment einer Partikel in bezug auf eine beliebige, durch ihren Schwerpunkt gehende Achse, multipliziert mit der Anzahl der Partikel in einer Volumeneinheit.

Savin: We have in this theory of elasticity two cases. For each there is a number of corresponding boundary conditions. In the first case we have five, in the second case we have six. I did not hear about boundary conditions.

Nowacki: In dieser Theorie erfordert das Differentialgleichungssystem sechs Randbedingungen. Man kann an der Oberfläche entweder drei Komponenten des Verschiebungsvektors $\boldsymbol{u}$ und drei Komponenten des Rotationsvektors $\boldsymbol{\omega}$ oder drei Komponenten des Spannungsvektors und drei Komponenten des Momentenspannungsvektors vorschreiben. In der Cosseratschen Theorie haben wir, wegen der Einschränkung $\boldsymbol{\omega} = {}^1/_2 \operatorname{curl} \boldsymbol{u}$, nur 5 Randbedingungen zu befriedigen.

Kaliski: The medium has six degrees of freedom. The six degrees of freedom are important for very short waves or very high frequencies. In this case, however, I think that taking the viscosity of the body into account may also be important.

Nowacki: Man kann analog, wie in der symmetrischen Elastizitätstheorie, auch hier eine asymmetrische Viskoelastizität aufbauen.

Kaliski: Only for very short waves the terms may be important.

On Thermodynamics of the Differential Type Material

By

W. Olszak and **P. Perzyna**
Warsaw (Poland)

1. Introduction

The basic object of the present paper is the analysis of the thermodynamic restrictions for the constitutive equations describing simple materials of the differential type. Our approach is based on principles of determinism, local action, material objectivity, equipresence and the second law of thermodynamics.

We are interested in the special case of the general thermodynamic theory of simple materials presented by COLEMAN [3, 4]. This special case is characterized by the constitutive equations of the differential type.

The restrictions which the second law of thermodynamics imposes on these constitutive equations are analyzed. The reduced forms of the general constitutive equations of the differential type are given and some special cases are considered. Among others, we thus obtain the constitutive equations describing the materials investigated in detail by COLEMAN and MIZEL [2] and by COLEMAN and NOLL [1].

By considering the isothermal process, the constitutive equations of the differential type established first by NOLL [5] are obtained.

All the constitutive equations obtained are valid for finite deformations.

1.1. Notations and Definitions[1]

We shall denote vectors in the threedimensional vector space $\mathscr{V}$ and points in the Euclidean space $\mathscr{E}$ by boldface Latin minuscules: $\boldsymbol{a}, \boldsymbol{x}, \ldots$ The word tensor is regarded as a synonym for a linear transformation of $\mathscr{V}$ into itself. Tensors are denoted by boldface Latin majuscules: $\boldsymbol{A}$, $\boldsymbol{F}$, $\boldsymbol{T}$, ... For the transpose of $\boldsymbol{A}$, we write $\boldsymbol{A}^T$. The trace

[1] In general we shall use similar notations as in [7].

and determinant operators are denoted, respectively, by tr and det. The unit tensor is written $\mathbf{1}$. A tensor $\boldsymbol{Q}$ is said to be orthogonal if $\boldsymbol{Q}\boldsymbol{Q}^T = \boldsymbol{Q}^T\boldsymbol{Q} = \mathbf{1}$; $\boldsymbol{Q}$ is proper orthogonal if it is orthogonal and has a positive determinant. Lightface symbols always denote scalars.

We shall use the material and the spatial description. In the material description, the $\boldsymbol{X}$ and t are taken as independent variables, and in the spatial description $\boldsymbol{x}$ and t are assumed as independent variables.

It is often convenient to identify the material point X of the body $\mathscr{B}$, with its position $\boldsymbol{X}$ in a fixed reference configuration $\varkappa$. We shall use the symbol ∇ to indicate a gradient in the reference configuration $\varkappa$, i. e., a gradient computed taking $\boldsymbol{X}$ as the independent variable, whereas grad is used when the position $\boldsymbol{x}$ in the present configuration is taken as the independent variable. Similarly, the operator div will refer to spatial derivatives. A superimposed dot will denote the material time derivative, i. e., the derivative with respect to t keeping $\boldsymbol{X}$ fixed.

The motion of the body $\mathscr{B}$ with points $\boldsymbol{X}$ is described by the equation

$$\boldsymbol{x} = \chi\,(\boldsymbol{X}, t). \tag{1.1}$$

The gradient $\boldsymbol{F}$ of $\chi\,(\boldsymbol{X}, t)$ with respect to $\boldsymbol{X}$, i. e.,

$$\boldsymbol{F} = \boldsymbol{F}(\boldsymbol{X}, t) = \nabla\chi\,(\boldsymbol{X}, t), \tag{1.2}$$

is the deformation gradient at $\boldsymbol{X}$ relative to the configuration $\varkappa$. It is assumed that $\chi\,(\boldsymbol{X}, t)$ is always smoothly invertible in its first variable, i. e., that the inverse $\boldsymbol{F}^{-1}$ of $\boldsymbol{F}$ exists, or, equivalently, that $\det \boldsymbol{F} \neq 0$. By the polar decomposition of $\boldsymbol{F}$, viz.,

$$\boldsymbol{F} = \boldsymbol{R}\boldsymbol{U} \qquad \text{or} \qquad \boldsymbol{F} = \boldsymbol{V}\boldsymbol{R} \tag{1.3}$$

we define the orthogonal rotation tensor $\boldsymbol{R}$, and the positive definite and symmetric right and left stretch tensor $\boldsymbol{U}$ and $\boldsymbol{V}$, respectively.

Similarly, the relations

$$\boldsymbol{C} = \boldsymbol{F}^T\boldsymbol{F} = \boldsymbol{U}^2, \qquad \boldsymbol{B} = \boldsymbol{F}\boldsymbol{F}^T = \boldsymbol{V}^2 \tag{1.4}$$

define the right and left Cauchy-Green tensors, respectively.

Let us denote by

$$\boldsymbol{L} = \operatorname{grad} \dot{\boldsymbol{x}} \tag{1.5}$$

the velocity gradient. Then we have the following relations:

$$\dot{\boldsymbol{F}} = \boldsymbol{L}\boldsymbol{F}, \qquad \boldsymbol{L} = \dot{\boldsymbol{F}}\boldsymbol{F}^{-1}. \tag{1.6}$$

The mass density ϱ is determined by $\boldsymbol{F}$ through the equation

$$\varrho = \frac{1}{|\det \boldsymbol{F}|} \varrho_{\varkappa} \tag{1.7}$$

where $\varrho_{\varkappa}$ denotes the mass density in the reference configuration $\varkappa$.

For a scalar field over $\mathscr{B}$, such as temperature ϑ, it is easily shown that

$$\nabla \vartheta = \boldsymbol{F}^T \operatorname{grad} \vartheta. \tag{1.8}$$

Let $\boldsymbol{F}(\tau)$ and $\vartheta(\tau)$ be the deformation gradient and temperature, respectively, at time τ, at a fixed material point $\boldsymbol{X}$. Then the functions $\boldsymbol{F}^t(s)$ and $\vartheta^t(s)$ defined by

$$\boldsymbol{F}^t(s) = \boldsymbol{F}(t-s), \qquad \vartheta^t(s) = \vartheta(t-s), \qquad 0 < s \leqq \infty, \tag{1.9}$$

are called the histories of the deformation gradient and temperature at $\boldsymbol{X}$ up to time t.

If $\boldsymbol{F}$ is replaced by the relative displacement gradient $\boldsymbol{F}_{(t)}$, then the notations $\boldsymbol{R}_{(t)}$, $\boldsymbol{U}_{(t)}$, $\boldsymbol{V}_{(t)}$, $\boldsymbol{C}_{(t)}$ and $\boldsymbol{B}_{(t)}$ are used, respectively, for the corresponding relative rotation tensor, relative stretch tensors, and relative Cauchy-Green tensor (cf. TRUESDELL and NOLL [7]).

The tensor

$$\boldsymbol{D}(t) = \dot{\boldsymbol{U}}_{(t)}(t) = -\dot{\boldsymbol{U}}^t_{(t)}(0) \tag{1.10}$$

is called the stretching tensor of motion.

The n-th acceleration gradient is defined by

$$\boldsymbol{L}_n(t) = \overset{(n)}{\boldsymbol{F}}(t) = (-1)^n \overset{(n)}{\boldsymbol{F}^t_{(t)}}(0). \tag{1.11}$$

Similarly, the n-th stretching $\boldsymbol{D}_n$ is defined by

$$\boldsymbol{D}_n(t) = \overset{(n)}{\boldsymbol{U}}_{(t)}(t) = (-1)^n \overset{(n)}{\boldsymbol{U}^t_{(t)}}(0). \tag{1.12}$$

For $n = 0, 1, 2, \ldots$, where we assume

$$\boldsymbol{L}_0 = \boldsymbol{D}_0 = \boldsymbol{1}, \qquad \boldsymbol{L}_1 = \boldsymbol{L}, \qquad \boldsymbol{D}_1 = \boldsymbol{D}.$$

The n-th Rivlin-Ericksen tensor $\boldsymbol{A}_n$ is defined by

$$\boldsymbol{A}_n(t) = \overset{(n)}{\boldsymbol{C}}_{(t)}(t) = (-1)^n \overset{(n)}{\boldsymbol{C}^t_{(t)}}(0). \tag{1.13}$$

We shall use a more compact notation introduced by COLEMAN [3]. Let us denote by $\mathbf{\Lambda}$ the ordered pair $(\boldsymbol{F}, \vartheta)$, where $\boldsymbol{F}$ is any tensor and ϑ any scalar:

$$\mathbf{\Lambda} = (\boldsymbol{F}, \vartheta). \tag{1.14}$$

For any scalar α we define the product $\alpha\mathbf{\Lambda} = \mathbf{\Lambda}\alpha$ as follows:

$$\alpha\mathbf{\Lambda} = (\alpha\boldsymbol{F}, \alpha\vartheta). \tag{1.15a}$$

Let $\mathbf{\Lambda}_1 = (\boldsymbol{F}_1, \vartheta_1)$ and $\mathbf{\Lambda}_2 = (\boldsymbol{F}_2, \vartheta_2)$. The sum $\mathbf{\Lambda}_1 + \mathbf{\Lambda}_2$ is defined as

$$\mathbf{\Lambda}_1 + \mathbf{\Lambda}_2 = (\boldsymbol{F}_1 + \boldsymbol{F}_2, \ \vartheta_1 + \vartheta_2). \tag{1.15b}$$

The scalar-multiplication and addition so defined make the set of ordered pairs (1.14) a vector space $\mathscr{A}$ of dimension ten. The natural definition of an inner product in the space $\mathscr{A}$ is

$$\mathbf{\Lambda}_1 \cdot \mathbf{\Lambda}_2 = \operatorname{tr}\{\boldsymbol{F}_1\boldsymbol{F}_2^T\} + \vartheta_1\vartheta_2. \tag{1.16}$$

Let $\mathbf{\Lambda}(\tau) = \big(\boldsymbol{F}(\tau), \vartheta(\tau)\big)$, where $\boldsymbol{F}(\tau)$ and $\vartheta(\tau)$ are, respectively, the deformation gradient and the temperature at time τ at a fixed material point $\boldsymbol{X}$. The function $\mathbf{\Lambda}^t(s)$, defined by

$$\mathbf{\Lambda}^t(s) = \mathbf{\Lambda}(t-s) = (\boldsymbol{F}^t(s), \vartheta^t(s)), \quad 0 \leqq s < \infty, \tag{1.17}$$

is called the total history of $\boldsymbol{X}$ up to time t (cf. COLEMAN [5]).

2. Thermodynamic Processes

Consider a body $\mathscr{B}$ with material points $\boldsymbol{X}$. A thermodynamic process in $\mathscr{B}$ is described by eight functions of $\boldsymbol{X}$ and the time t. The values of these functions have the following physical interpretation (see COLEMAN and NOLL [4]):

1. The spatial position $\boldsymbol{x} = \chi(\boldsymbol{X}, t)$. The function χ describes a motion of the body and is called the deformation function (cf. eq. (1.1)).
2. The symmetric stress tensor $\boldsymbol{T} = \boldsymbol{T}(\boldsymbol{X}, t)$.
3. The body force $\boldsymbol{b} = \boldsymbol{b}(\boldsymbol{X}, t)$ per unit mass.
4. The specific internal energy $\varepsilon = \varepsilon(\boldsymbol{X}, t)$ per unit mass.
5. The specific entropy $\eta = \eta(\boldsymbol{X}, t)$ per unit mass.
6. The local absolute temperature $\vartheta = \vartheta(\boldsymbol{X}, t)$, which is assumed to be always positive, $\vartheta > 0$.
7. The heat flux vector $\boldsymbol{q} = \boldsymbol{q}(\boldsymbol{X}, t)$.
8. The heat supply $r = r(\boldsymbol{X}, t)$ per unit mass and unit time (absorbed by the material and furnished by radiation from the external world).

A set of eight functions

$$\{\chi, \boldsymbol{T}, \boldsymbol{b}, \varepsilon, \eta, \vartheta, \boldsymbol{q}, r\} \tag{2.1}$$

is called a thermodynamic process if, and only if, it is compatible with the condition for the balance of linear momentum[1] (Cauchy's first law of motion)

$$\operatorname{div} \boldsymbol{T} - \varrho \ddot{\boldsymbol{x}} = -\varrho \boldsymbol{b}, \tag{2.2}$$

and the law of balance of energy (the first law of thermodynamics)

$$\operatorname{tr}\{\boldsymbol{T}\boldsymbol{L}\} - \operatorname{div} \boldsymbol{q} - \varrho \dot{\varepsilon} = -\varrho r. \tag{2.3}$$

We now introduce the specific free energy function which we define by

$$\psi = \varepsilon - \vartheta \eta. \tag{2.4}$$

In order to define a thermodynamics process, it suffices to prescribe the six functions

$$\{\chi, \boldsymbol{T}, \psi, \eta, \vartheta, \boldsymbol{q}\}. \tag{2.5}$$

The two remaining functions $\boldsymbol{b}$ and r are then uniquely determined by eqs. (2.2) and (2.3), and the function ε by eq. (2.4).

A thermodynamic process in $\mathscr{B}$, compatible with the constitutive equations at each point $\boldsymbol{X}$ of $\mathscr{B}$, and all time t, will be called an admissible process in $\mathscr{B}$.

2.1. *Principle of Material Objectivity*

Two thermodynamic processes $\{\chi, \boldsymbol{T}, \psi, \eta, \vartheta, \boldsymbol{q}\}$ and $\{\chi^*, \boldsymbol{T}^*, \psi^*, \eta^*, \vartheta^*, \boldsymbol{q}^*\}$ are equivalent if they are related by a change of frame (see Noll [5]).

We assume that the constitutive equations describing the actual physical properties of the material should satisfy the following principle of material objectivity:

If the process $\{\chi, \boldsymbol{T}, \psi, \eta, \vartheta, \boldsymbol{q}\}$ is compatible with a system of constitutive equations, then also all processes $\{\chi^*, \boldsymbol{T}^*, \psi^*, \eta^*, \vartheta^*, \boldsymbol{q}^*\}$ equivalent to it must be compatible with the same system of constitutive equations.

The physical meaning of this principle is simply that the material properties of a body should not depend on the observer, irrespective of how he moves (cf. Noll [5], Truesdell and Noll [7]).

---

[1] See Truesdell and Toupin [6]. Cauchy's second law of motion requires $\boldsymbol{T}^T = \boldsymbol{T}$, which we assumed in advance. Couple stresses and body couples are assumed to be absent.

2.2. *Second Law of Thermodynamics*

Let us define for each thermodynamic process the production of entropy in the body $\mathscr{B}$ as follows (cf. Coleman and Noll [1]):

$$\Gamma = \frac{d}{dt}\int_{\varphi} \eta \, dm - \int_{\varphi} \frac{r}{\vartheta} \, dm + \int_{\partial\varphi} \frac{1}{\vartheta} \boldsymbol{q} \cdot \boldsymbol{n} \, da, \tag{2.6}$$

where dm denotes the element of mass in the body $\mathscr{B}$ in the configuration at time t, i. e., in χ_t; $\partial\varphi$ denotes the surface and φ the volume of $\mathscr{B}$, $\boldsymbol{n}$ is the exterior unit normal vector to $\partial\varphi$ in the configuration χ_t, and da denotes the element of surface area in the configuration χ_t.

Under suitable smoothness assumptions we can write:

$$\Gamma = \int_{\varphi} \gamma \, dm \tag{2.7}$$

where

$$\gamma = \dot{\eta} - \frac{r}{\vartheta} + \frac{1}{\varrho\vartheta} \operatorname{div} \boldsymbol{q} - \frac{1}{\varrho\vartheta^2} \boldsymbol{q} \cdot \operatorname{grad} \vartheta$$

is the specific production of entropy.

We assume the following postulate: For every admissible thermodynamic process in a body $\mathscr{B}$, the following inequality must hold for all times:

$$\Gamma \geqq 0. \tag{2.8}$$

The inequality (2.8) is called the Clausius-Duhem inequality, and the postulate above is a statement of the second law of thermodynamics.

In order that (2.8) may hold in $\mathscr{B}$, it is sufficient that

$$\gamma \geqq 0$$

at all material points $\boldsymbol{X}$ of $\mathscr{B}$. Using (2.3), (2.4) and (2.7), we see that the condition (2.8) is equivalent to

$$\vartheta\gamma = -\dot{\psi} + \frac{1}{\varrho} \operatorname{tr} \{\boldsymbol{F}^{-1} \boldsymbol{T} \dot{\boldsymbol{F}}\} - \eta\dot{\vartheta} - \frac{1}{\varrho\vartheta} \boldsymbol{q} \cdot \operatorname{grad} \vartheta \geqq 0. \tag{2.9}$$

Let us denote by (cf. Coleman [3])

$$\boldsymbol{\Sigma} = \left(\frac{1}{\varrho}(\boldsymbol{F}^{-1}\boldsymbol{T})^T, -\eta\right) \tag{2.10}$$

the generalized stress at $\boldsymbol{X}$ and t. It is obvious that the function $\boldsymbol{\Sigma} = \boldsymbol{\Sigma}(x, t)$ is an element of the tendimensional vector space $\mathscr{A}$.

Thus the inequality (2.9) can be written as follows:

$$\vartheta\gamma = -\dot{\psi} + \boldsymbol{\Sigma} \cdot \dot{\boldsymbol{\Lambda}} - \frac{1}{\varrho\vartheta}\, \boldsymbol{q} \cdot \operatorname{grad} \vartheta \geqq 0. \tag{2.11}$$

2.3. *Principle of Equipresence*

We shall use the principle of equipresence (see TRUESDELL and TOUPIN [6]) which in its present form (cf. COLEMAN and MIZEL [2], COLEMAN [3], TRUESDELL and NOLL [7]) may be stated as follows: An independent variable present in one constitutive equation of a material should be present in all, until its presence is shown to be in direct contradiction to the assumed symmetry of the material, the principle of material objectivity, or the laws of thermodynamics.

3. Material of the Differential Type

According to COLEMAN's thermodynamic theory of simple materials (cf. COLEMAN [3, 4]) the specific free energy $\psi(t)$, the generalized stress $\boldsymbol{\Sigma}(t)$, and the heat flux $\boldsymbol{q}(t)$ at point $\boldsymbol{X}$ of a body $\mathscr{B}$ are determined when the total history $\boldsymbol{\Lambda}^t(s)$ and the present value of the temperature gradient grad ϑ are known at $\boldsymbol{X}$, i.e.,

$$\psi(t) = \mathop{\mathfrak{p}}_{s=0}^{\infty} (\boldsymbol{\Lambda}^t(s);\ \operatorname{grad} \vartheta), \tag{3.1}$$

$$\boldsymbol{\Sigma}(t) = \mathop{\mathfrak{S}}_{s=0}^{\infty} (\boldsymbol{\Lambda}^t(s);\ \operatorname{grad} \vartheta), \tag{3.2}$$

$$\boldsymbol{q}(t) = \mathop{\mathfrak{q}}_{s=0}^{\infty} (\boldsymbol{\Lambda}^t(s);\ \operatorname{grad} \vartheta). \tag{3.3}$$

For some thermodynamic processes and for some materials it may happen that only a very short part of the history of the pair $\boldsymbol{\Lambda}^t(s)$ has an influence on the specific free energy $\psi(t)$, the generalized stress $\boldsymbol{\Sigma}(t)$, and the heat flux $\boldsymbol{q}(t)$. If $\boldsymbol{\Lambda}$ has sufficiently many continuous time derivatives, then $\boldsymbol{\Lambda}^t(s)$ may be approximated, for s near $s = 0$, by its Taylor expansion up to some order r. This expansion is determined by the value of $\boldsymbol{\Lambda}^t(s)$ and its derivatives up to the order r at $s = 0$. A material in which the actual behaviour depends on only a finite number of time derivatives

$$\boldsymbol{\Lambda}^t(o) = \boldsymbol{\Lambda}(t), \quad \dot{\boldsymbol{\Lambda}}^t(o) = \dot{\boldsymbol{\Lambda}}(t), \ldots, \quad \overset{(r)}{\boldsymbol{\Lambda}^t}(o) = \overset{(r)}{\boldsymbol{\Lambda}}(t) \tag{3.4}$$

is called a material of the differential type[1].

[1] A material of the differential type for the isothermal process was first discussed by NOLL [5] (cf. TRUESDELL and NOLL [7]).

Thus, the general constitutive equations of the differential type may be written as follows:

$$\psi(t) = \hat{p}(\mathbf{\Lambda}, \dot{\mathbf{\Lambda}}, \ldots, \overset{(r)}{\mathbf{\Lambda}}; \quad \text{grad}\,\vartheta), \tag{3.5}$$

$$\Sigma(t) = \hat{\mathbf{S}}(\mathbf{\Lambda}, \dot{\mathbf{\Lambda}}, \ldots, \overset{(r)}{\mathbf{\Lambda}}; \quad \text{grad}\,\vartheta), \tag{3.6}$$

$$\boldsymbol{q}(t) = \hat{\boldsymbol{q}}(\mathbf{\Lambda}, \dot{\mathbf{\Lambda}}, \ldots, \overset{(r)}{\mathbf{\Lambda}}; \quad \text{grad}\,\vartheta), \tag{3.7}$$

where $\hat{\boldsymbol{p}}$, $\hat{\mathbf{S}}$ and $\hat{\boldsymbol{q}}$ are constitutive functions and

$$\overset{(r)}{\mathbf{\Lambda}} = \big(\overset{(r)}{\boldsymbol{F}}(t), \overset{(r)}{\vartheta}(t)\big), \quad \hat{\mathbf{S}} = \left(\frac{1}{\varrho}\,(\boldsymbol{F}^{-1}\hat{\boldsymbol{T}})^T, \; -\hat{h}\right). \tag{3.8}$$

4. Constitutive Restrictions

4.1. Thermodynamic Restrictions

The material time derivative of the free energy $\psi(t)$ is given by

$$\dot{\psi} = \sum_{k=0}^{r} \psi_{,\underset{\Lambda}{(k)}} \cdot \overset{(k+1)}{\mathbf{\Lambda}} + \psi_{,\boldsymbol{g}} \cdot \dot{\boldsymbol{g}} \tag{4.1}$$

where $\boldsymbol{g} = \text{grad}\,\vartheta$, $\psi_{,\underset{\Lambda}{(k)}}$ denotes the k-th gradient of ψ with respect to $\overset{(k)}{\Lambda}$ and

$$\sum_{k=1}^{r} \psi_{,\underset{\Lambda}{(k)}} \cdot \overset{(k+1)}{\mathbf{\Lambda}} = \sum_{k=0}^{r} \text{tr}\,\{\psi_{,\underset{F}{(k)}} \overset{(k+1)}{\boldsymbol{F}}\} + \sum_{k=0}^{r} \psi_{,\underset{\vartheta}{(k)}} \overset{(k+1)}{\vartheta}. \tag{4.2}$$

The thermodynamic postulate requires that

$$\vartheta\gamma = (\hat{\mathbf{S}} - \psi_{,\Lambda}) \cdot \dot{\mathbf{\Lambda}} - \sum_{k=1}^{r} \psi_{,\underset{\Lambda}{(k)}} \cdot \overset{(k+1)}{\mathbf{\Lambda}} - \frac{1}{\varrho\vartheta}\,\hat{\boldsymbol{q}} \cdot \boldsymbol{g} - \psi_{,\boldsymbol{g}} \cdot \dot{\boldsymbol{g}} \geqq 0. \tag{4.3}$$

The inequality $(4.3)_2$ represents the basic restriction imposed on the constitutive equations describing a material of the defferential type.

Due to the remark proved by COLEMAN [3], we know that, at a given material point $\boldsymbol{X}$ and t we can arbitrarily choose the total history $\mathbf{\Lambda}^t(s)$ and the vectors $\boldsymbol{g}(t)$ and $\dot{\boldsymbol{g}}(t)$ and be certain that there exists at least one thermodynamic process corresponding to this choice.

Thus, the postulate $(4.3)_2$ requires

$$\psi_{,\boldsymbol{g}} = 0. \tag{4.4}$$

It means that the function $\hat{p}$ is independent on $\boldsymbol{g}$, i. e.,

$$\psi = \hat{p}(\mathbf{\Lambda}, \dot{\mathbf{\Lambda}}, \ldots, \overset{(r)}{\mathbf{\Lambda}}). \tag{4.5}$$

The basic inequality $(4.3)_2$ reduces to the following form

$$(\hat{\boldsymbol{S}} - \psi,_{\Lambda}) \cdot \dot{\boldsymbol{\Lambda}} - \sum_{k=1}^{r} \psi,_{\overset{(k)}{\Lambda}} \cdot \overset{(k+1)}{\boldsymbol{\Lambda}} - \frac{1}{\varrho\vartheta} \hat{\boldsymbol{q}} \cdot \boldsymbol{g} \geqq 0. \tag{4.6}$$

Due to the same COLEMAN's remark we know that at a given time t we can arbitrarily specify $\boldsymbol{\Lambda}, \dot{\boldsymbol{\Lambda}}, \ldots, \overset{(r)}{\boldsymbol{\Lambda}}, \boldsymbol{g}$ and $\dot{\boldsymbol{g}}$ at a point $\boldsymbol{X}$ and be sure that there exists at least one admissible thermodynamic process corresponding to this choice. If we assign $\boldsymbol{\Lambda}, \dot{\boldsymbol{\Lambda}}, \ldots, \overset{(r)}{\boldsymbol{\Lambda}}$ and $\boldsymbol{g}$, the postulate requires

$$\Theta(\boldsymbol{\Lambda}, \dot{\boldsymbol{\Lambda}}, \ldots, \overset{(r)}{\boldsymbol{\Lambda}}, \boldsymbol{g}) - \psi,_{\overset{(r)}{\Lambda}} \cdot \overset{(r+1)}{\boldsymbol{\Lambda}} \geqq 0, \tag{4.7}$$

where Θ is a finite number. The inequality (4.7) holds for all choices of $\overset{(r+1)}{\boldsymbol{\Lambda}}$ only if

$$\psi,_{\overset{(r)}{\Lambda}} = 0. \tag{4.8}$$

By (4.4), (4.8) and (2.4) we can write

$$\psi = \hat{p}(\boldsymbol{\Lambda}, \dot{\boldsymbol{\Lambda}}, \ldots, \overset{(r-1)}{\boldsymbol{\Lambda}}), \tag{4.9}$$

$$\eta = h(\boldsymbol{\Lambda}, \dot{\boldsymbol{\Lambda}}, \ldots, \overset{(r-1)}{\boldsymbol{\Lambda}}), \tag{4.10}$$

$$\varepsilon = \hat{e}(\boldsymbol{\Lambda}, \dot{\boldsymbol{\Lambda}}, \ldots, \overset{(r-1)}{\boldsymbol{\Lambda}}). \tag{4.11}$$

Thus, the basic inequality (4.6) takes now the form

$$(\hat{\boldsymbol{S}} - \psi,_{\Lambda}) \cdot \dot{\boldsymbol{\Lambda}} - \sum_{k=1}^{r-1} \psi,_{\overset{(k)}{\Lambda}} \cdot \overset{(k+1)}{\boldsymbol{\Lambda}} - \frac{1}{\varrho\vartheta} \hat{\boldsymbol{q}} \cdot \boldsymbol{g} \geqq 0. \tag{4.12}$$

4.2. Reduced Forms of Constitutive Equations

The principle of material objectivity, which states that an admissible process must remain admissible after a change of frame, imposes certain restrictions on the constitutive equations of the differential type. Let us introduce the following pairs

$$\boldsymbol{\Lambda}^* = (\boldsymbol{U}, \vartheta), \quad \overset{(k)}{\boldsymbol{\Lambda}^*_k} = (\boldsymbol{D}^*_k, \vartheta), \quad k, = 1, 2, \ldots, r. \tag{4.13}$$

where $\boldsymbol{D}^*_k = \boldsymbol{R}^T \boldsymbol{D}_k \boldsymbol{R}$ is co-rotational streching.

We may write the constitutive equations of the differential type in the following reduced set

$$\begin{aligned}\psi &= \bar{p}(\boldsymbol{\Lambda}^*, \boldsymbol{\Lambda}_1^*, \ldots, \boldsymbol{\Lambda}_{r-1}^*),\\ \eta &= \bar{h}(\boldsymbol{\Lambda}^*, \boldsymbol{\Lambda}_1^*, \ldots, \boldsymbol{\Lambda}_{r-1}^*),\\ \boldsymbol{R}^T\boldsymbol{T}\boldsymbol{R} &= \bar{\boldsymbol{T}}(\boldsymbol{\Lambda}^*, \boldsymbol{\Lambda}_1^*, \ldots, \boldsymbol{\Lambda}_{r-1}^*, \boldsymbol{\Lambda}_r^*;\ \boldsymbol{R}^T\boldsymbol{g}),\\ \boldsymbol{R}^T\boldsymbol{q} &= \bar{\boldsymbol{q}}(\boldsymbol{\Lambda}^*, \boldsymbol{\Lambda}_1^*, \ldots, \boldsymbol{\Lambda}_{r-1}^*, \boldsymbol{\Lambda}_r^*\ \boldsymbol{R}^T\boldsymbol{g}).\end{aligned} \tag{4.14}$$

If we introduce the following pairs

$$\boldsymbol{\Lambda}^\# = (\boldsymbol{C}, \vartheta), \quad \boldsymbol{\Lambda}_k^\# = (\boldsymbol{A}_k^*, \overset{(k)}{\vartheta}), \quad k = 1, 2, \ldots, r, \tag{4.15}$$

where $\boldsymbol{A}_k^* = \boldsymbol{R}^T\boldsymbol{A}_k\boldsymbol{R}$, then the reduced set of the constitutive equations of the differential type has the form

$$\begin{aligned}\psi &= \bar{\bar{p}}(\boldsymbol{\Lambda}^\#, \boldsymbol{\Lambda}_1^\#, \ldots, \boldsymbol{\Lambda}_{r-1}^\#),\\ \eta &= \bar{\bar{h}}(\boldsymbol{\Lambda}^\#, \boldsymbol{\Lambda}_1^\#, \ldots, \boldsymbol{\Lambda}_{r-1}^\#),\\ \boldsymbol{R}^T\boldsymbol{T}\boldsymbol{R} &= \bar{\bar{\boldsymbol{T}}}(\boldsymbol{\Lambda}^\#, \boldsymbol{\Lambda}_1^\#, \ldots, \boldsymbol{\Lambda}_{r-1}^\#, \boldsymbol{\Lambda}_r^\#;\ \ \boldsymbol{R}^T\boldsymbol{g}),\\ \boldsymbol{R}^T\boldsymbol{q} &= \bar{\bar{\boldsymbol{q}}}(\boldsymbol{\Lambda}^\#, \boldsymbol{\Lambda}_1^\#, \ldots, \boldsymbol{\Lambda}_{r-1}^\#, \boldsymbol{\Lambda}_r^\#, \boldsymbol{R}^T\boldsymbol{g}).\end{aligned} \tag{4.16}$$

5. Special Cases

5.1. Isothermal Process

In the case of an isothermal process, the pairs (4.13) and (4.15) reduce to

$$\boldsymbol{\Lambda}^* = \boldsymbol{U}, \qquad \boldsymbol{\Lambda}_k^* = \boldsymbol{D}_k^*, \tag{5.1}$$

and

$$\boldsymbol{\Lambda}^\# = \boldsymbol{C}, \qquad \boldsymbol{\Lambda}_k^* = \boldsymbol{A}_k^* \tag{5.2}$$

respectively. By eqs. $(4.14)_3$ and $(4.16)_3$ we have the respective constitutive equations

$$\boldsymbol{R}^T\boldsymbol{T}\boldsymbol{R} = \bar{\boldsymbol{T}}(\boldsymbol{U}, \boldsymbol{D}_1^*, \ldots, \boldsymbol{D}_r^*), \tag{5.3}$$

$$\boldsymbol{R}^T\boldsymbol{T}\boldsymbol{R} = \bar{\bar{\boldsymbol{T}}}(\boldsymbol{C}, \boldsymbol{A}_1^*, \ldots, \boldsymbol{A}_r^*), \tag{5.4}$$

which were first introduced by Noll [5].

5.2. Caloric Equations of State

If we assume that the constitutive equations of the differential type material are independent of $\overset{(k)}{\vartheta}$ for $k = 1, 2, \ldots$, then

$$\begin{aligned} \psi &= \psi(\boldsymbol{F}, \dot{\boldsymbol{F}}, \ldots, \overset{(r-1)}{\boldsymbol{F}}, \vartheta), \\ \boldsymbol{T} &= \boldsymbol{T}(\boldsymbol{F}, \dot{\boldsymbol{F}}, \ldots, \overset{(r-1)}{\boldsymbol{F}}, \overset{(r)}{\boldsymbol{F}}, \vartheta, \boldsymbol{g}), \\ \boldsymbol{q} &= \boldsymbol{q}(\boldsymbol{F}, \dot{\boldsymbol{F}}, \ldots, \overset{(r-1)}{\boldsymbol{F}}, \overset{(r)}{\boldsymbol{F}}, \vartheta, \boldsymbol{g}), \end{aligned} \tag{5.5}$$

$$\eta = -\frac{\partial \psi}{\partial \vartheta}, \tag{5.6}$$

$$\operatorname{tr}\left\{\frac{1}{\varrho}(\boldsymbol{F}^{-1}\boldsymbol{T})^T \dot{\boldsymbol{F}}\right\} - \operatorname{tr}\{\psi_{,\boldsymbol{F}} \dot{\boldsymbol{F}}\} - \sum_{k=1}^{r-1} \operatorname{tr}\{\psi_{,\overset{(k)}{\boldsymbol{F}}} \overset{(k+1)}{\boldsymbol{F}}\} - \frac{1}{\varrho\vartheta} \hat{\boldsymbol{q}} \cdot \boldsymbol{g} \geqq 0\,. \tag{5.7}$$

In the more restricted case let us assume that the present values of ε, η, $\boldsymbol{T}$ and $\boldsymbol{q}$ are functions of the present values of $\boldsymbol{F}$, $\dot{\boldsymbol{F}}$, ϑ and $\boldsymbol{g}$, i. e.,

$$\begin{aligned} \psi &= \psi(\boldsymbol{F}, \dot{\boldsymbol{F}}, \vartheta, \boldsymbol{g}), \\ \eta &= \eta(\boldsymbol{F}, \dot{\boldsymbol{F}}, \vartheta, \boldsymbol{g}), \\ \boldsymbol{T} &= \boldsymbol{T}(\boldsymbol{F}, \dot{\boldsymbol{F}}, \vartheta, \boldsymbol{g}), \\ \boldsymbol{q} &= \boldsymbol{q}(\boldsymbol{F}, \dot{\boldsymbol{F}}, \vartheta, \boldsymbol{g}). \end{aligned} \tag{5.8}$$

This case was studied by Coleman and Mizel [2]. They assumed that each function ε, η, $\boldsymbol{T}$, $\boldsymbol{q}$ is jointly and continuously differentiable in all its variables $\boldsymbol{F}$, $\dot{\boldsymbol{F}}$, ϑ, $\boldsymbol{g}$. Under this assumption, a necessary and sufficient condition that the thermodynamic postulate holds for all admissible thermodynamic processes in $\mathscr{B}$ is that the following statements be true at each $\boldsymbol{X}$ in $\mathscr{B}$ (see [2]):

$$\psi_{,\boldsymbol{g}} \equiv \boldsymbol{0}, \qquad \psi_{,\dot{\boldsymbol{F}}} \equiv \boldsymbol{0}, \tag{5.9}$$

i. e., $(5.8)_1$ reduces to

$$\psi = \psi(\boldsymbol{F}, \vartheta). \tag{5.10}$$

The entropy is given by the relation (cf. eq. (5.6))

$$\eta = -\frac{\partial \psi}{\partial \vartheta}, \tag{5.11}$$

and hence is a function of $\boldsymbol{F}$ and ϑ alone. The "equilibrium stress" $\boldsymbol{T}^{(0)}$, defined by

$$\boldsymbol{T}^{(0)} = \boldsymbol{T}(\boldsymbol{F}, \vartheta, \boldsymbol{0}, \boldsymbol{0}) = \boldsymbol{T}^{(0)}(\boldsymbol{F}, \vartheta), \tag{5.12}$$

is given by the "equilibrium isothermal stress relation":

$$\boldsymbol{T}^{(0)} = \varrho \boldsymbol{F}\, [\psi,_{\boldsymbol{F}}(\boldsymbol{F}, \vartheta)]^T. \tag{5.13}$$

The "extra-stress" $\boldsymbol{T}^{(e)}$, defined by $\boldsymbol{T}^{(e)} = \boldsymbol{T} - \boldsymbol{T}^{(0)}$, and the heat flux $\boldsymbol{q}$, obeys the inequality

$$\operatorname{tr}\{\boldsymbol{T}^{(e)} \dot{\boldsymbol{F}} \boldsymbol{F}^{-1}\} - \frac{1}{\vartheta} \boldsymbol{q} \cdot \boldsymbol{g} \geqq 0 \tag{5.14}$$

for all values of $\boldsymbol{F}$, $\dot{\boldsymbol{F}}$, ϑ and $\boldsymbol{g}$.

So, the reduced form of the constitutive equations is as follows:

$$\begin{aligned} \psi &= \overline{\psi}(\boldsymbol{U}, \vartheta), \\ \eta &= \overline{\eta}(\boldsymbol{U}, \vartheta), \\ \boldsymbol{R}^T \boldsymbol{T} \boldsymbol{R} &= \overline{\boldsymbol{T}}^{(0)}(\boldsymbol{U}, \vartheta) + \overline{\boldsymbol{T}}^{(e)}(\boldsymbol{R}^T \boldsymbol{D} \boldsymbol{R}, \boldsymbol{U}, \vartheta, \boldsymbol{R}^T \boldsymbol{g}), \\ \boldsymbol{R}^T \boldsymbol{q} &= \boldsymbol{q}(\boldsymbol{R}^T \boldsymbol{D} \boldsymbol{R}, \boldsymbol{U}, \vartheta, \boldsymbol{R}^T \boldsymbol{g}). \end{aligned} \tag{5.15}$$

If in (5.15) we assume that $\boldsymbol{T}^{(e)}$ is a linear function with respect to $\boldsymbol{D}$ and independent on $\boldsymbol{g}$ and that $\boldsymbol{q}$ is independent on $\boldsymbol{D}$, then we obtain the case which was discussed by Coleman and Noll [1].

References

[1] Coleman, B. D., and W. Noll: The Thermodynamics of Elastic Materials with Heat Conduction and Viscosity. Arch. Ratl Mech. Anal. **13**, 167—178 (1963).
[2] Coleman, B. D., and V. J. Mizel: Existence of Caloric Equations of State in Thermodynamics. J. Chem. Phys. **40**, 1116—1125 (1964).
[3] Coleman, B. D.: Thermodynamics of Materials with Memory. Arch. Ratl Mech. Anal. **17**, 1—46 (1964).
[4] Coleman, B. D.: On Thermodynamics, Strain Impulses and Visco-elasticity. Arch. Ratl Mech. Anal. **17**, 230—254 (1964).
[5] Noll, W.: A Mathematical Theory of the Mechanical Behaviour of Continuous Media. Arch. Ratl Mech. Anal. **2**, 195—226 (1958).
[6] Truesdell, C., and R. Toupin: The Classical Field Theories. Encyclopedia of Physics, vol. III/1, 226—793. Berlin-Göttingen-Heidelberg: Springer. 1960.
[7] Truesdell, C., and W. Noll: The Non-linear Field Theories of Mechanics. Encyclopedia of Physics, vol. III/3. Berlin-Göttingen-Heidelberg: Springer. 1965.

Discussion

Truesdell: In the earlier work by Coleman and Noll on simple materials necessary and sufficient conditions were found. Now with materials of differential type it is not clear what the sufficient conditions are. You have found necessary conditions. It is not clear that there are any non-trivial functions that will satisfy the dissipation inequality identically. In special cases we know there are none, for instance in a quadratic theory of Reiner-Rivlin type. You have not really shown us that there exist any constitutive equations of the type you are talking about.

OLSZAK: We have shown that the constitutive equations presented earlier by COLEMAN and NOLL[1] and by COLEMAN and MIZEL[2] may be obtained as particular cases of the constitutive equations of differential type. These both cases may be treated as good examples of non-trivial response functions for differential type materials.

BRUN: I think that as a result of this theory you could find that the work of stresses along an isothermal closed path of deformation is necessarily non-negative.

OLSZAK: Yes, because this result is valid for a more general material, namely for a simple material.

KALISKI: There are derivatives with respect to time in your constitutive equations. I see also time derivatives of the temperature. I think that this must introduce some restrictions on the thermodynamic processes.

OLSZAK: We assume that our constitutive equations of the differential type are valid for certain types of materials and for certain processes. It means that there are some restrictions imposed on the thermodynamic processes. We believe that the constitutive equations of the differential type may describe properties of some materials in processes in which the thermodynamic parameters deviate only little from their original equilibrium values.

---

[1] Arch. Ratl Mech. Anal. **13**, 167—178 (1963).

[2] J. Chem. Physics **40**, 1116—1125 (1964).

The Notion of State and Its Implications in Thermodynamics of Inelastic Solids

By

E. T. Onat
New Haven, Conn. (U.S.A.)

Summary

A definition of state based on the observable histories of displacement gradients, temperature, internal energy and stress is introduced. When the state space is finite dimensional, thermomechanical behavior of the material in the range of finite deformations can be represented by differential equations. The nature of this representation and the restrictions placed upon it by the second law of thermodynamics are discussed.

1. Introduction

A typical experiment in the phenomenological study of thermomechanical behavior of inelastic solids would involve application of, say, given displacement gradients $\boldsymbol{D}(\tau)$ and temperature $\theta(\tau)$ to a given specimen over the time interval $0 \leq \tau \leq T$ and observation of the resulting stress $\sigma(\tau)$ and heat supply $r(\tau)$ over the same time interval. Different $\boldsymbol{D}(\tau)$ and $\theta(\tau)$ may be applied to different, initially identical specimens to obtain a set of stimulus-response pairs: $\left(\boldsymbol{D}^i(\tau), \theta^i(\tau); \boldsymbol{\sigma}^i(\tau), r^i(\tau)\right)$.

Since the experimental information is in the form of histories of relevant quantities, it would be natural to expect that these histories would play a critical role in the formulation of the thermodynamics of inelastic solids.

Indeed, the starting point of some recent work on thermodynamics has been the assumption that a thermodynamic quantity such as free energy is a functional of the histories of deformation and temperature. Studies of this type in linear viscoelasticity have been reported by STAVERMAN and SCHWARZL [1], HUNTER [2] and BREUER and ONAT [3].

COLEMAN's important work [4] on thermodynamics of material with memory is also based on the functional dependence of the stress, internal energy and entropy on the histories of displacement gradients, temperature and temperature gradients.

However, some workers in the field would prefer a formulation which starts from the notion of state and state variables. The present paper is concerned with the consolidation of these two points of view.

The paper starts with a discussion of a series of thought experiments where the displacement gradients, temperature, stress and heat flow are considered as measurable quantities.

In Sections 3 and 4 there is introduced the definition of state which is based on the observable stimulus-response pairs. The starting point of this definition is the possibility that the behavior of the material for $\tau \geq t$ may not depend on all the details of $\boldsymbol{D}(\tau)$ and $\theta(\tau)$ on $[0, t]$. Thus a set of distinct histories on $[0, t]$ may produce material elements which behave in identical manner for $\tau \geq t$ under identical future stimuli. These elements are said to be in the same state and orientation. In Section 4 we pass from the notion of state and orientation to that of state.

After a brief discussion of non-aging materials in Section 5, we consider the space composed of distinct states. The case where this space can be mapped onto a finite dimensional metric space is of special importance. As shown in Section 7 this case gives rise to the differential equation representation of thermo-mechanical behavior in the presence of finite deformations.

To the present author's knowledge, such a representation has not been available in the literature. Much is known on the integral representation of the mechanical behavior (see, for instance, RIVLIN [5]). However, as it was shown in [6], this mode of representation becomes inconvenient for some important materials. An advantage of the differential equation representation of behavior is that it is quite suitable to handle such cases.

The paper closes with a discussion of the restrictions imposed by the second law on the representation of mechanical behavior.

2. A Multiple Experiment

We assume that there exist at time $\tau = 0$ a number of test specimens composed of a given material. These specimens carry no stress at $\tau = 0$ and have the temperature $\theta = \theta_0$. The specimens are identical and are identically oriented with respect to a given fixed rectangular cartesian frame. In each test the specimen of interest is subjected to a homogeneous deformation defined by the equation

$$x_i = D_{ij}(\tau)\, X_j, \qquad 0 \leq \tau < \infty \tag{2.1}$$

where x_i and X_j are the coordinates of a material point at times τ and 0, respectively. Note that

$$D_{ij}(0) = \delta_{ij} \qquad \text{and} \qquad \det(D_{ij}(\tau)) > 0. \tag{2.2}$$

In addition, the specimen is subjected to a homogeneously distributed time dependent temperature

$$\theta = \theta(\tau), \qquad 0 \leq \tau < \infty \tag{2.3}$$

where $\theta(0) = \theta_0$.

The displacement gradients $D_{ij}(\tau)$ or, for short, $\boldsymbol{D}(\tau)$, and $\theta(\tau)$ are varied from test to test.

A deformation and temperature history $(\boldsymbol{D}(\tau), \theta(\tau))$ on $0 \leq \tau < \infty$ will cause the development of the stress $\boldsymbol{\sigma}(\tau)$ in the specimen. There will also be heat exchange with the surrounding medium, the rate of heat flow into the unit mass of the solid being denoted by $r(\tau)$.

A typical test therefore would involve application of

$$\boldsymbol{D}(\tau) \text{ and } \theta(\tau), \quad \text{on} \quad 0 \leq \tau < \infty \tag{2.4}$$

and observation of the resulting stress and heat flow:

$$\boldsymbol{\sigma}(\tau) \text{ and } r(\tau), \quad \text{on} \quad 0 \leq \tau < \infty. \tag{2.5}$$

We shall refer to (2.4) as *the stimulus* on $[0, \infty]$ and to (2.5) as *the response* on $[0, \infty]$. A *multiple experiment* would involve tests conducted with n different but identical specimens and would produce the stimulus-response pairs

$$(\boldsymbol{D}^{(1)}, \theta^{(1)}; \boldsymbol{\sigma}^{(1)}, r^{(1)}), \ldots, (\boldsymbol{D}^{(n)}, \theta^{(n)}; \boldsymbol{\sigma}^{(n)}, r^{(n)}). \tag{2.6}$$

For reasons of economy we dropped the argument τ in (2.6) so that $\boldsymbol{D}$, for instance, stands for a tensor-valued function defined on $[0, \infty]$.

It will be convenient to replace $r(\tau)$ in (2.5) and (2.6) by the internal energy e per unit mass of the solid. This can be done by using the first law of thermodynamics which for the present case has the form

$$\dot{e} = \frac{1}{\varrho} \sigma_{ij} V_{ij} + r. \tag{2.7}$$

Here ϱ is the density and V_{ij} is the tensor of rate of deformation:

$$V_{ij} = \frac{1}{\varrho} \left(\frac{\partial v_i}{\partial x_j} + \frac{\partial v_j}{\partial x_i} \right) = \text{Symmetric part of } (\dot{\boldsymbol{D}} \boldsymbol{D}^{-1}). \tag{2.8}$$

In (2.8) v_i is the velocity and the dot indicates the material rate of change of the dotted quantity; $\boldsymbol{D}^{-1}$ is the inverse of the matrix $\boldsymbol{D}$ and $\dot{\boldsymbol{D}} \boldsymbol{D}^{-1}$ denotes matrix multiplication of $\dot{\boldsymbol{D}}$ and $\boldsymbol{D}^{-1}$.

By using (2.7) we can in each test determine $e(\tau)$, with $e(0) = e_0$, from the observed values of $\boldsymbol{\sigma}$ and r and from the applied $\boldsymbol{D}$.

We shall henceforth consider the pair

$$(\boldsymbol{\sigma}, e) \quad \text{on} \quad 0 < \tau < \infty, \tag{2.9}$$

as the response of the material. We shall assume that $\boldsymbol{D}$ and θ are continuous[1] on the time interval $[0, \infty]$ and we restrict the discussion to solids for which the response $(\boldsymbol{\sigma}, e)$ to a *continuous* stimulus $(\boldsymbol{D}, \theta)$ is also *continuous*. This restriction which is introduced for reasons of simplicity is an important one and leaves some important solids, such as Voigt solid, outside of the present considerations. It may be noted that for the class of solids considered here $r(\tau)$ may be discontinuous whenever V_{ij} or $\theta^{\cdot}$ is discontinuous. It is for this reason that we replaced r by e in (2.5).

From a mathematical point of view a given solid acts as an operator which assigns to a pair of continuous functions $(\boldsymbol{D}, \theta)$, the pair $(\boldsymbol{\sigma}, e)$ on $0 \leq \tau < \infty$:

$$(\boldsymbol{\sigma}, e) = F[\boldsymbol{D}, \theta]. \tag{2.10}$$

The operator F is *causal*. This means that $(\boldsymbol{\sigma}(t), e(t))$ depends only on $(\boldsymbol{D}, \theta)$ on $0 \leq \tau < t$.

In order to study the nature of the operator F one must conduct a suitable multiple experiment. Such an experimental study must be accompanied by theoretical work aimed at a mathematical representation[2] of F.

The present paper is devoted to a discussion of the differential equation representation of (2.10). This mode of representation is intimately connected with the notion of state which we develop next.

3. State and Orientation

We consider a deformation and temperature history and focus the attention on times $\tau \geq t$. We introduce displacement gradients measured with respect to the configuration at time t. These are defined as

$$D^t_{ij}(\tau) = \frac{\partial x_i}{\partial y_j}, \quad \tau \geq t \tag{3.1}$$

where x_i and y_j are the coordinates of a material point in the fixed coordinate frame at times τ and t, respectively. It is easily seen that

$$\boldsymbol{D}^t(\tau) = \boldsymbol{D}(\tau)\, \boldsymbol{D}^{-1}(t), \qquad \tau \geq t \tag{3.2}$$

where $\boldsymbol{D}(\tau)$ has the meaning in (2.1).

---

[1] **D** is said to be continuous if each of its components is continuous.

[2] For a discussion of these points see [6] .

We introduce the reduced time ξ,

$$\xi = \tau - t, \qquad \tau \geq t \tag{3.3}$$

and the notation

$$\boldsymbol{D}^t (t + \xi) = \boldsymbol{D}^* (\xi), \tag{3.4}$$

so that $D^* (\xi)$ measures the deformation which took place in the interval $[t, t + \xi]$ with respect to the configuration at time t. An observer who enters the laboratory at time t will be able to measure only $\boldsymbol{D}^* (\xi)$ if the test specimen does not carry any markings associated with time $\tau = 0$.

We next introduce the reduced temperature

$$\theta^* (\xi) = \theta (\xi + t) - \theta (t) \tag{3.5}$$

which measures the temperature change which took place in the interval $[t, t + \xi]$.

Note that

$$\boldsymbol{D}^* (0) = \boldsymbol{I}, \qquad \text{and} \qquad \theta^*(0) = 0 \tag{3.6}$$

where $\boldsymbol{I}$ denotes the identity matrix.

The response of the material for $\tau > t$ is denoted by

$$\boldsymbol{\sigma}_t (\xi) = \boldsymbol{\sigma} (t + \xi) \qquad \text{and} \qquad e_t (\xi) = e (t + \xi). \tag{3.7}$$

$(\boldsymbol{\sigma}_t (\xi), e_t (\xi))$ on $0 \leq \xi < \infty$ will depend on the stimulus $(\boldsymbol{D}, \theta)$ on $[0, t]$ and also on the stimulus on $[t, \infty]$. The latter stimulus may be referred to the configuration and temperature at time t by means of (3.2), (3.4) and (3.5):

$$(\boldsymbol{\sigma}_t, e_t) = \underset{[0,t]}{\boldsymbol{F}} [\boldsymbol{D}, \theta; \boldsymbol{D}^*, \theta^*]. \tag{3.8}$$

We shall refer to $(\boldsymbol{\sigma}_t, e_t)$ and $(\boldsymbol{D}^*, \theta^*)$ as the future response and the future stimulus, respectively.

We may now conduct tests where we keep $(\boldsymbol{D}, \theta)$ on $[0, t]$ fixed, say, $(\boldsymbol{D}^{(1)}, \theta^{(1)})$, but vary $(\boldsymbol{D}^*, \theta^*)$ from test to test.

We may repeat this type of tests with a different stimulus on $[0, t]$, say, $(\boldsymbol{D}^{(2)}, \theta^{(2)})$.

There may be occasions where the response of the material on $0 \leq \xi < \infty$ in these two types of tests may be the same under identifical future stimuli $(\boldsymbol{D}^*, \theta^*)$. If such is the case then the observer who enters the laboratory at time t and who is able to measure the deformations and temperature only with respect to the configuration and temperature at time t will not be able to distinguish the first type of tests from the second type. This remark suggests the introduction of the following definition:

We shall say that the stimuli $(\boldsymbol{D}^{(1)}, \theta^{(1)})$ and $(\boldsymbol{D}^{(2)}, \theta^{(2)})$ on $[0, t]$ produce the *same state and orientation* at time t if the relation

$$F\underset{[0,t]}{[\boldsymbol{D}^{(1)}, \theta^{(1)}}; \boldsymbol{D}^*, \theta^*] = F\underset{[0,t]}{[\boldsymbol{D}^{(2)}, \theta^{(2)}}; \boldsymbol{D}^*\theta^*] \tag{3.9}$$

holds for any $(\boldsymbol{D}^*, \theta^*)$, where F has the meaning introduced in (3.8).

Note that in the above definition we *do not* demand that

$$\boldsymbol{D}^{(1)}(t) = \boldsymbol{D}^{(2)}(t), \quad \text{and} \quad \theta^{(1)}(t) = \theta_2(t).$$

We now consider the set H_t composed of all continuous stimuli $(\boldsymbol{D}, \theta)$ on $[0, t]$. We decompose this into subsets S_t^i such that any two stimuli belonging to S_t^i produce the *same state and orientation* at time t. S_t^i may be considered as elements of a set Σ_t. A generic element of this set will be denoted by S_t. A material element which has been subjected to a stimulus $(\boldsymbol{D}, \theta)$ on $[0, t]$ which belongs to S_t is said to be in the state and orientation S_t.

If the number of elements in the set Σ_t is much "less" than those in the set H_t then there would be advantage in working with the elements of Σ_t rather than the distinct stimuli[1] $(\boldsymbol{D}, \theta)$.

Our experience with linear viscoelastic or plastic materials suggests that for many materials of interest this will indeed be the case. Elastic materials constitute an important limiting case where a subset S_t contains all stimuli on $[0, t]$ which have the same terminal values $\boldsymbol{D}(t)$ and $\theta(t)$. Thus for an elastic material the set Σ_t can be mapped onto a 10 dimensional manifold whereas the set H_t occupies a function space.

With the above introduced notation, the operational relation (3.8) takes the form

$$(\boldsymbol{\sigma}_t, e_t) = F[S_t; \boldsymbol{D}^*, \theta^*], \tag{3.10}$$

or more explicitly

$$\begin{aligned} \boldsymbol{\sigma}_t &= \boldsymbol{F}[S_t; \boldsymbol{D}^*, \theta^*] \\ e_t &= E[S_t; \boldsymbol{D}^*, \theta^*], \end{aligned} \tag{3.11}$$

where F, $\boldsymbol{F}$ and E represent causal operators acting on $\boldsymbol{D}^*$ and θ^*.

We next consider two state and orientations, S_t and $S_{t'}$, at times t and t', respectively. We shall say that

$$S_t = S_{t'}$$

when

$$F[S_t; \boldsymbol{D}^*, \theta^*] = F[S_{t'}; \boldsymbol{D}^*, \theta^*]. \tag{3.12}$$

---

[1] The precise meaning of this comparison will become clearer in this and in the following sections.

(3.12) means that under identical future stimuli on $0 \leq \xi < \infty$, material elements which are in S_t and $S_{t'}$, produce identical future responses:

$$\boldsymbol{\sigma}^{(1)}(t+\xi) = \boldsymbol{\sigma}^{(2)}(t'+\xi); \qquad e^{(1)}(t+\xi) = e^{(2)}(t'+\xi). \tag{3.13}$$

For further considerations it will be essential to introduce the notion of *distance* between two state and orientations. Let S_t^α and $S_{t'}^\beta$ denote two state and orientations at times t and t', respectively. In the case of $t = t'$, S_t^α and S_t^β would represent two state and orientations at time t.

The distance between S_t^α and $S_{t'}^\beta$ may be considered large if the material elements in S_t^α and $S_{t'}^\beta$ produce widely different future responses under identical future stimuli. A convenient definition of distance based on the comparison of future responses is the following one:

$$d(S_t^\alpha, S_{t'}^\beta) = \max_{\boldsymbol{D}^*, \theta^*} \max_{\xi} \left(\{\mathrm{tr}\,(\boldsymbol{F}[S_t^\alpha; \boldsymbol{D}^*, \theta^*] - \boldsymbol{F}[S_{t'}^\beta; \boldsymbol{D}^*, \theta^*])^2\}^{1/2} + \right.$$
$$\left. + |E[S_t^\alpha; \boldsymbol{D}^*, \theta^*] - E[S_{t'}^\beta; \boldsymbol{D}^*, \theta^*]|\right). \tag{3.14}$$

The bracketed expression on the right hand side of (3.14) may be written in the following more explicit form

$$[(\sigma_{ij}^{(1)}(t+\xi) - \sigma_{ij}^{(2)}(t'+\xi))(\sigma_{ij}^{(1)}(t+\xi) - \sigma_{ij}^{(2)}(t'+\xi))]^{1/2} +$$
$$+ |e^{(1)}(t+\xi) - e^{(2)}(t'+\xi)|, \tag{3.15}$$

where $(\sigma_{ij}^{(1)}, e^{(1)})$ and $(\sigma_{ij}^{(2)}, e^{(2)})$ denote the responses of the material elements in S_t^α and $S_{t'}^\beta$ to the same future stimulus $\boldsymbol{D}^*(\xi)$, $\theta^*(\xi)$, respectively.

In order to determine the distance $d(S_t^\alpha, S_{t'}^\beta)$ one must find the maximum value of (3.15) over the interval $0 \leq \xi < \infty$ for a given (D^*, θ^*). One must then repeat this search for other (D^*, θ^*) until one arrives at the maximum possible value of (3.15). It can be shown that the distance function defined in (3.14) satisfies the three conditions [7] imposed on the distance in metric spaces. Thus the set Σ composed of all state and orientations at all times is a metric space.

4. The Notion of State

In order to arrive at the notion of state from the above considerations we must discuss briefly the role played by the superimposed rigid-body rotations in the study of thermo-mechanical phenomena.

We shall make the usual assumption that superimposed rigid body rotations do not affect the physical phenomena which are taking place within the material, but they alter the stresses measured in a fixed frame; more precisely, if the stimulus $(\boldsymbol{D}, \theta)$ causes the response $(\boldsymbol{\sigma}, e)$ on $[0, \infty]$;

$$(\boldsymbol{\sigma}, e) = F[\boldsymbol{D}, \theta], \tag{4.1}$$

then the stimulus $(\boldsymbol{R}\boldsymbol{D}, \theta)$ will cause the response $(\boldsymbol{R}\boldsymbol{\sigma}\boldsymbol{R}^T, e)$:

$$(\boldsymbol{R}\boldsymbol{\sigma}\boldsymbol{R}^T, e) = F[\boldsymbol{R}\boldsymbol{D}, \theta], \tag{4.2}$$

where $\boldsymbol{R}(\tau)$ is a time dependent, continuous proper orthogonal transformation, $\boldsymbol{R}^T$ is the transpose of it and $\boldsymbol{R}(0) = \boldsymbol{I}$.

We now consider the stimuli which belong to a given state and orientation S_t:

$$(\boldsymbol{D}, \theta) \in S_t \tag{4.3}$$

We derive from each stimulus in the above set, the following stimuli

$$(\boldsymbol{R}\boldsymbol{D}, \theta) \qquad \text{on} \qquad [0, t], \tag{4.4}$$

where $\boldsymbol{R}$ is an arbitrary continuous orthogonal transformation which satisfies the requirements

$$\boldsymbol{R}(0) = \boldsymbol{I} \qquad \text{and} \qquad \boldsymbol{R}(t) = \boldsymbol{Q} = \text{constant}. \tag{4.5}$$

A stimulus in (4.4) differs from $(\boldsymbol{D}, \theta)$ only by the superimposed rigid body rotation $\boldsymbol{R}(\tau)$. If one now considers the set composed of the stimuli (4.4) obtained from *all* $(\boldsymbol{D}, \theta)$ in S_t, then one can show using (4.2) and the definition (3.9) that this set is equal to the set of stimuli belonging to a state and orientation at time t, say, $S_t^{(1)}$:

$$(\boldsymbol{R}\boldsymbol{D}, \theta) \in S_t^{(1)}. \tag{4.6}$$

It may be useful to point out that the detailed proof of this important fact is based on the following result which is almont a direct consequence of (4.2):

If

$$(\boldsymbol{\sigma}_t, e_t) = \underset{[0,t]}{F}[\boldsymbol{D}, \theta; \boldsymbol{D}^*, \theta^*] \tag{4.7}$$

then

$$(\boldsymbol{Q}\boldsymbol{\sigma}_t\boldsymbol{Q}^T, e_t) = \underset{[0,t]}{F}[\boldsymbol{R}\boldsymbol{D}, \theta; \boldsymbol{Q}\boldsymbol{D}^*\boldsymbol{Q}^T, \theta^*], \tag{4.8}$$

where the proper orthogonal transformation $\boldsymbol{R}(\tau)$ is subject to the requirements (4.5).

We observe that the derived state and orientation $S_t^{(1)}$ in (4.6) depends only on S_t and on the fixed rotation $\boldsymbol{Q}$ at time t. It will be convenient to express this observation with the notation

$$S_t^{(1)} = \boldsymbol{Q} S_t. \tag{4.9}$$

Material elements in S_t and in $S_t^{(1)} = \boldsymbol{Q} S_t$ are said to be in the *same state*. The basis of this definition, is of course, the assumption that the superimposed rigid body rotations do not alter the physics of the situation. Now let us keep S_t fixed but vary $\boldsymbol{Q}$ in (4.9) to obtain new state and orientations[1]

$$S_t^{(2)} = \boldsymbol{Q}_2 S_t, \quad S_t^{(3)} = \boldsymbol{Q}_3 S_t, \ldots \tag{4.10}$$

In view of the preceding definition we shall say that the set composed of the elements $S_t^{(1)}$, $S_t^{(2)}$, $S_t^{(3)}$, ... defines the state $\mathfrak{S}_t$.

Two states $\mathfrak{S}_t$ and $\mathfrak{S}_{t'}$ at times t and t' are said to be equal,

$$\mathfrak{S}_t = \mathfrak{S}_{t'} \tag{4.11}$$

if one can find $S_t \in \mathfrak{S}_t$ and $S_t' \in \mathfrak{S}_t'$ such that

$$S_t = S_t'. \tag{4.12}$$

The *distance* between two states $\mathfrak{S}_t^\alpha$, $\mathfrak{S}_{t'}^\beta$ is defined as

$$\delta(\mathfrak{S}_t^\alpha, \mathfrak{S}_{t'}^\beta) = \operatorname*{Inf}_{\boldsymbol{Q}} d(S_t^\alpha, \boldsymbol{Q} S_{t'}^\beta) \tag{4.13}$$

where $S_t^\alpha \in \mathfrak{S}_t^\alpha$ and $S_{t'}^\beta \in \mathfrak{S}_{t'}^\beta$.

It can be shown by using the relation

$$d(S_t, S_{t'}) = d(\boldsymbol{Q} S_t, \boldsymbol{Q} S_{t'}), \tag{4.14}$$

which follows from the definitions of the distance function d and of $\boldsymbol{Q} S_t$ that the above defined distance is independent of the choice of S_t^α.

It can also be shown, using (4.14), (4.13) and (3.14), that the distance defined by (4.13) satisfies the usual requirements placed on a metric.

In the discussion of the state space it is convenient to regard the states which are equal to each other as a single entity. Thus the states which satisfy

$$\mathfrak{S}_t = \mathfrak{S}_{t'} = \mathfrak{S}_{t''} = \ldots \tag{4.15}$$

---

[1] We are interested here in materials for which

$$\boldsymbol{Q} S_t \neq S_t$$

when $\boldsymbol{Q} \neq \boldsymbol{I}$. It may be remarked that the case where $\boldsymbol{Q} S_t = S_t$ for any $\boldsymbol{Q}$ occurs for non-viscous gases.

will be denoted by the single generic symbol $\mathfrak{S}$. The set composed of distinct states $\mathfrak{S}$ is called *the state space*. The material will be able to assume the state $\mathfrak{S}$ at times $t, t', t'', \ldots$ For many purposes it will be necessary to give the set of times $(t, t', t'', \ldots)$ associated with a given state $\mathfrak{S}$ as a function of $\mathfrak{S}$.

Similarly, the state and orientations which satisfy

$$S_t = S_{t'} = S_{t''} = \ldots \tag{4.16}$$

will be denoted by the generic symbol S. On the other hand $\boldsymbol{Q} S$ will stand for $(\boldsymbol{Q} S_t, \boldsymbol{Q} S_{t'}, \boldsymbol{Q} S_{t''}, \ldots)$ where $(S_t, S_{t'}, S_{t''}, \ldots)$ satisfy (4.16). A state $\mathfrak{S}$ will be composed of state and orientations $\boldsymbol{Q} S$ where $\boldsymbol{Q}$ is an arbitrary proper orthogonal transformation:

$$\boldsymbol{Q} S \in \mathfrak{S}. \tag{4.17}$$

The distance between two states $\mathfrak{S}_1$ and $\mathfrak{S}_2$ will be defined as

$$\delta(\mathfrak{S}_1, \mathfrak{S}_2) = \delta(\mathfrak{S}_t^{(1)}, \mathfrak{S}_{t'}^{(2)}) \tag{4.18}$$

where $\mathfrak{S}_t^{(1)} \in \mathfrak{S}_1$, $\mathfrak{S}_{t'}^{(2)} \in \mathfrak{S}_2$.
Similarly

$$d(S_1, S_2) = d(S_t^{(1)}, S_{t'}^{(2)}) \tag{4.19}$$

where

$$S_t^{(1)} \in S_1, \quad S_{t'}^{(2)} \in S_2.$$

According to the remarks which followed (4.14) the *state space* is a *metric space*.

We now associate with each state $\mathfrak{S}$ a state and orientation S_r which belongs to $\mathfrak{S}$. This association is one-to-one, since a given S defines by (4.17) a distinct state $\mathfrak{S}$. A material element which is in S_r is said to be in the reference orientation and in the state $\mathfrak{S}(S_r)$. On the other hand an element in $\boldsymbol{Q} S_r$ is in the state $\mathfrak{S}(S_r)$ and has the orientation[1] $\boldsymbol{Q}$. The space composed of S_r will be called the *oriented state space*. This space is a *metric space* with the distance function as defined in (4.13). We have then here the correspondence

$$\mathfrak{S} \leftrightarrow S_r \tag{4.20}$$

which constitutes a one-to-one mapping of the state space onto the *oriented state space*.

---

[1] We shall henceforth use the symbol $\boldsymbol{Q}$ only to describe the orientation of an element with respect to an element in the reference orientation.

We shall, in addition, assume that this mapping is continuous. Namely if a sequence of states, $\mathfrak{S}^{(N)}$, converges to $\mathfrak{S}$ then the corresponding sequence, $S_r^{(N)}$, converges to $S_r(\mathfrak{S})$. Note that we have considerable freedom in the choice of S_r. The above assumption asserts that we can make the choice of S_r in such a way that the one-to-one mapping (4.20) is also continuous.

When such a mapping is possible, then there is no essential difference between the state space and oriented state space. For instance if one of these is finite dimensional then so is the other one.

Before going into a discussion of the oriented state space we would like to pause and introduce the notion of non-aging materials.

5. Non-aging Materials

Let us denote by S_0 the state and orientation of the material at $\tau = 0$. Let us consider an element subjected to the stimulus

$$\boldsymbol{D}(\tau) = \boldsymbol{I}, \quad \theta = \theta_0 \quad \text{on} \quad [0, t]. \tag{5.1}$$

This means that the element remains undeformed and its temperature remains constant on $[0, t]$.

The above stimulus will belong to a state and orientation S_t. If

$$S_t = S_0 \tag{5.2}$$

then the material is said to be non-aging. (5.2) means that a virgin element remains virgin under the absence of deformation and temperature changes.

It can easily be shown that non-aging materials have the following property: In the state space of a non-aging material one can find for any state $\mathfrak{S}_t$ at time t, a state $\mathfrak{S}_{t'}$ at time $t' \geq t$ such that

$$\mathfrak{S}_{t'} = \mathfrak{S}_t . \tag{5.3}$$

The consequences of this important property will be discussed in the next section.

It may be useful to introduce here briefly the notion of *strongly aging materials*. We shall say that a material is strongly aging if the state space of the material has the property that

$$\mathfrak{S}_{t'} \neq \mathfrak{S}_t \tag{5.4}$$

whenever $t' \neq t$.

Materials considered here are assumed to have the following additional properties:

If

$$(\boldsymbol{D}^{(1)}, \theta^{(1)}), (\boldsymbol{D}^{(2)}, \theta^{(2)}), \ldots \in S_t \in \mathfrak{S}_t \tag{5.5}$$

then

$$\theta^{(1)}(t) = \theta^{(2)}(t) = \ldots = \theta(\mathfrak{S}_t), \tag{5.6}$$

and

$$v^{(1)}(t) = v^{(2)}(t) = \ldots = v(\mathfrak{S}_t) \tag{5.7}$$

where $v(t)$ denotes the volume per unit mass at time t.
Note that

$$v^{(i)}(t) = v_0 \det\left(\boldsymbol{D}^{(i)}(t)\right) \tag{5.8}$$

where v_0 is the volume per unit mass at $t = 0$. (5.6) and (5.7) mean that two material elements which are in the same state at time t have the same volume and temperature at that time.

If the material is, in addition, non-aging then we have

$$\theta(\mathfrak{S}_t) = \theta(\mathfrak{S}_{t'}) \quad \text{and} \quad v(\mathfrak{S}_t) = v(\mathfrak{S}_{t'}), \tag{5.9}$$

whenever $\mathfrak{S}_t = \mathfrak{S}_{t'}$, so that each distinct state is associated with a given temperature and volume:

$$\theta(\mathfrak{S}), \quad v(\mathfrak{S}). \tag{5.10}$$

6. Structure of the Oriented State Space

We consider the oriented state space of a non-aging material. If the material can assume the state S_r at time t then, according to (5.3), it can assume the same state for times $\tau \geq t \geq 0$. Therefore, there exists a time,

$$t_{\min} = f(S_r), \tag{6.1}$$

below which the material cannot assume the given state S_r.
Note that the initial state of the material satisfies

$$f(S_r^0) = 0. \tag{6.2}$$

Let us now consider the states S_r which satisfy the inequality

$$t \geq f(S_r). \tag{6.3}$$

In view of (6.1) and (5.3) we decide that these states can exist at time t. We shall assume that the subset of the oriented state space defined by (6.3) is a *closed set* and its boundary is the points defined by $t = f(S_r)$. We shall further assume that the boundary is connected.

We observe that the set defined by

$$t_1 \geq f(S_r), \tag{6.4}$$

is contained in the preceding one if $t_1 < t$.

Consider a state S_r^* which belongs to the subset defined by (6.3). Assume that a material element is in the state S_r^* at time t. Let $t^* = f(S^*)$. Clearly $t^* \leq t$. We would like now to raise the following question: What can be said about the states S_r^{**} which the above material element can assume at time $t + \Delta t$, $(\Delta t > 0)$? Obviously we must have

$$t + \Delta t \geq f(S_r^{**}). \tag{6.5}$$

A stronger statement than the above one can be made, however, by considering the equivalent states S_r^* at times t^* and t. It follows from the definition of state that for non-aging materials the set Y composed of all states which can be created at $t^* + \Delta t$ starting with the state S_r^* at t^* is equal to the similar set created at $(t + \Delta t)$ starting with the state S_r^* at t. Therefore the stronger inequality follows

$$t^* + \Delta t \geq f(S_r^{**}). \tag{6.6}$$

This inequality which can also be written in the following form

$$\Delta t \geq f(S_r^{**}) - f(S_r^*), \tag{6.7}$$

will play a somewhat important role in the next section.

We now introduce the assumption that the state space can be mapped, one-to-one and continuously, onto a *finite dimensional* metric space. We further assume, in view of (5.10), that the temperature θ and the specific volume v of the solid can be taken as two of the coordinates of the latter space. The above assumptions imply that each distinct state $\mathfrak{S}$ and therefore S_r can be associated with the set of numbers $(\theta, v, q_1, \ldots q_n)$ which represent the coordinates of the finite dimensional space:

$$\mathfrak{S} \leftrightarrow S_r \leftrightarrow (\theta, v, q_1, \ldots, q_n), \tag{6.8}$$

$(\theta, v, q_1, \ldots, q_n)$ are called *state variables*.

The materials for which the correspondence (6.8) holds are endowed with a peculiar memory; For these materials the future response of a material element cannot depend on all the details of the past stimulus. Otherwise the distinct past stimuli would give rise to distinct states and therefore the state space would be a function space.

We are now ready to discuss the representation of thermo-mechanical behavior of non-aging solids.

7. Representation of Thermo-mechanical Behavior

The results of the previous sections can be used to obtain a mathematical representation of the dependence of $(\boldsymbol{\sigma}(t), e(t))$ on the stimulus $(\boldsymbol{D}, \theta)$ on $[0, t]$.

For this purpose we first study the relationship which exists between $(\boldsymbol{\sigma}(t), e(t))$ and the state and orientation of the material.

It follows from (3.11) and from the assumed continuity of $\boldsymbol{\sigma}(\tau)$ and $e(\tau)$ on $[0, \infty]$ that the response at time t must depend only on the state and orientation at time t:

$$\boldsymbol{\sigma}(t) = \boldsymbol{F}(S_t), \tag{7.1}$$

$$e(t) = E(S_t).$$

In view of (4.8), (4.6) and (4.9) the above relations have the following properties of invariance

$$\boldsymbol{F}(\boldsymbol{R} S_t) = \boldsymbol{R} \boldsymbol{F}(S_t)\, \boldsymbol{R}^T \quad \text{and} \quad E(\boldsymbol{R} S_t) = E(S_t), \tag{7.2}$$

where $\boldsymbol{R}$ is an arbitrary proper orthogonal transformation. If S_t belongs to the state $\mathfrak{S}(S_r)$, i. e.

$$S_t = \boldsymbol{Q} S_r,$$

then we have by (7.1) and (7.2)

$$\sigma(t) = \boldsymbol{Q} \boldsymbol{F}(S_r) \boldsymbol{Q}^T \quad \text{and} \quad e(t) = E(S_r). \tag{7.3}$$

For the class of materials considered in the previous section the correspondence (6.8) holds so that $\boldsymbol{F}$ and E can be considered as *functions* of the state variables θ, v, q_α $(\alpha = 1, \ldots, n)$ and hence

$$\sigma(t) = \boldsymbol{Q} \boldsymbol{F}(\theta, v, q_\alpha) \boldsymbol{Q}^T,$$

$$e(t) = E(\theta, v, q_\alpha). \tag{7.4}$$

The above relations have the following meaning: If at time t a material element is in the state (θ, v, q_α) and if it has the orientation $\boldsymbol{Q}$ then the stress and internal energy of the element are given by (7.4).

The state variables in (7.4) are subject to no requirements except that they must satisfy the inequality

$$t \geq f(\theta, v, q_\alpha), \tag{7.5}$$

which is equivalent to (6.3).

In order to complete the representation of thermo-mechanical behavior we must inquire as to how the state variables θ, v, q_α and the orientation $\boldsymbol{Q}$ depend on the stimulus $(\boldsymbol{D}, \theta)$ on $[0, t]$. We begin this inquiry with an observation concerning the state and orientations S_τ and S_t of a deforming material element at times τ and t, respectively $(\tau \geq t)$. By using the definition of state and orientation at time τ together with (3.10) we find that

$$S_\tau = G[S_t; \boldsymbol{D}^*(\xi), \theta^*(\xi)], \quad 0 \leq \xi \leq \tau - t. \tag{7.6}$$

The above relation has the meaning that the state and orientation of a material element at time $\tau \geq t$ depends only on its state and orientation at time t and on the stimulus $(\boldsymbol{D}^*, \theta^*)$ on $[0, \tau - t]$. The relation (7.6) has the following two invariance properties which are direct consequences of (4.2) and the definition (4.9):

(i)

$$\boldsymbol{R}(\tau - t) S_\tau = G[S_t; \boldsymbol{R}(\xi)\, D^*(\xi), \theta^*(\xi)], \tag{7.7}$$

where S_τ is given by (7.6) and $\boldsymbol{R}(\tau)$ is a proper orthogonal transformation which fulfills the requirement

$$\boldsymbol{R}(0) = \boldsymbol{I}. \tag{7.8}$$

(ii)

$$\boldsymbol{R} S_\tau = G[\boldsymbol{R} S_t; \boldsymbol{R}\boldsymbol{D}^*(\xi)\boldsymbol{R}^T, \theta^*(\xi)], \quad 0 \leq \xi \leq \tau - t, \tag{7.9}$$

where S_τ is defined by (7.6) and $\boldsymbol{R}$ is a *constant* proper orthogonal transformation.

We now observe from (7.6) that for a given S_t and for a given future stimulus $(\boldsymbol{D}^*, \theta^*)$ on $0 \leq \xi < \infty$, S_τ becomes a function of time. If S_τ belongs to the state $\mathfrak{S}(S_r)$ and possesses the orientation $\boldsymbol{Q}$, then both S_r and $\boldsymbol{Q}$ will be functions of time:

$$S_\tau = \boldsymbol{Q}(\tau) S_r(\tau). \tag{7.10}$$

Substituting (7.10) into (7.6) we have

$$\boldsymbol{Q}(\tau)\, S_r(\tau) = G[\boldsymbol{Q}(t)\, S_r(t); \boldsymbol{D}^*(\xi), \theta^*(\xi)]. \tag{7.11}$$

On the other hand if in (7.9) $\boldsymbol{R}$ is replaced by $\boldsymbol{Q}^T(t)$ and (7.10) is used one obtains, in view of $\boldsymbol{Q}\boldsymbol{Q}^T = \boldsymbol{I}$,

$$\boldsymbol{Q}^T(t)\, \boldsymbol{Q}(\tau)\, S_r(\tau) = G[S_r(t); \boldsymbol{Q}^T(t)\, \boldsymbol{D}^*(\xi)\, \boldsymbol{Q}(t), \theta^*(\xi)], \tag{7.12}$$

where $\boldsymbol{Q}(\tau)\, S_r(\tau)$ is defined by (7.11).

For the class of materials considered in the previous section, (7.12) has the following meaning: if $q_\alpha(\tau)$ and $\boldsymbol{Q}(\tau)$ are the state variables and orientation associated with (7.11) then

$$q_\alpha(\tau) = \boldsymbol{F}_\alpha[\theta, v, q_\beta;\ \boldsymbol{Q}^T(t)\,\boldsymbol{D}^*(\xi)\,\boldsymbol{Q}(t), \theta^*(\xi)],$$

$$\boldsymbol{Q}^T(t)\,\boldsymbol{Q}(\tau) = \boldsymbol{G}[\theta, v, q_\beta;\ \boldsymbol{Q}^T(t)\,\boldsymbol{D}^*(\xi)\,\boldsymbol{Q}(t), \theta^*(\xi)], \tag{7.13}$$

where $\boldsymbol{F}_\alpha$ and $\boldsymbol{G}$ are *functions* of the state variables $\boldsymbol{Q}$, v, q_α at time t and causal operators of the argument functions $\boldsymbol{Q}^T(t)\,\boldsymbol{D}^*(\xi)\,\boldsymbol{Q}(t)$ and $\theta^*(\xi)$.

Strictly speaking we must add to the first relations in (7.13) the following obvious ones which deal with temperature and volume:

$$\theta(\tau) = \theta(t) + \theta^*(\tau - t), \tag{7.14}$$

$$v(\tau) = v(t)\det(\boldsymbol{D}^*(\tau - t)).$$

We wish now to consider the time derivatives of $q_\alpha(\tau)$ and $\boldsymbol{Q}(\tau)$ at $\tau = t$. These will be denoted by $\dot{q}_\alpha(t)$ and $\dot{\boldsymbol{Q}}(t)$. In view of (7.13) $\dot{q}_\alpha(t)$ and $\boldsymbol{Q}^T(t)\,\dot{\boldsymbol{Q}}(t)$ will depend on θ, v, q_α at time t and also on the first and possibly higher derivatives (at $\xi = 0$ and with respect to ξ) of the argument functions $\boldsymbol{Q}^T D^*(\xi)\boldsymbol{Q}$ and $\theta^*(\xi)$.

The latter part of the above observation follows from the fact that these derivatives[1] determine the argument functions in the neighborhood of $\xi = 0$.

Here we introduce the assumption that only the first derivatives of the argument functions will appear in the above mentioned relations. This assumption is intimately connected with the "smoothness" of the operators $\boldsymbol{F}_\alpha$ and $\boldsymbol{G}$ and will be discussed further.

We next observe from (3.1) and (3.5) that

$$\dot{D}^*_{ij}(0) = \omega_{ij} + V_{ij} \quad \text{and} \quad \dot{\theta}^*(0) = \dot{\theta}(t), \tag{7.15}$$

where ω_{ij} and V_{ij} are the components, in the fixed frame and at time t, of the rate of rotation and the rate of deformation tensors, respectively.

Therefore, in view of the assumption introduced above, we have from (7.13)

$$\dot{q}_\alpha(t) = f_\alpha(\theta, v, q_\beta);\ \boldsymbol{Q}^T(\boldsymbol{\omega} + \boldsymbol{V})\boldsymbol{Q}, \dot{\theta}),$$

$$\boldsymbol{Q}^T(t)\,\dot{\boldsymbol{Q}}(t) = \boldsymbol{g}(\theta, v, q_\beta;\ \boldsymbol{Q}^T(\boldsymbol{\omega} + \boldsymbol{V})\boldsymbol{Q}, \dot{\theta}), \tag{7.16}$$

---

[1] Since $\boldsymbol{Q}^T\boldsymbol{D}^*(0)\boldsymbol{Q} = \boldsymbol{I}$, $\theta^*(0) = 0$ the values of the argument functions at $\xi = 0$ are of no importance.

where f_α and $\boldsymbol{g}$ are functions of the indicated arguments, these being evaluated at time t, and $\boldsymbol{g}$ is antisymmetric.

We now return to (7.12) and apply to it the invariance property (7.7) to obtain

$$q_\alpha(\tau) = F_\alpha[\theta, v, q_\beta; \boldsymbol{R}(\xi)\, \boldsymbol{Q}^T(t)\, \boldsymbol{D}^*(\xi)\, \boldsymbol{Q}(t), \theta^*(\xi)],$$

$$\boldsymbol{R}(t-\tau)\, \boldsymbol{Q}^T(t)\, \boldsymbol{Q}(\tau) = \boldsymbol{G}[\theta, v, q_\beta; \boldsymbol{R}(\xi)\, \boldsymbol{Q}^T(t)\, \boldsymbol{D}^*(\xi)\, \boldsymbol{Q}(t), \theta^*(\xi)], \quad (7.17)$$

where, again, $q_\alpha(\tau)$ and $\boldsymbol{Q}(\tau)$ are the state variables and orientation associated with (7.11).

We wish now to consider the time derivative of (7.17). In view of

$$\boldsymbol{R}(0) = \boldsymbol{D}^*(0) = \boldsymbol{I} \text{ and } \boldsymbol{Q}^T(t)\, \boldsymbol{Q}(t) = \boldsymbol{I}, \text{ we obtain}$$

$$\dot{q}_\alpha(t) = f_\alpha(\theta, v, q_\beta; \dot{\boldsymbol{R}}(0) + \boldsymbol{Q}^T(\boldsymbol{\omega} + \boldsymbol{V})\boldsymbol{Q}, \dot{\theta}), \quad (7.18)$$

$$\dot{\boldsymbol{R}}(0) + \boldsymbol{Q}^T(t)\, \dot{\boldsymbol{Q}}(t) = \boldsymbol{Q}(\theta, v, q_\beta; \boldsymbol{R}(0) + \boldsymbol{Q}^T(\boldsymbol{\omega} + \boldsymbol{V})\boldsymbol{Q}, \dot{\theta}).$$

Since $\dot{\boldsymbol{R}}(0)$ is an *arbitrary* antisymmetric tensor we can assign it the value,

$$\dot{\boldsymbol{R}}(0) = -\boldsymbol{Q}^T \boldsymbol{\omega} \boldsymbol{Q}. \quad (7.19)$$

Using (7.19), (7.18) becomes

$$\dot{q}_\alpha(t) = f_\alpha(\theta, v, q_\beta; \boldsymbol{Q}^T \boldsymbol{V} \boldsymbol{Q}, \dot{\theta}), \quad (7.20)$$

$$\boldsymbol{Q}^T(t)\, \dot{\boldsymbol{Q}}(t) = \boldsymbol{g}(\theta, v, q_\beta; \boldsymbol{Q}^T \boldsymbol{V} \boldsymbol{Q}, \dot{\theta}) + \boldsymbol{Q}^T \boldsymbol{\omega} \boldsymbol{Q}.$$

We add to (7.20) the time derivative of (7.14)

$$\dot{\theta}(t) = \dot{\theta}, \quad \dot{v}(t) = v\, tr\, \boldsymbol{V}, \quad (7.21)$$

to complete the final step of our discussion.

(7.21) and the first equation (7.20) are called *the growth law for the state variables*. The second equation in (7.20), on the other hand, may be called *the growth law for the orientation*. These laws represent a system of differential equations for the determination of $v(\tau)$, $q_\alpha(\tau)$ and $\boldsymbol{Q}(\tau)$ on $[0, \infty]$ for a given stimulus $(\boldsymbol{D}, \theta)$ and hence a given $(V, \omega, \dot{\theta})$ on $[0, \infty]$. The initial conditions for the system follows from the known values of (θ, v, q_α) at $\tau = 0$.

The growth laws and (7.4) constitute a complete representation of thermomechanical behavior of the solids considered in the previous section.

It may be of interest to mention here the following interpretation of (7.20).

We consider the rotating rectangular frame f_r which has the orientation $\boldsymbol{Q}(\tau)$ with respect to the fixed frame. We also consider the rectangular frame f_r^t which coincides with the above frame at $\tau = t$ but remains fixed for further times. We observe that at time t the material is in the reference orientation with respect to the frame f_r.

It is clear that

$$\boldsymbol{V}_r^t = \boldsymbol{Q}^T \boldsymbol{V} \boldsymbol{Q}, \quad \omega_r^t = \boldsymbol{Q}^T \omega \boldsymbol{Q}, \tag{7.22}$$

are the rate of deformation and the rate of rotation tensors, respectively, measured in the fixed frame f_r^t.

On the other hand

$$\boldsymbol{\Omega}_r^t = \boldsymbol{Q}^T \boldsymbol{Q}^{\cdot}, \tag{7.23}$$

measures the rate of rotation of the frame f_r with respect to f_r^t.

With these interpretations (7.20) takes the following intrinsic form:

$$\begin{aligned} q_\alpha^{\cdot} &= f_\alpha(\theta, v, q_\beta; \boldsymbol{V}_r^t, \theta^{\cdot}) \\ \boldsymbol{\Omega}_r^t &= \boldsymbol{g}(\theta, v, q_\beta; \boldsymbol{V}_r^t, \theta^{\cdot}) + \boldsymbol{\omega}_r^t. \end{aligned} \tag{7.24}$$

This simple form is intuitively obvious and could have been written down at once. For instance, since the state is independent of rigid rotations, it would be natural to expect that $q_\alpha^{\cdot}$ will not depend on ω_r^t. On the other hand the additive dependence of $\boldsymbol{\Omega}_r^t$ on ω_r^t is a natural consequence of the definition of $\boldsymbol{Q}$.

We wish now to discuss a simpler class of materials. We assume that for these materials the Fréchet derivatives [8] of the causal operators $\boldsymbol{F}_\alpha$ and $\boldsymbol{G}$ in (7.13) exist. We further use the assumption which was implicit in our previous discussion that if $\boldsymbol{D}^*$, θ^* and their first derivatives are continuous on $0 \leq \xi < \infty$ so are $q_\alpha(\tau)$ and $Q(\tau)$ and their first derivatives on $t \leq \tau \leq \infty$.

We assert without proof that under that above assumptions, the growth laws (7.20) become linear in $\boldsymbol{V}$ and $\theta^{\cdot}$. Thus we have in a more explicit notation and with the addition of (7.21).

$$\begin{aligned} \theta^{\cdot} &= \theta^{\cdot}, \quad v^{\cdot} = v V_{ii}, \\ q_\alpha^{\cdot} &= f_\alpha + g_{\alpha ij} Q_{ki} V_{kl} Q_{lj} + h_\alpha \theta^{\cdot}, \\ Q_{li} Q_{lj}^{\cdot} &= k_{ij} + l_{ijkl} Q_{sk} V_{sp} Q_{pl} + m_{ij} \theta^{\cdot} + Q_{li} \omega_{lk} Q_{kj}, \end{aligned} \tag{7.25}$$

where

$$f_\alpha, h_\alpha, g_{\alpha ij} = g_{\alpha ij}, \quad k_{ij} = -k_{ji}, \quad l_{ijkl} = -l_{jikl} = -l_{jilk},$$

and $m_{ij} = -m_{ji}$ are functions of (θ, v, q_α).

To complete the representation we repeat (7.4) in a more explicit form:

$$\sigma_{ij} = Q_{ik} F_{kl}(\theta, v, q_\alpha) Q_{jl}, \qquad (7.26)$$
$$e = E(\theta, v, q_\alpha).$$

It may be argued that most materials of interest will belong to the above mentioned class so that the representation (7.25) and (7.26) is of considerable generality. However as remarked in [6] plastic solids do not fall within this class; In these solids $q_\alpha^{\cdot}$ will depend on $\boldsymbol{V}$ and $\theta^{\cdot}$ in a more complicated manner on account of the marked difference in behavior observed under conditions of "loading" and "unloading".

Before closing this section we would like to consider very briefly the restrictions imposed by (6.7) on the growth law (7.25). We assume that $f(S_r) = f(\theta, v, q_\alpha)$ possesses first derivatives so that (6.7) can be written in the form

$$1 - \left(\frac{\partial f}{\partial \theta}\theta^{\cdot} + \frac{\partial f}{\partial v} v^{\cdot} + \frac{\partial f}{\partial q_\alpha} q_\alpha^{\cdot}\right) \geq 0. \qquad (7.27)$$

If one now substitutes the second equation of (7.25) into (7.27) one obtains

$$\left(1 - f_\alpha \frac{\partial f}{\partial q_\alpha}\right) - \left(\frac{\partial f}{\partial \theta} + \frac{\partial f}{\partial q_\alpha} h_\alpha\right)\theta^{\cdot} - \left(\frac{\partial f}{\partial q_\alpha} g_{\alpha ij} Q_{ki} Q_{lj} + \right.$$
$$\left. + v \frac{\partial f}{\partial v}\delta_{kl}\right) V_{kl} \geq 0, \qquad (7.28)$$

where δ_{kl} denotes the Kronecker delta.

With the assumption that $\theta^{\cdot}$ and $\boldsymbol{V}$ can have arbitrarily large values, the above inequality yields

$$1 - f_\alpha \frac{\partial f}{\partial q_\alpha} \geq 0,$$
$$\frac{\partial f}{\partial \theta} + \frac{\partial f}{\partial q_\alpha} h_\alpha = 0, \qquad (7.29)$$
$$\frac{\partial f}{\partial q_\alpha} g_{\alpha ij} Q_{ki} Q_{lj} + v \frac{\partial f}{\partial v}\delta_{kl} = 0.$$

We shall not pause here to discuss (7.29) further.

8. Restrictions Imposed by the Second Law

In the previous sections we made little use of thermodynamics. We employed the first law to replace the rate of heat flow r by the internal energy e in the response of the material. As a preparation for the present section, we also introduced an assumption of thermodynamic nature, namely that temperature and specific volume are state variables. However the discussion which led to the results (7.4), (7.20) and (7.25), (7.26) is essentially free of thermodynamic considerations.

We wish now to investigate the restrictions that the second law of thermodynamics imposes on (7.25) and (7.26).

We start with the assumption that the materials and the stimuli $(\boldsymbol{D}, \theta)$ considered in this section are such that it is meaningful to speak of the *entropy* S (per unit volume) of the deforming solid[1]. We further assume that material elements which are in the same state (in the sense of section 4) have the same entropy and moreover distinct states are associated with distinct entropies. We have, therefore,

$$S = S(\theta, v, q_\alpha). \tag{8.1}$$

In the present case where there is no heat conduction the second law of thermodynamics takes the form

$$S^{\cdot} - \frac{r}{\theta} \geq 0, \tag{8.2}$$

where $S^{\cdot}$ denotes the rate of change of entropy, r the rate of heat flow, θ the temperature, at a given time and in any test of interest.

Combining (8.2) with the first law (2.7) we obtain the familiar inequality:

$$\frac{1}{\varrho}\,\sigma_{ij} V_{ij} - e^{\cdot} + S^{\cdot}\theta \geq 0. \tag{8.3}$$

With the introduction of the Helmholtz free energy F,

$$F = e - \theta S = F(\theta, v, q_\alpha), \tag{8.4}$$

(8.3) becomes

$$\frac{1}{\varrho}\,\sigma_{ij} v_{ij} - F^{\cdot} - \theta^{\cdot} S \geq 0. \tag{8.5}$$

Denoting partial derivatives of F (which are assumed to exist) by $F_{,\theta}$, $F_{,v}$ and $F_{,\alpha}$ (8.5) becomes, upon the substitution of (7.25)

$$\frac{1}{\varrho}\,(\sigma_{ij} - \varrho F_{,\alpha}\, g_{\alpha kl}\, Q_{ik} Q_{jl} - F_{,v}\delta_{ij})\, V_{ij} -$$

$$- (F_{,\alpha} h_\alpha + F_{,\theta} + S)\theta^{\cdot} - F_{,\alpha} f_\alpha \geq 0. \tag{8.6}$$

[1] The symbol S need not be confused with S which denoted a distinct state and orientation.

A set of conditions which are sufficient for (8.6) to hold are

$$F_{,\alpha} f_\alpha \leq 0,$$
$$S = -(F_{,\theta} + F_{,\alpha} h_\alpha), \tag{8.7}$$
$$\sigma_{ij} = Q_{ik}(F_{,v}\delta_{kl} + \varrho F_{,\alpha}\, g_{\alpha kl}) Q_{jl},$$

where the last two equations imply that the expressions in the parentheses in (8.5) are zero.

The above conditions become necessary in the case where V_{ij} $(= V_{ji})$ and $\theta^{\cdot}$ can assume arbitrary values (note that by previous results and assumptions $F_{,\alpha} f_\alpha$ and the terms in the parentheses in (8.6) are independent of V_{ij} and $\theta^{\cdot}$). Now from the point of view of the preceding sections there is no reason for V_{ij} $(= V_{ji})$ and $\theta^{\cdot}$ not to be arbitrary. However in the present section we are dealing with situations where it is meaningful to speak of entropy and these may very well preclude[1] stimuli for which $|V_{ij}|$ and $\theta^{\cdot}$ have arbitrarily large values, and therefore it would be impossible to assert the necessity of the last two conditions in (8.7).

Here we wish to retain (8.7) as a consequence of the second law. To achieve this we introduce an additional assumption which is similar to the one used earlier in [3]; we assume that for the solids considered here the rate of entropy production does not depend on V_{ij} and $\theta^{\cdot}$. This assumption implies that (8.7) are also necessary conditions for the second law to hold. If one now compares (8.7) with (7.26) by using (8.4) one finds that

$$F_{kl} = F_{,v}\delta_{kl} + \varrho F_{,\alpha}\, g_{\alpha kl},$$

and (8.8)

$$E = F - \Theta(F_{,\theta} + F_{,\alpha} h_\alpha).$$

(8.8) constitutes the contribution of the second law to the representation of thermo-mechanical behavior. This contribution is an important one, since it enables one to derive the seven functions F_{ij} and E in (7.26) from a single function F. However, it is worth noting that the second law does not produce any restrictions on the growth law (7.24) except the first inequality in (8.7).

References

[1] STAVERMAN, A. J., and F. SCHWARZL: Thermodynamics of Viscoelastic Behavior. Verh. Akad. Wet. (Amsterdam) B-**55**, 474—485 (1952).
[2] HUNTER, S. C.: Tentative Equations for the Propagation of Stress. Strain and Temperature Fields in Viscoelastic Solids. J. Mech. Phys. Solids **9**, 39—51 (1961).

---

[1] This remark is due to MEIXNER (see, for instance, [9]).

[3] Breuer, S., and E. T. Onat: On the Determination of Free Energy in Linear Viscoelastic Solids. ZAMP **15**, 184—191 (1964).
[4] Coleman, B.: Thermodynamics of Materials and Memory. Arch. Ratl Mech. Anal. **17**, 1—46 (1964).
[5] Rivlin, R. S.: Nonlinear Viscoelastic Solids. SIAM Review **7** (3), 323—340 (1965).
[6] Onat, E. T.: Description of Mechanical Behavior of Inelastic Solids. Proc. 5th U. S. National of Congress Applied Mechanics (Minneapolis, Minn.) 421—434 (1966).
[7] Simmons, G. F.: Introduction to Topology and Modern Analysis. p. 51. New York: McGraw-Hill. 1963.
[8] Vainberg, M. M.: Variational Methods for the Study of Nonlinear Operators. p. 40. Holden-Day. 1964.
[9] Meixner, J.: TIP has many faces. IUTAM Symposia Vienna 1966. p. 237. Wien-New York: Springer. 1968.

Discussion

Drucker: Would you care to comment on the inverse process, where you specify the stress history, which then determines the deformation gradients' history. Does this in any way restrict the state variables?

Onat: It may be, but the invertibility in the case of finite deformations is not trivial, one has to worry about the rigid body rotation question.

Kröner: You have defined state quantities here and one would like to know what they are physically.

Onat: If you have no previous knowledge at all about the material you could plan experiments to determine the number of state variables. This is done as in the control theory by just watching input and outputs. Now, if you have in addition physical information, for instance on dislocation distribution, by all means use it. In one of your papers you introduce a certain tensorial dislocation density as your state variable. I think, information on the number and nature of state variables would be coming from both microscopic and macroscopic considerations.

Kröner: Of course, if the number of state quantities is infinite the problem is terrible, but we can hope at least that only a few are important for macroscopic experiments.

Onat: Yes! In linear viscoelasticity, if you want to study slow phenomena, you could get away with a few state variables. But the constitutive equations used for this purpose would not really apply to wave propagation. If you want to discuss fast phenomena you must use another set of constitutive equations, another set of state variables.

Kestin: The same difficulty as Prof. Drucker pointed out in relation to the control of strains relates to the control of temperature. We have no control of the temperature of a specimen other than at the boundaries, as a rule. So why not start with the concept of state and work from there? Why invert the whole thing and make such questions as the interchangeability of stress and strain, which is really simple, appear so complex and so difficult to resolve?

Onat: My point is that the experimental information available is a set of histories and therefore one should base the definition of the concept of state on histories.

Rimrott: It appears to me there must be very strict limitations as to what may happen during a stress history.

Onat: One works with continuous functions and one could also work with bounded functions. For instance if you don't want to introduce complications then you would work with histories where the stresses never go beyond a certain limit.

Kestin: You wrote down, as many people do, the Clausius inequality in a particular form. Where does it come from?

Onat: I take the view that it looks very reasonable.

Truesdell: Coleman amd I yesterday, when we talked about phenomena of this kind, were asked how we define temperature. I think, the fact that Prof. Onat has labeled himself an engineer should not excuse his being asked by somebody how he defines temperature in problems with finite deformation.

Onat: In my hypothetical experiments I measure the temperature and heat exchange. These two quantities I assume as measurable and I shy away from situations where temperature is not defined, I shy away from very fast phenomena and so on. And, since in the earlier part of my discussion I did not mention entropy, I felt that I was on solid ground.

Perzyna: What restriction should be satisfied by your response functions?

Onat: Some restrictions which come from considerations of rigid-body rotation. It may also be that I may like to introduce the Onsager restrictions. In plasticity, the smoothness hypotheses I used in part of the paper are not valid.

Kluitenberg: I wonder how it is possible that the entropy explicitly depends on the time.

Onat: I wanted to include in my considerations the possibility that the material may change its behavior over the course of time. This happens in concrete for instance; concrete ages over the course of time even when the stresses are zero.

Composite Stress-Strain Relations for Elastoplastic Solids[1]

By

William Prager

La Jolla, Calif. (U.S.A.)

Summary

The paper explores the use of composite stress-strain relations for inviscid elastoplastic behavior. Their predictions are compared with the observed behavior of thin-walled tubes of mild steel in combined tension and torsion. It is shown that even very simple stress-strain relations of this type furnish predictions that are in markedly better agreement with experiments than those of the Prandtl-Reuss theory. Finally, the extent is discussed to which use of the considered stress-strain relations implies a unique solution to a typical boundary value problem in plasticity.

1. Introduction

In the typical discussion of linear viscoelastic behavior (see, for instance, [1]), the simple stress-strain relations of the linear viscous fluid (Newtonian fluid) and the linear elastic solid (Hookean solid) are used as building blocks to construct composite stress-strain relations that represent the behavior of linked Newtonian and Hookean elements. Composite stress-strain relations for inviscid elastoplastic behavior can obviously be constructed in a similar manner by considering linked elastic and rigid, perfectly plastic elements. This technique, however, has not been widely used in plasticity. Ivlev [2] has discussed one- and two-dimensional arrangements of elastic, viscous, and rigid, perfectly plastic elements, but his paper is primarily concerned with the classification of the media represented by these models, and his composite stress-strain relations for inviscid elastoplastic behavior do not seem to have been applied to specific problems. When subjected to uniform strain with components that increase monotonically and in proportion, Ivlev's elastoplastic solids exhibit an unlimited capacity to workharden, whereas

[1] This research was supported by the Advanced Research Projects Agency (Project DEFENDER) and was monitored by the U. S. Army Research Office-Durham under Contract DA-124-ARO-D-257.

the solids considered in the present paper will ultimately flow under constant stress.

In Section 2, two-dimensional models of composite elastoplastic behavior are discussed with special reference to combined tension and torsion of thin-walled circular tubes. A graphical method of integrating the stress-strain relations is presented in Section 3. It is shown that even the most elementary composite stress-strain relations furnish predictions that agree much better with experiments than those of the Prandtl-Reuss theory. In Section 4, it is briefly indicated how these stress-strain relations can be generalized to arbitrary three-dimensional states of stress and strain, and the extent is investigated to which use of the generalized relations implies a unique solution to a typical boundary value problem.

2. Two-Dimensional Models of Elastoplastic Behavior

Consider combined tension and torsion of a thin-walled circular tube made of an elastoplastic material that is incompressible in both the elastic and the plastic ranges. Young's modulus E and the shear modulus G of the tube material are then related by

$$E = 3G. \tag{1}$$

The axial and shearing stresses in the tube will be denoted by σ and τ, and the axial strain and the engineering shear strain by ε and γ. The states of stress and strain in the tube will be specified by the stress vector $\boldsymbol{s}$ and the strain vector $\boldsymbol{e}$; with respect to a system of rectangular axes, these vectors are to have the components

$$\boldsymbol{s} = (\sigma, \tau \sqrt{3}), \quad \boldsymbol{e} = (E\varepsilon, E\gamma/\sqrt{3}). \tag{2}$$

Primes and double primes will be used to indicate elastic and plastic strains, and the strain vector will be written as $\boldsymbol{e} = \boldsymbol{e}' + \boldsymbol{e}''$ with

$$\boldsymbol{e}' = (E\varepsilon', E\gamma'/\sqrt{3}), \quad \boldsymbol{e}'' = (E\varepsilon'', E\gamma''/\sqrt{3}). \tag{3}$$

Note that the scale factors in the shear components of the vectors (2) and (3) satisfy the following conditions:

1. the stress vector $\boldsymbol{s}$ and the elastic strain vector $\boldsymbol{e}'$ coincide:

$$\boldsymbol{s} = \boldsymbol{e}'; \tag{4}$$

2. the work dW of the stress $\boldsymbol{s}$ on the strain increment represented by $d\boldsymbol{e}$ equals

$$dW = \boldsymbol{s} \cdot d\boldsymbol{e}/E. \tag{5}$$

where the center dot denotes scalar multiplication;

3. if the tube material obeys the yield condition of von Mises [3],

$$\sigma^2 + 3\tau^2 = k^2, \tag{6}$$

the stress vector **s** has constant magnitude during plastic flow.

The models discussed below consist of springs and friction blocks. The blocks are supported by a plane horizontal surface and experience a frictional resistance when they slide over this surface. This resistance has the constant intensity k and the direction opposed to that of the instantaneous motion. A spring that joins two friction blocks is supposed to exert, on each of them, a force that is directed towards the other block and has an intensity proportional to the distance between the blocks.

As is readily verified, a single friction block can be made to represent the behavior of a tube consisting of a rigid, perfectly plastic material that obeys the yield condition and flow rule of von Mises [3]. To this end, the stress vector **s** is interpreted as a load acting on the block, while the displacement vector of the block from the position corresponding to the virgin state of the tube is interpreted as the strain vector $\boldsymbol{e}$. As long as the block remains at rest, the load **s** must have an intensity below the maximum frictional resistance k. When the block slides quasi-statically over its support, the load **s** must have the intensity k and the direction of the instantaneous velocity $\dot{\boldsymbol{e}}$. We thus have the relation

$$\mathbf{s} = k\dot{\boldsymbol{e}} / \sqrt{\dot{\boldsymbol{e}} \cdot \dot{\boldsymbol{e}}}, \tag{7}$$

which agrees with the theory of von Mises [3].

The simplest model of elastoplastic behavior consists of a friction block and a spring. In Fig. 1, the point B represents the friction block and the solid line BA represents the spring. In the virgin state, both A and B are at the coordinate origin O. In Fig. 1, the vector $\overrightarrow{OA}$ is the strain vector $\boldsymbol{e}$, and the vector $\overrightarrow{BA}$, which represents the spring force exerted on the block B, is the stress vector **s**, and hence the elastic strain vector $\boldsymbol{e}'$. It follows that the vector $\overrightarrow{OB}$ is the plastic strain vector $\boldsymbol{e}''$. During the quasi-static motion of the friction block B, the force **s**, which the spring exerts on the block, balances the frictional resistance of the block. The force **s** thus has the constant intensity k and the block instantaneously slides in the direction of this force. These two conditions enable us to determine the new position $A_1 B_1$ of the model, when its

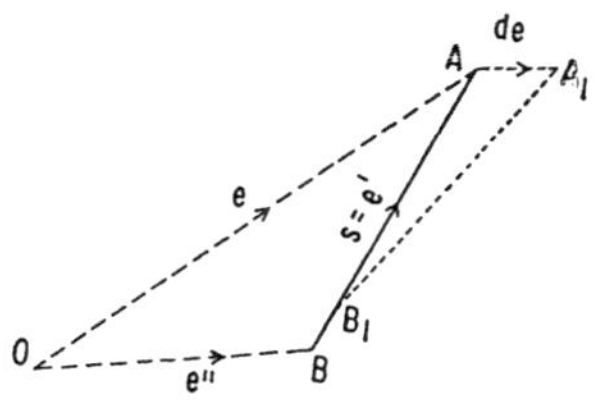

Fig. 1. Model with friction block and spring

current position AB and the increment $\boldsymbol{de}$ of the strain vector are given.

A model with two friction blocks and two springs is shown in Fig. 2, where the points B and C represent the friction blocks and the solid lines CB and BA the springs. To avoid overloading the figure, the coordinate origin O and the vectors from it to the points A, B, and C are not

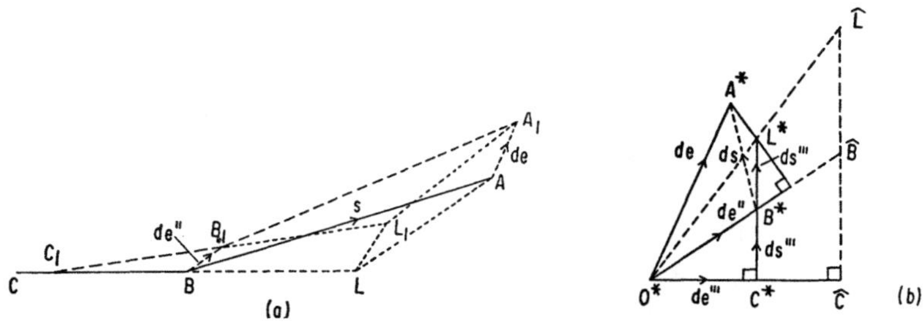

Fig. 2. (a) Model with two friction blocks and two springs; (b) displacement diagram

shown. The vectors $\overrightarrow{OA}$ and $\overrightarrow{OB}$ would again be the strain vector $\boldsymbol{e}$ and the plastic strain vector $\boldsymbol{e}''$. The vectors $\boldsymbol{s} = \overrightarrow{BA}$ and $\boldsymbol{s}''' = \overrightarrow{CB}$ will be interpreted as representing the states of macrostress and microstress. The first of these can be observed directly and completely, the second only to the extent to which it affects the subsequent behavior of the model. Together, these two stress vectors represent the influence of the loading history on this behavior. The vector $\boldsymbol{s}''' = \overrightarrow{CB}$ may also be interpreted as the vector of elastic microstrain, that is the strain that corresponds to the microstress by Hooke's law. Finally, the vector $\boldsymbol{e}''' = \overrightarrow{OC}$ will be interpreted as the vector of plastic microstrain.

Figure 2a is supposed to show a position of the model in which both blocks will move when the strain vector is given the increment $\boldsymbol{de}$ indicated in the figure. For simplicity, the frictional resistances experienced by the blocks B and C are assumed to have the same intensity k, though this assumption is not essential. The block C is under the action of the spring force $\overrightarrow{CB}$ and a resistance of the intensity k. The vector $\overrightarrow{CB}$ therefore has the magnitude k and indicates the direction of motion of the block C. The block B, on the other hand, is under the action of the spring forces $\overrightarrow{BC}$ and $\overrightarrow{BA}$ and a resistance of the intensity k. In Fig. 2a, the vector $\overrightarrow{LB}$ has been made equal to $\overrightarrow{BC}$. The vector $\overrightarrow{LA}$ therefore has the intensity and direction of the resultant of the spring forces that act on B. Since this resultant must be balanced by the frictional resistance of the block, the vector $\overrightarrow{LA}$ has the magnitude k and indicates the direction of motion of the block B.

Suppose now that a small displacement $\overrightarrow{CC_1}$ of the block C along CB is given as well as the direction of motion of the point A. The new position B_1 of the block B then follows from the requirements that B moves parallel to LA and that C_1B_1 has the length k. Extending C_1B_1 beyond B_1 by its own length, we obtain the point L_1. The new position A_1 of the point A then follows from the conditions that the point A moves in the given direction and that L_1A_1 has the length k. Note that the stress vectors $\overrightarrow{BA}$ and $\overrightarrow{B_1A_1}$ will not, in general, have the same magnitude.

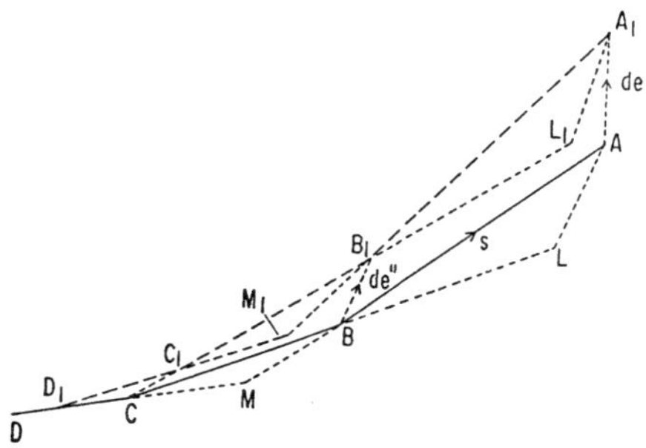

Fig. 3. Model with three friction blocks and three springs

Figure 3 shows a model with three friction blocks (B, C, D) and three springs (DC, CB, BA). The motion of this model can be discussed in a similar manner as that of the model with two friction blocks. The points D, C, M in Fig. 3 are collinear and $DC = CM$; similarly C, B, L are collinear and $CB = BL$. The vectors MB and LA have the lengths k and indicate the directions of motion of the blocks C and B; the block D moves along DC. We shall not discuss this model in greater detail, because it will be shown in the next section that the simpler model of Fig. 2a already gives satisfactory results.

3. Comparison with Experiments

The friction block-spring system in Fig. 1 models the stress-strain relations of Prandtl-Reuss. As was first observed by Hohenemser and Prager [5], the actual elastoplastic behavior of thin-walled tubes of mild steel in combined tension and torsion differs in important respects from that predicted by the Prandtl-Reuss relations. In their typical experiment, these authors first subjected the tubular specimen to plastic flow in simple tension and then twisted the tube while keeping the axial extension constant. Figure 4 illustrates the behavior of the friction block-spring model for this type of experiment. When the twisting of the tube is started, the model is in the position BA, the distance BA being k. As the endpoint A of the spring is moved normal to BA, the model successively assumes the positions B_1A_1, B_2A_2, . . ., indicating how the stress vector $\overrightarrow{BA}$ changes when the shear strain γ is increased while the axial strain ε is kept constant. The vectors $\overrightarrow{B_1^*A_1}$, $\overrightarrow{B_2^*A_2}$, . . . in

Fig. 4 are stress vectors predicted by a modified form of the stress-strain relations of HENCKY [6], which has been suggested by HOHENEMSER and PRAGER [5]. For the considered straining program, the loaded end of the spring moves in the direction of the vector AB while the specimen is stressed in tension, and B is the last position of the block for which this is the case. According to the modified stress-strain relations of Hencky, a typical later position of the block, say the position B_4^* that is associated with A_4, lies on BA_4 at the distance k from A_4. In Hencky's original theory, the coordinate origin O takes the role that is played by B in the modified theory.

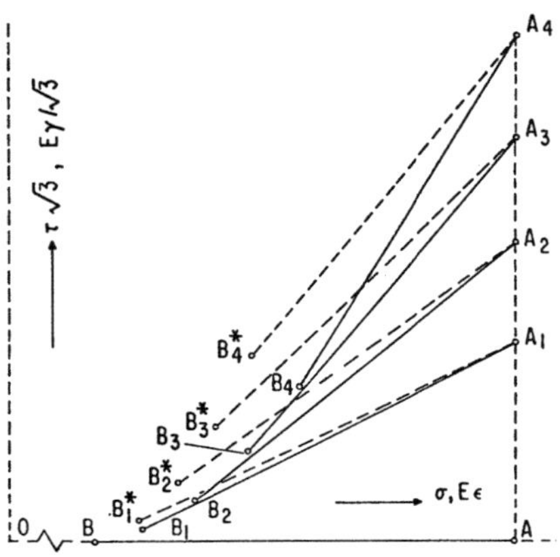

Fig. 4. Combined tension and torsion: Solid stress vectors correspond to Prandtl-Reuss theory, broken stress vectors to modified Hencky theory

As is seen from Fig. 4, for a given shear strain γ, the Prandtl-Reuss relations call for a greater shearing stress τ than the modified Hencky relations.

Stress vectors that represent experimentally determined axial and shearing stresses lie between the stress vectors predicted by the modified theory of HENCKY and the theory of PRANDTL-REUSS and usually agree better with the former than with the latter. This observation first reported in [5] has since been confirmed by many authors (see, for instance, FEIGEN [7]).

In Fig. 2b, the small displacements of the points of the model are shown at three times the scale used in Fig. 2a. When the displacement $de = \overrightarrow{O^*A^*}$ of the point A is given, the displacements $\overrightarrow{O^*B^*}$, $\overrightarrow{O^*C^*}$, and $\overrightarrow{O^*L^*}$ of the points B, C, and L are found as follows. The points B^* and C^* lie on the lines $O^*\hat{B}$ and $O^*\hat{C}$ that are parallel to LA and CB, which indicate the directions of motion of B and C. If $O^*\hat{B}$ and $O^*\hat{C}$ did represent the displacements of B and C, the triangle $O^*\hat{C}\hat{B}$ would have to have a right angle at $\hat{C}$. Indeed, the fixed length k of BC implies that the displacements of B and C have the same projection on BC. The displacement $O^*\hat{L}$ of L could now be found by extending $\hat{C}\hat{B}$ beyond $\hat{B}$ by $\hat{B}\hat{L} = \hat{C}\hat{B}$. This follows from the facts that the points C, B, and L remain collinear and the distances CB and BL remain equal during the motion of the model. Of course, the displacements $O^*\hat{B}$ and $O^*\hat{L}$ obtained from an assumed displacement $O^*\hat{C}$ will not, in general, be compatible with the given displacement $O^* A^*$. The line $O^* \hat{L}$, however, will indicate the direction of motion of the point L. To find the actual displacement $O^* L^*$ of L, we take account of the fact that LA, which has the direction of $O^* \hat{B}$, preserves its length k during the motion. This means, that $A^* L^*$ must be perpendicular to $O^* \hat{B}$. After L^* has been determined in this manner, C^* is found as the foot of the perpendicular from L^* onto $O^* \hat{C}$, and B^* is found as the intersection of this perpendicular with $O^* \hat{B}$. Note that the vectors $\overrightarrow{B^* A^*}$ and $\overrightarrow{C^* B^*}$ represent the changes $d\mathbf{s}$ and $d\mathbf{s}'''$ of the macrostress $\mathbf{s}$ and the microstress $\mathbf{s}'''$.

The construction in Fig. 2b is strictly valid only if the displacements are infinitesimal, but good approximations are obtained when it is applied to sufficiently small finite displacements. After the new positions A_1, B_1, C_1, L_1 of the relevant points of the model have been determined in Fig. 2a from the displacements found in Fig. 2b, the construction is applied to the next strain increment, and so on.

Figure 5, which corresponds to the same straining program as Fig. 4, shows a few positions of the model that have been obtained in this manner. For the strain represented by the radius vector of A_4, the

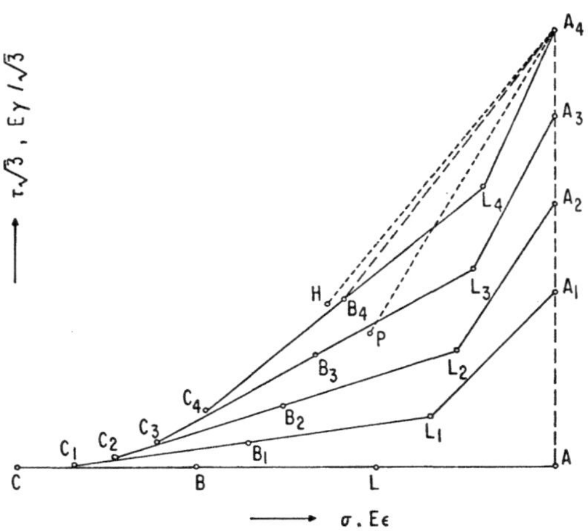

Fig. 5. Combined tension and torsion: Behavior of model of Fig. 2 agrees well with modified Hencky theory

stress vector $B_4 A_4$ lies between the stress vectors $H A_4$ and $P A_4$ furnished by the modified theory of Hencky and the theory of Prandtl-Reuss and is closer to the former. Thus, the behavior of the model roughly corresponds to the observed behavior of mild steel tubes in combined tension and torsion. If this appears desirable, the agreement between model and experiment can be improved by attributing frictional resistances of different intensities to the block B and C. The limiting case in which the friction at block C is infinite, so that this block is in fact fixed, has been discussed by KADASHEVICH and NOVOZHILOV [8].

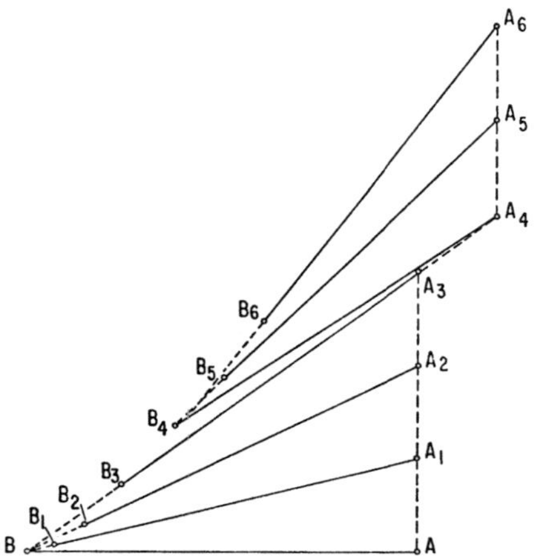

Fig. 6. Combined tension and torsion: Behavior of model of Fig. 2 after strain increment in direction of stress vector agrees well with modified Hencky theory

The straining program in Fig. 6 starts in the same way as that in Fig. 5, and the directions of the stress vectors $\overrightarrow{B_1A_1}$, $\overrightarrow{B_2A_2}$, and $\overrightarrow{B_3A_3}$ are very nearly those of the lines BA_1, BA_2, and BA_3. At A_3, however, the strain increment $\overrightarrow{A_3A_4}$ in the direction of the stress vector $\overrightarrow{B_3A_3}$ is introduced, which leads to the stress vector $\overrightarrow{B_4A_4}$. According to the modification of Hencky's theory that was proposed in [5], the subsequent stress vectors $\overrightarrow{B_5A_5}$, $\overrightarrow{B_6A_6}$, . . . should have the directions of B_4A_5, B_4A_6, . . . The stress vectors in Fig. 6, which correspond to the behavior of the model in Fig. 2a, very nearly satisfy this condition.

4. Uniqueness of Solution of Boundary Value Problem

Three-dimensional stress-strain relations are readily formulated that predict the same behavior as the models in Section 2 when they are applied to combined tension and torsion of thin-walled circular tubes. The simplest way of obtaining relations of this kind is to establish the vector equations for the motion of the models and then interpret these vectors as vectors in a nine-dimensional Euclidean space in which the rectangular components of a stress vector are the components of the corresponding stress deviation, symmetric components being formally treated as independent. Similarly, the rectangular components of a strain vector in nine-dimensional space are the components of the corresponding strain tensor multiplied by $2G$, where G is the shear modulus and symmetric components are again treated as independent. Since this derivation of three-dimensional stress-strain relations is straightforward, we refrain from presenting these relations here. We shall, however, discuss the extent to which use of these stress-strain relations implies a unique solution to the following boundary value problem. Given the instantaneous fields of stress and plastic strain throughout the volume of an elastoplastic body, the infinitesimal increments of the surface traction on part of its surface, and the infinitesimal increments of the displacement on the remainder of the surface, to determine the increments of stress and displacement throughout the body. It is well known [9, 10] that this problem has a unique solution if the stress-strain relation implies

$$(ds_{ij} - d\hat{s}_{ij})\,(d\varepsilon''_{ij} - d\hat{\varepsilon}''_{ij}) \geq 0, \tag{8}$$

provided the increments ds_{ij} and $d\hat{s}_{ij}$ of the stress deviation are respectively compatible with the plastic strain increments $d\varepsilon''_{ij}$ and $d\hat{\varepsilon}''_{ij}$.

For the model in Fig. 2a, the inequality corresponding to (8) would be

$$(d\boldsymbol{s} - d\hat{\mathbf{s}}) \cdot (d\boldsymbol{e}'' - d\hat{\mathbf{e}}'') \geq 0. \tag{9}$$

That this inequality is satisfied can be shown as follows. The direction of $d\boldsymbol{e}''$ in Fig. 2b is determined by the instantaneous configuration of the model, and $d\hat{\mathbf{e}}''$ must therefore have the same direction. Thus,

$$d\hat{\mathbf{e}}'' = \lambda d\boldsymbol{e}''. \tag{10}$$

It follows then from Fig. 2b that

$$d\hat{\mathbf{s}}''' = \lambda d\boldsymbol{s}''', \tag{11}$$

and

$$d\boldsymbol{s} \cdot d\boldsymbol{e}'' = d\boldsymbol{s}''' \cdot d\boldsymbol{e}'' = d\boldsymbol{s}''' \cdot d\boldsymbol{s}'''. \tag{12}$$

Writing the corresponding equation for $d\hat{\mathbf{s}} \cdot d\hat{\mathbf{e}}''$ and using (10) and (11), we find

$$d\hat{\mathbf{s}} \cdot d\boldsymbol{e}'' = \lambda d\boldsymbol{s}''' \cdot d\boldsymbol{s}''' = \lambda d\boldsymbol{s} \cdot d\boldsymbol{e}''. \tag{13}$$

Thus,

$$\begin{aligned}(d\boldsymbol{s} - d\hat{\mathbf{s}}) \cdot (d\boldsymbol{e}'' - d\hat{\mathbf{e}}'') &= (1 - \lambda)\,(d\boldsymbol{s} - d\hat{\mathbf{s}}) \cdot d\boldsymbol{e}'' \\ &= (1 - \lambda)^2\, d\boldsymbol{s} \cdot d\boldsymbol{e}'' = (1 - \lambda)^2\, d\boldsymbol{s}''' \cdot d\boldsymbol{s}''' \geq 0,\end{aligned} \tag{14}$$

by (10), (13), and (12). This establishes the inequality (9) and hence the uniqueness of the solution to the considered boundary value problem.

The inequality (8) can be established in the same manner by interpreting Figs. 2a and b as figures in nine-dimensional space. Note that the points A, B, C in the nine-dimensional version of Fig. 2a determine a two-dimensional subspace, which also contains the points L, C_1, B_1, and L_1, but not, in general, the point A_1. Similarly, in the nine-dimensional version of Fig. 2b, the points O^*, B^*, C^*, and L^* lie in a two-dimensional subspace which does not, in general, contain the point A^*.

A theorem of Drucker [11] provides an alternative uniqueness proof. According to this theorem, a sufficient condition for uniqueness is that no work can be extracted from the already stressed material by an external agency that first applies and then removes a set of infinitesimal stresses. Now, from the friction block-spring models in Figs. 2a and 3, work can clearly not be extracted by first applying an infinitesimal force to the point A and then removing this force. While Drucker [12] has described devices with dry friction for which this statement is not true, it is essential for these counterexamples that the amount of friction is changed during the cycle of loading and unloading. Since the frictional resistances of the blocks in the models considered here are constant, the use of stress-strain relations that are based on models of this kind ensures the uniqueness of the solution to the considered boundary value problem.

References

[1] Alfrey, T., Jr.: Mechanical Behavior of High Polymers. New York: Interscience Publishers. 1948.
[2] Ivlev, D. D.: The Theory of Complex Media. Doklady Ak. Nauk SSSR **148**, 64—67 (1963); Soviet Physics-Doklady **8**, 38—30 (1963).
[3] Mises, R. von: Mechanik der festen Körper im plastisch deformablen Zustand. Göttinger Nachrichten, math. phys. Kl. **1913**, 582—592.
[4] Prandtl, L.: Spannungsverteilung in plastischen Körpern. Proc. 1st Internatl Congr. Appl. Mech., pp. 43—54. Delft: 1924.
Reuss, E.: Berücksichtigung der elastischen Formänderungen in der Plastizitätstheorie. Z. angew. Math. Mech. **10**, 266—274 (1930).
[5] Hohenemser, K., and W. Prager: Beitrag zur Mechanik des bildsamen Verhaltens von Flußstahl. Z. angew. Math. Mech. 12, 1—14 (1932).
[6] Hencky, H.: Zur Theorie plastischer Deformation und der hierdurch im Material hervorgerufenen Nachspannungen. Z. angew. Math. Mech. **4**, 323—334 (1924).
[7] Feigen, M.: Inelastic behavior under combined tension and torsion. Proc. 2nd U. S. Natl Congr. Appl. Mech., pp. 469—476. New York: ASME 1955.
[8] Kadashevich, Iu. I., and V. V. Novozhilov: The Theory of Plasticity which Takes into Account Residual Microstresses. Prikl. Mat. Mekh. **22**, 78—89 (1958); J. Appl. Math. Mech. **22**, 104—118 (1958).
[9] Melan, E.: Zur Plastizität des räumlichen Kontinuums. Ing.-Arch. **9**, 116—126 (1938).
[10] Prager, W.: Recent Developments in the Mathematical Theory of Plasticity. J. Appl. Phys. **20**, 235—241 (1949).
[11] Drucker, D. C.: On Uniqueness in the Theory of Plasticity. Q. Appl. Math. **14**, 35—42 (1956).
[12] Drucker, D. C.: Some Implications of Work Hardening and Ideal Plasticity. Q. Appl. Math. **7**, 411—418 (1950).

Discussion

GRIOLI: Do you think that it is correct to use a linear theory of elasticity in the field of an elastic-plastic theory?

PRAGER: Yes, I don't see why not. If you take mild steel as a typical structural metal, the elastic deformations at the yield limit are sufficiently small to be treated as infinitesimal.

ODQVIST: I suppose that if you want to work out your theory in detail you must introduce certain new material constants or functions in order to express the actual deformation of the body.

PRAGER: The only material constants which enter in this very primitive theory are Young's modulus for the elastic part, the shear modulus, which follows from elastic incompressibility, and the dry friction σ_0, the yield limit. But there is some point I could make: The stress-vectors do not remain of constant length, the total stress does not satisfy the Mises yield condition, it's only the individual friction elements that satisfy something analogous to the Mises yield condition.

ONAT: Do you retain the notion of yield surface in this theory?

PRAGER: No, there is no yield surface.

ONAT: But loading and unloading?

PRAGER: Loading and unloading, yes! If I may speak in terms of this three-particle model — if you move the particle A in such a way, that you can have equilibrium without the particles B and C moving, the assumption is they will not move and this would be elastic unloading.

SETH: Plastic state is a totally ideal state. Should it not be treated as a limiting process through a transition rather than an exact process?

PRAGER: It is well known that, in the theory of finite elasticity, you can, by appropriate assumptions, produce a behavior which asymptotically, for very very large strains, shows something like deformation under constant stress, something which has been labeled yield. But in my opinion this has absolutely nothing to do with the physical phenomena, which occur in plastic deformation.

SETH: Don't you think, one should identify yield with a sort of an asymptotic phenomenon?

PRAGER: This is a different point of view. As I pointed out, I do not believe that it is justified.

DRUCKER: Would you care to make some comment on the Bauschinger-effect in this type of theory?

PRAGER: I have a chain of particles and even the last particle is free. If I attach that with a spring to the origin then I get strain hardening and a Bauschinger type of effect.

NAGHDI: You said that you don't have a yield surface but there is loading and unloading. Therefore, at some stage in your developments, there is an elastic domain. What would the boundary of this domain be?

PRAGER: This elastic domain is exactly the circle with the center B through the point A.

NAGHDI: You did not explicitly use it, but it is really there, isn't it?

PRAGER: Yes, if you want to distinguish between what strain increment produces plastic response and what produces purely elastic response, it's there. In terms of normality it's not there.

Kinetics of Creep and Creep Rupture

By

Yu. N. Rabotnov

Moscow (U.S.S.R.)

In the development of the modern mechanics of solids there are two distinct lines, not opposite but different. One of them is a trend to deduce the constitutive equations for different kinds of particular solids starting from few very general and very universal principles governing the behaviour of any media which might exist in reality or might be invented. Indeed it is only natural for a thinking human being to feel a great satisfaction when the external world is ordered according to the requirements of our mind.

But one must confess that the general considerations of tensor algebra and even of thermodynamics have a rather restrictive than creative power, their heuristic value being not so great.

On the other hand the aim of mechanics is not only the writing of equations, but also the integration of them. And that is possible only when these equations are sufficiently simple. Mechanics of solids needs a number of non-complicated models describing with a certain grade of accuracy the behaviour of real materials under certain load and temperature conditions enabling the formulation of basic boundary value problems.

Therefore, besides the very general approach developed by a number of modern authors, there exist a number of more particular approaches, considering a strictly limited circle of phenomena. Sometimes tentatives of this kind are said to be "engineering", but I think that is not a correct word. They are "mechanical" as was the theory of elasticity at its early stage, as is up to the present day the theory of plasticity of metals at room temperature.

The difficulty we meet in constructing phenomenological schemes is not the lack of possibilities, but rather the abundance of them. We have at our disposal too great a variety of analytical means to describe the behaviour of different materials as intricate as it might be. The problem is to make a choice between them and to find the most simple and adequate.

Surely, the only one way to say whether the theory is true or wrong is to compare its predictions with the results of the experiment. But on this way we meet some serious obstacles. First, the number of possible experimental programs is restricted and often the crucial experiment which could in principle make the choice between controversial theories definite appears to be technically impossible. On the other hand when we investigate the plastic and creep properties of metals and polymers we meet a great amount of divergence of the characteristics between the individual specimens. Therefore the predictions of almost every theory usually lie inside the band of scatter of the experimental data. So, the interpretation of the experiment is always ambiguous, the predictions of very different theories can be considered as equally good if we require nothing but the first approximation or equally bad if we are searching for the absolute. Often we speak about the physical meaning of theory and appreciate as good the theories which are physically meaningful. But one must confess that this definition seems to be somewhat uncertain, the present state of the physics of solids does not permit an adequate description of the macroscopic behaviour of metals starting from the considerations of the elementary microscopic mechanisms. The use of physics for the development of mechanical theories lies rather in its qualitative aspects then in quantitative predictions. Of prime interest for mechanics is the question: upon what does this or that quantity depend, by how many parameters is it defined. The question about the form of this dependence and its analytical approximation can always be solved by direct experiment.

After these preliminary remarks let us consider our main subject, the creep of metals at elevated temperatures. A very general statement meeting no objections will sound as follows:

The creep strain rate depends only on the stress and temperature at a given structural state.

Now we have to decipher the notion of the structural state. The structural state can be identified by different physical means, by microphotographs, X-ray patterns, measurements of electric conductivity and so on. These means give no numerical information, but, if we have two structural states, state A and state B, we can always say, whether they are "identical" ($A = B$) or "different" ($A \neq B$). Sometimes we have at our disposal much more information and we can say that A is before B ($A < B$) or A is after B ($A > B$). Now we introduce the concept of the one parameter material. At first this concept will be purely imaginary, and then we shall try to find some examples when it is applicable to real phenomena. We suppose that the specimen is subject to a certain program of loading, reloading, heating and cooling. This program will le called *program A*. In the course of this program we can fixe the structural states, for example take microphotographs of the structure. So, we obtain a sequence of microphotographs $A_0, A_1, \ldots A_n \ldots (A_{n+1} > A_n)$. Now let us consider another program B and a corresponding sequence of structural states $B_0, B_1, \ldots B_m, \ldots$ We call the process to be one-parameter irreversible if for any B there is a one to one correspondence between the elements in B and in A such that every $B_m = A_n$ and $B_{m+1} > B_m$ if $A_{n+1} > A_n$. The sequence A will be denoted as complete if any structural state possible for the material in processes under consideration belongs to this sequence.

Now one can take as a structural parameter defining the structural state the ordinal number n of the corresponding state in the reference sequence A, for example, or, naturally, an arbitrary monotonic function $q(n)$ of this number.

In a similar way one can define a two parameter, a three parameter and so on material. The problem is to learn to separate in an appropriate manner the corresponding subsequences of the structural states. Because the number of the experimental programs is finite, we can always characterize the structural state within any desired degree of accuracy with the aid of a finite number of parameters of state. The correct choice of these parameters permits to reduce their number.

We consider now the very simple problem of one-dimensional creep, say in tension. Supposing the process to be one-parameter, we put

$$\dot{e} = f(\sigma, T, q) \tag{1}$$

A very simple assumption will be to take q equal to the actual creep deformation or some function of it, so, we obtain the well known equation of state or the equation of the strain hardening theory. We write eq. (1) as follows:

$$\dot{p} = f(\sigma, T, p) \tag{2}$$

In contrast with (1) we introduce here not the total deformation but the creep deformation p, the instantaneous plastic deformation being subtracted. There is some physical and experimental evidence for it. Though it is always a result of slip in certain cristallographic planes, in the initial stage of the instantaneous plastic deformation it is always a coarse slip, localized in few atomic planes, the most part of the volume of the metal being not plastically deformed. On the other hand, the creep deformation is a result of fine slip equally distributed in the volume of the body. So, the instantaneous deformation in its earlier stage provides a very strong strain hardening in a very small part of the deformed volume which has practically no influence on the subsequent creep [1].

Therefore we arrive in a very natural way at the following decomposition of the total strain

$$e = e_e + e_0 + p$$

where the first term is the elastic strain, the second term is plastic strain and the third is due to creep. This idea can be developed somewhat farther. In some cases the first stage creep described by eq. (1) is followed by the secondary creep, when the creep rate becomes constant. The description of this transition from the first to the second stage of creep can be different. In the first case we assume that the strain hardening reaches its limit value, that is when p tends to infinity, the creep rate $\dot{p}$ tends to a certain value depending only on σ and T. The mechanism of creep in the first and second stage of creep is supposed to be the same. On the other hand we can assume that the creep deformation consists of two parts, transient creep deformation p, and steady creep deformation p_2, the first being governed by the equation of type (2) but giving zero strain rate when p_1 tends to infinity. The rate of the second part of the creep strain p_2 depends only on stress and temperature

$$\dot{p}_2 = v(\sigma, T).$$

Both points of view are self-consistent and non-contradictory, the difference is too small to make a direct experimental proof possible. But the mathematical formulation of the creep problem in these cases is quite different.

The greatest part of the creep deformation is irreversible, especially when the level of temperatures is high enough. Nevertheless creep recovery can play some role in the creep process even for metals at moderate temperatures; for high-polymers the recoverable part of the strain is predominant. To describe both the recoverable and the irreversible deformation of plastics BRYSGALIN [2], following the same course of reasoning split the creep deformation p into two parts, $p = p_1 + p_3$ where p_1, as formerly, is governed by the equation of the type (2) and p_3 is connected with the stress by the integral equation. As a first approximation we can take this equation to be linear, so we shall consider creep recovery as elastic aftereffect. BRYSGALIN found a suitable form of nonlinear hereditary equation for p_3 and used it to describe the behaviour of celluloid under alternating cycles of loading and unloading. The basic assumption of the foregoing reasoning is that mechanisms of recoverable and non-recoverable creep are different.

We return now to the one-parameter eq. (1) and try to make more precise the definition of the parameter q. It is easy to understand that even if we suppose the structural state to be fully determined by only one parameter, the identification of this parameter with the creep strain will be an oversimplification. Generally speaking we must admit that the rate of change of this parameter is governed by another kinetic equation. Really the experimental verification of the usual strain hardening rule showed the systematic deviation from the predictions of the theory. The papers [3], [4], [5] and many others should be mentioned. It is easy to show that, when taking for instance $dq = S(\sigma)\, dp$ we can fit the experimental data for the case of a stepwise changing load very well. The case $dq = \sigma\, dp$ was considered in more detail in [6], [7]. We obtain a more precise description of creep under changing loads if we take thermal softening into account and assume that

$$dq = S\, dp - \mu\, dt. \tag{3}$$

The simplest program to find the function μ is the following. The specimen is held a certain time under the stress σ, then it is unloaded. During the period of unloading $dp = 0$ and therefore the value of q diminishes according to the equation $dq = -\mu\, dt$. Therefore, after reloading of the specimen with stress of the same magnitude the creep rate shall be greater than immediately before unloading. A somewhat similar method was used by Japanese authors [5], as can be understood from their publications in English. The parameter we have considered may be called the strain hardening parameter even though we take thermal softening into account. As to the physical meaning of this parameter, it might be associated with the dislocation density. In some physical theories of creep, this density really enters in the constitutive equations like a parameter of state. But the general definition of the state parameter given above does not depend on the supposed mechanism of deformation, as we shall see later. The formalism may be the same even when the physical character of the process is profoundly changed.

Now we consider the third stage of creep, preceeding the rupture. At this stage another parameter appears, the parameter of embrittlement. It is well known that creep rupture is a result of the development of intercristalline cracking. At high temperatures the mechanism of the formation of cracks is well known — they appear as a result of diffusion of vacancies towards the grain boundaries. These cracks become visible in the third stage, the relative number of cracks must enter the kinetic equation of creep as a structural parameter defining the acceleration of creep before rupture. Some authors [8] believe that the formation of cracks begins at the very early stage of the creep process, the resulting weakening of the material compensating the strain hardening so that the creep rate remains constant. There is no direct experimental evidence for this two-parameter explanation of the second stage of creep, but for some materials it might be possible.

We introduce the parameter of embrittlement ω, we suppose that $\omega = 0$ in the virgin state and that rupture occurs when $\omega = 1$. It will be interesting to find such conditions when only one parameter ω enters in the equations. When temperature and stress level are high enough so that creep deformation becomes considerable within a few seconds or minutes, the creep curves for the most part of materials have no first stage, for some time the creep rate remains constant and then begins to increase.

The idea of separation of the instantaneous part of the irreversible deformation leads to the following equation, describing the early stage of the creep process:

$$\dot{p} = v(\sigma, T) \tag{4}$$

Some special experiments were undertaken to prove the non-dependence of the function v upon the previous instaneous and creep deformation (of course, within a certain limited range of strain values, say $1-2\%$) [9]. This function being known, we can take a stress-strain curve at some given stress rate and calculate for any point of this curve the corresponding creep deformations. Subtracting these calculated values from the abscissae of the experimental curve, we obtain the instantaneous stress-strain diagram.

Equation (4) describes only the initial parts of the creep curves. Taking crack formation into account we must improve this equation, putting

$$\dot{p} = f(\sigma, T, \omega), \qquad \dot{\omega} = g(\sigma, T, \omega) \tag{5}$$

The uniform elongation before rupture for short-time creep conditions is found to be approximately constant, therefore one can suppose that $p = p_* \omega$ and (5) turns into

$$\dot{p} = f(\sigma, T, p) \tag{6}$$

the value of $\dot{p}$ tending to infinity when p reaches the value p_*. If $\dot{p}$ is supposed to be a power function of the stress the same value of the exponent will determine the dependence of the creep rate on stress for the initial parts of the creep curves and the dependence of the time to rupture on stress. This result is very well confirmed experimentally. Having a stress-time-to-rupture curve we find the value of p_* and now we can calculate the shape of the creep curve including the accelerating part [9].

The idea that the parameter of the embrittlement is a monotonic function of the creep strain, that is, creep strain can be taken as a parameter of the embrittlement might seem to be oversimplified. Its real meaning is that we suppose eq. (2) to be valid not only in the first stage of creep, but in the second and even in the third, though the physical mechanism of the influence of the accumulated creep strain on the creep rate is not the same. Nevertheless, if we take into account the rather great divergence in the properties of individual specimens, we shall admit that this point of view can be valid as a first approximation.

The unified theory of creep and creep rupture, considering the formation of cracks at any stage of the creep process and the influence of these cracks on the creep rate has a certain practical interest. The acceleration of creep before rupture in hyperstatic systems results in the stress redistribution and therefore the value of the stress concentration factor at the moment of the formation of macroscopic crack is less than this value calculated for steady state creep conditions.

To make the theory more precise we must admit that different kinds of cracks may exist, not only the intergranular round-edge diffusion cracks, but also the sharp cracks due to the stress concentration on the boundaries. The more exact kinetic equations must take into account not only the growth of cracks, but also the closing of them.

The aim of this paper is not to give any accomplished theory or a set of equations of a sufficiently general character, but rather to point out the difficulties which may be encountered in the interpretation of the experimental data and the ambiguity of this interpretation. The possibilities of constructing models, non-violating the first and the second laws of thermodynamics, are too large, we have too many degrees of freedom. In mechanics we feel comfortable when we deal with a well established system of equations admitting the formulation of certain boundary value problems and solution of these problems by certain mathematical means. One must not forget that every small improvement of the constitutive equations leads to great mathematical difficulties.

Here, to make the basic idea more clear, we did not touch upon the problems of combined stress. There the uncertainty of the choice of defining parameters becomes much larger. Setting apart the difficulties connected with the instantaneous plastic deformation we must construct a reasonable generalization of eq. (1) supposing that the creep potential exists and that the strain hardening is isotropic and defined by a single scalar parameter which may be defined in different ways. Direct experiments show strain hardening to be essentialy anisotropic [10]. The existence of the creep potential for the non-steady phase of creep seems to be doubtful [11].

The situation in the theory of creep, as also in the theory of plasticity, can be estimated from the pessimistic and from the optimistic points of view. Pessimistic, because despite much work already done we are far away from knowing the absolute truth. Optimistic, because we have at our disposal a number of reasonable models, each of them having a certain well defined field of application.

References

[1] RABOTNOV, YU. N.: IXth International Congress on Theoretical and Applied Mechanics. Brussels: 1956.
[2] BRYZGALIN, G. I.: J. Prikl. Mech. Tech. Fisiki **1962** (2). [In Russian.]
[3] ODQVIST, F.: Symposium su la plasticitá nella scienza dalle costruzioni. 1956.
[4] NAMESTNIKOV, V. S., and A. A. CHVOSTUNKOV: J. Prikl. Mech. Tech. Fisiki **1960** (4). [In Russian.]
[5] TAIRA, S.: Creep in Structures. Berlin-Göttingen-Heidelberg: Springer. 1962.
[6] RABOTNOV, YU. N.: Progress in Applied Mechanics. New York: 1963.
[7] RABOTNOV, YU. N.: In: Proceedings of the Joint International Conference on Creep. 1963.
[8] ZHURKOV, S. N., and T. N. SAPFIROVA: J. Techn. Fisiki **28** (8), 1958. [In Russian.]
[9] RABOTNOV, YU. N.: J. Prikl. Mech. Techn. Fisiki **1965** (1). [In Russian.]
[10] NAMESTNIKOV. V. S.: Izvestia AN SSSR, OTN **1957** (10). [In Russian.]
[11] NAMESTNIKOV, V. S.: In: Proceedings of the Joint International Conference on Creep. 1963.

Discussion

ODQVIST: When you talk about the embrittlement, should the quantity ω be identified with $1 - \psi$ in the Kachanov theory or is it something still more general?

RABOTNOV: In a particular case you can identify this parameter ω with the relative area of the cross-section. But I don't want to do it. I prefer to introduce this parameter in a purely formal way. The difference between my theory and the theory of KACHANOV is that KACHANOV quotes that the creep-rate depends on the stress only. He does not take in account the influence of the embrittlement parameter on the creep. So he cannot describe the third stage of the creep-curve.

KESTIN: I think that one of the important outcomes of a conference like this is to recognize that very frequently we use the same words to mean different things or we use different words to mean the same thing. When you spoke of an equation of state, I became uncomfortable, because I do not think that you mean an equation of state here. I also have had in the past great difficulties in understanding what people meant by a constitutive equation. So I think that when the theory of creep is made to include the structural parameter, it should be recognized that what you called an equation of state, or what other people might call a constitutive equation, consists of two parts: one part which is the result of experiments on equilibrium states — and I would like to reserve the name of equation of state for this part — and a second part which consists of the results of experiments on essentially irreversible processes, that is on processes which express the interaction of elementary sub-systems within the system. In other words we should ask two types of question: (1) how does the material behave when it is always in equilibrium or when it performs only a reversible process, and (2) how does it behave when it performs an essentially irreversible process.

RABOTNOV: I did not invent the name "equations of state". I prefer terms like constitutive equations, kinetic equations. Equation of state is a term existing in the creep-theory for about 40 years, so I used it.

ONAT: The notions of state and state variables that you are speaking of here come close to the notion of state and state variables that I talked about in my lecture. There is one difference. You would like to decide about the state and the variables which control the state by taking photographs, by looking at the material at the instant of interest with a microscope. Now, this is probably the best way, but there is another possibility. One could watch a number of test-results, inputs — outputs, and from these one could obtain information about the number of state variables that one ought to use, and in fact about their nature. In the study of systems there are two methods: one of them is to open the black box, to look inside, and the other one is to watch how the black box behaves from the point of view of inputs and outputs.

Rabotnov: I think that really when we formulate a certain form of the assumed constitutive equations we deal always with the data of the mechanical experiments, its inputs and outputs.

The physical considerations might be useful to suggest us the number and the character of the defining parameters, which are introduced in the theory. You cannot obtain the equations looking in the microscope.

Drucker: You describe the structure as a whole with elements which creep, redistribute the stress, and so change their creep-rates. Have you taken this concept and applied it to the picture you see in the microscope where you have a block of material composed of microstructural elements. Does it give you the same or a different set of stress-strain relations and structural behavior?

Rabotnov: I think that in this way you can have all the different grades of approximation: the first approximation, when we introduce one parameter only, the change of the mean lenght. You then cannot take into account the particularities of the substructure. But one can go infinitely far on this way. You can introduce as a structural parameter for instance the size of grains. At first these grains may remain unchanged in the process or they can split, they can grow or change their shape, some change can be inside these grains — it is a question of the number of defining parameters, I think.

Seth: In your splitting up the strain into an elastic strain, a plastic strain and a creep strain what type of measure for elastic strain do you use? Do you use Hencky measure or a normal measure?

Rabotnov: You can generalize this theory for all the different types of strain, but for small strains only. I am sure that the splitting of the strain in a plastic part and a creep part has physical meaning only while the strain is no more than 1—2%. When the strain is greater, then the interdependence becomes very complicated. I don't know what to do with large deformation.

Seth: No, it's not large deformation. Don't you think, when you go into the micro-structure you should have a measure of strain which should correspond to the micro-structure and not to the ordinary classical measure?

Rabotnov: No, I don't think that.

The Influence of Dissipated Stresswork on the Rupture of Materials

By

Markus Reiner
Haifa (Israel)

Summary

When the stress deviator in a body upon which external forces act exceeds a certain limit, the body looses its continuity and breaks. This phenomenon of rupture implies the appearance of new surfaces to the several parts into which the body is converted, and therefore the increase of surface energy. The source of such additional surface energy is in the stored potential energy which has been produced by the elastic part of the stresswork (w_{se}). *Therefore that part of the stresswork which in the course of deformation has been dissipated by either plastic (frictional) or viscous resistance (w_{sd}) does not contribute to rupture.* Dissipation of the stresswork accordingly increases the breaking strength of the material. If a material such as a Newtonian liquid has no formelasticity, it will be sheared indefinitely, but never break. (Rupture under isotropic tension will occur in every material, due to the fact that every material has volume-elasticity.) Actually, any real liquid possesses some formelasticity. This has been proved to exist in lubricating oils, and such oil can suffer rupture in a journal bearing.

Let w_s be the stresswork, F the free intrinsic energy, D the dissipated energy, all per unit volume, then from the first law of thermodynamics

$$\frac{d}{dt}\left(\frac{w_s}{\varrho}\right) + \frac{d}{dt}\left(\frac{F}{\varrho}\right) + \frac{d}{dt}\left(\frac{D}{\varrho}\right) = 0 \tag{1}$$

while from the second law of thermodynamics

$$\frac{d}{dt}\left(\frac{D}{\varrho}\right) \geqq 0. \tag{2}$$

With ϱ constant in a distortion, we write for (1)

$$\dot{w}_s + \dot{w}_{se} + \dot{w}_{sd} = 0. \tag{3}$$

On the basis of this development Reiner and Weissenberg have postulated that rupture depends upon a maximum value of the intrinsic free energy. Rupture will occur at the time t_r, when

$$w_{se} = \int_0^{t_r} (w_s - \dot{w}_{sd})\, dt = R \tag{4}$$

where R, the resilience of the material is a material parameter.

This irreversible aspect in continuum mechanics is accompanied by certain properties, the most important of which is the wellknown fact that the strength of metals increases with the rate of strain. This is examined in the present paper for the case of the standard linear solid.

Let a metal rod be extended at constant rate of strain a. Let θ be the strain and ϑ the stress at which rupture occurs. The Reiner-Weissenberg criterion then yields that at infinitely slow rate of extension ($a = 0$), when there is no dissipation of energy, the rupture stress is $\vartheta_0 = \sqrt{2R\mu_H}$, where μ_H is the rigidity of the Hooke-component of the standard linear solid. When the rate of strain approaches infinity, the dissipated stress power is $\dot{D} = \frac{2R}{\eta} \frac{\mu_M^2}{\mu_H + \nu_M}$, where μ_M is the rigidity of the Maxwell component and η is its viscosity. The rupture stress in this case is $\vartheta_\infty = \sqrt{6R(\mu_H + \mu_M)}$. Therefore $\vartheta_\infty/\vartheta_0 = (1 + \mu_M/\mu_H)^{1/2}$ is the maximum ratio of the increased strength. In accordance with this theory the rupture stress increases to a maximum when the dissipated stress-power has increased to a maximum. This result is qualitatively in accordance with experimental findings. An experimental investigation for a quantitative examination is in hand.

1. When a material body is loaded with some forces, a strain (ε_{ij}) is produced accompanied by internal stresses (s_{ij}). Both strain and stress can be resolved into isotropic components ($\varepsilon_{\alpha\alpha}$ and $s_{\alpha\alpha}$) and deviatoric components ($\mathring{\varepsilon}_{ij}$ and $\mathring{s}_{ij}$). It may be assumed that under an isotropic stress a metal reacts simply in accordance with Hooke's law. If the isotropic stress is a tension, this may lead to rupture, but we shall not deal here with this kind of rupture, to which a fluid also is liable. Under a constant stress *deviator* ($\mathring{s}_{ij}$), the deformation either reaches a certain limit, when the material is solid, or it continues indefinitely, when the material is said to be fluid. In both cases there will, in general, be a certain limit (ϑ_{ij}) to the stress deviator when the body looses its continuity and *breaks*. This phenomenon of rupture implies the appearance of new surfaces to the several parts into which the one body is converted, and therefore the increase of *surface energy*. The source of such additional surface energy is in the *stored potential energy* which has been produced by the *elastic part of the stresswork* (w_{se}). Therefore that part of the stresswork which in the course of deformation has been dissipated by either plastic (frictional) or viscous resistance (w_{sd}) does not contribute to rupture. *Dissipation of the stresswork accordingly increases the breaking strength of the material.* If a material such as a Newtonian liquid has no formelasticity, it will be sheared indefinitely, but never break. Actually, any real liquid possesses some formelasticity. This has been proved to exist in lubricating oils [1], and such oil can suffer rupture in a journal bearing [2].

This irreversible aspect in continuum mechanics is accompanied by certain properties, the most important of which is the wellknown fact that the *strength of metals increases with the rate of strain*. This will be examined in detail in the present paper.

2. Let w_s be the *stresswork*, i. e. the work which the stresses perform in the deformation, F the *free intrinsic energy* and D the *dissipated energy* (all per unit volume); then the first law of thermodynamics requires

$$\frac{d}{dt}\left(\frac{w_s}{\varrho}\right) + \frac{d}{dt}\left(\frac{F}{\varrho}\right) + \frac{d}{dt}\left(\frac{D}{\varrho}\right) = 0 \tag{2.1}$$

where ϱ is the density and the free intrinsic energy is the total free energy minus the kinetic energy. The density must be introduced because the first law does not refer to unit volume, but to unit mass.

Now consider isothermal processes only, i. e. the heat into which the dissipated energy is converted is conducted away without delay. Then in accordance with the second law of thermodynamics, we have

$$\frac{d}{dt}\left(\frac{D}{\varrho}\right) \geqq 0 \tag{2.2}$$

where the sign $=$ stands for reversible, and the sign $>$ for irreversible processes.

From the fact that $d(D/\varrho)/dt$ vanishes for all reversible processes, it follows that the free intrinsic energy is equivalent to a generalized potential φ, which is the elastic part of the stresswork w_{se}. If we do not take into account the volumetric work, and the quantities in eqs. (2.1) and (2.2) refer to distortional work only, ϱ can be taken as constant and cancels out. Using Newton's dot for differentiation in respect of time, we accordingly write for (2.1)

$$\dot{w}_s + \dot{w}_{se} + \dot{w}_{sd} = 0 \tag{2.3}$$

where it is understood that w_{sd} can only be either positive or zero.

The total distortional work can be expressed in terms of stress and velocity of strain

$$d w_s = - s \dot{\varepsilon}\, dt. \tag{2.4}$$

The negative sign is used because w_s is imposed by external forces. From (2.4) we obtain for the distortional power

$$\dot{w}_s = - s \dot{\varepsilon}_2. \tag{2.5}$$

On the basis of this development Reiner and Weissenberg [3] have postulated that rupture depends upon a maximum value of the intrinsic free energy, i. e. the energy $\varphi = w_{se}$, which can be stored up elastically in the volume element of the material. Rupture will occur at the time t_r, when

$$w_{se} = \int_0^{t_r} (\dot{w}_s - \dot{w}_{sd})\, dt = R \tag{2.6}$$

where R, the *resilience* of the material is a material parameter. In an irreversible process where $\dot{D}$ is positive in accordance with (2.2) part of the power of distortion leaks out, and only part of it can be stored up. At certain states of stress and deformation, when $\dot{w}_s = \dot{w}_{sd}$ a body of material can be deformed without limit, as with the simple Newtonian liquid.

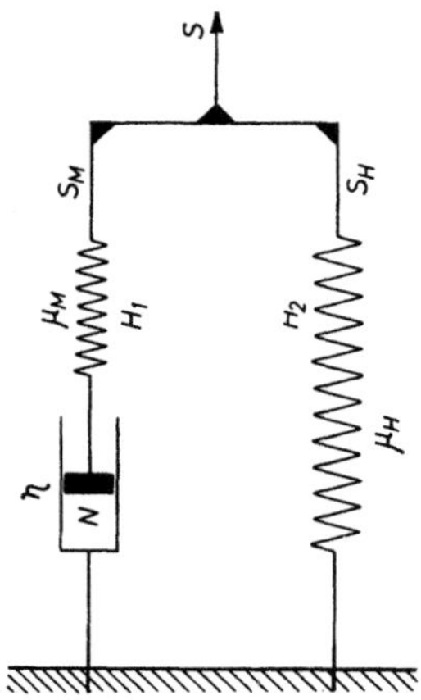

Fig. 1. Model of a standard linear solid

In order to evaluate eq. (2.6) for a special case, the following have to be known: (*i*) the rheological equation of state of the material, and (*ii*) the mechanical conditions of the tests. As an example we treat here *simple tension* at constant rate of deformation of a *standard linear solid.*

3. As shown by Zener [4], the rheological properties of metals are best represented by a model consisting of an elastic Hookean spring in parallel with a Maxwell element, the latter consisting of a spring and dashpot in series. This arrangement is shown in Fig. 1, originally proposed by Poynting and Thomson [5]. In every deformation there will be elastic responses of the two springs with stored potential energy and irreversible dissipation of energy in the dashpot. As said before, the model should be related to the *deviators of* stress and strain.

In the case of *simple tension*

$$s_{ij} = \sigma \begin{Vmatrix} 1 & 0 & 0 \\ 0 & 0 & 0 \\ 0 & 0 & 0 \end{Vmatrix}; \quad s_{\alpha\alpha} = \sigma \tag{3.1}$$

$$\varepsilon_{ij} = \varepsilon_l \begin{Vmatrix} 1 & 0 & 0 \\ 0 & -\nu & 0 \\ 0 & 0 & -\nu \end{Vmatrix}; \quad \varepsilon_{\alpha\alpha} = (1 - 2\nu)\, \varepsilon_l \tag{3.2}$$

where ν is Poisson's ratio, we have for the deviators

$$\overset{\circ}{s}_{ij} = s_{ij} - \frac{s_{\alpha\alpha}}{3}\,\delta_{ij} = \frac{\sigma}{3}\begin{Vmatrix} 2 & 0 & 0 \\ 0 & -1 & 0 \\ 0 & 0 & -1 \end{Vmatrix} \tag{3.3}$$

$$\overset{\circ}{\varepsilon}_{ij} = \varepsilon_{ij} - \frac{\varepsilon_{\alpha\alpha}}{3}\,\delta_{ij} = \frac{1+\nu}{3}\,\varepsilon_l\begin{Vmatrix} 2 & 0 & 0 \\ 0 & -1 & 0 \\ 0 & 0 & -1 \end{Vmatrix} \tag{3.4}$$

We start at $t = 0$ from a stressless and strainless state.

At any time t, both the Hooke and the Maxwell components proceed with the same rate of strain

$$\dot{\overset{\circ}{\varepsilon}}_H = \dot{\overset{\circ}{\varepsilon}}_M = \dot{\overset{\circ}{\varepsilon}} \tag{3.5}$$

while the stresses taken by the two elements are additive

$$\overset{\circ}{s} = \overset{\circ}{s}_H + \overset{\circ}{s}_M\,. \tag{3.6}$$

We now have from the rheological equation of the Hooke-body

$$\overset{\circ}{s}_H = 2\mu_H\,\overset{\circ}{\varepsilon}_H\,. \tag{3.7}$$

For the Maxwell element we draw upon REINER [6, p. 465] with

$$\overset{\circ}{s}_M = e^{-\frac{\mu_M}{\eta}t}\,2\mu_M\int\limits_0^t \dot{\overset{\circ}{\varepsilon}}\,e^{\frac{\mu_M}{\eta}t}\,dt \tag{3.8}$$

and therefore considering (3.3) and (3.4)

$$\overset{\circ}{s}_{11} = 2\sigma/3 = 4\mu_H\,\frac{1+\nu}{3}\,\varepsilon_{l,H} + 4\mu_M e^{-\frac{\mu_M}{\eta}t}\,\frac{1+\nu}{3}\int\limits_0^t \dot{\varepsilon}_l\,e^{\frac{\mu_M}{\eta}t}\,dt \tag{3.9}$$

from which

$$\frac{\sigma}{2(1+\nu)} = \mu_H\,\varepsilon_{l,H} + \mu_M e^{-\frac{\mu_M}{\eta}t}\int\limits_0^t \dot{\varepsilon}_l\,e^{\frac{\mu_M}{\eta}t}\,dt. \tag{3.10}$$

The stresswork expended upon the material is the sum of the stress-works expended upon its components, or

$$\overset{\circ}{w} = \overset{\circ}{w}_H + \overset{\circ}{w}_M. \tag{3.11}$$

For the H-part we have

$$\overset{\circ}{w}_H = \mu_H\,\overset{\circ}{\varepsilon}{}^2 \text{ (summed over nine terms)}. \tag{3.12}$$

This is entirely conserved. In contradistinction, the stress power of the M-component

$$\overset{\circ}{\dot{w}}_M = \overset{\circ}{s}_M \overset{\circ}{\dot{\varepsilon}} = \frac{\overset{\circ}{s}_M \overset{\circ}{\dot{s}}_M}{2\mu_M} + \frac{\overset{\circ}{s}{}^2_M}{2\eta} \text{ (summed over nine terms)} \tag{3.13}$$

is partly conserved and partly dissipated through viscous resistance [7]. In accordance with our theory only the first term of the right side of (3.13) has to be taken into account.

The total conserved stresswork is therefore

$$\overset{\circ}{w}_c = \overset{\circ}{s}{}^2_H / 4\mu_H + \overset{\circ}{s}{}^2_M / 4\mu_M . \tag{3.14}$$

Introducing $\overset{\circ}{s}_H$ from (3.7) and $\overset{\circ}{s}_M$ from (3.6), this is also

$$\overset{\circ}{w}_c = \mu_H \overset{\circ}{\varepsilon}{}^2 + (\overset{\circ}{s} - 2\mu_H \overset{\circ}{\varepsilon})^2 / 4\mu_M . \tag{3.15}$$

The dissipated power is from the second term of the right side of (3.13)

$$\overset{\circ}{w}_{sd} = \overset{\circ}{s}_M / 2\eta = (\overset{\circ}{s} - 2\mu_H \overset{\circ}{\varepsilon})^2 / 2\eta . \tag{3.16}$$

Let θ be the *rupture deformation* and ϑ the *rupture stress*, then rupture takes place in accordance with the Reiner-Weissenberg theory, when

$$\mu_H \overset{\circ}{\theta}{}^2 + (\overset{\circ}{\vartheta} - 2\mu_H \overset{\circ}{\theta})^2 / 4\mu_M = R . \tag{3.17}$$

Considering (3.3) and (3.4), this becomes, after appropriate summation, as the rupture criterion for simple tension

$$\mu_H \left(\frac{1+\nu}{3}\right)^2 \theta_l^2 \,(4+1+1) + \left(\frac{\vartheta_l}{3} - 2\mu_H \frac{1+\nu}{3}\,\theta_l\right)^2 (4+1+1)/4\mu_M = R$$

or (3.18)

$$4\mu_M \mu_H \theta_l^2 + \left(\frac{\vartheta_l}{1+\nu} - 2\mu_H \theta_l\right)^2 = \frac{6\mu_M}{(1+\nu)^2} R . \tag{3.19}$$

The dissipated power of rupture is from (3.15) after summation

$$\dot{D} = [\vartheta_l - 2\mu_H (1+\nu)\,\theta_l]^2 / 3\eta \tag{3.20}$$

4. We now carry out an experiment when the rate of strain is constant,

$$\dot{\varepsilon}_{l,H} = a \tag{4.1}$$

or

$$\varepsilon_{l,H} = a t . \tag{4.2}$$

Equation (3.10) now yields

$$\frac{\sigma}{2(1+\nu)a} = \mu_H t + \eta \left(1 - e^{-\frac{\mu_M}{\eta} t}\right) \tag{4.3}$$

and

$$\frac{\dot{\sigma}}{2(1+\nu)a} = \mu_H + \mu_M e^{-\frac{\mu_M}{\eta} t}. \tag{4.4}$$

Equations (4.3) and (4.4) are represented graphically in Fig. 2.

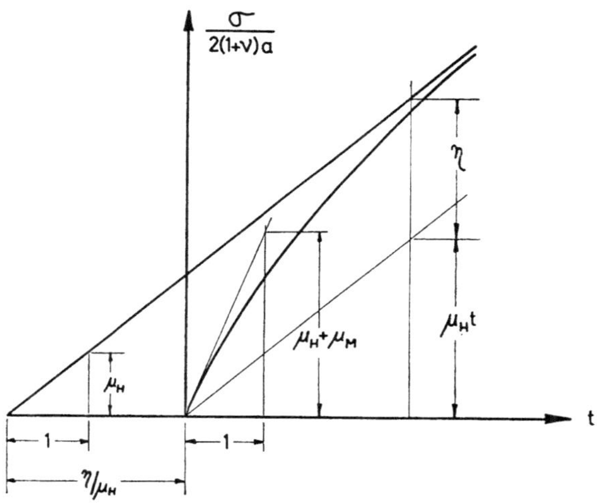

Fig. 2. Stress-time curve for constant rate of strain

Let τ be the time when rupture takes place, then the deformation at rupture is

$$\theta_l = a\tau \tag{4.5}$$

and the rupture stress is found from (4.3) as

$$\vartheta_l = 2(1+\nu)\, a \left[\mu_H \tau + \eta \left(1 - e^{-\frac{\mu_M}{\eta} t}\right)\right]. \tag{4.6}$$

Now we introduce

$$m = \tau \mu_M / \eta = \tau / \mathrm{T}_{\mathrm{rel}} \tag{4.7}$$

where $\mathrm{T}_{\mathrm{rel}}$ is the time of relaxation.

The yield criterion (3.19) then gives

$$a = \frac{1}{2(1+\nu)\,\mathrm{T}_{\mathrm{rel}}} \sqrt{\frac{6R}{\mu_H m^2 + \mu_M (1 - e^{-m})^2}}. \tag{4.8}$$

With this expression for a, eqs. (4.5) and (4.6) yield θ_l and ϑ_l as functions of m and the parameters of the material as follows:

$$\theta_l = \frac{m}{2(1+\nu)} \frac{\sqrt{6R}}{[\mu_H m^2 + \mu_M (1-e^{-m})^2]^{1/2}} \tag{4.9}$$

$$\vartheta_l = \frac{[\mu_H m + \mu_M (1-e^{-m})] \sqrt{6R}}{[\mu_H m^2 + \mu_M (1-e^{-m})^2]^{1/2}}. \tag{4.10}$$

Furthermore

$$\left.\begin{aligned} \frac{d\theta}{da} &= \frac{d\theta}{dm}\frac{dm}{da} \\ \frac{d\vartheta}{da} &= \frac{d\vartheta}{dm}\frac{dm}{da} \end{aligned}\right\} \tag{4.11}$$

With

$$\left.\begin{aligned} \frac{d\theta}{dm} &= \frac{\sqrt{6R}}{2(1+\nu)} \frac{\mu_M (1-e^{-m})(1-e^{-m}-me^{-m})}{[\mu_H m^2 + \mu_m (1-e^{-m})^2]^{3/2}} \\ \frac{d\vartheta}{dm} &= \sqrt{6R}\, \frac{\mu_H \mu_M [e^{-2m}(1+m) + e^{-m}(m^2-2) - m + 1]}{[\mu_H m^2 + \mu_M (1-e^{-m})^2]^{3/2}} \\ \frac{dm}{da} &= \left(\frac{da}{dm}\right)^{-1} = -\frac{2(1+\nu)\,\mathrm{T_{rel}}}{\sqrt{6R}} \frac{[\mu_H m^2 + \mu_M (1-e^{-m})^2]^{3/2}}{\mu_H m + \mu_M (1-e^{-m}) e^{-m}} \end{aligned}\right\} \tag{4.12}$$

we find

$$\left.\begin{aligned} \frac{d\theta}{da} &= -\frac{\mathrm{T_{rel}}\, \mu_M (1-e^{-m})(1-e^{m}-me^{-m})}{\mu_H m + \mu_M (1-e^{-m}) e^{-m}} \\ \frac{d\vartheta}{da} &= -2(1+\nu)\,\mathrm{T_{rel}} \frac{\mu_H \mu_M [e^{-2m}(1+m) + e^{-m}(m^2-2) - m + 1]}{\mu_H m + \mu_M (1-e^{-m}) e^{-m}} \end{aligned}\right\} \tag{4.13}$$

5. We now examine three special cases, two extreme and one intermediate.

(i) We assume $m = 0$. Then from (4.8) a $\rightarrow \infty$ and

$$\left.\begin{aligned} \theta_\infty &= \frac{\sqrt{6R}}{2(1+\nu)} \left[\mu_H + \mu_M \left(\frac{1-e^{-m}}{m}\right)^2\right]^{-1/2} = \frac{\sqrt{6R}}{2(1+\nu)} \frac{1}{(\mu_H+\mu_M)^{1/2}} \\ \vartheta_\infty &= \sqrt{6R} \left[\mu_H + \mu_M \left(\frac{1-e^{-m}}{m}\right)\right] \left[\mu_H + \mu_M \left(\frac{1-e^{-m}}{m}\right)^2\right]^{-1/2} \\ &= \sqrt{6R(\mu_H+\mu_M)} \\ D_\infty &= \frac{2R}{\eta} \frac{\mu_M{}^2}{\mu_H+\mu_M} \\ \left.\frac{d\theta}{da}\right|_\infty &= 0; \quad \left.\frac{d\vartheta}{da}\right|_\infty = 0 \end{aligned}\right\} \tag{5.1}$$

(*ii*) Assume $m \to \infty$, $a \to 0$

$$\left.\begin{aligned} \theta_0 &= \frac{1}{2(1+\nu)}\sqrt{\frac{6R}{\mu_H}} \\ \vartheta_0 &= \sqrt{6R\mu_H}; \quad \dot{D}_0 = 0; \\ \left.\frac{d\theta}{da}\right|_0 &= 0 \\ \left.\frac{d\vartheta}{da}\right|_0 &= 2(1+\nu)\,\mu_M\,\mathrm{T}_{\mathrm{rel}} = 2(1+\nu)\,\eta \end{aligned}\right\} \tag{5.2}$$

(*iii*) Assume $m = 1, \quad a = \dfrac{1}{2(1+\nu)\,\mathrm{T}_{\mathrm{rel}}}\,\dfrac{\sqrt{6R}}{(\mu_H + 0.37\mu_M)^{1/2}}$ (5.3)

$$\left.\begin{aligned} \theta &= a\,\mathrm{T}_{\mathrm{rel}} \\ \vartheta &= \frac{(\mu_H + 0.63\mu_M)\,\sqrt{6R}}{(\mu_H + 0.37\mu_M)^{1/2}} \\ \frac{d\theta}{da} &= -\frac{0.17\,\mathrm{T}_{\mathrm{rel}}\,\mu_M}{\mu_H + 0.23\,\mu_M} \\ \frac{d\vartheta}{da} &= 0.19\,(1+\nu)\,\mathrm{T}_{\mathrm{rel}}\,\frac{\mu_H\,\mu_M}{\mu_H + 0.23\mu_M} \end{aligned}\right\} \tag{5.4}$$

These relations are represented graphically in Fig. 3.

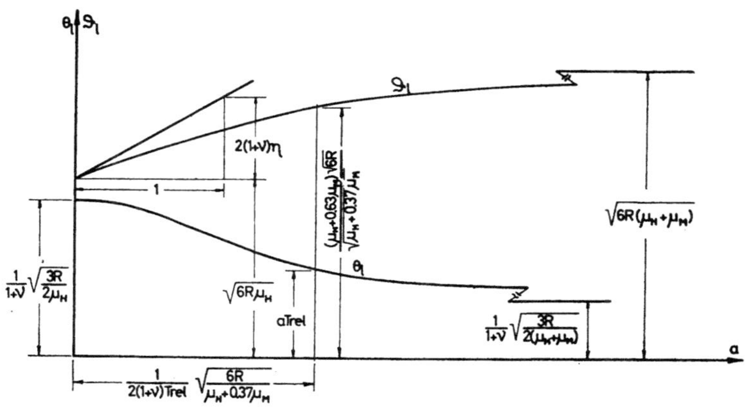

Fig. 3. Yield stress and yield strain as depending upon rate of strain

6. We see that in accordance with the Reiner-Weissenberg theory the yield stress ϑ_l, which is equal to $\sqrt{6R\mu_H}$ at infinitesimal slow rate of strain, increases to $\sqrt{6R(\mu_H + \mu_M)}$ at infinitely high rates, the ratio of

both being $\left(1+\frac{\mu_M}{\mu_H}\right)^{1/2}$. Under the same conditions the yield strain θ_l is reduced from $\frac{1}{1+\nu}\sqrt{\frac{3R}{2\mu_H}}$ to $\frac{1}{1+\nu}\sqrt{\frac{3R}{2(\mu_H+\mu_M)}}$. These extreme values are independent of the viscosity. However, viscosity determines the *rate* at which ϑ increases from ϑ_0 to ϑ_∞. In any case ϑ increases with increasing rate of strain. This is in accordance with experimental findings.

Qualitatively the result that the yield stress increases with the rate of strain, confirms that the greater the dissipated stresswork, the greater the strength of the material.

References

[1] Barlow, A. J., and J. Lamb: Proc. Roy. Soc. (London) **A-253**, 52–69 (1965).
[2] Reiner, M.: Z. oesterr. Ing. Arch. Ver. **63**, 803 (1911).
[3] Reiner, M., and K. Weissenberg: Rheol. Leaflet **10**, 12 (1939).
[4] Zener, C. M.: Elasticity and Anelasticity of Metals. Chicago: 1948.
[5] Poynting, J. H., and J. H. Thomson: Properties of Matter. London: 1929.
[6] Reiner, M.: In: Encyclopedia of Physics, ed. by S. Flügge, Vol. 6. Berlin-Göttingen-Heidelberg: Springer. 1959.
[7] Reiner, M.: ibid. p. 522.

Notations

a	rate of extension	$\dot{\varepsilon}$	rate of strain
D	dissipated energy	ε_{ij}	strain tensor
e	basis of natural logarithms	$\varepsilon_{\alpha\alpha}$	trace of ε_{ij}
F	free intrinsic energy	$\mathring{\varepsilon}_{ij}$	diviator of strain tensor
H	Hooke material	η	coefficient of viscosity
i, j	tensor indices	θ_{ij}	tensor of limit of strain
l	longitudinal	ϑ_{ij}	tensor of limit of stress
M	Maxwell material	μ	shear modulus
m	non-dimensional factor	ν	Poisson's ratio
R	resilience	ϱ	density
s_{ij}	stress tensor	σ	tensile stress
$\mathring{s}_{ij}$	deviator of stress tensor	$\dot{\sigma}$	rate of stress
$s_{\alpha\alpha}$	trace of s_{ij}	φ	potential
t	time		
w_s	stresswork		
w_{sd}	dissipated part of stresswork		
w_{se}	elastic part of stresswork		

Discussion

Olszak: Professor Reiner introduces the concept of the stored energy as being responsible for the phenomena of rupture and yielding. This idea is well confirmed by some experimental evidence but I think that this theory can easily be generalized. To represent linear phenomena, we generally use Boltzmannian elements, i. e. linear springs (E) and linear dashpots (V). For post-linear effects we have to include at least one of the non-Boltzmannian elements, for instance a "slider" (limiting the stress) or a "regulator" (limiting the strain rate), etc. And now the question is, what is the criterion of passing from the linear (Boltzmannian) range to the non-linear range? (In what follows, this passage will be referred to as the "transition" phenomenon or the "critical" stage.) For, say, the elastic-plastic body we generally suppose that a limit value of the conserved energy is decisive for attaining this critical state. But we know bodies which, for instance, exhibit nearly no possibility of storing energy; for them we observe, during the deformation process, nearly exclusively energy dissipation. There is, of course, quite a class of bodies with intermediate rheological properties. So the most general criterion of attaining the transition stage would be represented by a function of both the conserved energy and the power of dissipation, $f(W_E, W_D) = 0$. The most simple function which can be imagined is the sum of stored energy and the dissipation power (dimensionally "reduced" in an appropriate way).

Everything depends on the "interior organization" of the body. We can imagine three different levels of organization:

Level of interior organization	(E)	(V)
III "total"	all together	
II "partial"	blocks	
I "individual"	springs	dashpots

In the first one, (I) we can consider that there is quite an independence of the constitutive elements: the (first) critical state is attained in the relatively "weakest" (individual) (a) elastic element or (b) viscous element. In the second, (II) all the springs form one block and all the dashpots form another one: the critical state is attained in the relatively "weaker" block of elements constituting (a) the elastic component (or phase) or (b) the viscous component (or phase). In the third case, (III) the critical state is reached in the whole model, considered to be the symbol of one viscoelastic unit.

The "lowest" type of structural organisation ("individual" failure) hardly seems to reflect real material properties. On the contrary, the concept of "partial" organization will propably find many adherents. However, "total" resistance against critical phenomena (defined in the above sense as the transition from the linear to the non-linear range) in adequately organized materials seems to correspond to their real behaviour.

The simplest criterion may be expressed in the form

$$\xi_E W_E + \xi_D W_D = \xi_E k_E^2 + \xi_D k_D^2 = k^2.$$

If we consider the stored energy to be responsible for attaining the transition stage, we may assume $\xi_D = 0$ and we find the criterion of Professor REINER. On the other hand, if we consider the other extreme case, e. i. the viscous component to be responsible for this transition, we may introduce $\xi_E = 0$ (like, e. g., for "abnormal" fluids). Also many intermediate cases are possible: for these we may assume $0 \leq \begin{Bmatrix} \xi_E \\ \xi_D \end{Bmatrix} \leq 1$.

So, I think, this kind of approach is able to explain various classes of phenomena known from experiments, also those treated by Professor REINER. (For details cf.: W. OLSZAK, Progress in Solid Mechanics, Macmillan 1963; Z. BYCHUWSKI and W. OLSZAK, Arch. Mech. Stos., 1967.)

REINER: This is to assume that the dissipated energy itself would contribute to rupture. That would contradict the fact that a broken body because of the increased surfaces must have an increased surface tension and must show an increase of energy. In accordance with my theory only the conserved energy causes rupture.

Variational Methods of Constructing Models of Continuous Media

By

L. I. Sedov
Moscow (U.S.S.R.)

At present in plasma mechanics, in mechanics of plastic solids, of polymers, in the continuum theory of dislocations, when taking account of electromagnetic properties and effects in various media, when taking account of quantum laws in macroscopic phenomena, and in many other cases, it is necessary to construct new models of continuous media.

The classical models of the perfect viscous fluid and of linear theory of elasticity are continuing to play an important role in mechanics but by their nature they obviously came to be insufficient.

As important trends in the theory of new models for liquid and gaseous media let us mention the complicated models required to investigate motions and various processes in plasma, averaged motions and processes in turbulent motions of various media, in the theory of motion of multicomponent mictures, fluids, gases in which phase and chemical transformations and interactions with deformable bodies are taken into consideration, in particular, in porous media. A significant importance is attaining the theory of constructing models of non-Newtonian fluids in connection with biological problems, with problems of chemical technology, with the theory of liquid helium at very low temperatures etc.

Lately, interesting theoretical models of fluids and solids (ferromagnets) have been considered, in which the small particles possess significant magnetic fields and mechanical angular momentum i. e., spins.

In recent years, a great deal of papers have been published which involve complicated equations of motion and state for physical media. However, in many cases such papers have a particular character, sometimes, they are connected essentially only with simplest particular problems and with empirical formal approach. Because of difficulties, in essence, a large number of these papers poorly or in no way coordinate with universal or specific physical approachs. At best, only certain general restrictions imposed by the second law of thermodynamics on formally "ad hoc" introduced relations are considered.

In the following I will dwell on the highly general variational method of constructing models of continuous media, on the method which is based on and uses fundamental physical ideas and the knowledge of modern theoretical constructions. I mean the rational treatments of the field theory and physical media in theoretical mechanics and in relativity theory, in thermodynamics and statistical physics. Undoubtedly, it is time to introduce in daily practice a unified treatment of mechanical and electromagnetic phenomena. The importance of this statement is also connected with the fact that all the essential internal microscopic interactions in physical media are of electromagnetic nature. Taking into consideration the universal properties of physical or geometrical invariantness, otherwise, generalized properties of symmetry, is also very useful and fruitful [2, 5, 12].

Lately, it became clear that not only in microscopic, but also in many macroscopic laws and effects, the quantum nature of interactions (which is not described in the limits of Newtonian mechanics) is essential.

If the influence of electromagnetic effects are essential, than it is necessary to introduce the characteristics of the electromagnetic field, to add the Maxwell equations and to take account of ponderomotoric interaction forces between the field and physical medium.

If one takes into account the ponderomotoric forces in the most simple form, than the straightforward generalization of the Euler or Navier-Stokes equations leads to magnetohydrodynamic equations, which in some important problems turn out to be sufficient. In the general investigation of plasma motions the magnetohydrodynamic equations are not sufficient, therefore, other, generally more complex models, are to be introduced [21].

Following the magnetohydrodynamics, as a next complication one often uses directly the kinetic theory. Kinetic theories are also theories of continuous media (it is worthwhile to note specially, that the considerations connected with the discrete atomic-molecular structure of bodies and underlying the derivation of the kinetic equations have only heuristic significance) in which the physically infinitely small elements of the continuous medium have an infinitely great number of degrees of freedom.

In the classical models and in magnetohydrodynamics the physically infinitely small particles have a finite number of degrees of freedom; only finite bodies have an infinite number of degrees of freedom.

The variational methods developed in the following are well suited for constructing models with an increased but finite number of degrees of freedom, models, which by their nature, are intermediate between the kinetic models with an infinite number and the classical models with a very small number of degrees of freedom of the physically infinitely small particles.

It is to be noted, that in particular cases, when the solution of very difficult problems for the kinetic equations are possible, as a rule, one uses approximate methods which are, in essence, equivalent to transition to simpler models with elements having finite degrees of freedom. In this connection the general direct physical theory of models, with small particles having a great but finite number of degrees of freedom, have also a physical meaning for kinetic theories and, in essence, is always sufficient, when the aforementioned approximate methods of solution of the kinetic equations prove to be satisfactory.

Parallel with such concepts as the strain tensor with components $\varepsilon_{ij} = \frac{1}{2} (\hat{g}_{ij} - \mathring{g}_{ij})$ (where $\hat{g}_{ij}$ are the components of the metric tensor in the co-moving system of coordinate and $\mathring{g}_{ij}$ in the "initial state"), the stress tensor with components p^{ij}, the density ϱ and the temperature T, for multicomponent media it is necessary to introduce the specific densities $\varrho_1, \varrho_2, \ldots$, in some cases of vital importance are the characteristics of the electromagnetic field, the components of the electrical and magnetic tensions E^i and H^i, the components of the vectors of the electric current j^i, of electrification P^i and magnetization M^i the macroscopic characteristics $\chi_1, \chi_2, \ldots$ of the internal moments and of the various properties and structures of the disposition of atoms, molecules or small particles, the characteristics of dislocations etc.

If x^i $(i = 1, 2, 3, 4)$ are the coordinates of the points of the medium in the system of coordinates connected with the observer (it may be an inertial system or ony other specially introduced system of coordinates) and ξ^i the coordinates in the co-moving system, than the equations of the law of motion

$$x^i = x^i(\xi^1, \xi^2, \xi^3, \xi^4), \qquad i = 1, 2, 3, 4 \tag{1}$$

may serve as a basis for introducing various functions and, in particular, quantities with tensor indexes of various order, in the system connected with the observer and in the co-moving system, as for example, the functions

$$x^i(\xi^1, \xi^2, \xi^3, \xi^4 - s) \qquad -\infty \leq s \leq 0$$

or the quantities

$$x^i_j = \frac{\partial x^i(\xi^m)}{\partial \xi^j}, \qquad x^i_{je} = \frac{\partial^2 x^i(\xi^m)}{\partial \xi^e \, \partial \xi^j}$$

etc, etc ... Individual particles are distinguished by fixing ξ^1, ξ^2, ξ^3.

For fixed ξ^1, ξ^2, ξ^3, varying the coordinate ξ^4 in the four-dimensional space (space-time), gives a world line for the given individual particle.

In order to describe body and surface interactions inside or on the boundary of the body one must use generalized concepts of energy fluxes and of stress tensors of a higher order.

For example, for the first law of thermodynamics for the small particle in the co-moving system of coordinates along a world line, one must use an equation of the following form

$$dU = dA^{(e)} + dQ^{(e)} + dQ^{**} \tag{2}$$

where U is the internal energy, $dA^{(e)}$ the elementary work done by the macroscopic usual external mass and surfaces forces, $dQ^{(e)}$ — the external heat input, and dQ^{**} the macroscopic, external to the particle, energy input other than heat, due to complicated microscopic properties of the medium and to internal interactions. The input dQ^{**} is not zero, in particular, owing to interactions of the given medium with external electromagnetic field. The presence of dQ^{**} may be conditioned by internal interactions in the body, due to stresses of a higher order, in particular, to the work done by external surface moments on the boundary of the particle[1].

Let us specially note, that just as for the equations of kinetic energy in theoretical mechanics, the equation for the heat input (2), as applied to more general mental processes (not along the world line) may involve additional non-zero terms (the work done by the reactions of constraints) in comparison with the real processes. In particular, the assumption that the space is euclidian or any other restrictions on the curvature of the space may be regarded as examples of constraints.

Lately, in physics of elementary particles (and in mechanics of continuous media in a number of papers) a very wide use has already been made of spinors as characteristic parameters which, as shown in the paper [8, 9], may be considered as an ensemble of tensors of various order, having special properties of symmetrie.

The characteristic feature of modern theories is the introduction as determining parameters, besides the aforementioned quantities, their derivatives of various order with respect to coordinates and time. For example, $\frac{\partial T}{\partial x^i}$; $\frac{\partial \varepsilon_{ij}}{\partial \xi^k}$, x^i_{kj}, etc.

The natural appearance among the determining parameters of the consecutive derivatives follows directly from the statistical kinetic theories. The success, in practical applications, of the theory of heredity also

[1] For more details on the energy input dQ^{**} see [20, 22].

may serve as an indication for the expediency of introducing consecutive derivatives with respect to time.

Nevertheless, let us mention, that the introduction of functional constraints in kinetic theories and in the theory of heredity at once greatly complicates the system of equations for processes of general form. When passing to this theories the intermediate cases between the classical models of continuous media and models of continuous media introduced in the kinetic theories and the theory of heredity are excluded. In the latter theories the relations with the usual thermodynamics, where small particles with a finite number of degrees of freedom are considered, become difficult.

Of importance is the problem of constructing models of continuous media with the explicit formulation of a minimum number of simple hypotheses which are related with the universal physical concepts and laws and with the specific properties of the bodies considered.

Practice in physics, in the theory of relativity and in the theory of conservative media when two forces or forces of a higher order are considered [1, 3, 4, 6, 7, 10, 11, 14, 16, 22, 23], shows that the application of the variational equations is useful and convenient.

The consecutive derivatives with respect to time being absent among the arguments of the varied functionals, in order to obtain a closed system of differential equations, parallel with the variational principles, one may use the equation of energy (2), just as it is done in the classical theory of elasticity [15, 17, 18, 19].

However, I have shown [20] that in the presence of consecutive derivatives of the determining quantities with respect to time in the arguments of the internal energy, the eq. (2) is insufficient to obtain a closed sytem of equations.

I have shown [1, 22, 23] that in the general case, by means of the generalized variational, generally speaking, non-holonomic equation, it is convenient to obtain a closed system of differential equations or finite equations including the equation of state and the kinetic equations. This same variational equation provides, in specific problems, the boundary and initial conditions. In particular, by this means we obtain also the conditions on the strong and weak discontinuities in essentially new and complex cases, when among the determining parameter we have consecutive derivatives and other characteristics of the additional internal degrees of freedom, subject to discontinuities. The problem of the limit and boundary conditions is very important, since the orders of the differential equations increase.

In the journal PPM a paper [13] will soon be published by M. LOURIÉ, dedicated to the derivation of the conditions on the shocks from the variational equations.

The methods we developed in order to obtain a system of equations and additional conditions from the variational equations, may be applied to the construction of models with both reversible and irreversible processes.

Let us take the basic equation in the form

$$\delta \int_V \Lambda_1 \, d\tau + \delta W + \delta W^* = 0, \tag{3}$$

where $d\tau$ is the element of the four-dimensional Riemannian space-time "volume" V limited by the three-dimensional surface $\sum$; ($d\tau = \sqrt{-g} \cdot \cdot dx^1 \, dx^2 \, dx^3 \, dx^4$, $g = |g_{ij}|$, where $ds^2 = g_{ij} \, dx^i \, dx^j$ is the square of the element of length), $\Lambda_1 = \Lambda(K^C(x^i), L^B(\xi^i), x^i_j, g_{ij}, \mu^A, \nabla_e \mu^A) + \frac{1}{2\varkappa} R$ — generalized Langrangian functions, $\varkappa$ — a scalar constant, K^C and L^B are given functions of then arguments-physical "constants", scalar or tensor complexes, $B = 1, 2, \ldots, N_B$, $C = 1, 2, \ldots, N_c$. The parameters μ^A ($A = 1, 2, \ldots N$) and their corresponding four-dimensional gradients $\nabla_e \mu^A$, form a system of determining parameters, R is the scalar curvature of the four-dimensional Riemannian space.

Further, we set

$$\delta W^* = -\int_V (Q_i \delta x^i + D^{ij} \delta g_{ij} + M_A \delta \mu^A) \, d\tau - \int_\Sigma (Q^j_i \delta x^i + M^j_A \delta \mu^A) \, n_j \, d\sigma \tag{4}$$

and after some transformations and considering (4) we obtain

$$\delta W = \int_\Sigma (\Theta^{ij} \delta \eta_i + P^j_i \delta x^i + \mathscr{M}^j_A \delta \mu^A + Q^{pqj} \delta g_{pq} + G^{pqrj} \nabla_p \delta g_{qr}) \, n_j \, d\sigma \tag{5}$$

where n_j are the components of the unit normal vector, external to $\sum$, and $d\sigma$ the three-dimensional element of the surface $\sum$[1].

The parameters μ^A ($A = 1, 2, \ldots, N$) may involve several complexes which are responsible, in general, for the spin tensors of different valency[2].

When formulating the eq. (3) and its subsequent consequences, one may use fixed systems of coordinates x_i and ξ_i. From the point of view of subsequent derivations the nature of the quantities Λ_1 and of the coeffi-

---

[1] The more complicated case, when among the arguments of the function Λ_1 there are second derivatives $\varkappa^i_{jk}$ and $\nabla_j \nabla_e \mu^A$ is considered by BERTICHEVSKI [1]. Further complication, arising when gradients of higher order are introduced, is easy to realize, but is associated with the cumbersome form of an increased number of formulae.

[2] The case of spin arguments is investigated in the paper of GOLOUBIATNIKOV [6].

cients in the expression (4) for δW^* is not essential. However, eq. (3) and its consequences will have some essential, and from the point of view of physics, important features when the quantities Λ_1 and δW^* are invariant with respect to certain or arbitrary transformations of coordinates x^i and ξ^i or to certain groups of transformations, i. e. symmetry groups in the same coordinate system.

The basic equation (3) may be considered both in the general theory of relativity and in Newtonian mechanics.

In the general theory of relativity the elementary volume $d\tau$, Λ_1, and δW^* may be regarded as four-dimensional scalars, the arguments μ^A, $\nabla_e \mu^A$, in general, as four-dimensional tensors or spintensors. It is well known that the operator ∇_e is determined in connection with the parallel transfer of the objects μ^A by means of the connectivity coefficients Γ^i_{je}, which for the transfer of tensors are determined by the formula:

$$\Gamma^i_{je} = \frac{1}{2} g^{is} \left(\frac{\partial g_{is}}{\partial x^e} + \frac{\partial g_{se}}{\partial x^j} - \frac{\partial g_{ej}}{\partial x^s} \right).$$

The problem of the parallel transfer of spinors and of the operator ∇_e for the spin quantities is investigated by different authors, see, for example [6].

Newtonian mechanics admits the existence of the absolute time $\xi^4 = t$, the element of distance ds and the connectivity coefficients being determined by the three-dimensional euclidian fundamental tensor $g_{\alpha\beta}$, the μ^A complexes being also considered as three-dimensional tensors.

In many cases in Newtonian mechanics we may assume that

$$\Lambda = \Lambda^* = \frac{\varrho v^2}{2} - \varrho V,$$

where ϱ is the mass density, v — the speed, and V — the internal energy of the physical medium which depends on the parameters of the internal state. It is obvious that the quantity Λ^* is a three-dimensional scalar with respect to coordinate transformations which do not depend on time.

The expression for Λ^* does not conserve its form as the difference between the specific kinetic and internal energies, even in the case of Galilei-Newtonian transformations. Hence, the quantity Λ^* is independent of the choice of the coordinate system.

In the basic equation (3) and in the formulae (4) and (5) the variations are determined for individual particles, they are arbitrary and, in general, non-zero on the boundary $\sum$.

When carrying out the variations in the eq. (3), the quantity Λ_1 being a scaler and every term in the formula (4) being a scalar, it is essential from the physical point of view, that in the expression under the integral sign in formula (5), every term be obtained also as a scalar. In this case, the coefficients of the variations will form the components of the spintensors, if the corresponding variations are determined as components of spintensors. Defining the variations as components of spintensors we may at once obtain in an invariant form the equations of motion and all the other relations in an arbitrary curvilinear system of coordinates and, in particular, in the co-moving system of coordinates. In addition, in this case the components of the spintensors Θ^{ij}, P^j_i, M^j_A may be considered as physical characteristics.

The definition of variations used in the following involves some innovations [23] which are included in the following formulae:

$$\begin{aligned}
\delta x^i &= x'^i(\xi^1, \xi^2, \xi^3, \xi^4) - x^i(\xi^1, \xi^2, \xi^3, \xi^4),\\
\partial \mu^A &= \mu'^A(x^1, x^2, x^3, x^4) - \mu^A(x^1, x^2, x^3, x^4),\\
\delta \mu^A &= \partial \mu^A + \delta x^e \nabla_e \mu^A,\\
\delta g_{ij} &= \partial g_{ij} = g'_{ij}(x^e) - g_{ij}(x^e) = \partial^* g_{ij} + \nabla_i \partial \eta_j + \nabla_j \partial \eta_i,
\end{aligned}$$

where $\partial \eta_1$, $\partial \eta_2$, $\partial \eta_3$, $\partial \eta_4$ are the components of an arbitrary, infinitely small vector which depends on the coordinates x_i. If the space is pseudo-euclidian, than by definition we set $\partial^* g_{ij} = 0$, in this case the quantities $\partial \eta_i$ give the displacements when the varied coordinate system connected with the observer is deformed.

In addition, let us make use also of the following formulae:

$$\delta \Lambda = \partial \Lambda + \delta x^e \nabla_e \Lambda$$

and

$$\partial x^i_j = x^e_j \nabla_e \delta x^i - \delta x^e \nabla_e x^i_j$$

where

$$\nabla_e x^i_j = \frac{\partial x^i_j}{\partial x^e} + x^s_j \Gamma^i_{se}.$$

Further, use is made also of the formula:

$$\delta d\tau = \left(\frac{\partial \sqrt{-g}}{\sqrt{-g}} + \nabla_e \delta x^e\right) d\tau \quad \text{where} \quad g = |g_{ij}| \text{ and } \frac{\partial \sqrt{-g}}{\sqrt{-g}} = \frac{1}{2} g^{ij} \delta g_{ij}$$

Considering also the well known formula [11]

$$\partial R = -R^{ij} \delta g_{ij} + \nabla_e W^e, \quad \text{where} \quad \nabla_e W^e = g^{ij} \partial R_{ij},$$

$$W^e = (g^{pr} g^{eq} - g^{pq} g^{er}) \nabla_p \delta g_{qr}$$

after varying (3) and using the Gauss-Ostrogradski transformation, we obtain the Euler equations:

$$R^{ij} - \frac{1}{2} g^{ij} R = -\varkappa T^{ij} \tag{6}$$

$$\nabla_j T^{ij} = 0 \tag{7}$$

$$\frac{\partial \Lambda}{\partial x^s_j} \nabla_i x^s_j + \nabla_s \left(\frac{\partial \Lambda}{\partial x^i_j} x^s_j \right) = -Q_i - M_A \nabla_i \mu^A - \frac{\partial \Lambda}{\partial K^B} \nabla_i K^B - \frac{\partial \Lambda}{\partial L^C} \nabla_i L^C \tag{8}$$

$$\frac{\partial \Lambda}{\partial \mu^A} - \nabla_e \left(\frac{\partial \Lambda}{\partial \nabla_e \mu^A} \right) = M_A . \tag{9}$$

The following abbreviated designation is introduced in the eqs. (6) and (7):

$$T^{ij} = -\frac{2}{\sqrt{-g}} \left[\frac{\partial \sqrt{-g} \Lambda}{\partial g_{ij}} + \frac{\partial \sqrt{-g} \Lambda}{\partial \nabla_e \mu^A} \frac{\partial \nabla_e \mu^A}{\partial g_{ij}} - \right.$$
$$\left. - \frac{\partial}{\partial x^s} \left(\sqrt{-g} \frac{\partial \Lambda}{\partial \nabla_e \mu^A} \frac{\partial \nabla_e \mu^A}{\partial g_{ij,s}} \right) \right] - \mathscr{Q}^{ij} \tag{10}$$

Equations (6), (7), (8) and (9) form a complete system of equations for $x^i(\xi^j)$, g_{ij} and μ^A, $\left(\frac{\partial g_{ij}}{\partial x^s} = g_{ij,s} \right)$.

If the choice of the system connected with the observer is not essential, than the eqs. (7) depend on the eqs. (8) and (9).

The tensor T_{ij} is called the Einstein's momentum energy tensor. According to the basic equation (3) we obtain the following formulae for the quantities denoted in the expression (5):

$$\Theta^{ij} = T^{ij} + \frac{1}{\varkappa} \left(R^{ij} - \frac{1}{2} g^{ij} R \right) \tag{11}$$

$$P^j_i = -\Lambda \delta^j_i - \frac{\partial \Lambda}{\partial x^i_s} x^j_s + \frac{\partial \Lambda}{\partial \nabla_j \mu^A} \nabla_i \mu^A - \frac{1}{2\varkappa} R \delta^j_i + Q^j_i \tag{12}$$

$$\mathscr{M}^j_A = -\frac{\partial \Lambda}{\partial \nabla_j \mu^A} + M^j_A \tag{13}$$

$$Q^{pqj} = -\frac{\partial \Lambda}{\partial \nabla_e \mu^A} \frac{\partial \nabla_e \mu^A}{\partial g_{pq,j}} \tag{14}$$

$$G^{pqre} = \frac{1}{2\varkappa} (g^{pr} g^{qe} - g^{pq} g^{re}) . \tag{15}$$

In the special theory of relativity the space is pseudo-euclidian and the Riemann tensor, zero. In this case, we may assume that varying the metric, reduces to the deformation of the coordinate system connected with the observer, and therefore $\partial^* g_{ij} = 0$.

Thus, in the special theory, according to (10), we may introduce the Einstein's tensor $T^{ij} = \Theta^{ij} \neq 0$ and the corresponding eqs. (7), but in this case the equations of the field theory (6) will be absent.

The tensor P_i^j in the four-dimensional space may be considered as a direct generalization of the usual three-dimensional stress tensor to four-dimensional space. The tensor P_i^j determines the energy fluxes through $\sum$, due to variations of the law of motions δx^j.

In the general case, the tensor P^{ij} is not symmetric. Even when there is symmetry $P^{ij} = P^{ji}$, the tensor P^{ij} and T^{ij} may be different not only in the general theory of relativity, but also in the special theory. However, one may show that for classical models of continuous media and in several other more general examples $T^{ij} = P^{ij}$.

The eqs. (11—15) may be considered as generalized equations of state of the medium and of the gravitational field, defining the characteristics of the internal stresses, i. e. the interactions.

It follows from (14) and (15) that for the gravitational field one may introduce the tensors Q^{pqj} and G^{pqre}, characterizing the internal interactions and defining the corresponding energy fluxes on arbitrary mental variations of the metric.

Experience in mechanics of continuous media shows that the explicit introduction of internal stresses and of corresponding energy fluxes through the surface $\sum$ is useful and fruitful both in theory and applications and, in particular, when formulating the initial and boundary conditions.

In order to elucidate the character of the basic equation (3) and its consequences (6—9) let us note, that under additional particular assumptions about Λ and δW^* from eq. (3) we may obtain the basic equations of mechanics and dynamics of the system in the Newtonian theory.

In particular, we may obtain the basic equations of angular momentum when the elementary particles have their proper internal angular moments (proper spins). We may easily be convinced that in this case the corresponding equation of angular momentum is independent of the momentum equation in the Newtonian mechanics.

The equations of Einstein's general theory of relativity and Maxwell's equations, when properly defining the function Λ, follow from eq. (3) and are contained in eqs. (6—9).

It may be also easily shown, that Dirac's equations and their generalization to Riemannian space and to spin equations with partial differentials of second or higher order, when properly defining the function Λ, follow from (3); in curved space, in addition, the definition of parallel transfer and covariant differentiation for spiners are required.

It is easy to indicate also the arguments and properties of the function Λ and the expression for δW^* for classical models of ideal fluids, in the linear or non-linear theory of elasticity, in the theory of viscous fluid [3, 22].

In particular, in this way we may obtain the Navier-Stokes equations in Newtonian mechanics or the generalized equations for viscous medium with relativistic effects [3].

It is also obvious, that the problem of constructing models of continuous media in continuum theory of dislocations with irreversible effects, may be considered by means of the variational equation (3).

In order to apply the developed variational formalism, it is necessary to advance hypotheses of physical nature concerning the properties of the generalized Lagrangian function Λ and of the given, generally non-holonomic, functional for δW^*. Obviously, the problem of determining the function Λ for physical media reduces to determining the internal energy.

In many specific problems of practical interest the enumeration of physical and geometrical arguments on which depends the Lagrangian function Λ is of great importance. It is well known, that often the small range of variation of the characteristic parameters is the essential physical feature of the considered processes. In these cases, after introducing small parameters, the functions Λ may be regarded as a polynomial with respect to this parameters as arguments, in many cases practice and mathematical theories show that it is sufficient to consider polynomials of degree not higher than second.

When determining δW^* one may introduce entropy as one of the parameters $\mu' = S$ and in the formula (4) pick out a term of the form

$$\delta W^* = \int_V \varrho T \, \delta S + \ldots \tag{16}$$

where ϱ is the mass density and T, for equilibrium states, by definition the absolute temperature.

As for other coefficients of the variations in the expressions under the integral sign for δW^* in the formulae of the type (4), these quantities are closely connected with the dissipation laws of various forms of energy.

In order to investigate the problems of entropy variations, from the Euler equations we may derive the following equality:

$$\varrho T DS = \frac{1}{\sqrt{\hat{g}_{44}}} \nabla_j F^j + \frac{\partial \Lambda}{\partial K^B} DK^B + \frac{1}{\sqrt{\hat{g}_{44}}} \frac{\partial \Lambda}{\partial L^C} \frac{\partial L^C}{\partial \xi^4} + \\ + Q_i u^i + M_A D\mu^A. \tag{17}$$

where $u^i = \left(\frac{dx^i}{ds}\right) \xi^\alpha = \text{const}$ are the components of the four-speed.

The operator $D = u^i \nabla_i$ determines the differentiation with respect to S along the world line. Equation (17) shows that the derivatives with respect to time of K^B and L^C may lead to entropy variation.

The components of the vector F_j are expressed by means of the Lagrangian. In the simplest case considered above the following formula holds:

$$F^j = \frac{\partial \Lambda}{\partial x_s^i} x_4^i x_s^j \tag{18}$$

For more general cases the expresison for F^j is given in the paper [3] by V. Bertichevski.

It may be shown that in classical cases $F^j = 0$. However, in general $F^j \neq 0$, and in a Lagrangian system it may lead to an entropy increase.

Equation (17) and certain additional considerations may be assumed as basis for determining laws for δW^*, in particular, we may take advantage of the general considerations about the small deviations of the considered processes from reversible ones with the possible application of Onsager's principle.

References

[1] Бердичевский, В. Л.: Построение моделей сплошных сред при помощи вариационного принципа. ПММ **30**, **3** (1966).

[2] — О нелинейных тензорных функциях в теории относительности. ПММ **1966** (5).

[3] — Вариационные методы построения моделей сплошных сред с необратимыми процессами в специальной теории относительности. ПММ **1966** (6).

[4] Casal, P.: Capillarité interne en mécanique des millieux continus. Compt. Rend. **256** (3), (1961).

[5] Голубятников, А. Н.: Нелинейные спинорные функции. ДАН **165** (2), (1965).

[6] — Сплошная среда со спинорными и векторными характеристиками. ДАН **169** (2), (1966).

[7] Ericksen, J. L.: Anisotropic Fluids. Arch. Ratl Mech. Anal. **5** (4), 231—237

[8] Желнорович, В. А.: Представление спиноров в n-мерном пространстве системой тензоров. ДАН **169** (2), (1966).

[9] — Спинор как инвариантный объект. ПММ **30** (5), (1966).

[10] Kondo, K.: Memoirs of the Unifying Study of the Basic Problems in Engineering Science by Means of Geometry. Vols 1—3. Tokyo: Gakujutsu Bunken Fukyu-Kai. 1955—1962.

[11] Ландау, Л. Д., и Е. М. Лифшиц: Теория поля. Физматгиз 1962.

[12] Лохин, В. В., и Л. И. Седов: Нелинейные тензорные функции от нескольких тензорных аргументов. ПММ **27** (3), 1963.
Lokhin, V. V., and L. I. Sedov: Nonlinear Tensor Functions of Several Tensor Arguments. [Translation from Prikl. Mat. Mech. **27** (3), (1963).] New York: Pergamon Press.

[13] Лурье, М. В.: Применение вариационного принципа для исследования разрывов в сплошной среде. ПММ **30** (4), 1966.

[14] MINDLIN, R. D.: Micro-Structure in Linear Elasticity. Arch. Ratl Mech. Anal. No. 1 **16** (1), (1964).
[15] ROY, M.: Mécanique des millieux continus et déformables. Vol. 1—2 G. V. 1950.
[16] SCHÖPF, H. G.: Allgemeinrelativistische Prinzipien der Kontinuumsmechanik. Ann. Phys. **7** (12), 7—8 (1963).
[17] СЕДОВ, Л. И.: Об основных концепциях механики сплошной среды. В книге „Некоторые проблемы математики и механики". Новосибирск: Сибирское отдел. АН СССР. 1961.
[18] — Об основных принципах механики сплошной среды. Москва: Изд. МГУ. 1961.
[19] — Введение в механику сплошной среды. Москва: Физматгиз. 1962.
[20] SEDOV, L. I.: Some Problems of Designing New Models of Continuum Media. Proceedings of the 11th International Congress of Applied Mechanics, Munich, 1964. Berlin—Heidelberg—New York: Springer. 1966.
[21] СЕДОВ, Л. И.: О пондеромоторных силах взаимодействия электромагнитного поля и ускоренно движущегося континуума с учетом конечности деформаций. ПММ **1965** (1).
[22] — Математические методы построения новых моделей сплошных сред. Успехи Математ. Наук **20** (5), 125 (1965).
[23] — О тензоре энергии и импульса и о макроскопических внутренних взаимодествиях в гравитационном поле и в материальных средах. ДАН **164** (3), (1965).

Discussion

ROY: I want to express my admiration for the continued progress of the Sedov-theory. I believe that it is expecially interesting to see how you search now to combine some microscopic aspects of the media with a macroscopic view of the same media. I will ask a question. I read $x_i(\xi^1, \xi^2, \xi^3, \xi^4 - s)$. What is s?

SEDOV: s is a parameter for taking into account the effects of memory, the hereditary effects.

ROY: Volume V is bounded by a surface Σ. Is this surface a material surface or not? Is the media which is contained in this volume V always the same?

SEDOV: There is no material flux through the surface Σ.

SETH: How does the microscopic concept, which has been brought in, work at the boundary?

SEDOV: The macroscopic effects are considered in the function Λ, the dissipation, internal and on the surface, through the terms in δW^*, the surface forces $Q_i^j M_A^j$ are the terms corresponding also to the electromagnetic force. All these terms are macroscopic, but they are determined from the microscopic mechanism. In field theory this is the usual method. The only new things here are two terms δW and δW^* taking into account and the non-zero variations on the surface Σ.

KLUITENBERG: You have added a term D^{ij} to the momentum energy tensor. I should like to ask if that is due to irreversibility? Has it something to do with the entropy production. Is it just a new physical assumption?

SEDOV: This term is stipulated by δW^*; it is determined by the irreversible processes and some external forces and influences. The presence of the term $D^{ij}\,\delta g_{ij}$ does not affect the change of entropy. In particular, the other terms in δW^* are required for the model of viscous fluid, and they contain the energy flux connected with the viscosity and heat transport in viscous fluid.

Irreversible Transition in Continuum Mechanics

By

B. R. Seth

Dibrugarh (India)

Summary

There is hardly any branch of science or technology in which we do not come across transition into irreversible processes. Elasticity, plasticity, visco-elasticity, creep, boundary layers, shocks and oscillations are well-known examples. For elastic-plastic deformation, the ideal model is to assume some type of yield condition. For creep, a number of creep strain laws have to be taken. For boundary layers, the pressure from the irrotational flow has to be used in the layer. In shocks a number of jump conditions are utilized.

Irreversibility, and hence, non-linearity, is involved in all such problems; but generally they are linearized with the result that discontinuities, singular surfaces (non-differential regions) have to be introduced over which two successive states of a medium are matched together with the help of some empirical laws. Analytically this involves complex constitutive equations.

It is shown that irreversible transition may be identified with the asymptotic solutions at the transition points of the field equations of the system. Combined with the concept of a generalized measure, which provides for the intrinsic non-linearity in a condensed form, the use of semi-empirical laws is eliminated and a number of new and interesting results are obtained.

1. Introduction

The concept of transition has been very little treated by both micro- and macro-field theorists. Exchange and irreversibility occur at transitions where thermodynamic and other effects can be expected to be significant. Shock waves and turbulence are well known examples. When a medium in a state A goes into a state$B$ both A and B can in general be treated as linear and isothermal. But the transition state T between A and B need not behave similarly, and its mechanism needs to be investigated. Continuum model makers like to discuss only A or B and want to replace T by some conditions of state or jump conditions. Analytically, T can provide an interesting challenge and its solution can replace the *ad hoc* conditions of state whose use is so much on the increase. For example if we make Prandtl's assumptions to get the thickness of the boundary layer, δ, we get $\delta \sim R^{-1/2}$, R being the Reynold's number. Treated as a transition problem, this assumption is discarded and we get $\delta \sim R^{-k/(2+\lambda)}$, k and λ depending upon the vorticity permisible on the boundary and the free stream flow. This can include turbulent and other effects. For $\lambda = {}^1/_2$, $k = {}^1/_2$, we get $\delta \sim R^{-1/5}$, a well known result [1].

It will be shown that transitions may be identified with the asymptotic solutions at the turning point of the differential system defining a medium. They need not be represented by *ad hoc* laws, as is currently done. If these laws exist they will come out of the equations. In some cases, like work-hardening, they need not be there. Composite constitutive equations do not solve transition problems. They bring together adjacent states in some sense.

It should be noted that transition points may be multiple points, so that when complete or partial irreversibility sets in there may be a number of branches having distinguishing analytical and physical features. For example, an initially elastic state may go into a creep state or an elastic-plastic state.

2. Measure of Deformation

In his attempt to explain complicated natural phenomenon the worker of today has to discard simple models in favour of complex ones. High speed problems, multiphase systems, non-equilibrium states, all demand more and more attention. If any phenomenon be represented by a spectrum, its normal aspects may be linear and can be represented by the middle of the spectrum; the ends are non-linear, irreversible and lead to change of states. In such cases ordinary measures of deformations are insufficient; if used the constitutive equations become very complicated and the order to which they may be used cannot be determined in advance. Abstract measure theory is well developed but very little use has been made of it. Problem of initial stress, elastic plastic deformations, creep, fatigue, and turbulence lead one into micro-fields, and require the use of generalized measures. A large number of measures are already in use for particular problems. They can be shown to be only particular cases of weighted integral measures [2]. A solitary attempt or two has been made to use Lebesgue measures. But it is still to be recognized that suitable measures can help us to fix 'a priori' the order of the constitutive equations to be taken and to do way with semi-empirical laws like that of Norton and Andrade in creep problems. Any scientific advance should attempt to reduce the number of such laws, if they cannot be totally done away with.

Any isotropic measure $\boldsymbol{\varepsilon}$ of rank two can be expressed in terms of any other measures $\boldsymbol{e}$ of the same rank by the relation,

$$\boldsymbol{\varepsilon} = \boldsymbol{F}_0 + \boldsymbol{F}_1 \boldsymbol{e} + \boldsymbol{F}_2 \boldsymbol{e}^2, \tag{2.1}$$

where F's are function of the invariants of e. We can now choose ε in a closed form with parameters which may vary for different states of the medium. These parameters may be adjusted in conformity with experimental data, so that only the first power of ε is sufficient. Any constitutive equation defining the state of the medium then becomes linear and can be readily dealt with. The question of the order of the terms to be taken in F's does not arise. In other words such a measure respresents the intrinsic non-linearity of irreversible phenomenon in a condensed form, and can help to do away with semi-empirical laws. To give an example let us see how such laws are contained in generalized measures for problems in creep. A generalized measure in cartesian coordinates which contains most of the known measures is given by [3],

$$\varepsilon = \frac{1}{n^m} [1 - (1 - 2e^{m/2}]^m. \tag{2.2}$$

where m, n are parameters and ε and e are principal components of the generalized measure and Almamsi measure, respectively.

For the uni-axial case it reduces to

$$\varepsilon = \frac{1}{n}\left[1 - \left(\frac{l_0}{l}\right)^n\right] = \alpha t \text{ (say)}. \tag{2.3}$$

Let A be the cross-section of a rod of length $\boldsymbol{l}$, and A_0 and l_0 the initial values of these quantities. If P be the load we have the relations

$$A l = A_0 l_0, \tag{2.4}$$

$$P = T_y A = T_0 A_0. \tag{2.5}$$

Differentiating (2.3) with respect to t and combining the result with (2.4) and (2.5) we get

$$\gamma = \text{rate of creep strain} = \frac{1}{l}\frac{dl}{dt} = \alpha\left(\frac{T_y}{T_0}\right)^n \tag{2.6}$$

which is Nortons' law.

In like manner the well known Andrade's law which holds good for a variety of metals at different temperatures may be deduced from (2.2) by adding together the two creep-strain rates obtained by putting $m = 1$, $n = o$, and $m = 3$, $n = 1$ in (2.2) [4]. In fact the law which is reported to represent the most precise measurements at high temperatures corresponding to constant stress-creep is given by [5]

$$\gamma' = \sum_{p=1} a_p t^{n_p - 1} + \sum_{q=1} h_g t,^{n_q - 1} \quad 0 < n_p < 1, \quad n_q > 1, \tag{2.7}$$

where $1/m = n_p$ for the transient part and $1/m = n_q$ for the fracture part.

This can be readily deduced from (2.2). In like manner, for problems in fluid mechanics we may use a generalized measure of the type

$$\varepsilon = \frac{k}{m^p n^p} [1 - (1 - 2me)^{n/2}]^p, \tag{2.8}$$

where e is the ordinary strain velocity measure and m, k help to keep the dimensions correct [6].

3. Identification of Transition

In solid mechanics the transition points may be geometrically identified with the limiting form of the deformation ellipsoid when its principal axes tend to become zero or infinite. In other words the ellipsoid reduces to a cylinder or a plane according as the change in the medium is due to infinite extension or contraction. If G_{ii} are the principal metric tensors

$$ds^2 = G_{ii} dx^i dx^i, \tag{3.1}$$

and the transition conditions becomes $G_{ii} \to \infty$ or 0. Analytically, the transition points may be turning points, critical points or singular points of the differential equations defining the medium. Thus such points, though not obvious in the original plane, may be readily determined in another stateplane. Since all such transformations are continuous it only amounts to saying that some topological space should clearly indicate them. This may be illustrated with a few examples [7].

A. One-dimensional Steady Compressible Flow

In one-dimensional steady viscous compressible flow the equations of motion are

$$\varrho u = m,$$

$$\varrho u \frac{\partial u}{\partial x} = -\frac{\partial p}{\partial x} + \frac{4}{3} \frac{\partial}{\partial x} \left(\mu \frac{\partial u}{\partial x}\right). \tag{3.2}$$

Assuming $p = f(\varrho)$ and putting $P = \partial u / \partial x$ we get

$$mP\left(1 - \frac{c^2}{u^2}\right) = \frac{4}{3} P \frac{d(\mu P)}{du}, \tag{3.3}$$

where $c^2 = dp/d\varrho$ = the local sound velocity.

Since $P \neq 0$, as otherwise $u = 0$, we have, if μ is taken as constant

$$\frac{dP}{du} = \frac{3m}{4\mu}\left(1 - \frac{c^2}{u^2}\right) = \frac{3m}{4\mu}\left(1 - \frac{1}{M^2}\right). \tag{3.4}$$

The three points, which can be interpreted as the transition points, are

$$\text{(I) } \mu \to 0, \quad \text{(II) } \mu \to \infty \quad \text{(III) } u \to \pm c \quad (M \to \pm 1).$$

These are obvious, in the (P, u) plane. The first, $\mu \to 0$ ($R \to \infty$, being the Reynolds number) corresponds to the case when the fluid tends to become non-viscous at the edge of a boundary layer. Over this transition the pressure-density relation is

$$p = k - \frac{m^2}{\varrho}, \tag{3.5}$$

which may be expected to hold good for gases with small viscosity.

The second $\mu \to \infty$ $(R \to 0)$ corresponds to the slow motion of highly viscous liquids.

The third $u \to \pm c$ indicates the transition from supersonic to subsonic when the Mach number is 1. It is significant that μ has not been taken to be zero and no particular specifying condition in the form $p = f(\varrho)$ has been taken. Over the shock,

$$\frac{\partial p}{\partial \varrho} \sim u^2 \sim \frac{m^2}{\varrho^2}, \tag{3.6}$$

and hence $p \sim k - m^2/\varrho$, a relation which can be expected to be true for weak shocks where the entropy change can be neglected.

From (5.3) we see that if $\mu \to 0$ and $u \to \pm c$ simultaneously, dP/du is indeterminate. This shows that viscosity is significant over shock-transition region and should not be taken equal to zero.

Again, over transition $dP/du \to 0$. Putting this condition in (3.2) we get

$$m \frac{\partial u}{\partial x} = - \frac{\partial p}{\partial x},$$

$$m(u_2 - u_1) = p_1 - p_2,$$

which is the momentum equation. This has been obtained without taking $\mu = 0$, as is done in current literature.

B. Two-dimensional Symmetrical Flow

In like manner for two-dimensional symmetrical flow we get

$$2\pi\varrho r u = m, \quad r u = U, \quad \varrho = m/2\pi U,$$

$$\varrho u \frac{\partial u}{\partial r} = -\frac{\partial p}{\partial r} + \frac{4}{3}\frac{\partial}{\partial r}\left[\mu\left(\frac{\partial u}{\partial r} + \frac{u}{r}\right)\right]. \tag{3.7}$$

These give

$$\frac{m}{2\pi r}\left(\frac{1}{r}\frac{\partial U}{\partial r} - \frac{U}{r^2}\right) = \frac{c^2 m}{2\pi U^2}\frac{\partial U}{\partial r} + \frac{4}{3}\mu\frac{\partial U}{\partial r}\frac{\partial}{\partial U}\left(\frac{\mu}{r}\frac{\partial U}{\partial r}\right),$$

which in a (P, U) plane can be re-written as

$$\frac{dP}{dU} = \frac{d}{dU}\left(\int\frac{U}{r^3}dr + \frac{8\pi\mu}{3mr}\frac{dU}{dr}\right) = \frac{1}{r^2} - \frac{c^2}{U^2} = \frac{1}{r^2}\left(1 - \frac{1}{M^2}\right). \tag{3.8}$$

Once again we get transition at $M = \pm 1$.

C. One-dimensional Ideal Gas Flow

Similarly we may construct more difficult cases. For example, for the one-dimensional flow of an ideal gas

$$p = gR\varrho T, \quad \frac{\partial p}{\partial \varrho} = gRT = c^2, \tag{3.9}$$

$$\frac{4}{3}\frac{\mu}{m}\frac{\partial u}{\partial x} = u + \frac{gRT}{u} - c_1 = u + \frac{c^2}{u} - c_1,$$

$$\frac{k}{m}\frac{\partial T}{\partial x} = c_1 u - c_2 - \frac{1}{2}u^2 + \frac{g\varrho}{\gamma - 1}T$$

$$= c_1 u - c_2 - \frac{1}{2}u^2 + \frac{c^2}{\gamma - 1}. \tag{3.10}$$

These give

$$\frac{4}{3}\frac{\mu}{m}P\frac{\partial P}{\partial u} = P - \frac{gRT}{u^2}P + \frac{gR}{u}\frac{\partial T}{\partial u}P$$

$$= P\left[1 - \frac{c^2}{u^2} + \frac{mgR}{uPk}\left(c_1 u - c_2 - \frac{1}{2}u^2 + \frac{c^2}{\gamma - 1}\right)\right.,$$

where $P = \partial u/\partial x$, and c_1, c_2 being constants of integration.

This may be re-written as

$$\frac{1}{3m}\frac{\partial Q}{\partial u} = \left(1 - \frac{c^2}{u^2}\right)\left(pk + \frac{mgRU}{\gamma - 1}\right),$$

where $Q = 2\mu k p^2 - 3m^2 g R\left(c_1 u - c_2 \log u - \frac{1}{4}u^2 - \frac{1}{2}\frac{u^2}{\gamma - 1}\right)$, (3.11)

which again shows transition at $M = \pm 1$.

D. Elastic-plastic Transition in Shells and Tubes under Pressure

In spherical polar coordinates we can take

$$u = r(1 - \beta), \quad v = 0, \quad w = 0 \tag{3.12}$$

where β is a function of $r = (x^2 + y^2 + z^2)^{1/2}$ only.

The finite components of strain are now given by

$$e_{rr} = \frac{1}{2}[1 - (r\beta' + \beta)^2], \quad e_{\theta\theta} = e_{\varphi\varphi} = \frac{1}{2}(1 - \beta^2), \quad e_{r\theta},\ e_{\theta\varphi},\ e_{\theta r} = 0 \tag{3.13}$$

We take the stress-strain relation as

$$T_{ij} = \lambda I_1 \delta_{ij} + 2\mu e_{ij}, \tag{3.14}$$

where λ, μ are elastic constants δ_{ij} is KRONECKER's delta and I_1, is the first strain invariant.

The corresponding stresses are

$$\tau_{rr} = \lambda I_1 + \mu - \mu(r\beta' + \beta)^2, \quad \tau_{\theta\theta} = \tau_{\varphi\varphi} = \lambda I_1 + \mu - \mu\beta^2, \tag{3.15}$$

$$\tau_{r\theta},\ \tau_{\theta\varphi},\ \tau_{r\varphi} = 0,$$

where $\beta' = d\beta/dr$, $I_1 = \frac{3}{2} - \beta^2 - \frac{1}{2}(r\beta' + \beta)^2$.

The stress-equations of equilibrium are all satisfied excepting

$$\frac{\partial T_{rr}}{\partial r} + \frac{2(T_{rr} - T_{\theta\theta})}{r} = 0. \tag{3.16}$$

Substituting from (3.15) we get the non-linear differential equation satisfied by β as

$$(\lambda + 2\mu)[2\beta^2 + (r\beta' + \beta)^2] + 4\mu \int r\beta'^2 dr = k, \tag{3.17}$$

where K is a constant of integration.

Putting

$$c = 2\mu/(\lambda + 2\mu) = (1 - 2\sigma)/(1 - \sigma), \quad r\beta' = \beta P,$$

and differentiating (3.17) with respect to r we get

$$[P + (2 + c)P + 3]\frac{d\beta}{dP} + \beta\,(1 + P) = 0, \tag{3.18}$$

which shows readily the transition points in the (P, β) plane as $P = -1$ and $P = \pm\infty$.

These may be identified with elastic-plastic and creep transitions by suitably choosing the transition stress-functions.

Similar results should hold good for tubes under pressure [8].

A number of other examples have been worked out which verify, both geometrically and analytically the method indicated to find the transition points [9, 10]. For deformation of solids a very general treatment can be given, which includes a number of irreversible phenomena. One can therefore conclude that a large number of irreversible problems involve transition, which should be treated as asymptotic problems intimately connected with the field equations.

References

[1] SETH, B. R.: Synthetic Method for Boundary Layer Thickness, In: Proc. IUTAM Symposium. On boundary layer research. pp. 47–58. Freiburg: 1957.
— Boundary layer thickness. J. Sci. Engng Res. **4**, 1–6 (1960).

[2] SETH, B. R.: Measure-concept in Mechanics, Inter. J. Non-linear Mech. **1**, 35–40 (1966).

[3] SETH, B. R.: Generalized Strain Measure with Applications to Physical Problems. In: IUTAM Symposium: On Second Order Effects in Elasticity, Plasticity and Fluid Mechanics. Haifa: 1962.

[4] SETH, B. R.: Generalized Strain and Transitions Concepts for Elastic-Plastic Deformation, Creep and Relaxation. In: Proc. XIth Inter. Congr. App. Mech., Munich. 1965.

[5] FINNIE, I., and W. R. HELLER: Creep of Engineering Materials. Ch. 2. New York: McGraw Hill. 1959.

[6] SETH, B. R.: Continuum Concepts of Measure. Presidential Address, Xth Congr. App. Mech., Madras, India, 1965.

[7] SETH, B. R.: Transition Phenomenon in Physical Problems. Bull. Calcutta Math. Soc. **56**, 83–89 (1964).

[8] SETH, B. R.: Elastic-Plastic Transition in Shells and Tubes under Pressure. ZAMM **43**, 345–351 (1963).

[9] SETH, B. R.: Transition Theory of Elastic-Plastic Deformation. Extension Lectures, pp. 1–64. Hyderabad: Osmania University, 1964.

[10] PURUSHOTHAMA, C. M.: Elastic-plastic Transition. ZAMM **45**, 401–408 (1965).

On Conditions of Yielding and Fracture for Viscoelastic-Plastic Bodies

By

G. S. Shapiro

Moscow (U.S.S.R.)

Summary

The conditions of transition from one state (visco-elastic for instance) into another state (plastic state or a state of fracture for instance) are discussed for bodies exhibiting elastic, viscous and plastic properties. It is found that criteria previously suggested by various authors (constancy of elastic distortion energy, constancy of dissipation power or constancy of the sum of elastic distortion energy and dissipation power) are inadequate to describe some phenomena observed in experiments. In particular, by using these criteria it is not possible to explain delayed yielding for mild steel nor an increase in the work required to produce fracture for various vulcanizates. An assumption of a more general nature is made according to which "critical" quantities defining the transition from one state to another are not constants but are functions depending on strain rates and stress rates. The examples considered are the stretching of a bar of viscoelastic material at a constant strain rate and the impact of a semi-infinite viscoplastic bar with constant velocity. The yield limit of the material of the bar is assumed to be a linear function of strain rate. It is shown that in the latter case the problem reduces itself to the solution of Stefan's problem for the heat conduction equation. A similar problem for a bar whose yield limit is independent of strain rate was considered by Sokolovsky.

General Considerations

The conditions of transition of a body from one state (A) to another state (B) are discussed in the present paper from a purely phenomenological point of view. Temperature changes are disregarded. It is apparent that the conditions of transition are closely related to the constitutive equations of a body in states (A) and (B). Let, for instance, the body be elastic in state (A) and perfectly plastic in state (B). The constitutive equation for state (A), Hooke's law, then makes it possible to express strains in terms of stresses and to represent the condition of transition, the yield function, in terms of stresses. The yield function, in turn, is taken to be the plastic potential and appears in the constitutive equation for state (B).

Let the body now be viscoelastic in state (A) and viscoplastic in state (B). Suppose that in state (A) the constitutive equation contains strains, strain rates, stresses and stress rates. It is natural then to assume that the condition of transition from state (A) into the plastic state must contain stresses, strain rates and stress rates. This condition must, in turn, appear in the constitutive law for state (B).

The theory of deformation of complex media exhibiting elastic, viscous and plastic properties has been developed in recent years by Freudenthal [1], Landau, Weiner and Zwicky [2], Ivlev [3], Perzyna [4], Naghdi and Murch [5], Berezhnoi and Ivlev [6]. These authors did not, however, aim at formulating the conditions of transition of a body from the viscoelastic state into the plastic state.

The question of fracture of incompletely elastic materials was considered by Ishlinsky [7] by using a model governed by the equation of the standard linear body during deformation. The model represented a spring in series with a combination of another spring (of different stiffness) and a dashpot moving in a vessel with viscous liquid. Two modes of fracture were considered, external or internal, according to whether the first or second spring of the model was ruptured. In both cases it was assumed that the rupture of the springs occurred when the forces induced in them attained certain limiting values. On the assumption of internal fracture it was possible to describe experimentally observed phenomena of fracture of materials some time after the application of loads or deformation.

In the investigation of the transition of viscoelastic bodies into the plastic state Reiner [8] took as a basis the Huber-Mises criterion, assuming that plastic yielding occurred when the elastic distortion energy accumulated in a body attained a limiting value. By using the model of the standard linear body, Reiner could illustrate two qualitative effects observed in experiments: delayed yielding when a constant load is applied and an increase in the yield limit when a constant strain rate is involved.

A more general approach is due to Olszak [9] who used two groups of models, linear and non-linear. Considering "critical" states of viscoelastic bodies as transition states from linear to non-linear behaviour, Olszak proceeded from the conception of "equivalence" of both components, viscous and elastic, of the viscoelastic model. Various possibilities of partial or complete fracture of the model were under consideration. It was assumed that the critical state in an elastic element was set up when the elastic distortion energy per unit volume attained a certain value, and the critical state in a viscous element occurred when the dissipation power per unit volume took a limiting value.

In all the afore-mentioned papers [7—9] it is assumed that there exist some constants which characterize the transition of a body from one state into another. This restriction does not permit an explanation of some experimental facts. For example, the condition that the elastic distortion energy is constant for a viscoelastic body leads to an increase in the yield strength (or ultimate strength) with increasing strain rate and at the same time it requires a decrease in the corresponding critical strain. The latter point is in contradiction with the well-known delayed yielding effects observed for some metals (mild steel for instance) and with experimental data for some polymers (vulcanizates for instance [10]).

In the present paper it is assumed that "critical" parameters of a body are not constants but depend on stresses, strain rates and stress rates. Two cases are considered.

In the first case the static extension of a viscoelastic body at a constant strain rate is considered. The model of the standard linear body is explored. The limiting value of elastic distortion energy is assumed to be a function of strain rate. In the second case the dynamic problem of the impact of a semi-infinite viscoplastic body with constant velocity is studied. The yield limit of the material is assumed to be a linear function of strain rate. A similar problem for a bar whose yield limit is independent of strain rate has been previously considered by SOKOLOVSKY [11].

1. Static Extension of a Viscoelastic Bar at a Constant Strain Rate

The equation of deformation for the model of the standard linear body representing a combination of the Maxwell body and the Hooke body in parallel has the form [8]

$$\sigma + T_{\mathrm{rel}}\dot{\sigma} = M(\varepsilon + T_{\mathrm{ret}}\dot{\varepsilon}), \tag{1.1}$$

where

$$\sigma = \sigma_H + \sigma_M, \quad \varepsilon = \varepsilon_H = \varepsilon_M, \quad T_{\mathrm{rel}} = \frac{\eta}{G_M},$$

$$T_{\mathrm{ret}} = \eta\,\frac{G_H + G_M}{G_H\,G_M}.$$

For elastic strain energy we have [8]

$$w_C = G_H\varepsilon^2 + (\sigma - 2G_H\varepsilon)^2/4G_M. \tag{1.2}$$

Suppose that plastic yielding occurs when this expression attains a value R depending on strain rate and stress rate

$$R = R(\dot{\varepsilon}, \dot{\sigma}).$$

In the simplest case, for small changes in $\dot{\varepsilon}$ and $\dot{\sigma}$ we have the relation

$$R = R_0 + a\dot{\varepsilon} + b\dot{\sigma}. \tag{1.3}$$

In the case of deformation at a constant strain rate there must be [8]

$$\sigma = 2G_H\varepsilon + 2G_M T_{\mathrm{rel}}[1 - \exp(-t/T_{\mathrm{rel}})]\dot{\varepsilon}_C \quad (\dot{\varepsilon} = \dot{\varepsilon}_C). \tag{1.4}$$

Combining (1.2), (1.3) and (1.4) and taking $b = 0$ for simplicity, we find for the limiting strain ε^*

$$\varepsilon^* = \frac{1}{2}\left[a/\tau + \sqrt{(a/\tau)^2 + 4R_0(G_H + G_M\alpha^2)}\right](G_H + \alpha^2 G_M)^{-1} \tag{1.5}$$

where τ is the time required for yielding (or fracture) to occur

$$\alpha = \frac{T_{\mathrm{rel}}}{\tau}[1 - \exp(-\tau/T_{\mathrm{rel}})].$$

At $a = o$ we obtain for ε^* the value

$$\varepsilon^* = \sqrt{\frac{R_0}{G_H + \alpha^2 G_M}}$$

found by REINER [8]. It can easily be proved that the values of ε^* given by (1.5) and the corresponding values of the yield limit determined according to (1.4) will increase with increasing $\dot{\varepsilon}$ for a suitable choice of a.

2. Dynamic Problem of Impact of a Semi-infinite Viscoplastic Bar with Constant Velocity

Suppose that the material of a semi-infinite $(x \geq 0)$ bar obeys the law

$$\mu\dot{\varepsilon}_p = s(|\sigma| - \sigma_s) \quad (|\sigma| > \sigma_s), \tag{2.1}$$

where $s = \operatorname{sign}\sigma$, which is valid under loading, i. e., when $\dot{\sigma} \geq 0$. Under unloading the material is assumed to be rigid.

Following [11], we introduce the dimensionless quantities

$$\xi = \frac{\sqrt{a\varrho\sigma_s}}{\mu}x, \qquad \tau = \frac{a\sigma_s}{\mu}t, \qquad X = \frac{\xi}{2\sqrt{\tau}},$$
$$\Sigma = \frac{|\sigma|}{\sigma_s}, \qquad \varepsilon = sa\varepsilon, \quad V = s\sqrt{\frac{a\varrho}{\sigma_s}}\,v, \tag{2.2}$$

and obtain the equation of the problem in the form of the heat conduction equation

$$\frac{\partial^2 V}{\partial \xi^2} = \frac{\partial V}{\partial \tau}. \tag{2.3}$$

The initial data is

$$V = 0 \quad \text{for} \quad \tau = 0. \tag{2.4}$$

The boundary condition is

$$V = V_0 \quad \text{for} \quad \xi = 0. \tag{2.5}$$

The solution of eq. (2.3) is sought in the form

$$V = A + B \operatorname{erf}(X), \qquad \operatorname{erf}(X) = \frac{2}{\sqrt{\pi}} \int_0^X e^{-z^2} dz.$$

From condition (2.4) it follows that $A = -B$ or

$$V = A[1 - \operatorname{erf}(X)]. \tag{2.6}$$

At the initial moment after impact infinitely large stresses occur at the end of the bar, and subsequently the stresses fall off. It is apparent that the bar is then divided into two parts, rigid and deformed one. The length of the rigid portion adjacent to the end of the bar will increase all the time. The position of the non-stationary boundary between the rigid and deformed parts of the bar has to be determined. We arrive at a peculiar problem of heat conduction with a moving boundary, which is similar to the problem of freezing. For the determination of the boundary we have the condition $\frac{\partial \Sigma}{\partial \tau} = 0$. Since

$$\Sigma = 1 + \frac{\partial V}{\partial \xi} = 1 + B\,(\pi \tau)^{-1/2} \exp\left(-\xi^2/4\tau\right),$$

the condition $\frac{\partial \Sigma}{\partial \tau} = 0$ gives $\tau = \xi^2/2$. Consequently, at the boundary $X = 0.707$. The condition of continuity of velocity at the moving boundary gives $A = 3{,}15\, V_0$. Thus, we have finally

$$\left.\begin{aligned} V &= 3{,}15\, V_0 [1 - \operatorname{erf}(X)] \\ \Sigma &= 1 - \frac{3{,}15}{\sqrt{\pi \tau}} V_0 \exp(-X^2) \end{aligned}\right\} \quad (X \geq 0{,}707). \tag{2.7}$$

For comparison we present the solution obtained in [11] for a material with the same law (2.1) under loading and unloading. The corresponding values of velocities and stresses are marked by asterisks. Then

$$V^* = V_0 [1 - \operatorname{erf}(X)],$$
$$\Sigma^* = 1 - \frac{V_0}{\sqrt{\pi \tau}} \exp(-X^2). \tag{2.8}$$

It is apparent that $V/V^* = 3.15$. Since $V_0 < 0$, the stress Σ exceeds Σ^* by the quantity $\frac{2{,}15}{\sqrt{\pi \tau}} \exp(-X^2)$, $X \geq 0{,}707$.

References

[1] FREUDENTHAL, A., and H. GEIRINGER: The Mathematical Theories of the Inelastic Continuum. Berlin—Göttingen—Heidelberg: Springer. 1958.

[2] LANDAU, H., J. WEINER and E. ZWICKY: Thermal Stresses in a Viscoelastic-plastic Plate with Temperature-dependent Yield Stress. J. Appl. Mech. **27**, (2), 297—302 (1960).

[3] IVLEV, D. D.: Problemy mekhaniki sploshnoi sredy. [On the Theory of Transient Creep.] Izd. AN SSSR (Moscow) **1961**, 157—160 .

[4] PERZYNA, P.: The Constitutive Equations for Rate Sensitive Plastic Materials. Q. Appl. Math. **20**, (4) 321—332 1963.

[5] NAGHDI, P., and S. MURCH: On the Mechanical Behavior of Viscoelastic-plastic Solids. Trans. ASME, E **30** (3), 321—328 (1963).

[6] BEREZHNOI, I. A., and D. D. IVLEV: On the Influence of Viscosity on the Mechanical Behaviour of Plastic Media. Dokl. AN SSSR **163** (3), 595—598 (1965). [In Russian.]

[7] ISHLINSKY, A. JU.: Failure of Incompletely Elastic Materials. Uchenye zapiski MGU, 117. Mekhanika **1**, 109—126 (1946). [In Russian.]

[8] REINER, M.: Plastic Yielding in Anelasticity. J. Mech. Phys. Solids **8** (4), 255—261 (1960).

[9] OLSZAK, W.: On Critical States in Viscoelasticity. In: Progr. Appl. Mech. pp. 333—348. New York: Macmillan Co. 1963.

[10] GUL, V. E.: Strength of Polymers. Moscow—Leningrad: Izd. Khimia. 1964. [In Russian.]

[11] SOKOLOVSKY, V. V.: One-Dimensional Unsteady Motion of a Visco-plastic Medium. Prikl. Matem. Mekhan. **13** (6), 623—632 (1949). [In Russian.]

Thermodynamics for Beginners

By

C. Truesdell
Baltimore, Md. (U.S.A.)

Antecedents

The title of this lecture is not intended as an insult. You are far from being beginners in thermodynamics; unfortunately, I, too, have suffered instruction in it. I believe that the recent work of COLEMAN and his associates has made thermodynamics, for the first time, a clear, definite theory that *can* be taught to beginners. COLEMAN treats deformable media subject to non-uniform temperature fields. He combines mechanical and thermodynamic principles at the highest level of difficulty. I wish to explain the central thermodynamic idea without these complications. The intended range of application is just that of classical thermodynamics: homogeneous blocks of matter which can be assigned a temperature, volume, and other parameters at each time. There results a theory that is simple, explicit, and embracing, in the same sense as the elementary "Newtonian" mechanics of "bodies" is simple, explicit, and embracing. It is not thermo*statics*, a theory of equilibria; rather, it is the simplest kind of true thermo*dynamics*. I wish I had been taught thermodynamics in this way; it would have saved me years of perplexity, frustration, and disgust. I am speaking to you, not as beginners yourselves, but as teachers, because I expect that anyone who is shown this way of looking at thermodynamics will adopt it at once and will use it to explain the theory of temperature and heat to his students.

Since none of us here is a beginner, each has his views and prejudices on thermodynamics which must be met by any new presentation. Therefore, as a preface to the formal theory I shall state what now seem to me the central, fatal faults of the older approaches. I say "approaches" because those are what I am criticizing, not the facts or final equations. The *results* of nineteenth-century thermodynamics have proved their value through myriad applications to specific problems and through their power of suggestion to theorists. We need to see a position for them in a framework of greater ideas, since otherwise their limitations, and hence also their validity, are obscure. I shall frequently draw upon the history of mechanics for parallels. Rational dynamics made immeasurable progress in the seventeenth century through the solutions, based on special concepts or, as we should now call them, devices, of a number of central special problems: descent along an inclined plane, ballistic motion, cycloidal oscillation, vibration of a simple spring, and some others. The solutions were correct, though to us they now seem awkward, indirect, and long, because we have learned from NEWTON and his successors a universal approach to the problems of elementary mechanics. We solve the older problems now in a few lines, by applying the general principles to appropriate special cases. That the seventeenth-century geometers could not do this was not so much from failure to know what the general laws of elementary mechanics assert as from failure to see that they were indeed general. Much was known that was true, in some cases more than enough, but there was no *order*, no *rank* of ideas. Nineteenth-century thermodynamics is similar. The thermodynamic substance was never allowed to diverge essentially from the ideal gas, and the thermodynamic process was never allowed to divagate far from equilibrium. Therefore, the basic theory was never disentangled from its main applications. To some people, this is a virtue. These same people, if they were consistent, ought to condemn NEWTON as the destroyer of "physical" mechanics, for it was he who told us that force is something bigger than gravity and elasticity and the few then known, measurable forces — force, *any* force, is something we can *imagine*, whether or not it exists in nature, and what we now teach beginners in mechanics first of all is to dream about all sorts of forces and the effects they would produce if they existed. One look into any textbook of thermodynamics shows that the theory of heat and temperature never was lifted to the level of abstraction NEWTON gave the theory of motion and force in 1687.

First, the classical authors insisted on *defining* temperature and heat in terms of something else, preferably something mechanical or the result of some primitive or imaginary experiment. As I have said for many years, disregarding the scoffs of those endowed with physical intuition, temperature and entropy join mass and place and time as primitive, undefined variables, described only by such properties as are laid down for them in mathematical terms. Whenever anyone tries to do anything rational with temperature and entropy, some physicist rises to ask whether a temperature always exists, and how to define entropy. The same physicist if asked whether forces exist will reply that he is not interested in philosophical questions, and if asked how to define mass and force is likely to explain that mass is a measure of the quantity of matter in a body, while force represents a push or a pull in a given direction. Such a person ought to be satisfied if I answer that temperature is a measure of how hot a body is, while entropy represents how much heat has gone into a body at a given temperature. Just as we expect that the student in his elementary courses in physics has already acquired regarding mechanics some truths we need to reinforce and some errors we need to correct, so also we may presume that he has had the benefit of the vocabulary and confusion imparted by an elementary course in thermodynamics.

I am not saying that temperature and entropy do not need to be connected with experience and experiment. Of course they do. So do time and mass and place and force. Such motivation, such experiment, belong in works on laboratory technique or the philosophy of physics; they are not in the province of the theorist who wishes to get something done. Just as NEWTON could look back on a century of sound if incoherent success by his predecessors in rational mechanics, we today may look back on a century of special successes in thermodynamics. This is our experience for theory. This is what makes it worthwhile to attempt a general formulation. In his great treatise, NEWTON writes not one word to tell us what a force is or how to measure one. His greatest gift to mechanics, a gift all his major contemporaries and many of his successors for nearly a century rejected, is an *a priori* concept of force. After describing a number of special forces, NEWTON[1] wrote in explanation, "And these quantities of forces we may for brevity call absolute, accelerating, and motive forces ... The concept is mathematical in any case. For I do not now ponder the causes and physical seats of forces." In the middle of his book, after presenting his brilliant solution of the problem of two bodies, NEWTON felt obliged to submit a new defense of his purely mathematical approach[2]: "In the same general sense I use the word impulse, since in this treatise I do not ponder the kinds and physical qualities of forces, but [their] mathematical quantities and proportions, as I explained in the definitions. In mathematics we are to investigate the quantities of forces and those reckonings which may follow from whatsoever conditions be laid down; hence, when we descend into physics, these reckonings are to be compared with the phenomena ..."

Nowadays the true nature of NEWTON's work is often mistaken. He put the word "mathematical" in the title of his book, in explanatory passages he used the words "mathematics" or "mathematician" frequently, and he always began them with a capital letter. It was NEWTON

[1] *Principia*, 1st ed., p. 4.
[2] *Principia*, 1st ed., pp. 191—192.

above all who taught us that the way to solve real physical problems is to shut off the physical talk and get down to equations. It was for this reason that the physicists of his own day largely refused to accept his views. His book was widely criticized as being a set of mathematical recipes, passing in silence over the real physics — what *is* space, what *are* forces? Time has judged for NEWTON's recipes, and against the cogitations of his physically-minded critics. Nineteenth-century thermodynamics had no NEWTON to give it a recipe for problem-solving. Instead, it churned over and over again the physical basis of what can be seen now to be a single, special problem of thermodynamics but what was then mistakenly esteemed the essence of the subject, indeed a theory of the "universe", that favorite term of gloomy thermodynamic prophets.

In any given, special motion, force can be defined, but such a definition is useless. Mechanics as we know it today is the science of forces given *a priori*; we use the forces to determine the motion, not the motion to calculate the forces. Similarly, thermodynamics is the science of temperature and heat given *a priori*, subject only to certain general axioms or laws, like NEWTON's laws in mechanics. NEWTON's laws are meaningless tautologies without a prior concept of force. The laws of thermodynamics, likewise, are relations among temperature, heat, and specifying parameters given *a priori*.

Second, Newtonian mechanics divides the theory into two parts. First there are the *general laws* of mechanics, relating forces and motions. By themselves, these laws are empty. Certainly they do not suffice alone to solve a single one of the problems NEWTON's predecessors, with their special theories, had solved, and solved correctly. Rather, they furnish us a framework upon which statements of special problems may be set. The "Newtonian equations" state common properties of all bodies as far as they are described by elementary mechanics. Particular ideal bodies are then defined by *constitutive equations*, which specify the way in which the forces are determined by aspects of the motion occurring. These constitutive equations, of course, are required to be consistent with the general laws. For example, the heavy body is defined to be one subject to a constant force, the linear spring as one subject to a force proportional to the distance from a fixed point, *etc.* Three or four constitutive equations exhaust mechanics before NEWTON's immense generalization of the subject. After his time, the distinction between the laws of mechanics and the constitutive equations of particular bodies came to be understood, and many more constitutive equations were suggested and studied. In nineteenth-century thermodynamics, as in pre-Newtonian mechanics, the general principles and the constitutive equations are stated or implied together. In fact, only two or three thermodynamic constitutive equations have been studied to any extent, and the classical books treat essentially only one.

My design here is to present the thermodynamics of homogeneous processes in two parts: first the general laws, and second the theory of constitutive equations. After this long preamble, I trust you will be patient enough to listen to what is indeed the elementary theory, fit for beginners.

The beginner does not ask questions of principle. Rather, he first learns the subject as it is, and only later turns back to probe its foundations. Our problem in thermodynamics is to teach the beginner something as clear, sound, and above all *specific* as he is taught in elementary mechanics. Those few beginners who continue their studies will then be better qualified to reconsider the foundations. Search of the foundations of Newtonian mechanics remained sterile so long as the searchers knew only Newtonian mechanics. NEWTON's own approach, which satisfied few students of the foundations, did the job as far as solving specific problems was concerned, and it still does the job today. It fails only in broader contexts such as theories of deformation and space-time structure. Likewise, the thermodynamics I shall present here fails by itself to enlighten us on deeper questions: theories of deformation or molecular structure. It inherits this failure from the nineteenth-century muddle it replaces. Unlike that muddle, but like Newtonian mechanics, it delivers the goods fairly and squarely, with clean, elementary mathematics, and a minimum of talk. That is all it is for. Let us return to it now.

Temperature and Calory

Experience makes it plain that mechanical action does not always give rise to mechanical effects alone. Doing work upon a body may make it hotter, and heating a body may cause it to do work. This fact suggests that alongside the mechanical working or *power* P should be set a *heat working* Q, such that the total working is $P + Q$. If K is the kinetic energy of motion, we recognize that not all work leads to motion and allow for storage of the remainder as *internal energy* E. Thus the *balance of energy* is asserted in the equation

$$\dot{K} + \dot{E} = P + Q, \tag{1}$$

where the dots denote time rates.

Much as the concept of place abstracts the idea of where a body is, the concept of *temperature* abstracts the idea of how hot it is. The temperature θ is a real number indicated on a thermometer, just as distance is a real number indicated on a ruler. Experience suggests that no matter what thermometer is used, there is a temperature below which no body can be cooled. That is, the temperature is bounded below. If this greatest lower bounded is assigned the temperature 0, the temperature is said to be "absolute":

$$\theta > 0, \tag{2}$$

no matter what unit of temperature be selected. We shall always adopt (2) for convenience.

Third, for any given body experience shows that there is an upper limit to the rate at which energy may increase without work's being done. That is, there exists an assignable *bound* B for $\dot{K} + \dot{E} - P$:

$$\dot{K} + \dot{E} - P \leqq B. \tag{3}$$

Equivalently, by (1), the heat working Q is bounded by B. Unlike the lower bound for the temperature, the upper bound for the heat working varies from body to body and depends on the circumstances in which the body finds itself.

In terms of the bound B, the *calory* H is defined by

$$H = \int \frac{B}{\theta}\, dt. \tag{4}$$

Thus the calory is defined only to within an arbitrary constant. The calory is of importance for the theory only through $\dot{H}$, the heating. Since $B = \theta \dot{H}$, (3) assumes the form

$$Q \leqq \theta \dot{H}. \tag{5}$$

A body stores calory, then, at least as fast as the heat working divided by the temperature. The hotter the body is, the greater must be the heat working in order to reach the maximum, which is determined by the heating.

Intended Application

The intended application of elementary thermodynamics is to blocks of material suffering no local deformation or variation of temperature. We describe the theory, therefore, as one of homogeneous processes in that all quantities occurring are *functions of time only*. Place plays no part. It is customary in the older studies to neglect motion and hence to set $K = 0$, but for later use we shall retain K, noting, however, that it occurs in the theory of homogeneous processes only in the combination $P - \dot{K}$. Since this quantity is the excess of the mechanical working over the kinetic energy rate, we shall call it the *work storage* and write it as W:

$$W \equiv P - \dot{K}. \tag{6}$$

In terms of it, the balance of energy (1) reads as

$$\dot{E} = W + Q. \tag{7}$$

In other presentations, what I call *calory* H is called "entropy"; (1) and (5), respectively, are among the various statements that physical writers call the "first and second laws of thermodynamics"; specifically, (5) is the *Clausius-Planck inequality*. This latter is just another way of writing our assumption (3) that Q is bounded. In any particular case, (3) would seem empty, since it is satisfied trivially by taking Q as its own upper bound. That is not what the law means. Rather, it will be interpreted as stating that we can assign a *universal* bound for all thermo-energetic charges in a given material. We shall make this idea precise a little later, but I mention it now in explanation of why we introduce H and use the equivalent statement (5) rather than the original one. Namely, it is more convenient in applications to specify the calory H rather than the bound B.

The Thermodynamic State

As in mechanics the concept of place occupied at a given time is primitive, in thermodynamics we consider a *state* at a given time. The state is a set of $k + 1$ real-valued functions of time:

$$\theta(t),\ \Upsilon_1(t), \ldots, \Upsilon_k(t),$$

or, briefly,

$$\theta(t),\ \Upsilon(t).$$

The temperature $\theta(t)$ is a positive function: $\theta(t) > 0$. The k *parameters* $\Upsilon_j(t)$ also are quantities given *a priori*. In a particular case, Υ_1 might be the volume of a body, and $\Upsilon_2, \ldots, \Upsilon_k$ might be the masses of $k-1$ constituents, but for the theory, no specific interpretation is necessary. We agree to consider only unconstrained states in the sense that the functions $\theta(t)$, $\Upsilon(t)$ may be arbitrary, except that $\theta(t) > 0$. In particular, at any one time Υ, $\dot{\theta}$, and $\dot{\Upsilon}$ may assume any real values whatever, and θ may assume any positive value. (I specifically do *not* demand that $\theta = 0$ be excluded from particular applications as a limit case, but there is no need to consider that case here.)

Thermodynamic Processes

A *thermodynamic process* is a set of $k + 5$ functions of time,

$$\theta(t),\ \Upsilon(t),\ W(t),\ E(t),\ Q(t),\ H(t),$$

that satisfy the *first and second axioms of thermodynamics:*

$$\dot{E} = W + Q, \tag{7}$$

$$\theta > 0, \tag{2}$$

If all but the last two of these functions are given, $Q(t)$ is uniquely determined by (7). Hence it is equivalent to say that a thermodynamic process is a set of $k + 4$ functions

$$\theta(t),\ \Upsilon(t),\ W(t),\ E(t),\ H(t)$$

such that (2) holds.

A process in which

$$\left.\begin{array}{l} Q = 0 \\ \dot{H} = 0 \end{array}\right\} \quad \text{is called} \quad \left\{\begin{array}{l} \text{adiabatic} \\ \text{isentropic.} \end{array}\right.$$

A process in which equality holds in (5) at all times is called *reversible*. All other processes are called *irreversible*. In a reversible process, $Q = \theta\dot{H}$; in reversible processes "adiabatic" and "isentropic" have the same meaning. In an irreversible adiabatic process, the calory increases: $\dot{H} > 0$. In an irreversible isentropic process, the heat working is negative: $Q < 0$.

The Reduced Dissipation Inequality

If we eliminate Q between (7) and (5), we obtain the *reduced dissipation inequality:*

$$\dot{E} - W \leqq \theta\dot{H}. \tag{8}$$

The *free energy* Ψ is defined by

$$\Psi \equiv E - \theta H. \tag{9}$$

In terms of it, the reduced dissipation inequality becomes

$$\dot{\Psi} + H\dot{\theta} - W \leqq 0. \tag{10}$$

This is the form of the reduced dissipation inequality we shall use for proving theorems of thermodynamics.

Thermodynamic Constitutive Equations

No problems can be solved by means of the thermodynamic axioms alone. In particular, they do not suffice to determine thermodynamic processes. Different natural materials react in different ways to the application of heat and temperature. The general theory affords us a means of representing this diversity. Namely, we lay down for study some particular, simple relation among the functions making up the process. Such a relation must be consistent with the axioms. If it is, it gives us an example of how a material may behave. With some help from experiment, we may even come close to describing how real bodies do behave, to the extent that they can be idealized as homogeneous blocks.

If $f(t)$ is a function of time, its *history* $f^t(s)$ up to time t is defined by $f^t(s) \equiv f(t-s), 0 \leq s < \infty$. The *state history* is the set of $k+1$ functions $\theta^t(s)$, $\Upsilon^t_k(s)$. The *principle of thermodynamic determinism* asserts that the state history determines the work storage, the free energy, and the calory. That is,

$$\left.\begin{array}{l} W \\ \Psi \\ H \end{array}\right\} \text{ are functionals of } \left\{\begin{array}{l} \theta^t \\ \\ \Upsilon^t \end{array}\right. . \tag{11}$$

Such a set of three relations is called a *thermodynamic constitutive equation*.

Not all functionals are admissible, however. In order to define a *thermodynamic material*, we require that the constitutive equation be compatible with the reduced dissipation inequality (10) *for all thermodynamic processes*.

Example 1: The Classical Caloric Equation of State

The standard applications of thermodynamics concern only, or almost only, one extremely special class of constitutive equations. The work storage is thought of as being purely mechanical and also linear and homogeneous in the "velocities" $\dot{\Upsilon}_a$:

$$W = -\sum_{a=1}^{k} \varpi_a(\theta, \Upsilon)\, \dot{\Upsilon}_a, \tag{12}$$

where the coefficients ϖ_a are called *thermodynamic pressures*, and the free energy and calory are given by *caloric equations of state:*

$$\begin{aligned} \Psi &= \Psi(\theta, \Upsilon), \\ H &= H(\theta, \Upsilon). \end{aligned} \tag{13}$$

Substitution of (12) and (13) into the reduced dissipation inequality (10) yields

$$(H + \partial_\theta \Psi)\,\dot{\theta} + \sum_{a=1}^{k} (\varpi_a + \partial_{\Upsilon_a} \Psi)\,\dot{\Upsilon}_a \leqq 0. \tag{14}$$

The coefficient of $\dot{\theta}$ and of $\dot{\Upsilon}_a$ are functions of θ and Υ. Since for any given values of $\theta(t)$ and $\Upsilon(t)$ at a fixed time t we may construct a process in which $\dot{\theta}(t)$ and $\dot{\Upsilon}(t)$ have any values we wish, (14) holds if and only if each summand vanishes:

$$\begin{aligned} H &= -\partial_\theta \Psi, \\ \varpi_a &= -\partial_{\Upsilon_a} \Psi. \end{aligned} \tag{15}$$

We have proved, then, that the $k + 2$ functions ϖ_a, Ψ, and H are not independent in a thermodynamic constitutive equation. The calory and the thermodynamic pressures are determined from the free-energy function $\Psi(\theta, \Upsilon)$ as a potential. Substitution back into (14) shows that equality holds always there. That is, *there are no irreversible processes*, and a process is adiabatic if and only if it is isentropic.

The contents of classical books on thermodynamics consist mainly in ringing the changes on (15).

Parenthesis

If I were really explaining thermodynamics to beginners, I should turn aside here to the use and interpretation of other thermodynamic potentials and to proof of inequalities and identities concerning them. Since this body of relations is essentially classical, once we are this far along, there is no need to go over it in this lecture. I remark only that the variants may be obtained either directly by an argument of the kind I have just given, or by transformation of the results already obtained.

Example 2: Linear Friction

The existence of the classical potentials, as we have just seen, is a consequence, not of any general thermodynamic ideas, but of the very special constitutive equations (12) and (13) taken as the starting point. If we generalize these assumptions just a little, we no longer get the classical results. Namely, if we include the effect of "linear friction", the work storage W is taken as a quadratic instead of a linear function of the parameter rates $\dot{\Upsilon}_a$:

$$W = -\sum_{a=1}^{k} \varpi_a^0(\theta, \Upsilon)\,\dot{\Upsilon}_a - \sum_{a,b=1}^{k} \varpi_{ab}(\theta, \Upsilon)\,\dot{\Upsilon}_a \dot{\Upsilon}_b, \tag{16}$$

while the caloric equations of state (13) are retained unchanged. Here $\varpi_a^0(\theta, \Upsilon)$ is the "equilibrium pressure", and the coefficients $\varpi_{ab}(\theta, \Upsilon)$ may be called "frictional pressures". Substitution into the reduced dissipation inequality (10) yields now

$$(H + \partial_\theta \Psi)\,\dot{\theta} + \sum_{a=1}^{k} (\varpi_a^0 + \partial_{\Upsilon_a} \Psi)\,\dot{\Upsilon}_a + \sum_{a,b=1}^{k} \varpi_{ab} \dot{\Upsilon}_a \dot{\Upsilon}_b \leqq 0\,. \tag{17}$$

We shall analyze this inequality somewhat more carefully. Since the coefficients of $\dot{\theta}$, $\dot{\Upsilon}_a$, and $\dot{\Upsilon}_a \dot{\Upsilon}_b$ are functions of θ and Υ only, and since we may construct a thermodynamic process in which the $2k + 2$ quantities θ, Υ_a, $\dot{\theta}$, $\dot{\Upsilon}_a$ have any values we choose to assign at the time t, provided only $\theta > 0$, the inequality (17) is of the type

$$\sum_i A_i x_i + \sum_{i,j} A_{ij} x_i x_j \leqq 0 \tag{18}$$

for all real x_k, where the A_i and A_{ij} are constants. First, let the x_k be particular numbers; then (18) must hold if they are replaced by αx_k, where α is any constant. Therefore for all α

$$\alpha \sum_i A_i x_i + \alpha^2 \sum_{i,j} A_{ij} x_i x_j \leqq 0. \tag{19}$$

This quadratic inequality is satisfied for all α if and only if the coefficient of α is 0 while the coefficient of α^2 is non-positive. That is,

$$\sum_i A_i x_i = 0, \quad \sum_{i,j} A_{ij} x_i x_j \leqq 0. \tag{20}$$

These relations are to be satisfied for all x_i. Hence

$$A_i = 0. \tag{21}$$

Conversely, if (21) and $(20)_2$ hold, (18) is satisfied, Applying these results to (17) shows that that inequality is equivalent to

$$\begin{aligned} H &= -\partial_\theta \Psi, \\ \varpi_a^0 &= -\partial_{\Upsilon_a} \Psi, \\ \sum_{a,b=1}^{k} \varpi_{ab}(\theta, \Upsilon)\, \dot{\Upsilon}_a \dot{\Upsilon}_b &\leqq 0. \end{aligned} \tag{22}$$

Collecting our results, we see that $\Psi(\theta, \Upsilon)$ is now a potential for H and the equilibrium pressures ϖ_a^0 but stands in no relation to the frictional pressures ϖ_{ab}. The free-energy function, then, no longer determines all constitutive equations of the material, since the frictional pressures ϖ_{ab} are independent of it. The frictional pressures, however, are not arbitrary. For each value of the thermodynamic state, the quadratic form of ϖ_{ab} must be non-positive definite. Conversely, if (22) holds, the assumed constitutive equations satisfy the reduced dissipation inequality.

If we substitute $(22)_{2,3}$ and (16) back into (7), we find that

$$\dot{H} = \frac{Q}{\theta} + \varDelta, \tag{23}$$

where

$$\theta \varDelta = - \sum_{a,b=1}^{k} \varpi_{ab}(\theta, Y)\, \dot{\Upsilon}_a \dot{\Upsilon}_b. \tag{24}$$

The positive-definite quadratic form $\varDelta(\theta, \Upsilon, \dot{\Upsilon})$ is called the *heat gain*. Comparison of (23) with (5) shows that in the present kind of material, a process is reversible if and only if the heat gain is zero. In particular, an adiabatic process is not generally isentropic; indeed, in an adiabatic process the calory generally increases, and an isentropic process generally is not adiabatic but requires negative heat working.

Looking back at the constitutive equation (16), which was our starting point, we see that the value of the work storage W is unaffected by the value of $\varpi_{ab} - \varpi_{ba}$. The heat gain (24) is likewise unaffected by this quantity. Thus there is no loss in generality in supposing that

$$\varpi_{ab} = \varpi_{ba}\,. \tag{25}$$

for all a and b. If we make this assumption and set

$$\mathcal{F}_a \equiv - \sum_{b=1}^{k} \varpi_{ab} \dot{\Upsilon}_b, \tag{26}$$

then not only do we have

$$\theta \varDelta = \sum_{a=1}^{k} \mathcal{F}_a \dot{\Upsilon}_a, \tag{27}$$

but also

$$\mathcal{F}_a = - \partial_{\dot{\Upsilon}_a}(\theta \varDelta)\,. \tag{28}$$

That is, $\theta\varDelta$ is a potential for $\mathcal{F}_a$. This potential is sometimes called "RAYLEIGH's dissipation function".

An assertion of symmetry of the form (25) is called a "reciprocity relation". In the theory of linear friction, as we have seen, it is trivially justified to demand that the reciprocity relations (25) shall hold.

Parenthesis for the Non-Beginner

Much has been made in the literature over reciprocity relations. It seems to me that all the fuss can be explained by the mere remark that up to now, thermodynamics was approached through detours in the fog of word-play. Once we agree that a thermodynamic constitutive equation specifies the work storage W, it becomes mathematically trivial and physically obvious that the magnitude of $\varpi_{ab} - \varpi_{ba}$ cancels out of every equation in which ϖ_{ab} appears, so that speculation about what value it may have can never be tested by comparing the results of the theory with experiment. We might as well set $\varpi_{ab} = \varpi_{ba}$, although if we do not wish to, we do not have to. It simply makes no difference in any result the theory is capable of predicting. Clearly this observation is independent of the choice of parameters Υ_a, so long as they correspond to a constitutive equation of the type considered here, namely (16). We may observe that if we replace the parameters by any linear combination,

$$\Upsilon_a^* = \sum_{b=1}^{k} A_{ab}\Upsilon_b + C_a, \quad a = 1, 2, \ldots k, \tag{29}$$

where the A_{ab} and C_a are constants, an equation of the same form results for W in terms of the $\dot{\Upsilon}_a^*$, equally trivially. The new coefficients ϖ_{ab}^* may be taken as symmetric if we please. This observation, too, has an imposing literature.

(Here I whisper aside to our colleagues who prefer to see everything in terms of expectations and fluctuations that a treatment of the corresponding problem in statistical thermodynamics likewise rests upon calculation of a differential change of energy and hence, trivially, *cannot* lead to anything more general than $\varpi_{ab} = \varpi_{ba}$, no matter how "forces" and "fluxes" be defined statistically, provided only that a constitutive equation of the form (16) somehow emerges.)

The observation that I have made is not new. That symmetry or lack of symmetry of the coefficients ϖ_{ab} is empty in terms of energy and entropy calculations has been remarked many times. This old fact takes on a new meaning here, because we have shown that considerations of energy and entropy alone, with no new axioms, *suffice* to get everything else we understand under the term "thermodynamics". If we desire more information about ϖ_{ab}, we have to introduce ϖ_{ab} itself from considerations *outside* pure thermodynamics. For example, we might suppose mechanics or electromagnetism already given, in which case we might get non-thermodynamic information about the ϖ_{ab}, which quantities then should be *proved* to enter a relation such as (16), not merely *defined* by laying down such a relation. I shall make this simple remark explicit in the appendix.

Closure

The part of classical thermodynamics that depends upon caloric equations of state has been separated from the general principles. The theory, both in general and in the two applications worked out, is clear, elementary, easy, and rigorous. By following it the teacher may clear away the mists of "intuition" from thermodynamics. If his interest is in the applications of particular equations of state, he is free to begin his applying after far less time and effort spent on foundations. If he wishes to develop new cases, he sees open before him the prospect of a general theory based on functionals, much like one-dimensional visco-elasticity, on the same foundation as the classical theory of equations of state. If he wishes to generalize, he will find that the new thermodynamics of deformable continua is just as easy a step from this thermodynamics of homogeneous processes as ordinary continuum mechanics is from the elementary mechanics of discrete systems, or easier. Nothing has to be unlearned. If, finally, he wishes to interpret gross thermodynamics in terms of molecular actions, he can see at once what is to be interpreted. Most important here is the separation between the general principles and the constitutive equations. In a case when statistical mechanics fails to deliver a caloric equation of state, no inconsistency with the general principles of gross thermodynamics need be implied. Statistical mechanics is set free from its traditional shackles[1] to the most primitive thermodynamics, and a great new range of interrelation between statistical and gross theories of heat and temperature opens before us.

Appendix: Extension to Fields

Nowadays we are interested primarily in fields, not merely functions of time. These occur in problems of heat conduction, deformation, and diffusion. Unfortunately, the simple properties of the constitutive equation (16) do not generalize to fields. While the thermodynamics of homogeneous processes takes account of mechanical actions *only through their effects on energy*, in field problems we have to add *force* itself as a

---

[1] "Statistical mechanics" and "statistical thermodynamics" mean different things to different people. Of course it has been known for a century that the thermodynamics of the text books is in some sense only a first approximation to what statistical mechanics can deliver, at least in principle. Writers on statistical mechanics often fail to see that, likewise, textbook thermodynamics is only a first approximation to what a real phenomenological theory of heat and temperature can deliver. Such a theory is presented in this lecture. In view of it, students of statistical mechanics can no longer justly describe the approximate status of equations of state as an approximate status for phenomenological thermodynamics as a whole.

primitive concept and discuss constitutive equations for forces. Such is mechanics. Now if we had somehow formulated a thermodynamics of homogeneous processes in which pressures such as ϖ_{ab} had entered from the start, as primitive quantities, not merely as coefficients in an equation defining a special material, we should have had to recast everything in such a way as to make ϖ_{ab} a quantity delivered by a constitutive equation. It would then have required a *special assumption*, not a mathematical triviality, to get the symmetry relation $\varpi_{ab} = \varpi_{ba}$, and there *would have been no other way* to get it. It would correspond, naturally, to a *special choice* of parameters Υ_a. All the numerous attempts to formulate symmetry relations for fields have failed for just this reason. They are not consistent as stated, and I do not know any way to make a consistent assertion of the kind writers on the linear "thermodynamics of irreversible processes" seem unanimously to desire. If more evidence of this failure be needed, it can be found in the publication of at least twenty papers per year whose expressed sole purpose is to "prove" or "justify" the "Onsager relations", despite the fact that "proofs" and "justifications" galore can be found in hundreds of earlier papers and in a dozen authoritative textbooks. In no other branch of classical physics is there evidence of need felt for such continuing reaffirmation.

The difficulty here reflects an old, old division between mechanics and thermodynamics. Thermodynamic writers have always tried to do everything with energy and in particular to define forces in terms of energy. This can be done for homogeneous processes. My approach in this lecture, while I hope clean and neat, has been of the same kind, and has yielded, by simple and rigorous mathematics, the same kind of results. In mechanics, which is the science of forces given *a priori*, this approach has never worked and, I think, never will. To expect it to do so is illusory.

Neither mechanics nor thermodynamics can buy out the other, but they can go into partnership. A combined thermodynamics rests upon use of the primitive concepts of both separate theories, and upon the basic laws of *both*. Such a thermodynamics has been constructed by COLEMAN and has been generalized by WANG and BOWEN. In it, of course, the same special constitutive equations are *allowed* as may occur in either of the older, separate theories, but they lose their privileged status. In particular, while symmetric arrays of coefficients may occur, either they result from some other axiom such as frame-indifference (*e. g.*, in the NAVIER-STOKES theory) or from some assumption of material symmetry (*e. g.*, in heat conduction), or they indicate a happy chance in the choice of variables, not any general principle.

The investigation reported here was sustained in part by a grant from the U.S. National Science Foundation to the Johns Hopkins University.

Discussion

KLUITENBERG: Can you say something more on the nature of these functionals? Is in this type of theory stress- and strain-relaxation included?

TRUESDELL: This lecture did not concern continuum mechanics but rather an elementary thermodynamics in which the role of a particular class of constitutive equations is clearly separated from the general principles of the subject. I don't think it's proper to talk about stress-relaxation and strain-relaxation here. Just as you teach to a student Newtonian mechanics in some form or other before you teach him continuum mechanics, you should teach him also some form of thermodynamics. Actually what I have presented here is a lecture on the principles which COLEMAN uses in his thermodynamics of continua, which is certainly general enough to include the effects that you wish. Perhaps one of the things that might be slightly new in this lecture is its remarking the possibility of a memory theory in pure thermodynamics. This has never appeared in the literature, as far as I know. Perhaps it is too easy.

KLUITENBERG: Have you any simple restrictions when you say functionals?

TRUESDELL: Yes. They must satisfy the principle of dissipation. Explicitly: in order to define a material, the functionals must be such that you can imagine no situation in which the material would disobey the second law. The question of invariance which you have in continuum mechanics does not appear in this zero-dimensional theory.

LEE: The standpoint you present things is, as you mentioned, in teaching thermodynamics. It seems to me it's worrying that the entropy is defined in terms of a bound B, which is simply defined by an inequality. Don't you think, one should make some more definite statement than this?

TRUESDELL: A certain amount of motivation is necessary, depending on the level at which you are teaching. When you teach Newtonian mechanics, you may or may not spend a lot of time telling people what forces and masses are. Likewise, in thermodynamics you may or may not spend a lot of time motivating temperature and entropy.

LEE: Well, one is used to working with equalities rather than inequalities, and an equality is needed to define a value precisely.

TRUESDELL: Is that really true, nowadays? I just followed a course in analysis, and an equality was the rarest thing in it.

KALISKI: It is not clear to me, what is the physical sense of temperature in this theory. For instance, if you consider a very high-frequency field with considerable times of relaxation in the period of the vibrations, then in classical thermodynamics the temperature loses its physical meaning.

TRUESDELL: I think the physical explanation of temperature is a thing which does not belong in a course on mathematical thermodynamics, any more than physical explanation of mass belongs in a course on mathematical mechanics. As for "physical meaning", I claim no physical applicability for anything I ever say. That is, the applications of the theory are left to the appliers. The theorist attempts to get the theory clear and to prove the theorems. Whether a theory applies to a given piece of material at a given time is something very important but something that the theorist cannot be expected to tell, in thermodynamics or any other theory.

MEIXNER: Could you explain what is θ? How can I use it? How can I measure it?

TRUESDELL: That's a question that people are always asking me. You open any book on thermodynamics, and there is a temperature. There is nothing new or different here.

KESTIN: The simple difference between the presentation which was given us today and the presentation which we give to students who begin their studies in thermodynamics, is the concept of an equilibrium state. We start with it as an axiomatic concept and assert that every elementary part of a large system obeys the same equation of state as the uniform system does. And we recognize then the appearance of a new class of quantities, which describe the interactions of the system elementary in the size with the surroundings in the generalized fluxes. And then we seek ways to describe these fluxes as was done in Professor MEIXNER's article in the Handbuch der Physik or in the book by MAZUR and DE GROOT and so on. I would like to end my statement with another axiom and that is, that that field which exceeds the normal field, in which we use the concept of state has no physical meaning.

COLEMAN: In a manuscript now in preparation, Professor MIZEL and I show that the thermodynamics of homogeneous systems can be presented as a consequence of a stability hypothesis on the equations governing the evolution of the system under consideration. Although the point of view is quite different, the results we obtain are in agreement with Professor TRUESDELL's presentation. An interesting feature of our new axiomatization is that it permits us to *prove* the existence of entropy- and temperature-functionals.

UHLHORN: Your system is called a body and I should like to know, does that mean what is called by thermodynamics a closed system.

TRUESDELL: I never got a clear idea what thermodynamicists mean by a closed system. I visualize the system as a finite portion cut out of an infinite homogeneous medium.

UHLHORN: This is called a closed system by thermodynamicists.

Thermomechanics of a Continuous System with Internal Structure

By

U. Uhlhorn

Lund (Sweden)

Summary

A motion of a three dimensional continuous body, admitting at each particle an internal structure of an unspecified nature, is represented in terms of a definite number of generalized co-ordinate fields which are given as functions of time and particle co-ordinates. With each generalized co-ordinate field is associated a generalized velocity, a generalized specific momentum, a generalized body force and a generalized contact force. Equations of motion are postulated in the form of balance equations for the generalized momenta. Balance of energy is postulated as an equation relating the rate of total energy change to the power of generalized contact and body forces and the supply of heat. The entropy principle is adopted in the form laid down by TRUESDELL and TOUPIN and a thermomechanical theory is developed in analogy to the scheme for simple materials given by COLEMAN and NOLL. Only an outline of the general theory is given in the present report.

A material continuum with internal structure may be considered as a generalization of the familiar COSSERAT continuum [1]. A particle in the continuum is characterized by its material co-ordinates X, Y, Z. The state of a particle is determined by a definite number N (the same for all particles) of generalized co-ordinates $Q^1, \ldots, Q^N$ (analogous to the case in analytical mechanics). The generalized co-ordinates are assumed to determine uniquely the position x, y, z of the particle in space by relations of the form

$$x = f(X, Y, Z, Q^1, \ldots, Q^N),$$

and also the values of certain adjoined physical parameters (in the sense of HELLINGER [2]) of a mechanical or non mechanical nature, e. g. the chemical state of the particle. Remark: In this approach we assume that there is only one kind of particles so that a point in space is at a given time occupied by at most one particle. The chemical state of a particle, represented e. g. by the concentrations of the chemical constituents, may however be variable. A continuous system in which diffusion and chemical reactions can take place may also be treated as a superposition of several constituents, each constituent being considered as a material continuum. This is the approach advocated and developed by TRUESDELL [3]. The two approaches are not equivalent but under certain special conditions the first approach may be considered a limiting case of the second. Both approaches have been studied by DUHEM [4].

The following general discussion is largely independent of the particular nature of the generalized co-ordinates.

We define a motion of the material continuum by prescribing the generalized co-ordinates as functions of the time and the material co-ordinates. In this formulation of the kinematics in terms of the set of variables $t, X, Y, Z; Q^1, \ldots, Q^N$, the time t and the material co-ordinates X, Y, Z are used as independent variables and the generalized co-ordinates $Q^1, \ldots, Q^N$ as dependent variables. We call this a material formulation of the kinematics in contrast to a field formulation, in which the set of variables t, X, Y, Z; $Q^1, \ldots, Q^N$ is replaced by another set t, x, y, z; q^1, $\ldots$, q^N, which is related to it by a non-singular transformation. In the field formulation the time t and the spatial particle co-ordinates x, y, z are used as independent variables and the generalized co-ordinates $q^1, \ldots q^N$ as dependent variables. It is obviously not possible to use the same set of generalized co-ordinates in the field formulation as in the material formulation.

For heuristical reasons dynamics is first introduced by an approach similar to that used by the brothers COSSERAT. Equations of motion are derived from a variational principle which can be given both a material formulation and a field formulation corresponding respectively to the usual procedure in continuum mechanics and in the classical field theories of physics (electromagnetic theory, gravitational theory etc.). The spatial form of the variational principle may be considered as an extended version of a principle for hyperelastic bodies, and in particular for elastic fluids, given by HÖLDER [5] which may also, in a slightly different form, be associated with the names of HERIVEL, LIN and SERRIN [6]. The field equations obtained as Lagrangean equations of the variational problem are the equations of motion of a generalized, hyperelastic material. The Cosserat continuum is included as a special case. The generalized hyperelastic material is said to be of the first grade if the corresponding Lagrangean density of the variational problem is a function only of the four independent variables, the generalized co-ordinates and their first order derivatives with respect the independent variables, but does not depend on the higher order derivatives of the generalized co-ordinates. If this condition is not satisfied the material is said to be of higher grade, the grade of the material being the order of the derivative of highest order in the Lagrangean density. Hyperelastic materials (with no internal structure) of grade two have been studied by TOUPIN [7]. Fluids which are of grade two in this sense have been considered long ago by KORTEWEG [8] in connection with a proposed continuum theory of capillary forces. It is interesting to note that superfluid helium II, which is frequently described as a quantum liquid or more properly as a Bose liquid, may be considered as a classical fluid of grade two. By using an approach of the kind described here it is in fact possible to derive equations of motion for an extended two-fluid model of liquid helium II in which the motion of the superfluid constituent is not artificially restricted to potential flow.

In the general case the Lagrangean field equations obtained from the variational principle are invariant under arbitrary changes of generalized co-ordinates and in fact even under much more general transformations. As in the special case of the Cosserat continuum it is sometimes convenient to use an anholonomic co-ordinate base for the representation of the gradients of the generalized co-ordinates with respect to the independent variables. The balance of energy is studied by means of the so called canonical tensor and certain expressions is the energy equation are unambigously identified as the power of generalized contact and body forces corresponding to the arbitrary assigned generalized velocity field.

Guided by the heuristic approach we then postulate a generalization of Cauchy's first law of motion which relates the flux of generalized momentum to the action of generalized contact and body forces. The generalized contact forces are determined by generalized stress tensors. Balance of energy is postulated as an equation relating the rate of total energy change (in a theory of the degree of generality presented here there is of course no natural decomposition of the total energy into kinetic and internal) to the power of generalized contact and body forces and the caloric power. The caloric power is determined by the flux of entropy and the local supply of entropy from external sources. The formula for the caloric power is similar to that for the mechanical power. The entropy flux and the local supply of entropy correspond to stress and body force respectively whereas the temperature plays the role of a generalized velocity, in agreement with some classical observations by HELMHOLTZ [9]. The entropy principle is adopted in the form of the Clausius-Duhem inequality laid down by TRUESDELL and TOUPIN.

By combination of the generalized Cauchy law of motion with the energy and entropy principles in their local forms, an inequality restricting the rate of change of the entropy is obtained, after elimination of the generalized *body* forces and the specific heat absorption. Several equivalent formulations of this dissipation-inequality can be obtained by Legendre transformations.

At this stage there are introduced constitutive equations by which the energy, the entropy, the heat flux and the generalized stress tensors are given as definite functions of the configuration, configuration gradient and temperature histories and the actual temperature gradient. As in the analysis of simple materials by COLEMAN and NOLL [10] there are some interesting particular cases. Thus under certain conditions it can be proved that the total energy is a thermodynamic and a kinetic potential, being also a function of the generalized momenta. This is a generalization of classical results due to HELMHOLTZ, NATANSON, and in particular to DUHEM [4].

Finally the formal generality of the present approach is restricted by consideration of invariance requirements. The balance equations for linear momentum and moment of momentum in their completely general form, presumably first given by the COSSERATS [11] (see also TRUESDELL [12]), are required to be consequences of the general equations of motion for the generalized co-ordinates. In order to satisfy this requirement it is necessary to prescribe the transformation properties of the generalized co-ordinates corresponding to an arbitrary *rigid displacement* of the body. For instance in the particular case of the Cosserat continuum the three-leg of mutually orthogonal unit vectors in each particle is assumed to rotate with the body in a rigid displacement. A detailed exposition will be given in a forthcoming paper.

References

[1] COSSERAT, E. et F.: Théorie des corps déformables. Paris 1909.
[2] HELLINGER, E.: Die Allgemeinen Ansätze der Mechanik der Kontinua. Encyklopädie der mathematischen Wissenschaften. Vol. 4, No. 30 (1913).
[3] TRUESDELL, C.: Rend. Lincei **22** (8), 33—38 (1957).
[4] DUHEM, P.: Trav. Mém. Fac. Lille **3** (2) (1893); Théorie thermodynamique de la viscosité du frottement et des faux équilibres chimiques. Paris 1896.
[5] HÖLDER, E.: Leipziger Berichte **97** (2) (1950).
[6] SERRIN, J.: Mathematical Principles of Classical Fluid Mechanics. In: Handbuch der Physik. Vol. VIII/1. Berlin—Göttingen—Heidelberg: Springer 1959.
[7] TOUPIN, R. A.: Arch. Ratl Mech. Anal. **17**, 85 (1964).
[8] TRUESDELL, C., and W. NOLL: The Non-Linear Field Theories of Mechanics. Handbuch der Physik. Vol. III/3, p. 513. Berlin—Göttingen—Heidelberg: Springer 1965.
[9] HELMHOLTZ, H. v.: Wissenschaftliche Abh. **3**, 203 (1895).
[10] COLEMAN, B. D., and W. NOLL: Arch. Ratl Mech. Anal. **13**, 167 (1963).
[11] COSSERAT, E. et F.: loc. cit. p. 167.
[12] TRUESDELL, C.: ZAMM **44**, 11 (1964).

Die Beeinflussung der Intensität des Transportprozesses durch das dynamische Verhalten der Phasengrenze Flüssigkeit-Gas oder Flüssigkeit-Dampf

Von

J. Zemánek
Běchovice (ČSSR)

Summary

The present scientific knowledge is evaluated and the basic physical models of transport processes at the phase boundary between liquid and gas (vapour) are mentioned. Attention is paid to the different transfer resistances appearing in the separate phases and at their boundary, and the significance of dynamic effects of the phase boundary for an intensification of the transfer process is pointed out.

A strong influence on mass transfer intensity in the gaseous phase is demonstrated on several generalized examples of original measurements concerning water evaporation (condensation) into turbulent air stream in film systems. Partial results in criterial form $Nu_{kor} = f(Re_g)$, $Nu_{kor} = f(Re_{gr})$ are given in Fig. 1, and by the relation $C_r = f(Re_f)$, eq. (3), in Fig. 2. The dynamic effect is characterized by the Reynolds number of the liquid phase Re_f and the modified pulsating Reynolds number $Re_{\sim} = 2fvd/v_g$. Particular attention is paid to the influence of liquid distribution and to the extent of wetting, especially in the region of low Reynolds number Re_f and by using hydrofobe plastic materials for film packings.

Eq. (5), the energetic equation, relates the intensification of transport processes to the pressure loss. In an evaluation of condensing coefficients k_a, which are characteristic of the phase boundary resistence R_s, attention is called to a considerable scatter of experimental and theoretical data, published in the scientific literature (Fig. 3 and 4). The favourable guiding data for k_a are given.

An equivalent transfer coefficient, eq. (21) and (22), is introduced in order to characterize the transfer resistance in the liquid phase.

In Table 1 and Fig. 6 a survey of modified formulas for this coefficient in the region of laminar, pseudolaminar and turbulent flow of liquid layer is given.

Fig. 6 presents a survey of various different theoretical and experimental systems and an illustrative demonstration of the high intensification of transport processes by two-phase systems, working with high hydraulic loading (Re_f).

For film systems the criterial expression $\chi \sim Re_f^s$ with values $s = 1{,}4 - 1{,}5$ is experimentally obtained in Graph 6. The results are of direct significance for the design of intensified contact exchangers.

Zusammenfassung

Übersicht und Beurteilung der jetzigen wissenschaftlichen Erkenntnisse und der physikalischen Modelle, die die Transportprozesse an der Phasengrenze Flüssigkeit-Gas oder Flüssigkeit-Dampf bei den Rieselfilm-, teilweise auch bei den Flachstrahlsystemen beschreiben. Die aus verschiedenen Modellvorstellungen abgeleiteten resultierenden Kennzahlbeziehungen ermöglichen den Vergleich mit Experimentalergebnissen und gleichzeitig auch vereinfachte Beschreibungsmethoden zu wählen. Dabei wird es auch möglich, Erkenntnisse über die Einflüsse der Phasengrenzendynamik zur Intensivierung des Transportprozesses zu gewinnen.

Die Abhandlung befaßt sich mit aktuellen Fragen aus der Thermodynamik und Hydrodynamik der Rieseleinbauten, die man bei Kontaktaustauschprozessen, d. h. bei der Verdunstung, Kondensation, Verdampfung, Absorption usw. mit Vorzug verwendet. Die Arbeit beschränkt sich auf Zweiphasensysteme Gas-Flüssigkeit und Sattdampf-Flüssigkeit. Am Rande werden auch die Eigenschaften der Flachstrahlen erwähnt.

Da sich die exakte theoretische Lösung des Problemes unseren Möglichkeiten entzieht, erscheint es als das Beste, die obenangeführte Problematik mittels vereinfachter Modellvorstellung zu verfolgen [1].

Zu solchen Modellvorstellungen gehört z. B. das Modell der drei additiven Teiltransportwiderstände, wonach der Gesamttransportwiderstand durch den Teilwiderstand in der Gasphase (Dampfphase), an der Phasengrenze und in der Flüssigkeitsschicht bestimmt wird. Dabei verstehen wir unter dem Begriff Transportwiderstand den Wärmeaustausch oder Stoffaustauschwiderstand.

Im Falle, daß wir den Teilwiderstand an der Phasengrenze vernachlässigen können, bekommen wir das bekannte Lewis-Whitmannsche Doppelfilmmodell [2] für das Gas-Flüssigkeitssystem. Die Zulässigkeit dieser Modellvorstellung wurde z. Z. von Szekely und King [3] überprüft und praktisch auch für den Fall bestätigt, daß in der Flüssigkeitsschicht Higbies Oberflächenerneuerungsmodell benützt wurde. Dieses Doppelfilm-Modell wurde weiter von Mickley [4], Trevissoi [5] und Tanaka [6] durchgearbeitet. Außerhalb des Rahmens dieser Abhandlung steht das bisher quantitativ nur teilweise bearbeitete Oberflächenerneuerungsmodell von Higbie oder Dankwert. Bei dem Zweiphasensystem Flüssigkeit-Gas beschreiben wir oft den gleichzeitig durchlaufenden Transportprozeß durch die Enthalpiedifferenz. Bei vernachlässigbarem Stoffaustauschwiderstand der Flüssigkeitsphase gilt die einfache Beziehung

$$\frac{1}{B_i} = \frac{1}{\beta_i} + \frac{m}{\alpha_f} \tag{1}$$

wobei

$$m = \frac{i_s - i_g}{t_s - t_g}.$$

Wenn wir noch den Wärmewiderstand der Flüssigkeitsphase vernachlässigen können, bekommen wir die Identität zwischen den Gesamtaustausch- und Teilaustauschkoeffizienten, d. h. $B_i = \beta_i$ für $\frac{m}{\alpha_f} = 0$. Diese einfache Beziehung findet Verwendung z. B. bei der Berechnung der Verdunstungs- und Kondensationsvorgänge bei dem Zweiphasensystem Luft—Wasser.

Die Modellvorstellung mit dem einzigen Transportwiderstand in der Gasphase wurde experimentell bestätigt im SFIM bei der Verdunstung des gekühlten Wassers in dem Luftstrom in Rieseleinbauten [1], und zwar bei gewöhnlichen atmosphärischen Parametern: Lufteintrittstemperatur 0 bis 37 °C u. Luftfeuchte 30% bis 100%; turbulente Luftströmung durch das System mit hydraulischem Durchmesser 10 bis 120 mm und relative Länge 10 bis 150; relative Wärmestromdichte an der Kontaktfläche bis 1300 kcal/m²h; Gesamttemperaturunterschied des Wassers bis 30 °C; nicht turbulente (pseudolaminare) Strömung der Wasserschicht d. h. $Re_f \leq 615\, Pr_f^{-0,6}$.

Auch bei Teilkondensation des Wasserdampfes aus der feuchten Luft an einer innen durch das Wasser gekühlten Platte [8] wurde indirekt die Gültigkeit dieser einfachen Modellvorstellung experimentell erwiesen, und zwar bei folgenden Parametern:

Luftgeschwindigkeit 20 bis 40 m/s d. h. Reynoldszahl $Re_g = 60 \cdot 10^3$ bis $500 \cdot 10^3$.
Lufttemperatur 75 bis 93 °C und Feuchte 31 bis 94%
Wandtemperatur 35 bis 70 °C.

Zur Korrelation der Meßergebnisse hat sich die Stefansche Korrektur, bezogen auf die Verhältnisse an der Flüssigkeitsoberfläche, als genügend genau erwiesen. Da dabei nur eine kleine hydraulische Belastung der Versuchsplatte entstand, wurde die Grenzschicht nicht dynamisch beeinflußt und hatte deshalb keine merkbare Wirkung auf die Intensität des Transportprozesses, siehe Abb. 1.

Ähnlich verhalten sich auch die Zweiphasensysteme Gas—Flüssigkeit in innen berieselten Versuchskolonnen, wie das die Analysen von GILLILAND, SHERWOOD, BARNET-KOBE, TREYBAL, KUZNECOV, VAN KREVELEN, UCHIDA, POSIN, BERMAN, MORRIS, JACKSON, CAIRNS-ROPER und anderer Forscher gezeigt haben [9]. In diesem Falle können wir zur Interpretation des Transportprozesses in der Gasphase mit Erfolg die sogenannte dreifache Analogie benützen [10, 11].

Diese Analogie beschreibt in CHILTON-COLBURNS Modifikation [12] die Ähnlichkeit des Strömungsfeldes mit dem Temperatur- und Konzentrationsfeld durch die Gleichheit der j_α-Kennzahl für den Wärmeaustausch mit der j_β-Kennzahl für den Stoffaustausch und der Fa-Kennzahl für den Impulsaustausch. Als Strömungsparameter wird dabei nur die Reynoldszahl der Gasphase Re_g bzw. ihr Wert Re_{gr} bezogen auf die Relativgeschwindigkeit benützt [1]. Die Kennzahlform der dreifachen Analogie ist die folgende [10]:

$$j_\alpha = j_\beta = Fa = C\,Re^{n-1}. \tag{2}$$

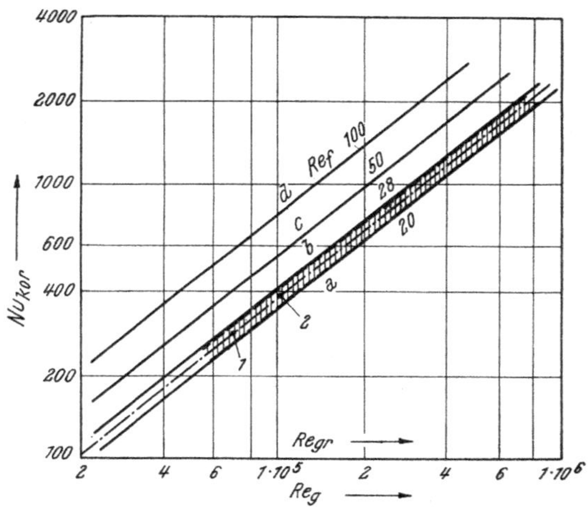

Abb. 1. Kurve 1 $Nu_{kor}\, Pr^{-1/3} = 0{,}0382\, Re_g^{0{,}8}$

Zone 2 $Nu_{kork} = \dfrac{Nu_\Sigma}{1 + \left(\dfrac{D}{a}\right)^{2/3} \dfrac{r}{\varepsilon_{gs}\gamma_c}\, \dfrac{\theta_c}{\theta_t}}$

$Nu_{kor} = f(Re_g)$ Meßergebnisse ↓↑

$Nu_{kor} = f(Re_{gr})$ Meßergebnisse ↓↑

Kurve	a	b	c	d
Re_g	20	28	50	100

Bei technischen Rieseleinbauten können wir die von der Dynamik der Phasengrenze entstandenen Zusatzeinflüsse der Formwiderstände und Pulsationseffekte nicht mehr unberücksichtigt lassen. Für die gegebene Konfiguration des Einbaues läßt sich das dynamische Verhalten der Flüssigkeitsoberfläche durch die Strömungskennzahl der Flüssigkeit $Re_f = \dfrac{w_f \delta_f}{\nu_f}$ bzw. durch ihre Modifikation (Pulsationskennzahl) $Re_{f\sim} = \dfrac{2fvd}{\nu_g}$ charakterisieren. Da bei dem Herabfließen der Wasserschicht die Amplitude $\dfrac{v}{2}$, sowie die Frequenz f durch Re_f eindeutig bestimmt wird, wie es die nähere Analyse der von Kapica [13], Brauer [14], Wilke [17], und Malewski [15] durchgeführten diagnostischen Messungen gezeigt hat, kann man die $Re_{f\sim}$ Zahl durch die Re_f Zahl ersetzen.

Abb. 2 zeigt die Resultate der im SFIM durchgeführten Versuchsmessungen, die in dem sogenannten unterkritischen Gebiet der Gasströmung durchgeführt wurden.

Das unterkritische Strömungsgebiet wird dadurch gekennzeichnet, daß das dynamische Verhalten der Phasengrenze nicht durch den Zusatzeinfluß der Re_g-Zahl der Gasphase gestört wird.

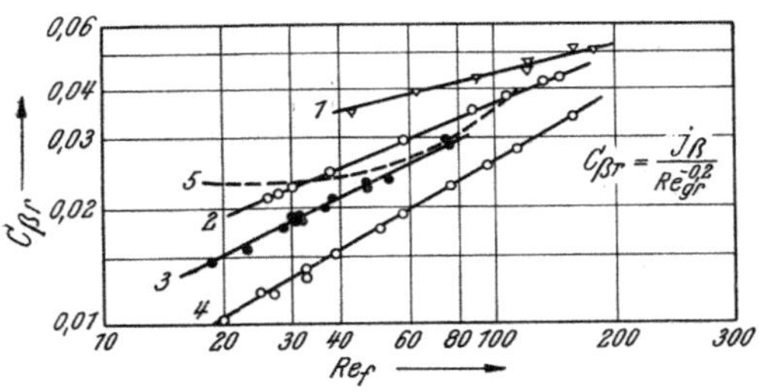

Abb. 2

Da die Rieseleinbauten überwiegend in diesem Strömungsgebiet arbeiten, wird auch der Koeffizient der Chilton-Colburns-Gleichung $C_{\beta r}$ nur durch die Re_f-Zahl der Flüssigkeit beeinflußt. Da die Dynamik der Phasengrenze mit der Re_f-Zahl gesteigert wird, wird analog auch die Intensität des Transportprozesses erhöht. Das haben unsere Versuchsmessungen (siehe Abb. 2) an folgenden Kühlsystemen bestätigt [1]

— an einem Rieselsystem mit Wirbeleinbauten (Kurve 1)
— an gut benetzbaren Plattensystemen mit zerstäubungslosen Flüssigkeitsverteilern (Kurve 2)
— an denselben Systemen mit gewöhnlicher Flüssigkeitszerstäubung (Kurve 3)
— oder an einem hydrophoben kompakten Kunststoffsystem mit Flüssigkeitszerstäubung (Kurve 4).

Deutlich bemerkbar ist der Einfluß der Oberflächenbeschaffenheit auf die Intensität des Transportprozesses. Bei gegebener Einbauform und Gasströmung läßt sich die Steigerung des Transportprozesses durch die Abhängigkeit

$$C_{\beta r} = \frac{j_\beta}{Re_{gr}^{n-1}} \sim Re_f^s \tag{3}$$

ausdrücken, wobei folgende Werte für den Exponenten s experimentell festgestellt wurden:

$s = 0{,}6$ bei schlecht benetzbaren (d. h. hydrophoben) Rieseleinbauten,

$s = 0{,}5$ bis $0{,}3$ bei normal benetzbaren Einbauten mit Zerstäubung der Flüssigkeit,

$s = 0{,}4$ bis $0{,}08$ bei gut benetzbaren Einbauten mit störungsfreier Flüssigkeitszufuhr.

Visuelle Beobachtungen haben dabei gezeigt, daß wir erst bei $Re_f \geq 20$ regelmäßige Benetzung der Rieselfläche erwarten können[1]. Ähnlicherweise beeinflußt die Dynamik der Phasengrenze die aerodynamischen Verluste. Die wellenförmige Bewegung des Rieselfilmes mit der Pulsationswirkung der Oberfläche führt aber dazu, daß die Druckverluste rascher zunehmen als die Übergangszahlen. Deshalb unterscheiden sich die Kennzahlbeziehungen der Zweiphasensysteme von den Einphasensystemen [1, 10, 11]:

$$j_{\alpha f} = j_{\beta f} = \eta F a_f \qquad F a_f = \varkappa_p F a. \tag{4}$$

In Gl. (4) bedeutet

η die relative Ausnutzung der aerodynamischen Widerstände zum Transportprozeß.

$\varkappa_p = \frac{F a_f}{F a}$ die relative Erhöhung der aerodynamischen Verluste des mit Flüssigkeit benetzten Rieseleinbaues gegenüber seinem Trockenzustand. Energetisch fällt die Steigerung des Transportprozesses durch die aerodynamischen Verluste ähnlich aus, wie die rauhen Austauschflächen oder Wirbeleinbauten der Einphasensysteme d. h.

$$\varkappa_\alpha \doteq \varkappa_p^\varrho; \quad \varkappa_\beta \doteq \varkappa_p^\varrho. \tag{5}$$

In dieser Beziehung bedeuten:

$\varkappa_\alpha$, $\varkappa_\beta$ die relative Erhöhung des Transportprozesses (Wärme- bzw. Stoffaustausches)

$\varrho = 0{,}60 \pm 0{,}05$ laut Meßergebnisse SFIM [1].

Zum Vergleich ist in Abb. 2 noch die Kurve 5 eingetragen. Ihr Verlauf wurde unter folgenden Voraussetzungen rechnerisch ermittelt:

— die Aerodynamik des Systems richtet sich nach der vereinfachten Lairds Vorstellung der Phasengrenzendynamik [17] (feste schwingende Wand mit bekannter Frequenz und Amplitude)
— Gültigkeit der oben erwähnten energetischen Beziehung $\varkappa_\beta = \varkappa_p^{2/3}$;
— die Bewegung der Phasengrenze richtet sich nach den diagnostischen Labormessungen Malewskis [15], die an einem von außen benetztem Rohr durchgeführt wurden. (Diese Voraussetzung wurde gewählt wegen des Mangels an genaueren Angaben für Plattensysteme).

---

[1] Das bestätigt auch Abb. 1, wo die Meßergebnisse in Kennzahlform $Nu_{kor} = f(Re_{gr})$ beigefügt sind, die bei der Verdunstung in Luftstrom aus dem Rieselsystem bei $Re_f = 20$, 50 u. 100 festgestellt wurden. Die Beeinflussung der Transportintensität durch Re_f ist bei $Re_f > 20$ ganz deutlich. Die Reynoldszahl Re_{gr} bezieht sich dabei auf die Höhe h des Gegenstromsystems.

Die Kurve 5 ist mit der Kurve 2 vergleichbar. Auffallend ist ihr steilerer Verlauf. Wie eigene aerodynamische Messungen an Plattensystemen gezeigt haben, wird diese kleine Abweichung durch die größeren Strömungserscheinungen an der Phasengrenze (d. h. auch Formwiderstände) im Gebiet $Re_f = 30 - 80$ verursacht.

Bei der Verwendung des Rotationsverteilers war die Steigerung der Verdunstungsintensität im Einklang mit der Modellvorstellung, die von uns [1] aus LAIRDS [17] aerodynamischen Messungen abgeleitet wurde.

Im weiteren beurteilen wir den Einfluß des Transportwiderstandes an der Phasengrenze. Aus der kinetischen Gastheorie haben HERZ und KNUDSEN die Intensität der Verdampfung einer Flüssigkeit mit Hilfe des sogenannten Kondensationskoeffizienten charakterisiert. Ausgehend von der HERZ-KNUDSEN-Gleichung hat SILVER [18] für den Transportwiderstand an der Phasengrenze folgende Beziehung abgeleitet

$$Re_f = \frac{15{,}1 \cdot 10^{-6}\, T_s^{3/2}}{k_a r_s^2 M_D^{1/2}}. \tag{6}$$

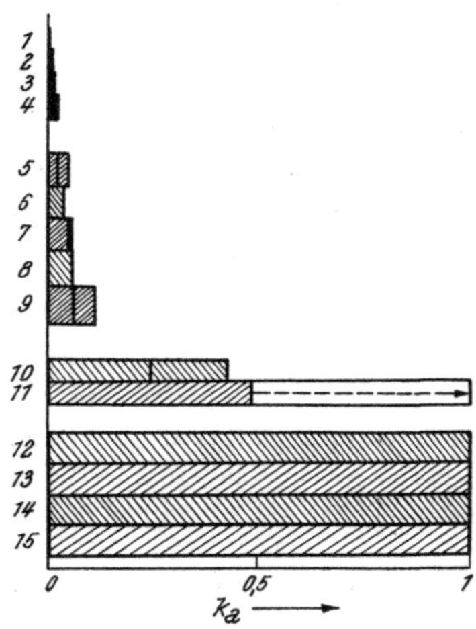

Abb. 3. Der Kondensationskoeffizient k_a von Wasserdampf. [1] *Heidrich*: Z. techn. Phys. **8**, 1953; [2] *Marselin*: C. R. hebd. 1964; [3] *Alty*: Proc. Roy. Soc. **104**, 1935; [4] *Pruger*: Z. techn. Phys. 1940; [5] *Knacke* und *Stranski*: Progr. Met. Phys. **6**, 1956; [6] *Barancov*: Z. f. Ch. 1939; [7] *Mackay* und *Alty*: Canad. J. Res. 1951; [8] *Delaney* und *Eagleton*: AICHE J. **3**, 1962; [9] *Kappler* und *Hammeke*: Forsch.-Ber. NRW 125, 1955; [10] *Hickmann*: Ie EC **46**, 1954; [11] *Kočurova*: IFŽ **3**, 1964; [12] *Hickman*: Ie EC **46**, 1954; [13] *Littlewood* and *Rideal*: Tr. Far. Soc. 1956; [14] *Madden*: AICHE 1959; [15] *Kočurova*: IFŽ **3**, 1964

In dieser Gleichung ist k_a der Kondensationskoeffizient, dessen Wert die relative Menge der Dampfmoleküle angibt, die nach dem Kontakt mit der Flüssigkeit aus der Gesamtmenge nicht mehr in den Dampfraum abgeprallt werden. Eine Übersicht der Kondensationskoeffizienten mit näheren Literaturangaben ist in Abb. 3 angegeben. Die Werte der Koeffizienten können wir in vier Bereiche teilen. Im ersten Bereich sind meistens experimentelle Angaben, die zu sehr kleinen Werten der Kondensationskoeffizienten geführt haben. Die zweite Kategorie enthält die Werte, die meistens in den Literaturangaben benützt wurden. Die dritte Kategorie wird durch mittlere Angaben der Kondensationskoeffizienten bestimmt. Die vierte Kategorie enthält die Angaben solcher Forscher, die behaupten, daß sämtliche Dampfmoleküle von der Phasengrenze verschluckt werden, d. h. $k_a = 1$. Irgendwelche Abweichungen von diesen Werten sollen durch unreine Beimischungen verursacht werden.

Als Grund für eine so große Streuung der Kondensationskoeffizienten sehen wir:

— die Meßtechnik, die zur objektiven und genauen Feststellung der Oberflächentemperatur bzw. des Partialdampfdruckes gewählt wird;
— die Feststellung der Kontaktfläche;
— die Nebeneinflüsse oder Zusatzeinflüsse.

Theoretisch ermittelte Werte Kočurovas führen zur folgenden empirischen Gleichung [19, 20].

$$k_a = 0{,}535 + 0{,}035 \log p_D \,[\mathrm{kp/cm^2}]. \qquad (7)$$

Der Vergleich des Phasengrenzwiderstandes R_s des Systems Wasserdampf—Wasser unter der Voraussetzung, daß $k_a = 0{,}036$ bzw. $k_a = f(p_D)$ laut Gl. (7), ist in Abb. 4 durchgeführt, und zwar im Intervall $p_D = 0{,}01$ bis $10\ \mathrm{kp/cm^2}$. In beiden Fällen nehmen die Werte des Phasengrenzwiderstandes R_s mit dem Dampfdruck p_D sehr rasch ab. Mit Hilfe der durchgeführten Überlegung kann man die Überhitzung der Oberfläche ausrechnen. Ist z.B. gesättigter Wasserdampf von 100 °C und die Wärmestromdichte 300000 kcal/m²h angenommen, so entsteht beim Kondensationskoeffizienten $k_a = 0{,}535$ nur eine Überhitzung von 0,081 °C, bei $k_a = 0{,}036$ aber schon eine beträchtliche Temperaturdifferenz von der Größe 1,2 °C.

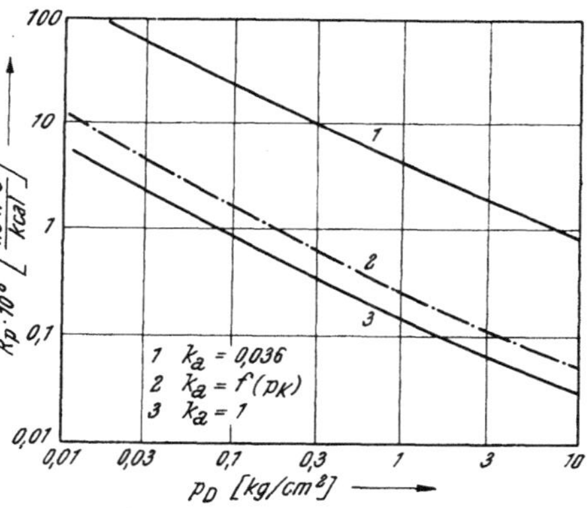

Abb. 4

Auch bei dem Zweiphasensystem Flüssigkeit—Gas kann der Transportwiderstand an der Phasengrenze R_s bedeutend sein, besonders im Gebiet von niedrigen Partialdrücken der Inertgase und größeren Durchflußgeschwindigkeiten. Im Falle, daß z. B. in der Gasphase ein intensiver Wärmeaustausch herrscht und wegen der Verschmutzung der Phasengrenze der Kondensationskoeffizient nur sehr klein ist, $k_a = 0{,}036$, siehe [21], kann die prozentuelle Abweichung beim Vernachlässigen des Transportwiderstandes an der Phasengrenze Gas—Flüssigkeit 5 bis 7% betragen.

Normalerweise ist aber der Wert des Kondensationskoeffizienten, siehe [19], ungefähr 15 × größer, so daß die Abweichung in dem erwähnten Falle nur $^1/_2$% beträgt. Berechtigung der höheren k_a-Werte wurde inzwischen indirekt von Hasson [39] experimentell bestätigt. Deshalb brauchen wir im Gegensatz zu Delaneys Vorschlag [40] bei dem Zweiphasensystem Flüssigkeit-Gas nur kleine Korrekturen am Phasengrenzwiderstand R_s zu machen und beim Zweiphasensystem Flüssigkeit-Sattdampf wird dieser Transportwiderstand im Gebiete der höheren Vakua nicht so groß, wie bisher vermutet wurde [21].

Dem Mechanismus des Transportprozesses im Innern der herabfließenden Schicht oder des Strahles wurde bisher nur unvollkommene Aufmerksamkeit gewidmet [12, 22 bis 27]. Im Falle, daß wir diesen Prozeß als nichtstationär ansehen und daß das Treibgefälle mit genügender Genauigkeit durch seinen logarithmischen Mittelwert ausgedrückt werden kann, kann man den Wärme- bzw. Stoffaustausch durch folgende Kennzahlbeziehung beschreiben

$$Nu_f = \frac{1}{Fo_\alpha} \ln \frac{1}{\Omega_\alpha}, \tag{8}$$

$$Sh_f = \frac{1}{Fo_\beta} \ln \frac{1}{\Omega_\beta}. \tag{9}$$

Der relative Temperatur- Ω_α bzw. Konzentrationsausgleich Ω_β wird dabei folgenderweise charakterisiert

$$\Omega_\alpha = \frac{t_{fs} - t_{f2}}{t_{fs} - t_{f1}}, \tag{10}$$

$$\Omega_\beta = \frac{c_{fs} - c_{f2}}{c_{fs} - c_{f1}}. \tag{11}$$

Weitere Kennzahlen sind wie folgt:

Nusseltsche Kennzahl $Nu_f = \frac{\alpha_f \delta_f}{\lambda_f}$, (12)

Sherwoodsche Kennzahl $Sh_f = \frac{B_f \delta_f}{D_f}$, (13)

Fouriersche Kennzahl für den Wärmeaustausch

$$Fo_\alpha = \frac{a h}{w_f \delta_f^2}, \tag{14}$$

Fouriersche Kennzahl für den Stoffaustausch

$$Fo_\beta = \frac{D_f h}{w_f \delta_f^2}. \tag{15}$$

Für das flache Geschwindigkeitsprofil kann man aus der Theorie der Wärmeleitung (siehe z. B. LYKOV [28]) folgende Beziehungen für den Fall ableiten, daß an der Flüssigkeitsoberfläche konstante Temperatur bzw. Konzentration herrscht

$$\Omega_\alpha = 2 \sum_{n=1}^{\infty} \frac{1}{N_n^2} e^{-N_n^2 Fo_\alpha}, \tag{16}$$

$$\Omega_\beta = 2 \sum_{n=1}^{\infty} \frac{1}{N_n^2} e^{-N_n^2 Fo_\beta}. \tag{17}$$

In beiden Gleichungen bedeutet

$$N_n = (2n - 1) \frac{\pi}{2}. \tag{18}$$

Für das parabolische Geschwindigkeitsprofil können wir VJAZOVOVS Gleichungen benützen, die in Reihenform angegeben werden

$$\Omega_\alpha = \sum_{n=1}^{\infty} k_{1n} e^{-k_{2n} Fo_\alpha}, \tag{19}$$

$$\Omega_\beta = \sum_{n=1}^{\infty} k_{1n} e^{-k_{2n} Fo_\beta}. \tag{20}$$

Beide Konstanten k_{1n}, k_{2n} sind in der Fachliteratur bestimmt [23, 24, 25].

Durch das dynamische Verhalten der Flüssigkeitsoberfläche sowie durch innere Durchmischung erhöht sich die Intensität des Transportprozesses. Diese beiden zusammenwirkenden Einflüsse führen zur Vergrößerung des effektiven Transportkoeffizienten im Vergleich zum Molekular-Transportkoeffizienten. Im weiteren führen wir deshalb folgende Definitionsgleichungen ein:

$$\Pi_e = \varkappa \Pi, \tag{21}$$

$$\chi = \varkappa - 1. \tag{22}$$

Dabei wird das Strömungsfeld durch die kinematische Zähigkeit $\Pi = \nu$, das Temperaturfeld durch die Temperaturleitfähigkeit $\Pi = a$, und das Konzentrationsfeld durch den Diffusionskoeffizienten $\Pi = D$ charakterisiert. Die effektiven Transportkoeffizienten sind allgemein von der Reynoldszahl Re, Prandtl-Zahl Pr, Schmidt-Zahl Sc und von der geometrischen Konfiguration abhängig.

Die Beziehungen $\chi = f(Fo)$ bei gewähltem Faktor $\varkappa = 1, 5, 10, 20$ und 100 sind für das flache und parabolische Geschwindigkeitsprofil

Tabelle 1. *Übersicht der wichtigsten*

Kurve Zone	Beziehung	Form der Kontaktfläche	Charakteristik des Transportprozesses
1	$\chi_{af} = f(Re_f)$	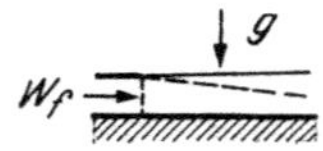	Kondensation des Sattdampfes an der Oberfläche des turbulenten Wasserstrahles
2	$\chi_{af} = f(Re_f)$	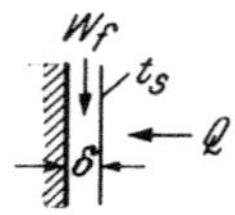	Kondensation des Sattdampfes an der herabrieselnden Flüssigkeitsschicht
3	$\chi_{af} = f(Re_f)$		Diffusion in die herabrieselnde Flüssigkeitsschicht
4	$\chi_{af} = f(Re_f)$	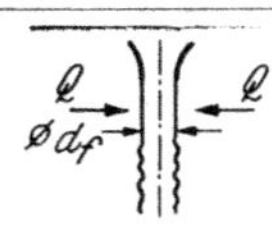	Kondensation des Sattdampfes an Rundstrahlen
5	$\chi_{af} = f(Re_f)$	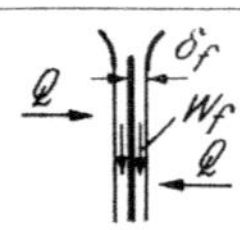	Kondensation des Sattdampfes an der Flüssigkeitsschicht
6	$\chi_{Dg} = f(Re_g)$	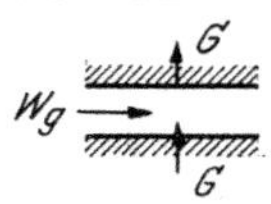	Stoffaustausch (gleichzeitige Verdunstung und Absorption) im Flachkanal bei turbulenter Gasströmung
7	$\chi_{af} = f(Re_f)$	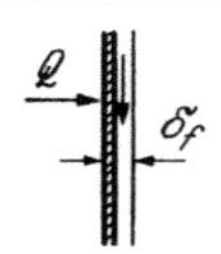	Erwärmen (Konvektion) der herabrieselnden Flüssigkeitsschicht (von der Wand)
8			
9	$\chi_{af} = f(Re_f)$	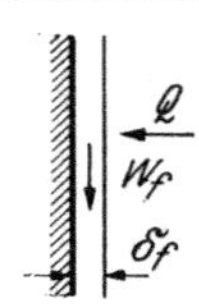	Kondensation des Sattdampfes an der Flüssigkeitsschicht
10	$\chi_{af} = f(Re_f)$	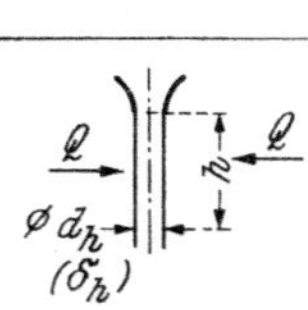	Kondensation des Sattdampfes am Flach- oder Rundstrahl

bisherigen Ergebnisse

Berechnungsunterlagen	Angenommen	Literatur
Theoretische Lösung $Nu_f = 250 Pe_f k^{1/3}$ für $k = 0{,}02 + 0{,}06$ $Nu_f = 200\, Pe_f\, k^{0,28} \quad k > 0{,}06$ $k = \frac{1}{Ku} = \frac{C \Delta t}{r}$	$Pr_f = 2{,}23$; ($t_f = 80\,°C$); $\Delta t = 20\,°C$ $Re_f = 5000 \div 50000$	[29]
Empirische Kennzahlbeziehungen $Re_f = 3{,}5\,\delta^+ + 2{,}5\delta^+ \ln \delta^+ - 64$ $\chi_{vf} = 0{,}084\delta^+ - 1$ nach Lin [30]	Einfluß der Zone $y^+ < 0{,}7\delta^+$ vernachlässigt	[41]
Meßergebnisse Pr_f nicht angegeben		[31]
Meßergebnisse: Pr_f nicht angegeben	$\varnothing\, d_f = 6u \cdot 10$ mm $R_s = 0\ m^2 h\,°C/kcal$	[32]
Ergebnisse SFIM	$Pr_f = 2{,}23$; $t_f = 80\,°C$; $\Delta t = 20\,°C$ R_s laut Gleichung (7) flaches Geschwindigkeitsprofil	[41]
Meßergebnisse		[33, 34]
$Nu_f = 0{,}0066\, Re_f^{0,933}\, Pr_f^{0,344}$	$Pr_f = 2{,}23$ ($t_s = 80\,°C$; $\Delta t = 20\,°C$) $R_s = 0\, m^2 h\,°C\,/\,kcal$; $Re_f > Re_{f\,\mathrm{krit}}$ $Ga_\delta^{1/3} = 0{,}436\, Re_f^{0,534}$ [14]	[7]
Meßergebnisse	$Re_f < Re_{f\,\mathrm{krit}}$	[7]
$Nu_f = 0{,}436 \frac{\delta^{+1/3}\, Re_f^{0,534}}{\int_0^{\delta^+} \frac{dy^+}{\chi_a + 1}}$ – $Ga_\delta^{1/3} = 0{,}436\, Re_f^{0,534}$ $Re_f = f(\delta^+; Pr_f)$ nach Lin [27, 30]	$Re_f > Re_{f\,\mathrm{krit}}$ $R_s = 0\ m^2 h\,°C/kcal$ $Pr_f = 2{,}23$ ($t_s = 80\,°C$; $\Delta t = 20\,°C$)	[27]
Theoretische Lösung $\chi_{af} = \frac{\varepsilon_*}{2} Re_f Pr_f$	$\varepsilon_n = 5 \cdot 10^{-4}$ nach Nikuradze [35] $Pr_T = \frac{\nu_T}{a_T} = 1$ $Pr_f = 1;\ 5;\ 10$	[35]

in Abb. 5 eingetragen. Diese Kurvenschar benützen wir im weiteren zur Analyse der im turbulentem Gebiet durchgeführten Versuchsmessungen oder Berechnungsmethoden.

Diese Analyse führte zu folgenden Kennzahlbeziehungen

$$\frac{\chi_a}{f(Pr)} = f(Re_f) \qquad \frac{\chi^v}{f(Pr)} = f(Re_f) \qquad \frac{\chi_D}{f(Sc)} = f(Re_f). \tag{23}$$

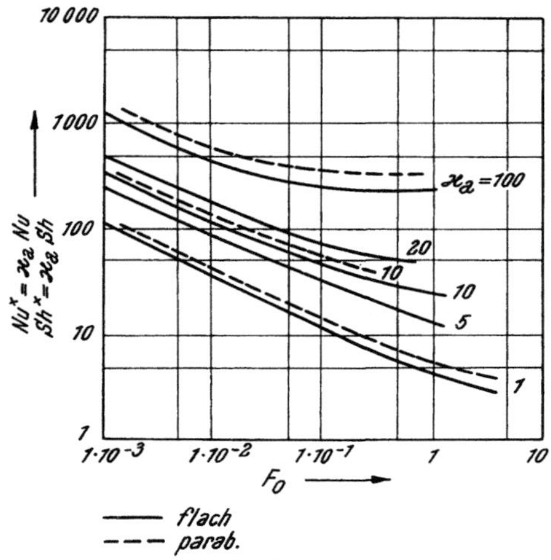

Abb. 5

SUBBOTINS Versuche [38] mit Luft, Wasser und Quecksilberströmung im Kanal führen z. B. zur einfachen Proportionalität $\chi_a \sim Pr$. Diese Messungen wurden bei folgenden Parametern durchgeführt $0 \leq Pr \leq 10$, $10000 \leq Re \leq 500000$.

Die Ergebnisse der Analyse sind in Abb. 6 zusammengestellt.

Die Bedeutung der einzelnen Kurven und Zonen ist aus Tab. 1 ersichtlich.

Aus dem Vergleich können wir folgende Schlußfolgerungen ziehen: ABRAMOVIČS Modellvorstellung [29] ist zu grob und führt zu einem unwahrscheinlich intensiven Transportprozeß, dessen Verlauf in Abb. 6 die Kurve 1 charakterisiert. Den Hauptgrund für das Versagen dieser Modellvorstellung sehen wir in der Annahme, daß der Wärmewiderstand des Flüssigkeitskernes vernachlässigt werden kann. Die weitere vereinfachende Annahme, daß der Transportwiderstand an der Phasengrenze unbedeutend ist, spielt dagegen bei atmosphärischem Druck nicht die entscheidende Rolle.

Umgekehrt führt KUTATELADZES Modellvorstellung [35] zu sehr kleinen Austauschkoeffizienten, wie näher in [20] diskutiert wurde. Als den Hauptgrund für die ungenügende Interpretation des wirklichen Kondensationsprozesses sehen wir, daß KUTATELADZES Lösung nicht die Beeinflussung der Intensität des Transportprozesses durch das dynamische Verhalten der Phasengrenze Flüssigkeit—Dampf berücksichtigt. Durch die Annahme $R_x = 0$ m²h °C/kcal und $F_{th} < F_{\text{wirkl.}}$ wird dagegen die Intensität des Prozesses der Rechnungsmethode noch teilweise erhöht. Interessant ist, daß die von KUTATELADZE übernommene Voraussetzung $a_T = 5 \cdot 10^{-4}\, W_x r_x$, die in die Kennzahlform $\chi_{af} = 2{,}5 \cdot 10^{-4}\, Re_f Pr_f$ umgebaut werden kann, für $Pr_f = 5$ mit WÖRTZS und

TOWLES Experimentalergebnissen praktisch identisch ist, siehe Vergleich in Abb. 6, Kurve $10_{(5)}$ und Kurve 6. Für $Pr_f = 1$ und 10 sind die aus KUTATELADZES Modellvorstellung folgenden abgeleiteten Beziehungen $10_{(1)}$ und $10_{(10)}$ beigefügt.

Die Anwendung des aus der Kanalströmung übernommenen Universalprofils [27, 30] und die Beziehung für den turbulenten Transportkoeffizienten (Kurve 9 bzw. 2) führt qualitativ zu einem ähnlichen Ein-

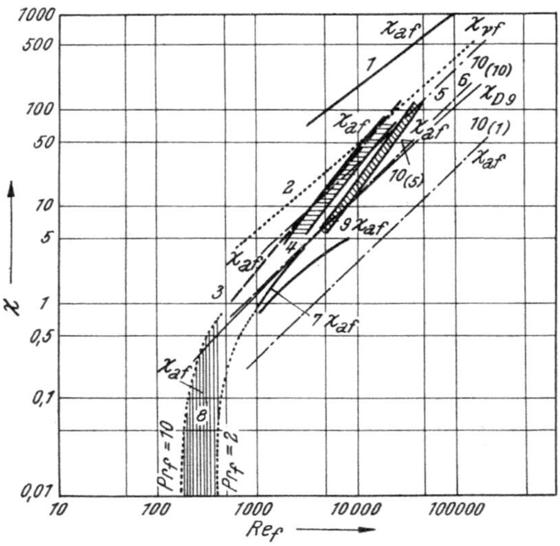

Abb. 6. $\chi_v = v_e - 1$ Impulsaustausch
$\chi_a = a_e - 1$ Wärmeaustausch
$\chi_D = D_e - 1$ Stoffaustausch

1 ABRAMOVIČ
2 KARMAN
3 BRÖTZ
4 ENDE
5 SFIM
6 WÖRTZ, TOWLE
7 WILKE ($Pr_f = 2{,}2$)
8 WILKE
9 LABUNCOV
10 KUTATELADZE ($Pr_f = 10;\ 5;\ 1$)

fluß, wie die Turbulenz am Diffusionsprozeß im Kanal (Kurve 6). In beiden Fällen gilt eine ungefähr lineare Beziehung zwischen χ und Re_f^s d. h. $\chi \sim Re_f^s$; Exponent $s \leq 1$. Das bedeutet, daß eine Steigerung des Transportprozesses durch die Durchflußgeschwindigkeit der Flüssigkeitsschicht nicht zu erreichen wäre. Dieses Resultat steht aber im Widerspruch mit dem wirklichen Mechanismus des Mischkondensationsprozesses, den wir in der an der Tragwand herabfließenden Flüssigkeitsschicht verfolgen.

Schon aus Kurve 7, die wir aus WILKE's Kriterialbeziehungen abgeleitet haben [7], läßt sich erkennen, daß bei dem Aufheizen der Flüssigkeitsschicht durch die Tragwand folgende Beziehung gilt: $\chi \sim Re_f^s$ wobei $s > 1$. Der Vergleich zwischen Kurve 9 und Kurve 7 läßt also deutlich den Zusatzeinfluß der Phasengrenzendynamik auf den Transportprozeß in der Flüssigkeitsschicht erkennen.

Für die Diffusion in der unbeschränkten Ebene gilt nach Brötz [31] die Beziehung (3) (siehe Abb. 6).

Bemerkenswert ist, daß auch die Meßresultate Endes [32], die bei der Kondensation des Sattdampfes an Rundstrahlen festgestellt wurden, einen ähnlichen Charakter haben. Die Bearbeitung von Endes Messungen führt zur Zone 4 in Abb. 6. Dabei handelte es sich um Kondensation an Wasserstrahlen von 6 und 10 mm Durchmesser.

Zum positiven Einfluß der Durchflußgeschwindigkeit bzw. der Re_f (Reynoldszahl der Flüssigkeit) auf die Intensität der Mischkondensation führen auch Bubeníčeks Messungen [1] bei der Kondensation des Sattdampfes an Rieselfilm, der mittels Schlitzmündung an der Tragwand gebildet wurde.

Eine nähere Analyse dieser Messungen hat gezeigt, daß wir den erwähnten Transportvorgang durch folgende Beziehung charakterisieren können

$$\chi_{af} \sim Re_f^s,$$

wobei $s = 1{,}4$ bis 1,5.

Da die gute Stabilität des Rieselsystems auch bei großen Durchflußbelastungen der Flüssigkeitsphase bewiesen wurde, haben diese Erkenntnisse große Bedeutung für die technische Ausnützung im Apparatebau [1, 36, 37, 41, 42].

In diesem Referat wurde kurz und nur übersichtlich die Problematik der Zweiphasensysteme Flüssigkeit—Gas bzw. Flüssigkeit—Dampf besprochen. Die einzelnen Teiltransportwiderstände wurden an Hand von Experimentalergebnissen verglichen und diskutiert. In der Gas- sowie in der Flüssigkeitsphase wurde eine starke Beeinflussung der Intensität des Transportprozesses durch das dynamische Verhalten der Phasengrenze nachgewiesen. Da die Kenntnis des verfolgten Mechanismus für den Kontaktapparatebau entscheidend ist, halten wir es für notwendig, die bisher überwiegend integralen experimentellen Forschungsmethoden durch diagnostische Messungen zu ergänzen und so die jetzigen Berechnungsmethoden mit genaueren Modellvorstellungen zu unterlegen.

Literatur

[1] Zemánek, J.: Výzkumné zprávy. Státniho výzkumného ústavu pro stavbu strojů (SVÚSS). Běchovice.

[2] Lewis und Whitmann: Ind. Engng Chemistry 16 (1924).

[3] Szekely: Chem. Engng Sci. 2 (1965).

[4] Mickley, H. S.: Chem. Engng Progress **45**, 739 (1949).

[5] Trevissoi, C.: Distributione delle temperature nel contatto diretto liquido-gas umido. Il Calore 5 (1957).

[6] TANAKA, K., T. TAKAMUTSU und M. HIROAKA: Simultaneous Heat and Mass Transfer between Gas and Liquid Phases. Int. J. Heat Mass Transfer **7**, 6 (1964).
[7] WILKE, W.: Wärmeübergang an Rieselfilme. VDI-Forschungsheft 490 (1962).
[8] STÝBLO, S.: Disertačni práce. Běchovice: Státní výzkumný ústav pro stavbu strojů. 1961.
[9] ZEMÁNEK, J.: Sdileni tepla a přenos hmoty. XVII kap. Vybrané statě z termomechaniky. Praha: Československá Akademie věd. 1965.
[10] ЗЕМАНЕК, Я.: О гидродинамической аналогии при тепло- и массообмене в двух фазных системах газ-жидкость Тепло-массоперенос, 3 1965 Изд. Минск: АН БССР. 1965.
[11] ZEMÁNEK, J.: Wärmetechnische und aerodynamische Eigenschaften der Rieseleinbauten bei Zweiphasensystemen Flüssigkeit-Gas. Int. Kongress CHISA 1965. Mariánske Lázně.
[12] SHERWOOD und PIGFORD: Absorption and Extraction. New York: McGraw Hill. 1952.
[13] КАПИЦА: Волновое течение тонких слоев вязкой жидкости. Журнал эксперементальной и теоретической физики **18**, 1 (1948).
[14] BRAUER, H.: Strömung und Wärmeübergang bei Rieselfilmen. VDI Forschungsheft 457 (1956).
[15] MALEWSKI: Zusammenhang zwischen Stoffübergang und Wellenstruktur beim welligen Rieselfilm. Chemie-Ingenieur-Technik **37**, 8 (1965).
[16] FEIND, K.: Strömungsuntersuchungen bei Gegenstrom von Rieselfilmen in lotrechten Rohren. VDI Forschungsheft 481 (1960).
[17] LAIRD, A. D.: Transaction ASME **76**, 10 (1954).
[18] SILVER, R. S.: Engineering 161 (1946).
[19] КОЧУРОВА, Н. Н.: О расчете коэффициента конденсации. Ннженерно физический журнал 6 (1965).
[20] ZEMÁNEK, J.: Übersicht über einige theoretische und experimentelle Arbeiten zur Mischkondensation von Sattdampf. In: Kolloquium Mischkondensation von Dämpfen. Dresden: Techn. Univ. 1965.
[21] КУТАТЕЛАДЗЕ, С. С.: Вопросы теплообмена при изменении агрегатного состояния вещества. Москва: Госэнергоиздат. 1953.
[22] РАММ, В. М.: Абсорбционные процессы в химической промышленности. Москва: Госхимиздат. 1951.
[23] КАФАРОВ: Анализ процессов массопередачи на основе представления о междуфазной турбулентности Тепло-и массоперенос, Том II. Минск: Издательство АН БССР. 1962.
Основы массопередачи. Москва: Госхимиздат. 1962.
[24] ВЯЗОВОВ, В. Е.: Теория абсорбции малорастворимых газов жидкими пленками. Журнал технической физики **10**, 18 (1940).
[25] ВЯЗОВОВ, В. В.: О жидкостном коэффициенте абсорбции. Журнал химическои промышленности **17**, 12 (1940).
[26] DUKLER, A. E.: Dynamic of vertical falling film systems. Chem. Engng Progress **55**, 10 (1959).
[27] ЛАБУНЦОВ, Д. А.: Теплообмен при конденсации пара на вертикальной поверхности в условиях турбулентного стекания пленки жидкости. Инженерно физический журнал 12 (1960).
[28] ЛЫКОВ, А. В.: Теория теплопроводности. Москва: Государственное издательство теоретической литературы. 1952.
[29] АБРАМОВИЧ, Г. Н.: Теория турбулентных струй. Москва: Физматгиз. 1960.

[30] Lin, C. S., R. W. Moulton und G. L. Puttman: Ind. Engng Chemistry **45**, 3 (1953).
[31] Brötz, W.: Die Vorausberechnung der Absorptionsgeschwindigkeit von Gasen in strömenden Flüssigkeitsschichten. Chemie-Ingenieur-Technik **26**, 8/9 (1954).
[32] Ende: Berechnungsunterlagen zur Auslegung thermischer Rieselentgaser. Dresden: Tech. Univ. 1964.
[33] Wörtz, B. B., and T. K. Sherwood: Ind. Engng Chemistry 31 (1939).
[34] Towle, W. L.: Ind. Engng Chemistry 31 (1939).
[35] Кутателадзе, С. С.: Основы теории теплообмена. Москва: Машгиз. 1962.
[36] Zemánek, J.: Improvements in and Relating to Fillings for Gas-Liquid Contact Apparatus. Brit. Patent Specification No. 973746 (March 1961).
[37] Heller und Forgo: Österr. Patentschrift Nr. 238231.
[38] Субботин, В. И., М. Х. Ибрагимов и Е. В. Номофилов: Обобщающая зависимость коэффициента турбулентного переноса тепла в потоке жидкости. Теплофизика высоких температур 3 (1965).
[39] Hasson, D., D. Luss and U. Navon: Theoretical Analyses of Vapour Condensation on Laminar Liquid Jets. An Experimental Study of Steam Condensation on Laminar Water Sheet. Int. J. Heat Mass Transfer **7**, 9 (1964).
[40] Delaney, L. J., and L. C. Eagleton: A Test of the Asumption of Interfacial Equilibrium in Measurements of the Gas Film Mass Transfer Coefficients. AICHE J. **8**, 3 (1962).
[41] Bubeníček, M.: Výzkumné zpárvy. Běchovice: Státni výzkumný ústav pro stavbu strojů.
[42] Schneller, J., und Zemánek, J.: Mischkondensator. Schweizer Patentschrift Nr. 393385 (1961).

A Possible Generalization of Onsager's Theory

By

Hans Ziegler
Zürich (Switzerland)

Summary

It is known that ONSAGER's symmetry relations for linear irreversible processes are without meaning unless a criterion for the proper choice of fluxes and forces is given. Author proposes such a criterion. He then establishes some extremum principles which are equivalent to ONSAGER's theory. These principles are compared with former ones, and it is shown that they provide a basis for the generalization of ONSAGER's theory to nonlinear processes.

1. Introduction

The title of this symposium indicates that deformation and flow of continuous media are, in principle, thermo-mechanical processes. It further accentuates that such processes are irreversible, reversibility being nothing more than a limiting case.

Once these ideas are accepted, the question arises how to apply the laws of thermodynamics to a continuum. It is surprising that this problem scarcely appears in the literature on thermodynamics although this literature is primarily concerned with gases, i. e., with a particularly simple continuum. It is only recently that the problem has been recognized; even so, a considerable amount of work has been devoted to it [1—8]. As a main result, general agreement has been reached with respect to the local and the global formulations of the first and second laws of thermodynamics. (These laws are independent of the choice of the state variables and their general formulation fortunately remains unaffected by any controversies concerning the admissibility of one or the other set of such variables).

However, in irreversible thermodynamics, the fundamental laws use to be supplemented by ONSAGER's symmetry relations [9, 10]. Their application to problems of continuum mechanics raises a few questions, and it is with respect to these that the opinions considerably diverge.

ONSAGER's theory is concerned with a set of „fluxes" and with the „forces" associated with them through the expression for the entropy production rate. It is assumed that there exists a linear relationship between fluxes and forces, and it is affirmed that, in the absence of Coriolis forces or magnetic fields, the matrix representing the linear connexion between fluxes and forces is symmetric.

ONSAGER's theory and, in particular, its application to continuum mechanics, have been severely criticized [11—13]. The reasons for this criticism stem from the fact that the theory, in its current form, is at the same time too general and too restricted.

In the first place, there does not exist a criterion for the proper choice of the fluxes and the corresponding forces. It has unfortunately been emphasized in the literature on irreversible thermodynamics that almost any quantities may be considered as fluxes or forces. On the other hand, COLEMAN and TRUESDELL [11] and more recently GAGE and KLINE [13] have shown that ONSAGER's symmetry relations become meaningless if partial interchange of fluxes and forces or mixture of the two sets of quantities by linear combination are admitted.

In the second place, ONSAGER's theory is restricted to linear relationships between fluxes and forces. However, in continuum mechanics, the actual interest is focused on nonlinear constitutive equations. Moreover, it is doubtful whether a statement confined to a particularly simple mathematical situation really has any physical content. In any case it seems necessary to recast the theory in such a way as to make it independent of details connected with its mathematical presentation.

There is no doubt that these objections are serious. One may be tempted under their impression to drop ONSAGER's theory. I do not think, however, that this is justified. It seems that, in physics and chemistry, the theory has been successful and is firmly established. Even in continuum mechanics there are situations where a theory of this type is needed. An example occurs in plasticity, where the so-called theory of the plastic potential requires a foundation.

It will be shown here that it is possible to account for the criticism mentioned above and to present the theory in a form which is at the same time more precise and more general. To achieve this, it is necessary to establish a criterion for the choice of fluxes and forces and to replace ONSAGER's symmetry relations by a physically significant principle [5, 6, 14—16] which is not restricted to the linear case.

2. Fluxes and Forces

Since we are mainly interested in continuum mechanics, let us consider an infinitesimal material element with strain components y_k $(k = 1, 2, \ldots, n)$ and temperature $\vartheta > 0$. The strain components are arbitrarily defined but independent; thus, n amounts to 6 or 9. For simplicity, let us confine ourselves to cases where the state of the element is completely determined by the actual values of y_k and ϑ. The state of a finite portion of the continuum then is described by the fields $y_k(x_l)$, $\vartheta(x_l)$, where $x_l\,(l = 1, 2, 3)$ are cartesian coordinates.

If we use the summation convention, the infinitesimal work done on the element in an infinitesimal deformation is given by an expression of the form

$$dW = Y_k dy_k. \tag{2.1}$$

The coefficients Y_k appearing on the right-hand side are called the stresses associated with the strain components y_k.

Part of the work (2.1) may be recoverable, while the remainder is transformed into heat. It has been shown elsewhere [5, 6, 14—16] how the two contributions are accounted for by distinguishing between reversible and irreversible stresses. Since we are not interested in the reversible part of the process, let us denote the *irreversible* stresses by Y_k. With this modification, (2.1) represents the irrecoverable work, also referred to as dissipation work;

$$L = Y_k \dot{y}_k \geqq 0, \tag{2.2}$$

is the rate of dissipation work and

$$\overset{\circ}{S} = \frac{1}{\vartheta} Y_k \dot{y}_k \geqq 0 \tag{2.3}$$

is the rate of entropy production due to deformation.

In analytical mechanics, an expression similar to (2.1) is used to define the generalized forces Y_k associated with the generalized coordinates y_k, which describe the configuration of the system. The only difference is that the *virtual* work is used instead of the real one. There, the problem of exchanging coordinates and forces would be absurd and has never posed itself. Starting from (2.2) it seems equally clear that, in ONSAGER's terminology, the time rates $\dot{y}_k$, which will henceforth be denoted by w_k, are the fluxes and the Y_k are the corresponding forces.

It is obvious that one might replace (2.2) by (2.3) and define the quotients Y_k/ϑ as the forces associated with the fluxes $\dot{y}_k = w_k$. This modification, which has been adopted by many authors, is unessential and need not be discussed here. The significant fact is that, in any monomer on the rigth-hand side of the expression

$$\vartheta \overset{\circ}{S} = Y_k w_k \tag{2.4}$$

one of the two factors, w_k, is a time rate and hence has the character of a flux[1].

In the case considered above the w_k are the strain rates and $\overset{\circ}{S}$ is the rate of entropy production by deformation. A similar expression can be set up for the rate of entropy production in more general situations. We are thus lead to the following

Criterion

Let the entropy production rate $\overset{\circ}{S}$ be given by an expression of the form (2.4). Then the time rates w_k are the fluxes and their coefficients Y_k are the associated forces.

If, in the case of a material element, the entropy production by heat exchange is included, $\overset{\circ}{S}$ contains an additional term

$$\left(\frac{1}{\vartheta}\right)_{,j} q_j, \tag{2.5}$$

where q_j is the heat flux vector and $(1/\vartheta)_{,j}$ the gradient of $1/\vartheta$. According to our criterion, q_j is a flux and $\vartheta(1/\vartheta)_{,j}$ is the associated force. The total rate of entropy production can still be expressed by (2.4), provided the notations

$$w_{n+j} = q_j, \quad Y_{n+j} = \vartheta\left(\frac{1}{\vartheta}\right)_{,j} = -\frac{\vartheta, j}{\vartheta} = (\ln \vartheta^{-1})_{,j} \tag{2.6}$$

are used and the summation is extended from $k = 1$ to $k = n + 3$.

The criterion given above excludes any interchange between fluxes and forces and hence disposes of the objections raised by COLEMAN and TRUESDELL [11] and by GAGE and KLINE [13]. On the other hand, the criterion does not prevent the formation of a new set of fluxes by linear combination of the original ones, accompanied by the appropriate transition to the associated forces. In the present context, such transformations are important (as far as the strain rates are concerned) in connection with the transition between different cartesian coordinate systems.

3. Extremum Principles

The second criticism mentioned in Section 1 is concerned with the fact that ONSAGER's theory is restricted to linear relationships between fluxes and forces. It is easy, however, to formulate it in such a way that a generalization becomes possible. As a rule, it is in the shape of an ex-

[1] There is no reason for connecting the distinction between fluxes and forces with the one between "intensive" and "extensive" variables. In a continuum theory we are interested in strain rates and stresses or in heat fluxes and temperature gradients. None of these quantities is "extensive" according to the current definitions.

tremum principle that a physical statement assumes its most invariant and hence its most general form. Let us therefore try to replace ONSAGER's symmetry relations first by a geometric statement and finally by an extremum principle. This has been done in various ways; unfortunately it has not always been made quite clear which quantities are to be considered as prescribed and which ones are varied.

Let us denote all of the fluxes by w_j $(j = 1, 2, \ldots, n)$ and the associated forces by Y_j. Since the w_j now may include heat flow and possibly other non-mechanical quantities, the expression

$$L = Y_j w_j = \vartheta \dot{S} \geqq 0, \tag{3.1}$$

need not be a rate of work and will henceforth be called the rate of dissipation. In those cases where the fluxes and forces are not defined beforehand, they follow from (3.1) with the aid of the criterion established in Section 2. Once the connection between fluxes and forces is known, the rate of dissipation can be expressed, by means of the dissipation function

$$L = D(w_j) \geqq 0, \tag{3.2}$$

in terms of the fluxes (and possibly the instantaneous values of the state variables) alone. The reverse is not true: the dissipation function does not completely determine the relationship between fluxes and forces unless an additional statement is made. In the linear case such a statement is provided by ONSAGER's theory which thus yields the forces in terms of the dissipation function.

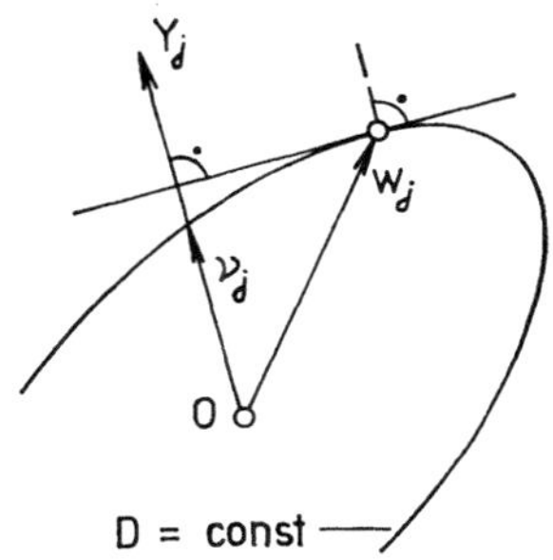

Fig. 1. Vector representation of fluxes and forces

A linear connection between fluxes and forces is given by

$$Y_j = a_{jk} w_k. \tag{3.3}$$

On account of (3.1) and (3.2) the corresponding dissipation function is

$$D(w_j) = a_{jk} w_j w_k \geqq 0, \tag{3.4}$$

where the matrix (a_{jk}) is at least positive semidefinite. If w_j and Y_j are interpreted as vectors in a euclidean space of n dimensions (Fig. 1), the surfaces $D = \text{const}$ are ellipsoids. The gradient of D is given by

$$\frac{\partial D}{\partial w_j} = (a_{jk} + a_{kj}) w_k. \tag{3.5}$$

ONSAGER's symmetry relations are

$$a_{kj} = a_{jk}; \tag{3.6}$$

on account of (3.3) and (3.5) they are equivalent to

$$Y_j = \lambda \frac{\partial D}{\partial w_j}, \qquad \lambda = \frac{1}{2}. \tag{3.7}$$

In geometrical terms (3.7) states that the force vector Y_j corresponding to a given flux vectoc w_j is normal to the D-surface passing through the end point of w_j and points to the exterior of this surface; its magnitude is determined by the condition that the two expressions (3.1) and (3.2) for L must be equal.

There are several possibilities of interpreting this normality condition in terms of an extremum principle. They are all based on the assumption that the structure of the dissipation function (i. e., the matrix (a_{jk})) is known. A first extremum principle [14—16, 5, 6] is the

Principle of Least Force

Let the value L of the dissipation rate and the direction ν_j of the force vector Y_j be prescribed. Then the actual flux vector w_j minimizes the magnitude m of Y_j subject to the side condition

$$D(w_j) = Y_j w_j = L \geqq 0. \tag{3.8}$$

Proof: According to (3.8) the value of L fixes a certain D-surface and the scalar product $Y_j w_j = m \nu_j w_j$. Comparing all vectors w_j with end points on the given D-surface we see that, on account of the normality condition (3.7) the projection of the actual one on ν_j is a maximum. The actual flux vector w_j thus maximizes the scalar product $\nu_j w_j$ and hence minimizes m.

An equivalent statement [5, 6 14—16,] is the

Principle of Maximum Dissipation Rate

If the force vector Y_j is prescribed, the actual flux vector w_j maximizes the dissipation rate L subject to side the condition

$$D(w_j) == Y_j w_j. \tag{3.9}$$

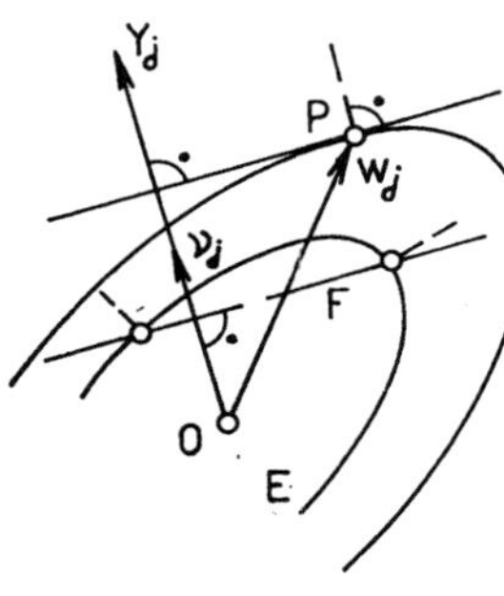

Fig. 2. Maximization of dissipation rate

Proof: For a given value of L the equation $D(w_j) = L$ represents an ellipsoid E (Fig. 2) with center O, and $Y_j w_j = L$ represents a plane F perpendicular to Y_j. The distance of F from O is proportional to L whereas the semi-axes of E are proportional to $\sqrt{L}$. The side condition (3.9) requires that E and F have at least one point in common; this condition is clearly satisfied for sufficiently small values of L. It is further obvious that the direction of the exterior normal of E differs from the direction ν_j of Y_j in the points of intersection with F but coincides with ν_j in the point of contact P. This point corresponds to the largest value of L for which (3.9) is satisfied.

According to (3.1) the rate of entropy production, $\mathring{S}$, differs from L only by the factor $\vartheta > 0$. It follows that the last principle can also be stated as a *principle of maximum rate of entropy production*. For this purpose it is only necessary to replace the expression "dissipation rate" by "entropy production rate".

Another principle has been formulated already by ONSAGER. To the best of my interpretation, his reasoning in [10] amounts to the following statement:

Onsager's Principle of Minimum Energy Dissipation

If the force vector Y_j is prescribed, the actual flux vector w_j maximizes the expression

$$Y_j w_j - \frac{1}{2} D(w_j). \tag{3.10}$$

Since, by carrying out the differentiation, (3.7) is obtained, this principle is equivalent to the ones stated above. It appears simpler insofar as it does not contain a side condition. On the other hand, the expression (3.10) is more complicated and physically less convincing than the functions minimized or maximized in the preceding principles. Finally, and in particular in view of the fact that, on account of (3.8), the maximized expression (3.10) is equal to $L/2$, there remains the question why ONSAGER refers to a "minimum" of the energy dissipation.

Still another extremum principle has been proposed by BIOT [17]:

Biot's Minimum Principle

If the force vector Y_j is prescribed, the actual flux vector w_j minimizes the dissipation function $D(w_j)$ subject to the side condition

$$Y_j w_j = L = \text{const.} \tag{3.11}$$

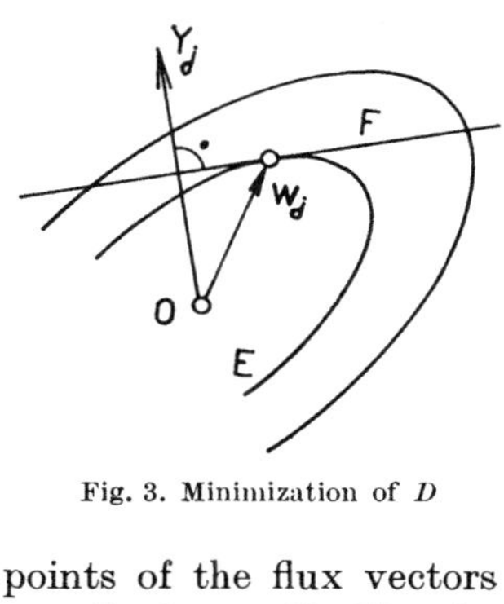

Fig. 3. Minimization of D

The proof follows from Fig. 3. The end points of the flux vectors w_j satisfying (3.11) lie on a given plane F perpendicular to Y_j. For the actual one, the ellipsoid E touches F; its value of D is a minimum compared to those of the ellipsoids intersecting F. Thus, BIOT's principle is equivalent to the preceding ones. Its content, however, is rather of a formally mathematical than of a physical nature. Fixing L and asking for the minimum of D one has to forget that L and D are merely two different expressions for one and the same physical quantity.

A last principle to be mentioned here is due to PRIGOGINE and has been extended by DE GROOT [1, p. 96]. To facilitate its formulation, let us notice that the linear relationship (3.3) between fluxes and forces can be inverted by setting

$$w_j = b_{jk} Y_k. \tag{3.12}$$

It is easy to see that ONSAGER's relations (3.6) also imply the symmetry of the matrix (b_{jk}) and that, therefore, all our principles remain valid if fluxes and forces are exchanged, provided the dissipation function now is written [16] in terms of the forces. Making use of this possibility, we obtain the

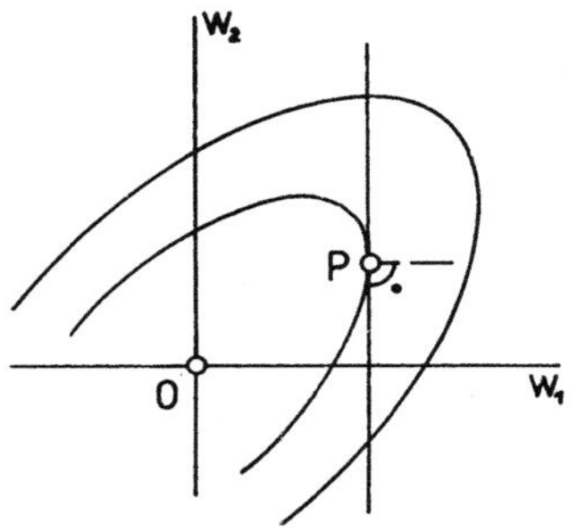

Fig. 4. Principle of PRIGOGINE and DE GROOT

Principle of Prigogine and de Groot

Let us fix part of the fluxes, w_j $(j = 1, 2, \ldots, k < n)$ and choose the remaining ones such that the dissipation function $D(w_j)$ is a minimum. Then the actual forces Y_j $(j = k + 1, \ldots, n)$ corresponding to the fluxes which have been fixed are zero.

Fig. 4, which strongly resembles Fig. 3, shows (in two dimensions and for w_1 fixed) that the statement of PRIGOGINE and DE GROOT is a consequence of the normality condition (3.7). Conversely, the Principle of PRIGOGINE and DE GROOT implies the normality condition, but only for certain points in w_j-space. It follows that the principle is a consequence of, but is not equivalent to, the ones discussed above.

4. Generalization

It has been pointed out that ONSAGER's relations become meaningless if the connection between fluxes and forces is nonlinear. On the other hand, the normality condition and most of the principles stated in Section 3 remain meaningful [16] and can still be used to express the force Y_j in terms of the dissipation function $D(w_j)$.

In the nonlinear case, the dissipation function is not a quadratic form of the velocities; the D-surfaces therefore are not ellipsoids. On account of the second law, however, the dissipation function is still positive definite. Equation (3.9) still holds; here, the right-hand side may be interpreted as the product of the projection of Y_j on w_j and the magnitude $|w_j|$. In order to exclude instabilities (in the form of self-excited oscillations) we have to require that this projection increases when the magnitude of w_j is increased. It then follows from (3.9) that $D(w_j)$ does not only increase on any ray emanating from the origin O in w_j-space, but increases faster than $|w_j|$. Thus, the D-surfaces are ordered, each one of them containing those with smaller values of D. Moreover, they become more dense with increasing D, in the sense that the distances between the points of intersection of an equidistant family of D-surfaces with a given ray from O decrease monotonically.

It is obvious that the normality condition discussed in connection with Fig. 1 can be applied to more general D-surfaces. Analytically, it then takes the form

$$Y_j = \lambda \frac{\partial D}{\partial w_j}, \qquad \lambda = D \left(\frac{\partial D}{\partial w_k} w_k \right)^{-1}, \tag{4.1}$$

which reduces to (3.7) in the linear case. In a similar manner, the principles of least force, of maximum dissipation rate, of maximum entropy production rate and BIOT's principle are still applicable. It can further be shown [16] that, subject to the stability condition discussed above, all these principles still are equivalent and that the D-surfaces must be convex if any given flux vector shall be associated with a force vector. For the Principle of PRIGOGINE and DE GROOT the generalizytion is possible under the condition that the D-surfaces are smooth; ONSAGER's extremum principle, however, cannot be generalized.

It appears that, physically as well as mathematically, the principle of maximum dissipation rate (or maximum rate of entropy production) is the simplest and most satisfactory of the principles enumerated. Subject to the stability condition, it supplies the generalization we have been looking for, and it even ensures the convexity of the D-surfaces.

5. Invariance

It has been mentioned at the end of Section 2 that linear transformations of the fluxes may be of practical importance. Such transformations do not violate the criterion set up in Section 2 for the choice of fluxes and forces. The transformed forces follow from the observation that the dissipation rate,

$$L = Y_j w_j, \tag{5.1}$$

is a scalar, i. e., an invariant.

It is to be expected and is easily verified that the principle of maximum dissipation rate (or any one of the equivalent principles) is invariant with respect to such transformations. In fact, if the w_j are considered as the contravariant components of the flux vector, the invariance of (5.1) requires that the Y_j are the covariant components of the flux vector. This is the case if the Y_j are defined by (4.1), i. e., by means of the gradient of D.

In order to show this in more detail, let a new set of fluxes be defined by the linear transformation

$$w'_j = \alpha_{jk} w_k. \tag{5.2}$$

Since (5.1) must remain valid in the transformed variables, we have

$$L = Y_k w_k = Y'_j w'_j = Y'_j \alpha_{jk} w_k. \tag{5.3}$$

Comparing the first and the last sums in this chain of equations, we obtain the transformation

$$Y_k = \alpha_{jk} Y'_j \tag{5.4}$$

for the forces. Since, on account of (5.2),

$$\frac{\partial D}{\partial w_k} = \frac{\partial D}{\partial w'_j} \frac{\partial w'_j}{\partial w_k} = \alpha_{jk} \frac{\partial D}{\partial w'_j}, \tag{5.5}$$

the forces defined by (4.1) actually transform according to (5.4).

The principle of maximum dissipation rate has another practically significant invariance property. Let some of the fluxes, denoted by $w_{m+1}, \ldots, w_n$, be identically zero. Then D becomes a function

$$D(w_1, \ldots, w_m, 0, \ldots, 0) = D'(w_i, \ldots, w_m), \tag{5.6}$$

of the remaining fluxes $w_1, \ldots, w_m$ alone. Since

$$\frac{\partial D}{\partial w_j}(w_1, \ldots, w_m, 0, \ldots, 0) = \frac{\partial D'}{\partial w_j} \quad (j = 1, \ldots, m) \tag{5.7}$$

and hence

$$\frac{\partial D}{\partial w_k}(w_1, \ldots, w_m, 0, \ldots, 0)\, w_k = \frac{\partial D'}{\partial w_k} w_k, \tag{5.8}$$

where the last sum extends only over the subscripts $k = 1, \ldots, m$, (4.1) yields, for $k = 1, \ldots, m$, the forces

$$Y_j = \lambda \frac{\partial D'}{\partial w_j}, \qquad \lambda = D' \left(\frac{\partial D'}{\partial w_k} w_k\right)^{-1}, \tag{5.9}$$

where the summation again is restricted to the first m products. It follows that the principle of maximum dissipation rate remains valid for any part of the considered process.

The practical importance of the last result lies in the possibility of applying the principle of maximum dissipation rate in cases where only a part of the irreversible process is being discussed. In the case of a material element, e. g., where the entropy production stems from deformation and heat exchange, the principle, applied to each one of the processes, supplies the relationship between strain rate and irreversible stress in the absence of heat flow as well as Fourier's law of heat conduction in the absence of a strain rate.

6. Flux Fields

DRUCKER [18] has criticized the principle of maximum dissipation rate on the basis of an example consisting in a series or parallel combination of two viscous elements. Assuming that each one of these elements obeys the principle, he concludes that this need not be true for the combined system. Drucker does not explain the structure of his system, and it is difficult to visualize a series or parallel combination of elements each one of which must have more than one degree of freedom, since otherwise the principle of maximum dissipation rate would remain meaningless. I therefore do not think that DRUCKER's argument disproves the principle of maximum dissipation rate. However, the argument might be considered to confirm the fact that, in a finite system, it may be impossible to give a global form of the principle even though it applies locally.

As a simple example Fig. 5 shows two particles m_1 and m_2 assumed to be rigidly connected and to move in a plane x, y, under the influence of the forces X, Y and the moment M acting at the centroid C. If we assume that each one of the two particles sustains a drag whose magnitude depends on, and whose direction is opposite to, the velocity, the principle of maximum dissipation rate is locally satisfied. It is even satisfied for the entire system provided the rigid connection of the two particles is dropped. With this connection present, however, it is easy to see that, in general, the relationship between the fluxes $\dot{x}$, $\dot{y}$, $\dot{\varphi}$ and the corresponding forces X, Y, M does not obey the principle.

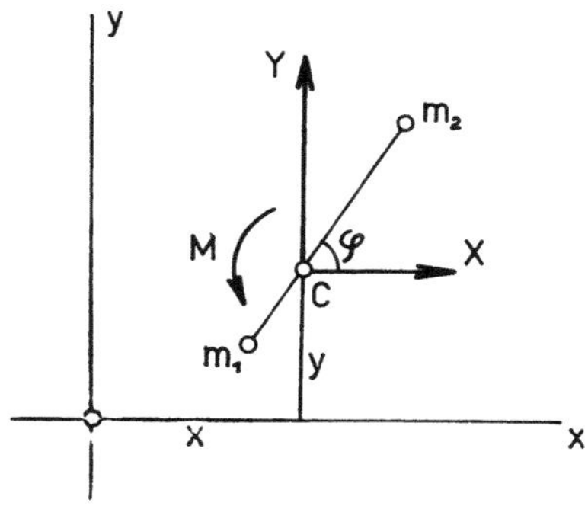

Fig. 5. Two-particle-model

This result is not surprising. Entropy is a measure for the disorder of a system containing a great number of elements. It is conceivable that, subject to given circumstances (i. e., under given macroscopic forces), the average development of such a system (representing its macroscopic response) corresponds to a maximum increase of disorder. A constraint in the form of a rigid connection within the system however represents an element of order which is apt to disturb this process.

Returning to the continuum, let us assume that the flux field is given by $w_j(x_l)$ and that $(Y_j(x_l)$ is the corresponding force field. The total dissipation rate within a finite domain may be defined by

$$L_{tot} = \int Y_j w_j \, dV. \tag{6.1}$$

Locally (3.9), i. e.,

$$D(w_j) = Y_j w_j, \tag{6.2}$$

still holds. We now will prove the following

Principle of Maximum Total Dissipation Rate

If, in a finite domain, the force field is prescribed by $Y_j(x_l)$ and if the flux field $w_j(x_l)$ is free from any restriction, the actual flux field maximizes the total dissipation rate (6.1) subject to the locally valid side condition (6.2).

Proof: The variational formulation of this principle reads

$$\delta L_{tot} = \delta \int [Y_j w_j - \mu(D - Y_j w_j)] \, dV = 0, \tag{6.3}$$

where $Y_j(x_l)$ is fixed and μ is a lagrangean multiplier. Carrying out the variation, we have

$$\int \left[(1 + \mu) \, Y_j - \mu \frac{\partial D}{\partial w_j}\right] \delta w_j \, dV = 0. \tag{6.4}$$

Since $w_j(x_l)$ is free, the variation $\delta w_j(x_l)$ is arbitrary. Thus, (6.4) is equivalent to

$$Y_j = \frac{\mu}{1 + \mu} \frac{\partial D}{\partial w_j}. \tag{6.5}$$

Setting $\mu/(1 + \mu) = \lambda$ and considering (6.2) we finally obtain (4.1), and since the local principle yields a maximum, this is also true for the global principle.

In a similar way, the principle of maximum rate of entropy production can be globalized. For this purpose it is only necessary to prescribe the temperature field in addition to the field $Y_j(x_l)$. The other principles stated in Section 2 apparently do not have simple global versions.

There is no contradiction between the principle just proved and the example consicered above. In our proof it is essential that the flux field be free from restrictions. On account of the rigid connection between the particles, this condition is not satisfied in the example.

It is obvious that the global principle, as well as the local one, is applicable to partial processes. It holds, e. g., for the process of heat conduction in a finite body, since the field of the heat flux is free from restrictions. For a linear relationship between heat flux and temperature gradient, a similar principle has been formulated by ONSAGER [9]. On the other hand, no corresponding global principle holds for the dissipation caused by deformation of a finite continuum, since the field of the strain rates connot be chosen arbitrarily but is subject to the compatibility conditions.

7. Conclusion

The purpose of these considerations was to show that ONSAGER's theory can be reformulated so as to take account of the objections mentioned in Section 1. I am not in a position to offer any proof for the general validity of the principles formulated here in place of ONSAGER's symmetry relations. It seems, however, that these principles are consistent. Moreover, there is some evidence in their favor: ONSAGER's relations in linear physics and chemistry, the significance of the theory of the plastic potential in plasticity, and maybe also the fact that the principles are satisfied by other nonlinear constitutive laws as, e. g., the creep theory of ODQVIST [19] or the theory of thixotropic substances developed by SLIBAR and PASLAY [20].

References

[1] DE GROOT, S. R.: Thermodynamics of Irreversible Processes. Amsterdam: North-Holland Publ. Co. 1952.
[2] PRAGER, W.: Introduction to Mechanics of Continua. Boston: Ginn and Co. 1961.
[3] DE GROOT, S. R., and P. MAZUR: Non-equilibrium Thermodynamics. Amsterdam: North-Holland Publ. Co. 1962.
[4] COLEMAN, B. D., and W. NOLL: Arch. Ratl Mech. Anal. **13**, 167 (1963).
[5] ZIEGLER, H.: Summer Course, Centro Internazionale Matematico Estivo. 1963. Roma: Ist. Matem. dell'Università.
[6] ZIEGLER, H.: 1964 Minta Martin Lecture, Mass. Inst. of Technology and American Inst. of Aeronautics and Astronautics.
[7] COLEMAN, B. D., and V. MIZEL: Arch. Ratl Mech. Anal. **13**, 245 (1963).
[8] COLEMAN, B. D., and M. E. GURTIN: Arch. Ratl Mech. Anal. **19**, 266 and 317 (1965).
[9] ONSAGER, L.: Phys. Rev. **37**, 405 (1931).
[10] ONSAGER, L.: Phys. Rev. **38**, 2265 (1931).
[11] COLEMAN, B. D., and C. TRUESDELL: J. Chem. Phys. **33**, 28 (1960).
[12] TRUESDELL, C.: Six Lectures on Modern Natural Philosophy. Berlin-Heidelberg-New York: Springer. 1966.

[13] Gage, D. H., and S. J. Kline: Report IT-1, Thermosciences Div., Dept. Mech. Engng, Stanford University.
[14] Ziegler, H.: Ing. Arch. **30**, 410 (1961).
[15] Ziegler, H.: Ing. Arch. **31**, 317 (1962).
[16] Ziegler, H.: In: Sneddon, I. N. and R. Hill, Progress in Solid Mechanics, **4**, 91 (1963).
[17] Biot, M. A.: J. Appl. Phys. **27**, 240 (1956).
[18] Drucker, D. C.: In: Reiner, M., and D. Abir, Proc. Intern. Sympos. on Second-order effects in Elasticity, Plasticity and Fluid Dynamics. New York-London: Pergamon Press. 1964.
[19] Odqvist, F. K. G.: Proc. 4th Intern. Congress of Applied Mechanics. p. 228. Cambridge, U. K.: 1934.
[20] Slibar, A., and P. R. Paslay: In: Reiner, M., and D. Abir, IUTAM-Symp. on Second-order effects in Elasticity, Plasticity and Fluid Dynamics. p. 314. New York-London: Pergamon Press. 1964.

Discussion

Meixner: Your principle is a generalization of the well understood linear case. There may be, of course, many other possible generalizations. So what one must do is to check the principle by a model. There is a very good model available. It is the so-called "Fowler-Ansatz" for chemically reacting gases. Have you tried to compare the consequences of your principle with the Fowler-Ansatz?

Ziegler: No, I have not.

Meixner: The Onsager relations have found so many proofs. One of the most interesting proofs has been given by Kubo. His formalism is quite general and can be used to investigate the existence of relations between the coefficients of higher order terms in the phenomenological equations. It would be interesting to see whether your principle is in accordance with the results of the Kubo formalism.

Ziegler: Thank you.

Koiter: When it comes to the use of the strain components as the state variables does not this imply that the consideration is restricted to elastic bodies for which the strain variables indeed specify the state completely, in addition to the temperature. But, for instance in a plastic body, is then a strain variable indeed a state variable?

Ziegler: The case I have been treating here was just the opposite of the elastic case; it was the case where all the work done on the element is transformed into heat. In other words, the process is completely irreversible. No elastic strain components were present. If they are present, what has to be done is to take the elastic strains, the plastic strains, and the temperature as the original state variables. It then turns out that the state functions (internal energy, entropy, etc.) depend on the elastic strains and the temperature alone whereas the *rate* of entropy production is determined by the plastic strain rates.

5000/198/67

GPSR Compliance

The European Union's (EU) General Product Safety Regulation (GPSR) is a set of rules that requires consumer products to be safe and our obligations to ensure this.

If you have any concerns about our products, you can contact us on ProductSafety@springernature.com

In case Publisher is established outside the EU, the EU authorized representative is:

Springer Nature Customer Service Center GmbH
Europaplatz 3
69115 Heidelberg, Germany

Batch number: 10577878

Printed by Printforce, the Netherlands